Advanced Drilling Engineering

Advanced Drilling Engineering

Principles and Designs

G. Robello Samuel, Ph.D.
Xiushan Liu, Ph.D.

Houston, Texas

Advanced Drilling Engineering: Principles and Designs

Gulf Publishing Company
2 Greenway Plaza, Suite 1020
Houston, TX 77046

Library of Congress Cataloging-in-Publication Data
Samuel, G. Robello.
Advanced drilling engineering : principles and designs / G. Robello Samuel and Xiushan Liu.
p. cm.
Includes bibliographical references and index.
ISBN-13: 978-1-933762-34-0 (alk. paper)
ISBN-10: 1-933762-34-9 (alk. paper)
1. Oil well drilling. 2. Gas well drilling. I. Liu, Xiushan. II. Title.
TN871.2.S284 2009
622'.3381—dc22

2009004742

Text design and composition by Ruth Maassen.

Transferred to Digital Printing in 2013

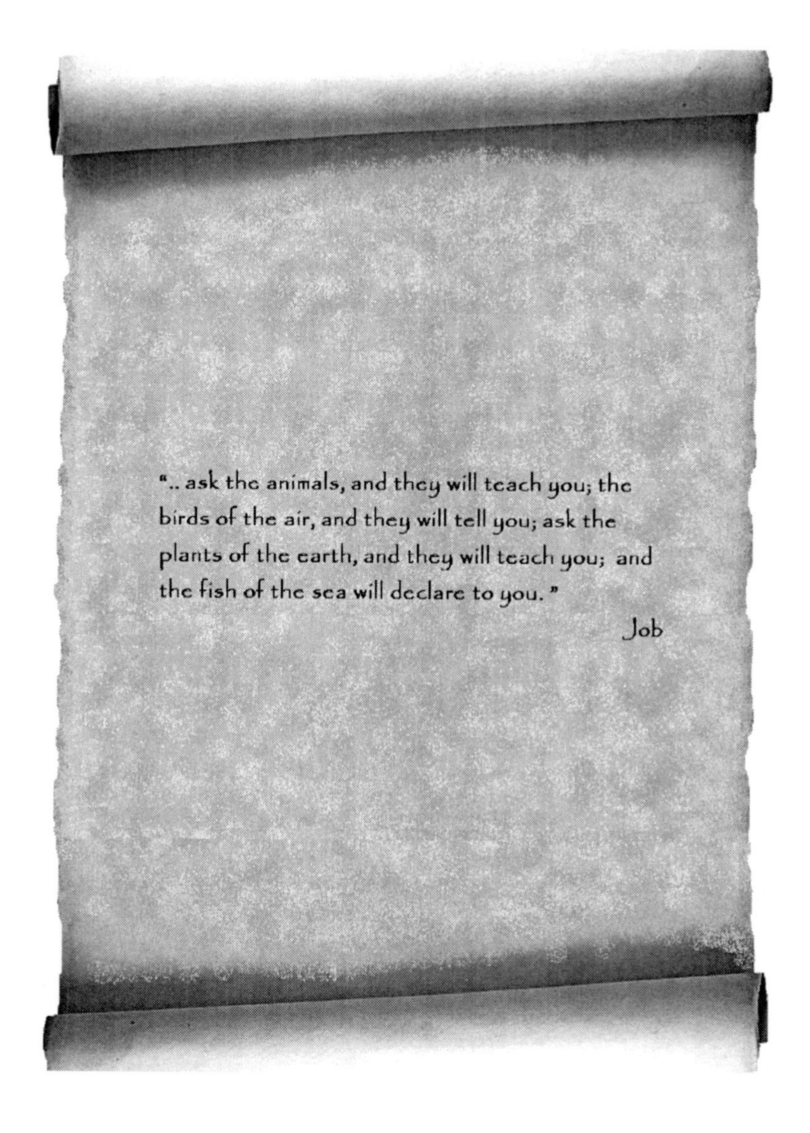
".. ask the animals, and they will teach you; the birds of the air, and they will tell you; ask the plants of the earth, and they will teach you; and the fish of the sea will declare to you."
Job

Contents

Preface

New engineers entering into the petroleum industry and enthusiastic students at the university were always my source of inspiration. During my long association with the industry coupled with my six years' teaching experience at universities, I have observed that the present curricula specialty courses related to drilling engineering lack in presentation of the basic principles and concepts required for designing complex wells. The recent advent of striking developments in new technologies and new techniques requires a broader knowledge and deeper understanding of the associated theory involved. This book is written to elucidate to the students and young practicing engineers a sound understanding of the engineering principles involved in designing complex wells. The analytical insights presented provide greater understanding of the design principles. Sample problems, including answers, are provided in between the topics, balancing between the theory and practical application, as well as providing educational value to the contents. Readers are also encouraged to work out the supplementary problems provided at the end of each chapter to increase their understanding of the theory presented. It is impossible to cover all the subjects and topics related to drilling and completing complex wells. We have selected and covered the topics that not only pave the way for basic understanding, but also provide useful underlying theory. This book can also be used as a supplementary textbook to a main textbook on drilling engineering for advanced-level undergraduate and graduate courses due to its balanced content of the topics with fundamentals and analytical methodologies.

The petroleum industry is at times crude, sometimes sweet, and many times slick, and therefore I advise my colleagues and students to "learn for the quadrants, not for the coordinate" at any point in life, in addition to following my LEAP^2 rule: Learn or Leave, Evaluate or Eliminate, Accomplish or Adios, Publish or Perish. I hope this book is a small part in providing an opportunity to achieve these objectives.

Claiming perfection is impossible even though painstaking efforts have been taken to present usable material for learning and applications; however, errors are inevitable in this complex endeavor. Readers are encouraged to send their suggestions and comments toward the improvement of this book.

—Robello Samuel
Houston, Texas

Acknowledgments

Writing a book is not an easy task and is like any other endeavor in the pursuit of attaining high standards.

Words fail to adequately thank Dr. Suresh Lee, David Engle, Andrea Engle, and Dongping Yao, who were not only able to proofread the chapters, but also were boundlessly helpful at various stages of the manuscript preparation. In addition, we would like to thank Dr. Jorge Sampaio, Jr., for sharing some of his class notes on well path design and survey tools.

We specially thank the following who have graciously shared their time and offered to contribute toward this book: Jerry Codling, Halliburton; Dr. Michael Economides, University of Houston; Angus Jamiesen and Patrick York, Weatherford.

Permission credits: Thanks are expressed to all the companies that have contributed their texts and images.

We particularly thank associate publisher Katie Hammon and production manager Sheryl Stone for their cooperative and untiring efforts to bring the book into print. In addition, we are greatly indebted to Jodie Allen and Ruth Maassen for the fine copy editing and composition on the manuscript.

Every possible effort has been taken to acknowledge and give appropriate credits in using the copyrighted materials. Should there be any omission, we sincerely apologize for the mistake, and suitable correction will be made at the first possible update.

We have dedicated our previous works to the members of our family, leaving this to our students as well as our professors who have crossed our paths during the past 25 years in this industry. Still, we cannot deny the fact that without the inspiration, patience, and unwavering love of our families, this work may not have been possible. In a real sense, they are also the coauthors of this book.

About the Authors

Dr. Robello Samuel has been a principal technical advisor (Drilling Engineering) in the Drilling, Evaluation division of Halliburton since 1998. He has over 20 years of experience in domestic and international oil/gas drilling and completion operations, management, consulting, research, and teaching. His skills include well planning, design, cost estimates, supervision of drilling and completion operations, personnel and technical review, project management, and creative establishment of project relationships through partnering and innovation. At present, he is a technical and engineering lead for a well planning application suite for drilling, completions, and well services operations. He has been an educator in the United States, Venezuela, Mexico, and India. Also, he is an adjunct professor for the past six years in the Petroleum program of Chemical and Biomolecular Engineering and Mechanical Engineering Technology Departments at the University of Houston, teaching advanced drilling engineering, downhole tools, and technologies, as well as complex well architecture courses.

Dr. Samuel has written or cowritten more than 65 journal papers, conference papers, and technical articles. He has given several graduate seminars at various universities. Dr. Samuel has been the recipient of numerous awards including the "CEO for a Day (Halliburton)" award. He is presently serving as a review chairman on the Society of Petroleum Engineers (SPE) Drilling and Completions Editorial Review Committee, Journal of Energy Resources Technology, reviewer of projects for the Natural Sciences and Engineering Research Council of Canada, editorial review chairman of Journal of Petroleum Science and Engineering, and co-chairman of SPE Multilateral Technical Interest Group. He also worked as a drilling engineer at the Oil and Natural Gas Corporation (ONGC) in India from 1983 to 1992.

Dr. Samuel holds B.S. and M.S. (mechanical engineering) degrees from the University of Madurai (Madurai) and the College of Engineering, Guindy, Anna University (Chennai), and M.S. and Ph.D. (petroleum engineering) degrees from Tulsa University. He is a member of ASME, SoR, and SPE. He is also the author of Downhole Drilling Tools—Theory and Practice for Students and Engineers (Gulf Publishing Company), Drilling Engineering Optimization—Techniques and Applications (manuscript under preparation), and coauthor of Drilling Engineering (Pennwell Publishers). Dr. Samuel can be reached via email at robellos@hotmail.com or on the Web at www.sigmaquadrant.com.

Dr. Xiushan Liu is a chief expert and professor of Sinopec Corporation's Petroleum Exploration & Production Research Institute, a visiting professor at CNPC's National Key Laboratory of Drilling Engineering, and an adjunct professor at Daqing Petroleum Institute in China. He has B.S. and M.S. degrees from Daqing Petroleum Institute and a Ph.D. from the Research Institute of Petroleum Exploration & Development of CNPC, all in petroleum engineering. Liu is a member of SPE, a member and secretary at Basic Theory Group of Chinese Petroleum Society, and an editorial board member for several of the world's highly respected journals, including SPE Drilling & Completion, Journal of Hydrodynamics, and Petroleum Drilling Techniques.

Dr. Liu has worked as a lecturer and associate professor for 11 years at Daqing Petroleum Institute in petroleum engineering, as a postdoctoral fellow for two years at Tsinghua University in engineering mechanics, and as a senior engineer and professor for eight years at Sinopec Corporation in drilling engineering. His primary research interests are related to key fundamental issues related to directional drilling, and downhole drilling tools especially in EM-MWD. He has presided over several key scientific and technological research projects in China, has applied for nine Chinese national invention patents, and has been the recipient of six research awards. Dr. Liu is the author of The Geometry of Wellbore Trajectory (Petroleum Industry Press–China) and Programming Techniques for Engineering Software (Petroleum Industry Press–China), the coauthor of Theory and Method for Designing and Describing Wellbore Trajectory (Heilongjiang Science and Technology Press) and Research and Application of New Techniques for Deep and Ultra-Deep

Drilling (China Petrochemical Press). He has written or cowritten more than 100 journal articles and international conference papers on wellbore trajectory planning and monitoring, multiphase flow in oil and gas drilling and production, drilling tubular mechanics, steerable drilling, and EM-MWD techniques. Dr. Liu can be reached via email at xliu@sina.com.

Advanced Drilling Engineering

1

General Introduction

Summary and Organization

The increased effort and quest to explore for oil and gas in complex reservoirs and deep waters have resulted in new technologies, methods, designs, and techniques. Significant challenges associated with these operations are very high, and therefore, the planning and well construction process is very important. The methods of drilling wells, as well as well construction and well engineering, have changed tremendously with the increased frontiers to explore and exploit the oil reserves. The associated downhole drilling tools, as well as downhole measurement tools, have also changed tremendously to cater to the needs of the new technology.

This book concentrates mainly on the topics involved in the design of complex wells with design and principles. General downhole drilling tools that can be used in bottomhole assembly (BHA) are discussed in detail in *Downhole Drilling Tools—Theory and Practice*.[1,2] Basic knowledge and understanding of the fundamental theory behind different tools are essential, so that they can effectively be used at optimum operating conditions. Bottomhole assembly with different tools is an important part of the overall well planning process. It is important to have good bit-BHA performance whether it is a simple design or a complex, multilateral well design.

The outline and the basis of the organization of various chapters presented in this book are given briefly. The book is divided into 11 chapters, and is organized in such a way that it provides continuity and flow to the study. This chapter introduces a few basic concepts, as well as a review of some of the mathematical fundamentals to help readers understand the mathematical formulation involved in the rest of the book. Specifically, vector algebra and the inductive definition of the geometric terms required for describing the well path in a plane or space are discussed. Practical problems are presented in between the topics to illustrate the usefulness of the theory. Substantial references and ample supplementary problems are included throughout the chapters to enable students and engineers to assimilate the topics for a deeper understanding. Readers are encouraged to work out the problems, thus gaining a thorough understanding of the theory presented. Both system of units, API and SI, are used wherever possible in the example problems.

Chapter 2 is focused toward the geodesy, a science describing the shape, size, and area of the Earth. This chapter is dedicated to providing a basic understanding of the geographic coordinate's datum, as well as map projections. The topics have been selected so that they are useful for well construction. A basic theory on the map projections and a few methods that are widely used in the petroleum industry are discussed. Detailed analytical treatment involved in the transformation of the map projections has not been presented and is beyond the scope of this book.

In Chapters 3 and 4, the fundamentals required for the profile design are discussed. For basic understanding of well path planning and for general reading, readers are referred to the book *Drilling Engineering*.[2] Procedures to facilitate the calculation of the trajectory shape, moving frame, borehole curvature, and wellbore torsion are also discussed in Chapter 3. To provide theoretical understanding on borehole torsion, different models, such as helical, circular arc, and natural curve, are presented with analytical discussions.

In Chapter 4, general and complex curve types for two-dimensional (2D) and three-dimensional (3D) well path planning are discussed for constructing directional, horizontal, and extended-reach wells. Models introduced in this chapter are used extensively in the later part of the book for planning wells to bypass obstruction, branched wells, and, in general, multilateral wells. Chapter 4 includes an exhaustive discussion with typical calculations and associated supplementary problems.

Chapter 5 covers the quantification of well paths in single, as well as multiple, well environments. Different types of well architecture, including multilateral, multibranch, and multilevel configurations, are discussed. Selection of well path designs based on geological and reservoir considerations are also covered.

In Chapter 6, the trajectory-monitoring survey section covers azimuth and azimuth conversion, grid coordinates system including magnetic declination, meridian convergence angle, and azimuth conversions. Survey calculation and trajectory deviation from the planned path is an important aspect of the path planning. Risk-based analysis and trajectory uncertainty that help to avoid anti-collision problems while planning multiple wells are discussed in Chapter 7. Tortuosity and different methods of estimating the tortuosity factors, which aid collision-avoidance and estimate the wellbore position uncertainty, are discussed.

Chapter 8 presents optimization methodology and calculation techniques to select the well path geometrical parameters, such as length, deviation angle, and well path profile. In this chapter, some new types of well profiles are also discussed, and the quantification based on the minimum-energy principle is introduced.

For wellbore deviation or hole inclination, wellbore torsion and downhole directional tools play an important part (there are several directional tools). Chapter 9 discusses some of the basic telemetry methods used in the measurement-while-drilling (MWD) tools. Special types of tools, such as hollow whipstock and rotary steerable tools, are described. Other downhole tools used in well construction are covered in *Downhole Drilling Tools—Theory and Practice for Students and Engineers*.[1] Rotary steerable tools have become an inherent part of completing complex well architecture. Borehole quality is generally related to the "smoothness" of the wellbore, which manifests itself in a number of ways, all adversely affecting the efficiency of the drilling process and increasing drilling and well-completion cost. The phenomenon of borehole spiraling or oscillation is being increasingly recognized, and techniques, such as point-the-bit and push-the-bit modes, with new, different drilling assemblies are being used to alleviate the problem.

In seeking to exploit deep and ultradeep wells, and in a more hostile environment, implementation and introduction of more downhole measurement tools and enabling technologies, such as tubular expansion, will continue to play a major role in combating costly problems. Chapter 10 provides the fundamental foundation for understanding how expandable products are used. As enabling a technology as expandable products are, they are not a panacea to solve every drilling or production challenge. Knowing how and when to apply expandable products is every bit as important as what the products are and what value they can add. This area of understanding, coupled with proper well designs, is critical to extracting the maximum benefit from this technology.

Chapter 11 covers some of the cost and economic analysis of various well paths and trajectory designs. This helps to design an optimized and energy-balanced well path. This chapter is provided to cover the economics of drilling directional wells as opposed to drilling near-vertical wells. A successful drilling and completion is the one that combines a good well plan profile with a satisfactory economic outcome. New techniques such as neural network, support vector machine, and econometrics type of risk-analysis models are also discussed.

Overview, Concepts, and Definitions

In this introductory chapter, some of the basic terms involved and required for the design and path planning are presented. Review of the fundamental concepts facilitates more understanding of the complex well path analysis. The explanations and descriptions of various terms are simple. Readers are advised, if needed, to read books where analyses are presented in considerable detail. The scope of this chapter is to present a concise and adequate summary of the commonly used concepts and definitions so that readers need not refer to other books immediately. A table of commonly used conversion factors between the unit systems is provided at the end of this book.

Spherical Coordinates

Spherical coordinates are also called spherical polar coordinates. For convenience of calculations, the following notations are used: Let ϕ be the direction or azimuth angle in the xy plane spanning from 0 to 2π radians and let α be the inclination spanning from 0 to π radians in the yz plane. Consider a point P with a distance r from the origin. These parameters are shown in Figure 1.1.

The position vector can be given as:

$$r = x\boldsymbol{i} + y\boldsymbol{j} + z\boldsymbol{k} \tag{1.1}$$

The coordinates x, y, and z can be given as:

$$\begin{cases} x = r \sin\alpha \cos\phi \\ y = r \sin\alpha \sin\phi \\ z = r \cos\alpha \end{cases} \tag{1.2}$$

where

x, y, z = Cartesian coordinates
$\boldsymbol{i}, \boldsymbol{j}, \boldsymbol{k}$ = Coordinate axes
r, α, ϕ = Cylindrical polar coordinates:
r = Distance from the origin
α = Inclination, degree

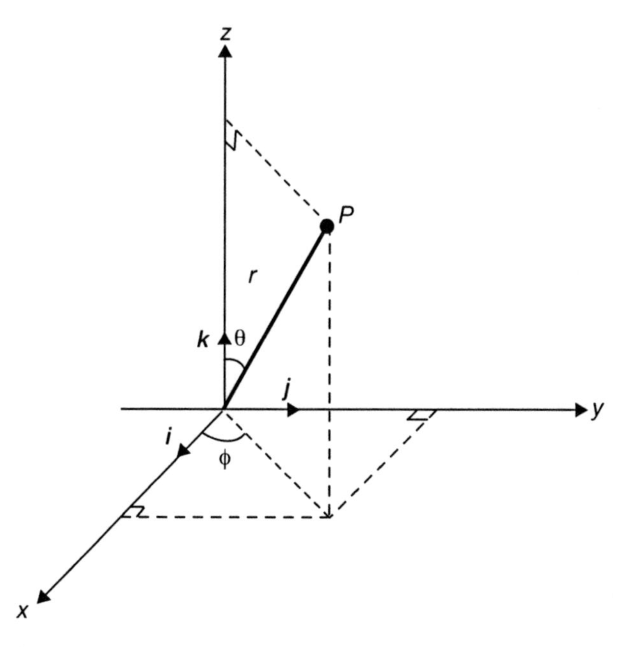

Figure 1.1 Representation of spherical coordinates.

ϕ = Azimuth or direction, degree

The length can be written as:

$$r = \sqrt{x^2 + y^2 + z^2} \tag{1.3}$$

The above equation can also be written in spherical coordinates as:

$$r = \sqrt{x^2 + y^2 + z^2}$$

$$\alpha = \cos^{-1}\left(\frac{z}{\sqrt{x^2 + y^2 + z^2}}\right) \tag{1.4}$$

$$\phi = \tan^{-1}\left(\frac{y}{x}\right)$$

Direction Cosines

A direction cosine of a vector represents the cosine of angles α, β, γ, namely $\cos\alpha$, $\cos\beta$, $\cos\gamma$, that makes with the coordinate x, y, z axes, respectively. Direction cosines are used for coordinate transformation and they form the elements of the matrix. The sum of the squares of the direction cosines is equal to unity.

PROBLEM 1.1

A directed line makes an angle of 90° with the x-axis and 45° with the z-axis. Calculate the angle at which it makes with the y-axis.

Solution

It is known that:

$$\cos^2\alpha + \cos^2\beta + \cos^2\gamma = 1 \tag{1.5}$$

Substituting the given values $\alpha = 90°$, $\beta = 45°$,

$$\gamma = 45° \text{ or } 135°$$

45° results in the line pointing upwards bisecting the negative y-axis and the z-axis.

Coordinate Transformation

It may be required to transform a point from one coordinate system to another coordinate system. For 2D transformation from one Cartesian coordinate system to another Cartesian coordinate system, it can be given in matrix form as:

$$\bar{r} = \begin{bmatrix} x \\ y \end{bmatrix} = \begin{bmatrix} a_1 \\ a_2 \end{bmatrix} + \begin{bmatrix} \cos\alpha & \sin\alpha \\ -\sin\alpha & \cos\alpha \end{bmatrix} \begin{bmatrix} x' \\ y' \end{bmatrix} \tag{1.6}$$

Alternatively, in equation form it can be written as:

$$\begin{aligned} x &= a_1 + x'\cos\alpha + y'\cos\alpha \\ y &= a_2 - x'\sin\alpha + y'\cos\alpha \end{aligned} \tag{1.7}$$

With translation, $\bar{a}$ between coordinates.

Similarly, transformation of x-y-z coordinates frames from x-y-z to x-y-z by rotating the coordinates x-y-z around the z-axis, and is given as:

$$\bar{r} = \begin{bmatrix} x \\ y \\ x \end{bmatrix} = \begin{bmatrix} a_1 \\ a_2 \\ a_3 \end{bmatrix} + \begin{bmatrix} \cos\alpha & \sin\alpha & 0 \\ -\sin\alpha & \cos\alpha & 0 \\ 0 & 0 & 1 \end{bmatrix} \begin{bmatrix} x' \\ y' \\ z' \end{bmatrix} \tag{1.8}$$

or in equation form:

$$\begin{aligned} x &= a_1 + x'\cos\alpha + y'\cos\alpha \\ y &= a_2 - x'\sin\alpha + y'\cos\alpha \\ z &= z' \end{aligned} \tag{1.9}$$

Similarly, coordinates rotated around the x-axis and y-axis are respectively given as:

$$\begin{bmatrix} x \\ y \\ z \end{bmatrix} = \begin{bmatrix} a_1 \\ a_2 \\ a_3 \end{bmatrix} + \begin{bmatrix} 1 & 0 & 0 \\ 0 & \cos\alpha & \sin\alpha \\ 0 & -\sin\alpha & \cos\alpha \end{bmatrix} \begin{bmatrix} x' \\ y' \\ z' \end{bmatrix} \tag{1.10}$$

and

$$\begin{bmatrix} x \\ y \\ z \end{bmatrix} = \begin{bmatrix} a_1 \\ a_2 \\ a_3 \end{bmatrix} + \begin{bmatrix} \cos\alpha & 0 & -\sin\alpha \\ 0 & 1 & 0 \\ \sin\alpha & 0 & \cos\alpha \end{bmatrix} \begin{bmatrix} x' \\ y' \\ z' \end{bmatrix} \tag{1.11}$$

Vectors

Cross Product

Cross product is also called the vector product. If the two vectors are given as:

$$\begin{aligned} \boldsymbol{A} &= \mathrm{a}_1\boldsymbol{i} + a_2\boldsymbol{j} + a_3\boldsymbol{k} \\ \boldsymbol{B} &= b_1\boldsymbol{i} + b_2\boldsymbol{j} + b_3\boldsymbol{k} \end{aligned} \tag{1.12}$$

the third vector lies perpendicular to the plane of these two vectors and can be obtained using the cross product of the two vectors.

The cross product results in a vector and can be written as:

$$\boldsymbol{A} \times \boldsymbol{B} = \begin{bmatrix} \boldsymbol{i} & \boldsymbol{j} & \boldsymbol{k} \\ a_1 & a_2 & a_3 \\ b_1 & b_2 & b_3 \end{bmatrix} = \left(a_2b_3 - a_3b_2\right)\boldsymbol{i} + \left(a_3b_1 - a_1b_3\right)\boldsymbol{j} + \left(a_1b_2 - a_2b_1\right)\boldsymbol{k} \tag{1.13}$$

Also, it can be seen that

$$\boldsymbol{A} \times \boldsymbol{B} = -\boldsymbol{B} \times \boldsymbol{A} \tag{1.14}$$

The three orthonormal vectors are schematically shown in Figure 1.2.

Dot Product

Dot product is also known as inner product and is a scalar quantity of the two vectors and is independent of the coordinates. If the two vectors are defined as in the cross product, the inner product (or the dot product) of the two vectors can be given as:

$$\boldsymbol{A} \bullet \boldsymbol{B} = \left(a_1\boldsymbol{i} + a_2\boldsymbol{j} + a_3\boldsymbol{k}\right) \bullet \left(b_1\boldsymbol{i} + b_2\boldsymbol{j} + b_3\boldsymbol{k}\right) = a_1b_1 + a_2b_2 + a_3b_3 \tag{1.15}$$

since:

$$\begin{aligned} \boldsymbol{i} \bullet \boldsymbol{i} &= \boldsymbol{j} \bullet \boldsymbol{j} = \boldsymbol{k} \bullet \boldsymbol{k} = 1 \\ \boldsymbol{i} \bullet \boldsymbol{j} &= \boldsymbol{j} \bullet \boldsymbol{k} = \boldsymbol{k} \bullet \boldsymbol{i} = 0 \end{aligned} \tag{1.16}$$

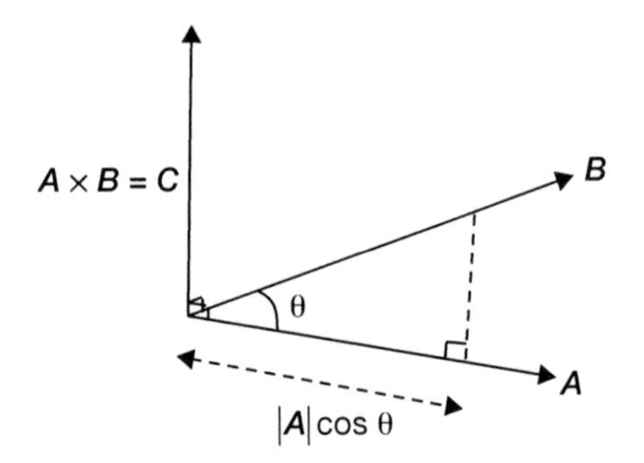

Figure 1.2 Vector representation.

Also, it can be seen that:

$$\boldsymbol{A} \bullet \boldsymbol{B} = \boldsymbol{B} \bullet \boldsymbol{A} \tag{1.17}$$

Angle between the Vectors

The angle between two vectors can be given in two different forms as:

$$\boldsymbol{A} \times \boldsymbol{B} = |\boldsymbol{A}||\boldsymbol{B}|\sin\theta \tag{1.18}$$

or

$$\boldsymbol{A} \bullet \boldsymbol{B} = |\boldsymbol{A}||\boldsymbol{B}|\cos\theta \tag{1.19}$$

Vector Norm

Vector norm is a scalar quantity that expresses the magnitude of a vector. It is usually denoted with single bar $|\ |$ or double bars $\|\ \|$.

Usually, the norm of a vector $x = (x_1, x_2, x_3, ..., x_n)$ is given as $|x| = \sqrt{\sum_{i=1}^{n} x_i^2}$.

Tangent-Normal-Binormal (TNB) Frame

A moving orthonormal frame called Frenet frame or Frenet trihedron is associated with a curve in a 3D space, as shown in Figure 1.3. As seen in the figure, the three unit vectors forming a right-hand system constitute the moving frame. Vectors are the tangent vector $\boldsymbol{t}$, pointing along the curve, and the other two orthogonal vectors are normal vector $\boldsymbol{n}$ and binormal vector $\boldsymbol{b}$. These three vectors are called Frenet vectors. There are three planes in the Frenet frame and they are osculating, rectifying, and normal planes. The $\boldsymbol{tn}$ plane with the unit tangent and normal vectors is called the osculating plane. The $\boldsymbol{tb}$ plane with the unit tangent and binormal vectors is called

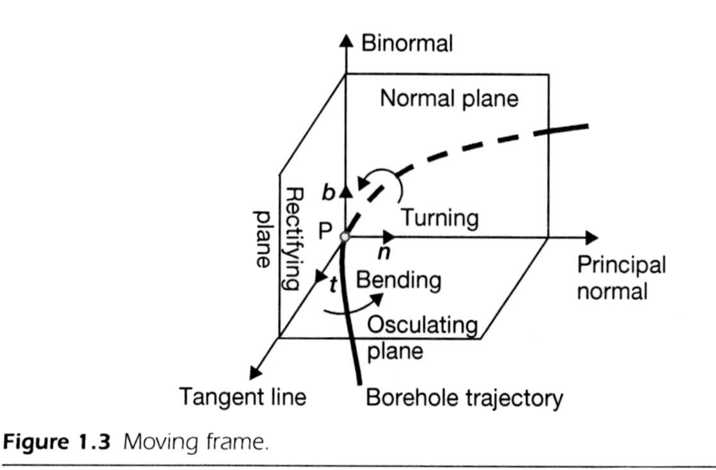

Figure 1.3 Moving frame.

the rectifying plane. The $\boldsymbol{nb}$ plane with the unit normal and binormal vectors is called the normal plane.

Frenet-Serret Formulas

Let $r = r(s)$ be the equation of a curve where r is the radius vector of a point A on a curve and s is the arc length, as shown in Figure 1.4. The vector $\boldsymbol{n}$ is orthogonal to $\boldsymbol{t}$ and lies in the osculating plane at point A.

The binormal vector $\boldsymbol{b}$, the product of the unit tangent, and the unit normal is given as:

$$\boldsymbol{b} = \boldsymbol{t} \times \boldsymbol{n} \tag{1.20}$$

where

$$\begin{cases} \boldsymbol{t} = \dot{\boldsymbol{r}} = \dfrac{d\boldsymbol{r}}{ds} \\ \boldsymbol{n} = \dfrac{\dot{\boldsymbol{t}}}{|\dot{\boldsymbol{t}}|} = \dfrac{\ddot{\boldsymbol{r}}}{|\ddot{\boldsymbol{r}}|} \end{cases} \tag{1.21}$$

Frenet equations[3] are given as:

$$\begin{cases} \dot{\boldsymbol{t}} = \kappa \boldsymbol{n} \\ \dot{\boldsymbol{n}} = -\kappa \boldsymbol{t} + \tau \boldsymbol{b} \\ \dot{\boldsymbol{b}} = -\tau \boldsymbol{n} \end{cases} \tag{1.22}$$

This equation can be given in matrix form as:

$$\begin{bmatrix} \dot{\boldsymbol{t}} \\ \dot{\boldsymbol{n}} \\ \dot{\boldsymbol{b}} \end{bmatrix} = \begin{bmatrix} 0 & \kappa & 0 \\ -\kappa & 0 & \tau \\ 0 & -\tau & 0 \end{bmatrix} \begin{bmatrix} \boldsymbol{t} \\ \boldsymbol{n} \\ \boldsymbol{b} \end{bmatrix} \tag{1.23}$$

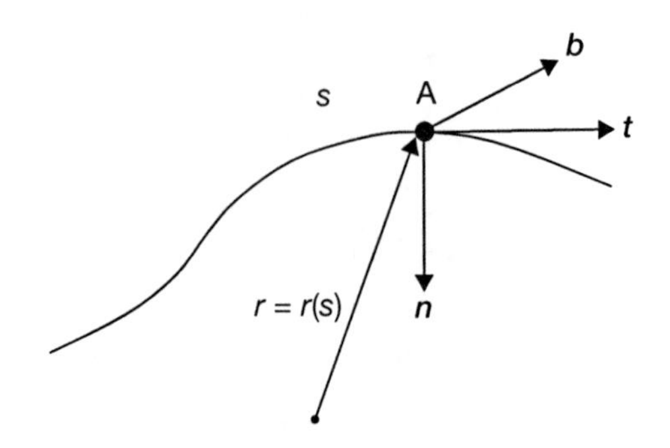

Figure 1.4 The tangent, normal, and binormal vectors.

Curvature and Torsion

Curvature

At a given point on a curve, R is the radius of the osculating circle at that point. By definition, it is the length of the normal vector. It is obtained by fitting a circle at the point of the curve. When the circle fits just right, the radius of the circle is defined as the radius of the curvature and its inverse is the curvature, κ (Figure 1.5) Therefore, the larger the circle of curvature, the smaller the curvature.

Consider a space curve in 3D space with a tangent vector ***t***. The curvature of the curve is the derivative of the tangent vector with respect to the arc length, and it provides a measure of how quickly the unit tangent changes with respect to arc length. Curvature is given as:

$$\kappa = \left|\frac{d\boldsymbol{t}}{ds}\right| = \left|\dot{\boldsymbol{t}}\right| = \left|\frac{d^2\boldsymbol{r}(s)}{ds^2}\right| \tag{1.24}$$

The relationship between the curvature and the unit normal vector is given as:

$$\boldsymbol{n} = \frac{1}{\kappa}\frac{d^2\boldsymbol{r}(s)}{ds^2} \tag{1.25}$$

or can be written as:

$$\kappa = \frac{\left|\dot{\boldsymbol{r}}(s) \times \ddot{\boldsymbol{r}}(s)\right|}{\left|\dot{\boldsymbol{r}}(s)\right|^3} \tag{1.26}$$

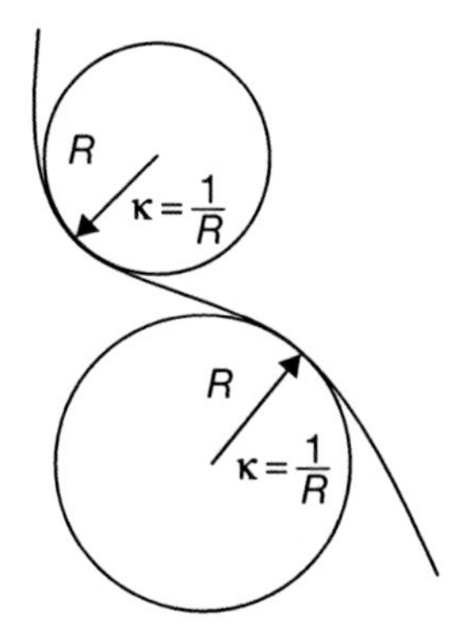

Figure 1.5 Radius of curvature.

Torsion

Torsion of a space curve gives a measure of the rate at which the osculating plane changes its direction. The torsion of a space curve in 3D space can be defined as:

$$\tau = -\boldsymbol{n} \bullet \dot{\boldsymbol{b}} \tag{1.27}$$

Torsion can be calculated using the following equation:

$$\tau = \frac{\left(\dot{\boldsymbol{r}}(s) \times \ddot{\boldsymbol{r}}(s)\right) \bullet \dddot{\boldsymbol{r}}(s)}{\left|\dot{\boldsymbol{r}}(s) \times \ddot{\boldsymbol{r}}(s)\right|^2} \tag{1.28}$$

Similar to radius of curvature, radius of torsion is inverse of the torsion.

Torsion has a sign convention and a right-handed curve has a positive torsion.

$$\tau = \begin{cases} +\left|\dot{\boldsymbol{b}}\right|, & \text{if } \dot{\boldsymbol{b}} \text{ and } \boldsymbol{n} \text{ is in the opposite direction} \\ -\left|\dot{\boldsymbol{b}}\right|, & \text{if } \dot{\boldsymbol{b}} \text{ and } \boldsymbol{n} \text{ is in the same direction} \end{cases} \tag{1.29}$$

where

$\dot{\boldsymbol{r}}$ = First derivative of coordinate vector
$\dot{\boldsymbol{t}}$ = First derivative of unit tangent vector
$\dot{\boldsymbol{n}}$ = First derivative of unit principal normal vector
$\dot{\boldsymbol{b}}$ = First derivative of unit binormal vector
$\ddot{\boldsymbol{r}}$ = Second derivative of coordinate vector
$\dddot{\boldsymbol{r}}$ = Third derivative of coordinate vector
κ = Curvature of curve, degree/ft
τ = Torsion of curve, degree/ft

PROBLEM 1.2

Equation of a space curve is given as $r(s)=(5\sin t, 3\cos t, 4\cos t)$. Calculate the curvature and torsion.

Solution

Taking the derivatives:

$$\begin{cases} \dot{r}(s)=(5\cos t,-3\sin t,-4\sin t) \\ \ddot{r}(s)=(-5\sin t,-3\cos t,-4\cos t) \\ \dddot{r}(s)=(-5\cos t,3\sin t,4\sin t) \end{cases}$$

Taking the cross product:

$$\dot{r}(s)\times\ddot{r}(s)=(0,20,-15)$$

Thus, $\dot{r}(s)\times\ddot{r}(s)=(0,20,-15)$

$$|\dot{r}(s)\times\ddot{r}(s)|=\sqrt{20^2+15^2}=5$$

$$|\dot{r}(s)|=\sqrt{25\cos^2 t+9\sin^2 t+16\sin^2 t}=5$$

$$\text{Curvature}=\kappa=\frac{|\dot{r}(s)\times\ddot{r}(s)|}{|\dot{r}(s)|^3}=\frac{5}{5^3}=\frac{1}{25}$$

Torsion is further calculated as:

$$\tau=\frac{\dot{r}(s)\times\ddot{r}(s)\bullet\dddot{r}(s)}{|\dot{r}(s)\times\ddot{r}(s)|^2}=\frac{60\sin t-60\sin t}{25^2}=0$$

Classification

The characteristics of the space curve can be summarized as follows:

- A necessary and sufficient condition that a curve be a straight line is that curvature $\kappa\equiv 0$ at all points.
- If the torsion is $\tau\equiv 0$, then the condition is satisfied only if the curve is a plane curve.
- If the κ = constant but > 0 and $\tau = 0$, then the condition is satisfied only if the curve is part of a circle.
- If the κ = constant but > 0 and τ = constant but > 0 , then the condition is satisfied only if the curve is part of a circular helix or $\frac{\tau}{\kappa}$ is constant.

Rotational Vector

The rotational vector, also called the Darboux vector,[4] of the Frenet frame moving along a curve of unit speed is given as:

$$\omega = \tau \boldsymbol{t} + \kappa \boldsymbol{b} \tag{1.30}$$

It is also called the angular velocity vector of a space curve with the first term being due to torsion and the second term due to curvature.

PROBLEM 1.3

Prove that $\dot{\boldsymbol{t}} \times \ddot{\boldsymbol{t}} = \omega\kappa^2$.

Solution

Using the Frenet equation:

$$\dot{\boldsymbol{t}} = \kappa \boldsymbol{n}$$

Taking the derivative:

$$\ddot{\boldsymbol{t}} = \kappa \dot{\boldsymbol{n}} + \dot{\kappa} \boldsymbol{n}$$

Using Eq. 1.22, the equation can be written as:

$$\ddot{\boldsymbol{t}} = \kappa\left(-\kappa \boldsymbol{t} + \tau \boldsymbol{b}\right) + \dot{\kappa} \boldsymbol{n}$$

which is

$$\ddot{\boldsymbol{t}} = -\kappa^2 \boldsymbol{t} + \kappa\tau \boldsymbol{b} + \dot{\kappa} \boldsymbol{n}$$

Thus,

$$\dot{\boldsymbol{t}} \times \ddot{\boldsymbol{t}} = \kappa \boldsymbol{n} \times \left(-\kappa^2 \boldsymbol{t} + \kappa\tau \boldsymbol{b} + \dot{\kappa} \boldsymbol{n}\right) = \left(-\kappa^3 \boldsymbol{n} \times \boldsymbol{t} + \kappa^2 \tau \boldsymbol{n} \times \boldsymbol{b}\right) = \kappa^3 \boldsymbol{b} + \kappa^2 \tau \boldsymbol{t}$$

Using the relationship $\omega = \tau \boldsymbol{t} + \kappa \boldsymbol{b}$,

$$\dot{\boldsymbol{t}} \times \ddot{\boldsymbol{t}} = \omega\kappa^2$$

Rotation Index

This is another way to define the characteristic of a space curve between two points on the curve, which can also be used to describe the well paths or trajectories. It is a geometric quantity estimating the number of loops in a curve and positive looping if the normal is oriented inward (i.e., the curve is counterclockwise).

It is defined as:

$$I = \frac{1}{2\pi}\int_a^b \kappa(s)\,ds \tag{1.31}$$

where

$$\kappa = \frac{d\alpha}{ds}$$

The term $\int_a^b \kappa(s)ds$ is called the total curvature, which is 2π times the rotation index of the curve.

For example, the rotation index of a plane curve, such as a circle or curves obtained by deforming a circle, will always be one.

More generally, for a space curve using the total curvature, it can be given as:

$$I = \frac{1}{2\pi}\int_a^b \left(\sqrt{\kappa(s)^2 + \tau(s)^2}\right)ds \tag{1.32}$$

Path versus Trajectory

The petroleum industry uses path and trajectory interchangeably. According to the conventional definition, *path* is the planned sequence of coordinates that can be used to construct a shape or a profile without any time element, whereas *trajectory* is the tracking of the planned path with respect to time. To be consistent with the basic definition, the same terms and meanings are used throughout the book. Well path is the sequence of wellbore course coordinates that can be obtained based on the design method and profile. Wellbore trajectory is the actual well path constructed using the actual survey data measured with the downhole surveying equipment, which includes positional uncertainty and tool errors.

Transition Curves

Transition curves, or bridge curves, are used in the well path design as a connecting curve between two sections of the well paths to change the inclination and direction, or azimuth, of the well paths. Usually transition curves are smooth curves and some of the notable curves are circular arc, helix, cubic spiral, catenary, clothoid, and parabola. Depending on the transition curve, the radius of the curvature and course coordinates of the well path change.

Covariance and Covariance Matrix

Variance is defined as a measure of the spread of a variable around its mean value. It is given as:

$$s^2 = \frac{\sum_{i=1}^{n}\left(X_i - \bar{X}\right)^2}{(n-1)} \tag{1.33}$$

where

X = Data set
$\bar{X}$ = Mean of the data set X
n = Number of elements in the data set X

Covariance is a measure that gives the variance between any two variables in an n variable data. It can be written as:

$$\operatorname{cov}(X,Y) = \frac{\sum_{i=1}^{n}\left(X_i - \bar{X}\right)\left(Y_i - \bar{Y}\right)}{(n-1)}$$

or

$$\sum[i,j] = \frac{\sum_{i=1}^{n}\left(X_i - \bar{X}\right)\left(Y_i - \bar{Y}\right)}{(n-1)} \tag{1.34}$$

If the value covariance is positive, both variables compared tend to increase together, whereas when negative, one increases as the other decreases. The other possible condition of zero covariance indicates that the two variables are independent of each other.

A covariance matrix is a collection of many covariances expressed in $n \times n$ matrix form for n data set and is always symmetric. As an example for three variables x, y, and z, the covariance matrix can be given as:

$$C = \begin{bmatrix} \operatorname{cov}(x,x) & \operatorname{cov}(x,y) & \operatorname{cov}(x,x) \\ \operatorname{cov}(x,y) & \operatorname{cov}(y,y) & \operatorname{cov}(y,z) \\ \operatorname{cov}(x,x) & \operatorname{cov}(z,y) & \operatorname{cov}(z,z) \end{bmatrix} \tag{1.35}$$

Splines

This is defined as a smooth, piecewise polynomial function forming a smooth curve. Splines are extensively used for interpolation of data. Increasing the order of polynomial increases the order of requirements to satisfy the continuity conditions across interval points. For example, the linear splines piecewise linear between the interpolation points, quadratic splines piecewise quadratic between the interpolation points, first derivative is continuous at the interpolation points, whereas cubic splines piecewise cubic between the interpolation points, first derivative and second derivative are continuous at the interpolation points, etc.

Splines-in-Tension

Splines-in-tension eliminate extraneous inflection points in curve-fitting by cubic splines.[5,6] A parameter λ, called tension parameter, determines the actual behavior of the function.

A linear function is any function $y(x)$ that satisfies the differential equation:

$$\frac{d^2y}{dx^2} = 0$$

On the other hand, a cubic function is any function $y(x)$ that satisfies the differential equation:

$$\frac{d^4y}{dx^4} = 0$$

From these two assertions, the following differential equation can be formed with the dependent parameter λ:

$$\frac{d^4y}{dx^4} - \lambda^2\frac{d^2y}{dx^2} = 0 \tag{1.36}$$

A general solution of this differential equation is called splines-in-tension or tension splines and is given by:

$$y(x) = A + Bx + C\sinh(\lambda x) + D\cosh(\lambda x) \tag{1.37}$$

It can be seen that when $\lambda \to 0$, the solutions of the differential equation are cubic functions (dependent on four boundary conditions).

When $\lambda \to \infty$, the solutions are linear functions.

Supplementary Problems

1. Express curvature of a curve in terms of $\dot{\kappa}, \tau, \dddot{r}(s), \boldsymbol{t}, \boldsymbol{n}, \boldsymbol{b}$.
2. Equation of a space curve is given as $r(s) = (5 + 5\sin\alpha, 4 + 4\cos\alpha, 3 + 3\cos\alpha)$. Calculate the curvature and torsion. Also, determine the radius of curvature, as well as radius of torsion.
3. Calculate the torsion of the parametric curve:

   $$p(s) = \left(a\cos\left(\frac{s}{c}\right), a\sin\left(\frac{s}{c}\right), \frac{bs}{c}\right),\ \text{where } c^2 = a^2 = b^2.$$

4. Find the Frenet frame, curvature, and torsion at time t of the following curve:

$$p(t) = \left(\sqrt{2}\sin(t), \sqrt{2}\sin(t), \cos(t)\right)$$

5. Calculate the Frenet apparatus (T, κ, N, B, and τ) of the following curves:
   a. $p(t) = \left(e^t \cos(t), e^t \sin(t), e^t\right)$
   b. $p(t) = \left(\cosh(t), \sinh(t), t\right)$

6. Defining the rotation vector (Darboux rotation vector) as $\omega = \tau \boldsymbol{t} + \kappa \boldsymbol{b}$, prove that

$$\begin{cases} \dfrac{d\boldsymbol{t}}{ds} = \omega \times \boldsymbol{t} \\ \dfrac{d\boldsymbol{n}}{ds} = \omega \times \boldsymbol{n} \\ \dfrac{d\boldsymbol{b}}{ds} = \omega \times \boldsymbol{b} \end{cases}$$

   Also, derive the equation for the length of the rotation vector.

7. Prove that $\dot{\boldsymbol{b}} \times \ddot{\boldsymbol{b}} = \omega\tau^2$.

8. For the catenary defined by the equation

$$y = a\cosh\left(\frac{t}{a}\right), x = t$$

   for some constant a, the curvature can be expressed as

$$\kappa = \frac{a}{a^2 + s^2}$$

   where s is the arc length from $t = 0$.

9. Show that the parametric curve given by the following equation has the curvature $\kappa = \dfrac{\pi}{a}$.

$$x(t) = \int_c^t \sin(\pi t)dt, \; y(t) = \int_c^t \sin(\pi t)dt$$

10. Find the Frenet-Serret frame, curvature, and torsion for the curve given by the parametric equations:

$$x(t) = \int_c^t \sin(\pi t)dt, \; y(t) = \int_c^t \sin(\pi t)dt, \; z(t) = \boldsymbol{b} \times \boldsymbol{t}$$

11. Show that $\left[\dot{r}, \ddot{r}, \dddot{r}\right] = 0$ is a necessary sufficient condition for a curve to be a plane.

12. An azimuth angle ϕ that can have values between 0° and 360° is used to specify the direction of a target. Express β, a quadrant angle (quadrature system), in terms of ϕ.
13. A rig is located at south y feet and west x feet. A proposed target is at L feet $S\theta E$ of the rig location. Determine the rectangular coordinates of target with respect to the rig.
14. Convert the given azimuth angles to quadrature equivalents or quadrature to azimuth equivalents:
    257°, 30°, 111°, 279°
    S10°E, S21°W, N71°W, N90°E
15. Show that the total signed curvature of a closed curve is a multiple of 2π.
16. Compute the total curvature and rotation index for the following curves:
    a. A circle oriented clockwise ($\cos t$, $\sin t$)
    b. A circle that is oriented counterclockwise ($\cos(-t)$, $\sin(-t)$)
    c. A curve with the parameters ($\cos t$, $\sin 2t$), $0 \leq t \leq 2\pi$
    d. A curve with the parameters ($\cos t$, $\sin 2t$), $0 \leq t \leq \pi$
17. Show that curves with constant curvature are either line segments or arcs of circles.
18. Find the direction of cosine of a line pointing upwards, bisecting the positive y and z-axis. Also, calculate if it bisects (1) negative y-axis and positive z-axis (2) positive y-axis and negative z-axis.
19. Prove that the covariance of two independent random variables is 0. Explain why the converse is not necessarily true.
20. What is the norm of a unit vector?

References

1. Robello, Samuel G. *Downhole Drilling Tools—Theory and Practice for Students and Engineers.* Houston: Gulf Publishing, 2007.
2. Azar, J. J., and Samuel G. Robello. *Drilling Engineering.* Tulsa, OK: Penwell Publishers, 2007.
3. Borisenko, A. I., and Tarpov, I. E. *Vector & Tensor Analysis with Applications.* New York: Dover, 1968.
4. Darboux, G. *Leçons sur la théorie générale des surfaces.* Paris: de L'École Polytechnique, 1914.
5. Schweikert, D. "An Interpolation Curve Using Splines in Tension." *J. Math. Phys.*, Vol. 45, 1966.
6. Pruess, S. "Properties of Splines in Tension." *Journal of Approximation Theory,* Vol. 17, 1976.

2

Geodesy

Studies for the location and placement of wells are important during the preplanning and planning phase of drilling operations. Such studies require the use of the principles of geodesy. This chapter discusses some principles of geodesy and explains the calculations needed for the placement of wells taking into account the Earth's curvature, and explains commonly used geodetic expressions, such as projection of the Earth's surface on maps, datum, datum corrections, and other terms. It provides a general overview of the geodetic concepts, which are of importance in well construction especially when complex architecture is involved, such as in wells with extended and ultra-extended reach. However, only the matter of relevance to well drilling studies is discussed. For more details, readers are advised to refer to other books (see the References at the end of this chapter).

Ellipsoid and Geoid

Geodesy[1,2] is the study of the shape, size, and geometrical surface of the Earth, including the definition of datum (reference frame for locating points on the Earth's surface). Defining the reference frame or the datum is important for surveying and mapping of places on the Earth. Several ideas have been put forward over the years regarding the shape and size of the Earth. Earlier, the Earth was approximated as a sphere. With increased knowledge about its shape, the Earth was adequately modeled as an oblate spheroid or ellipsoid—a shape obtained by rotating an ellipse about its minor (polar) axis. The coordinates x, y, z of any point on the surface of the ellipsoid, with origin at the center of the ellipsoid, satisfies the equation:

$$\frac{x^2 + y^2}{a^2} + \frac{z^2}{b^2} - 1 = 0 \tag{2.1}$$

where

a = Semi-major axis of the ellipse of rotation
b = Semi-minor axis of the ellipse of rotation
x, y, and z are the axes of the Cartesian coordinate system on which the polar z-axis coincides with the minor axis and the y-axis coincides with the major axis.

Important parameters characterizing the ellipse are:

Eccentricity = e
Flattening or compression = $f = \frac{a - b}{a}$ (2.2)

The eccentricity is defined in two ways as:

$$\text{First eccentricity} = e_1 = \frac{\sqrt{a^2 - b^2}}{a} \tag{2.3}$$

For convenience, the first eccentricity (Eq. 2.3) is referred to as e in the rest of the chapter.

$$\text{Second eccentricity} = e_2 = \frac{\sqrt{a^2 - b^2}}{b} \tag{2.4}$$

Further accuracy in the knowledge of the shape of the Earth resulted in the ellipsoidal model being improved by the introduction of a geoidal model to define the true shape of the Earth. The geoidal surface is an imaginary surface with equipotential of Earth's gravity, which closely approximates mean sea level. Since the Earth's mass is not distributed uniformly, the magnitude and direction of the gravity vector changes at the surface of the Earth, resulting in an undulation of the geoid. The surface of the reference ellipsoid, the geoid, the ocean, and the Earth are shown in Figure 2.1. Traditionally, reference ellipsoids have been defined locally and are not used to fit the whole Earth. Different ellipsoids are used to fit a particular part and surface of the Earth. The relationship between ellipsoid and geoid height at a particular point on the Earth's surface is given by:

$$H = h \pm N \tag{2.5}$$

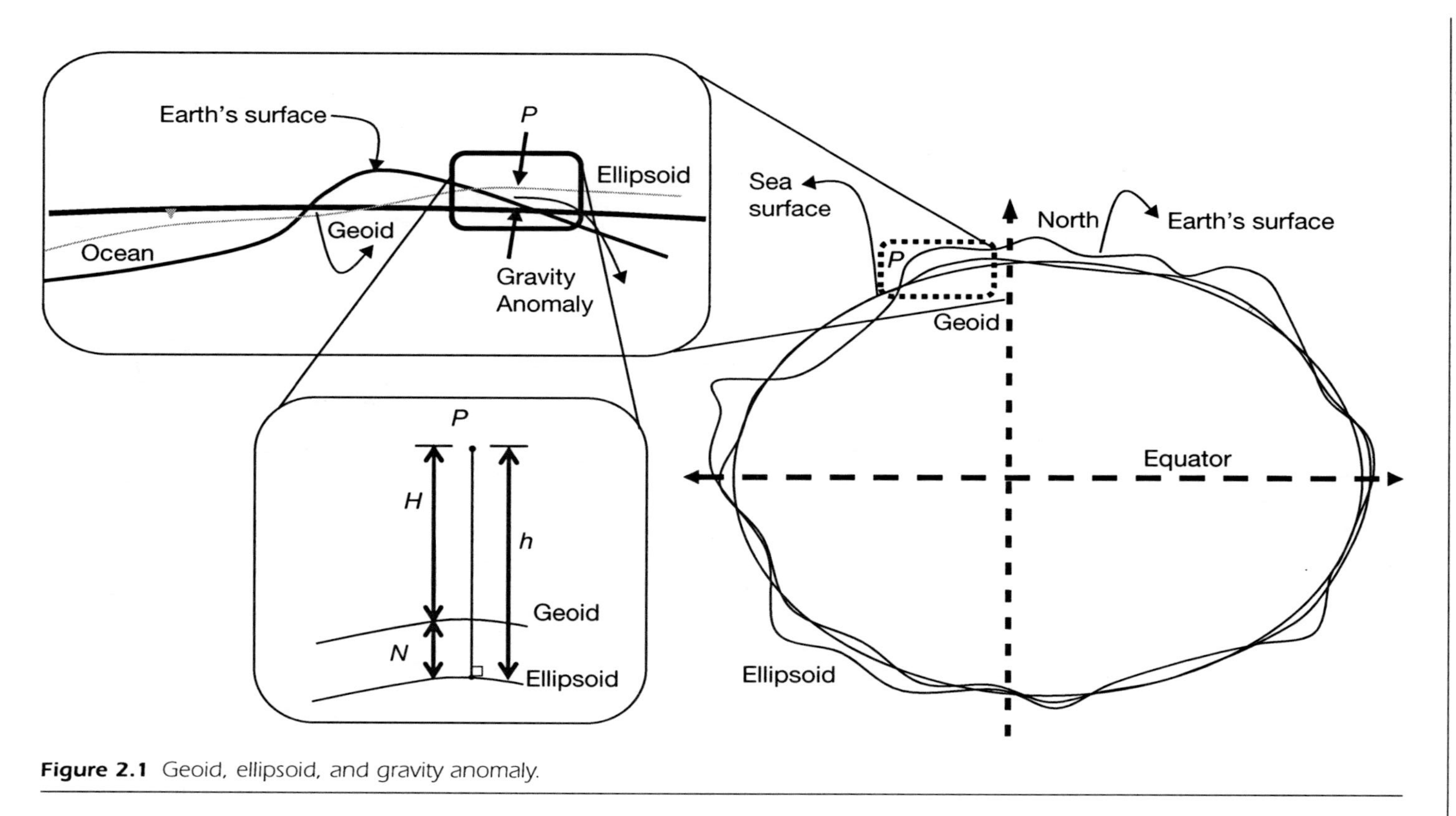

Figure 2.1 Geoid, ellipsoid, and gravity anomaly.

where, as indicated in Figure 2.1,

H = Orthometric height, which is the perpendicular distance from a point on the Earth's surface to the geoid
h = Ellipsoid height, which is the perpendicular distance from the point to the ellipsoid
N = Geoid height, which is the perpendicular distance from the ellipsoid to the geoid

In the United States, the geoid is below the ellipsoid, whereas at most other places the geoid is above the ellipsoid. Gravity anomaly or gravity undulation is the difference between the ellipsoidal surface and geoid.

Some of the terms associated with geodesy are reviewed below, as well as shown in Figures 2.1 and 2.2.

The Earth rotates about the primary polar axis, which passes through the center of the Earth and through the North and South Poles. The equatorial plane is equidistant from the poles and passes through the center of the Earth perpendicular to the primary polar axis. The intersection of the equatorial plane with the Earth's surface defines the equator. The latitude of a point on the Earth's surface is the angle between the radius from the Earth's center to that point and the equatorial plane, and is designated as ϕ. All

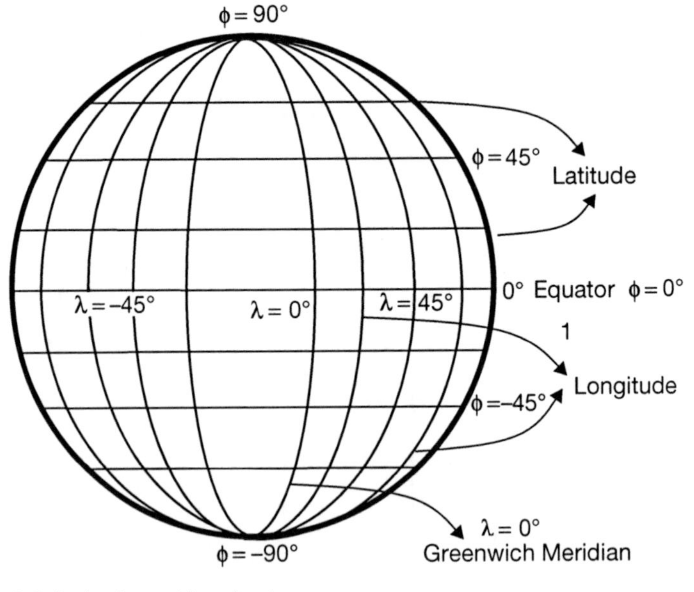

Figure 2.2 Latitude and longitude.

points of the same latitude lie on a line on the Earth's surface parallel to the equator, and these are called *parallels, east–west lines*, or the *lines of latitude.*

Meridian plane is any plane passing through the primary polar axis perpendicular to the equatorial plane. The intersection of a meridian plane with the Earth's surface defines a meridian. The meridian passing through Greenwich Observatory in England is taken as zero meridian.

Longitude of a point on the Earth's surface is the angle between the meridian passing through that point and the Greenwich or zero meridian and is designated as λ. All points with the same longitude lie on the same meridian, which are *north–south lines* or the *lines of longitude.*

The Greenwich meridian and the equator are both shown in Figure 2.3.

The radius of curvature in the prime vertical plane is given as:

$$N = \frac{a^2}{\sqrt{1 - e^2 \sin^2 \phi}} = \frac{a^2}{\sqrt{a^2 \cos^2 \phi + b^2 \sin^2 \phi}} \tag{2.6}$$

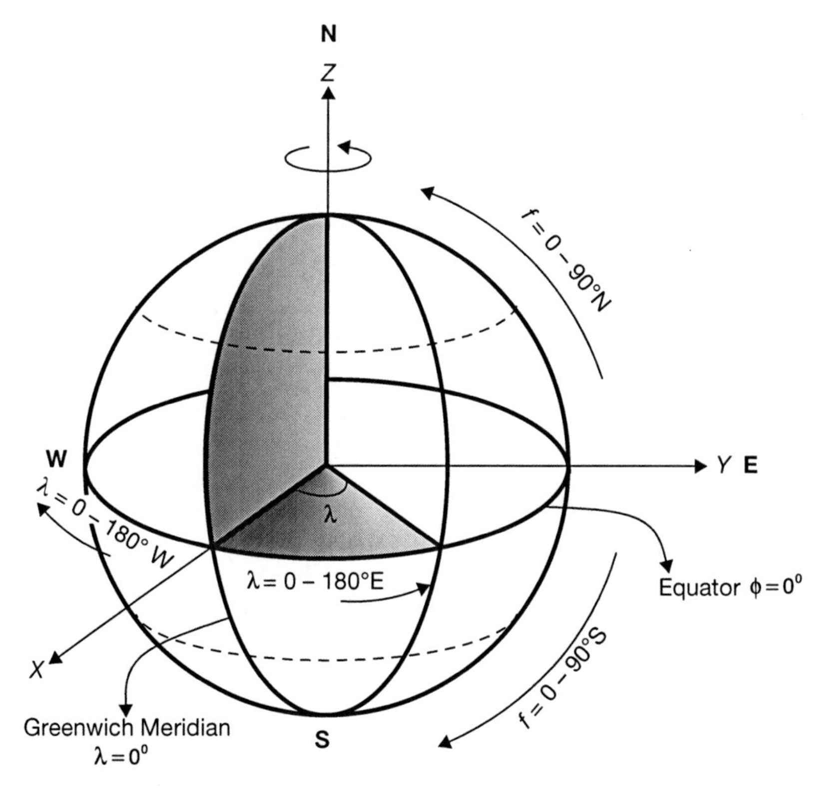

Figure 2.3 Meridians and equator.

The radius of curvature in the meridian is given as:

$$M = \frac{a\left(1-e^2\right)}{\left(1-e^2\sin^2\phi\right)^{\frac{3}{2}}} = \frac{a^2b^2}{\left(a^2\cos^2\phi + b^2\sin^2\phi\right)^{\frac{3}{2}}} \tag{2.7}$$

And the mean radius can be given as:

$$R_m = \sqrt{MN}$$

PROBLEM 2.1

Given the following data, calculate the mean radius of the Earth at 45°N latitude:

Semi-major axis = 6,378,137 m
Semi-minor axis = 6,356,752 m

Solution

a = 6,378,137 m and b = 6,356,752 m.

Radius of curvature in the prime vertical:

$$N = \frac{6,378,137^2}{\sqrt{6,378,137^2\cos^2 45 + 6,356,752\ ^2\sin^2 45}} = 6,386,976.3 \text{ m}$$

Radius of curvature in the meridian:

$$M = \frac{6,378,137^2 \times 6,356,752^2}{\left(6,378,137^2\cos^2 45 + 6,356,752^2\sin^2 45\right)^{\frac{3}{2}}} = 6,367,381.657 \text{ m}$$

$$\text{Mean radius} = R_m = \sqrt{6,139,320.055 \times 6,137,679.617}$$
$$= 6,374,383.53 \text{ m}$$

Types of Coordinates

The coordinates used for defining the spatial position of any point on the Earth's surface and the related variables are shown in Figures 2.4 and 2.5. The equatorial plane at any epoch becomes the mean equatorial plane.

1. Global Cartesian or geocentric coordinates are defined in terms of x, y, and z. It should be noted that z is the actual height of the point above the equatorial plane. The origin of the Cartesian system is the center of the ellipsoid with the primary polar axis as the semi-minor axis (z-axis). These Earth-centered/Earth-fixed coordinates are shown in Figure 2.4.

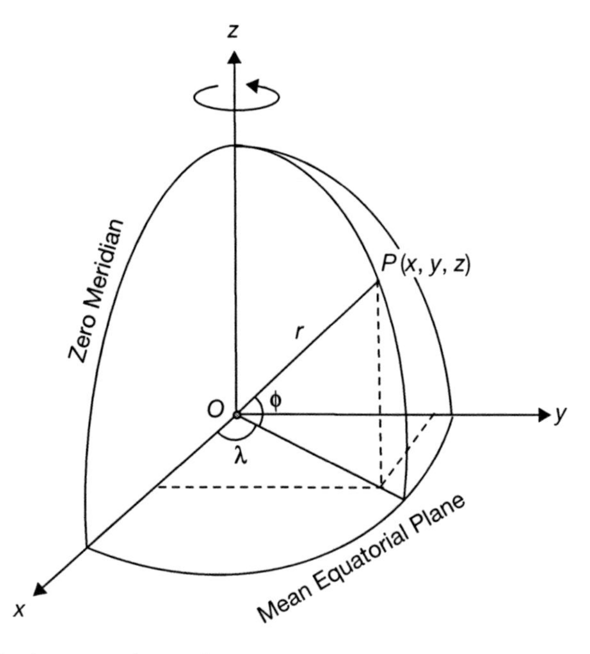

Figure 2.4 Earth-centered coordinates.

2. Geographical or geodetic coordinates, as shown in Figure 2.5, define the position of any point on the Earth's surface in terms of latitude, longitude, and height (ϕ, λ, h, respectively). The height h, called the ellipsoid height, is defined as the linear distance of the point above the ellipsoid.
3. Projected coordinates are plane coordinates used to map a point on the Earth's surface onto a tangent plane and expressed in terms of x, y, and z. A tangent plane is an orthogonal rectangular plane with the origin at any point on the Earth's surface. More details of map projections are explained in the Families of Map Projections section.

Using the Earth-center/Earth-fixed coordinates:

$$x_i = (h_i + N_i)\cos\phi_i \cos\lambda_i \tag{2.8}$$

$$y_i = (h_i + N_i)\cos\phi_i \sin\lambda_i \tag{2.9}$$

$$z_i = \left[h_i + (1 - e^2)N_i\right]\sin\phi_i \text{ or } z_i = \left[h_i + \frac{b^2}{a^2}N_i\right]\sin\phi_i \tag{2.10}$$

where

$$N = \frac{a^2}{\sqrt{1 - e^2\sin^2\phi}} = \frac{a^2}{\sqrt{a^2\cos^2\phi + b^2\sin^2\phi}} \tag{2.11}$$

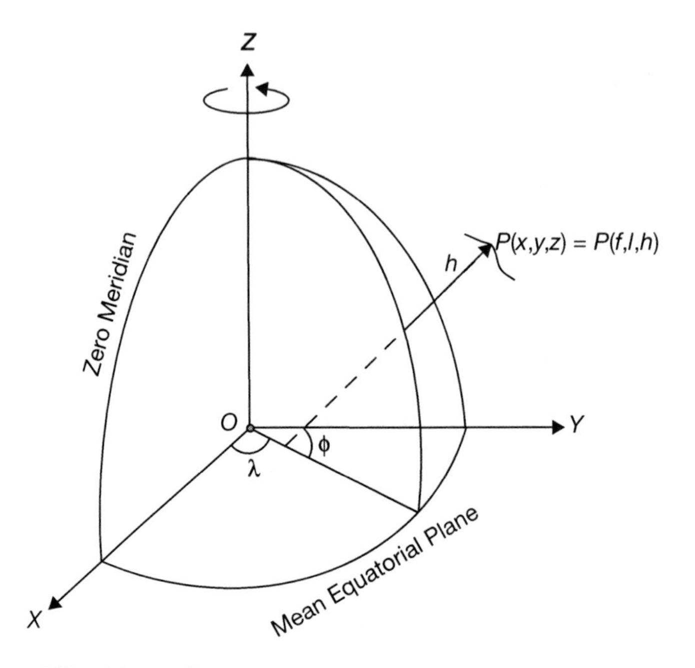

Figure 2.5 Ellipsoid coordinates.

Using Eqs. 2.8–2.11, the following equations can be derived:

$$\lambda_i = \arc\tan\left(\frac{y_i}{x_i}\right) \tag{2.12}$$

$$\phi_i = \arc\tan\left(\frac{z_i}{\sqrt{x_i^2 + y_i^2}\left(1 - e^2\frac{N_i}{N_i + h_i}\right)}\right) \tag{2.13}$$

$$h_i = \frac{\sqrt{x_i^2 + y_i^2}}{\cos\phi_i} - N_i \quad \text{or } h_i = \frac{z_i}{\sin\phi_i} - N_i\left(1 - e^2\right) \tag{2.14}$$

PROBLEM 2.2

Using the following data, calculate the tangent plane origin using the Earth-centered/Earth-fixed coordinates at the location 34°N latitude and 117.3°W longitude:

Semi-major axis = 6,378,300 m
Semi-minor axis = 6,356,860 m

Solution

$h = 232$ m.

Radius of curvature in the prime vertical:

$$N = \frac{6,378,300^2}{\sqrt{6,378,300^2 \cos^2 34 + 6,356,860^2 \sin^2 34}} = 6,384,823.308 \text{ m}$$

The coordinates are:

$$x_i = (232 + 6,384,823.308)\cos 34 \cos 11.3 = -2,427,838.83 \text{ m}$$
$$y_i = (232 + 6,384,823.308)\cos 34 \sin 117.3 = -4,703,851.56 \text{ m}$$
$$z_i = \left[232 + (1 - e^2)6,384,823.308\right]\sin 34 = 3,546,576 \text{ m}$$

Easting and Northing

Easting and northing are used to locate a particular point on a map using rectangular coordinates. Easting is the x value, the distance measured east of the origin, while northing is the y value, the distance measured north of the origin. The origin is selected in such a way that all easting and northing values are always positive, even when the point is south or west of the origin. For this, false origins with arbitrarily large positive values are used that will result in false easting and false northing values. Use of false easting and northing may not be necessary with modern computers.

Families of Map Projections

Projection involves the transformation of the approximately spherical surface of the Earth into a flat planar map surface. There is no perfect method to achieve this and, to some extent, all projections cause distortion or deformation in the areas, shapes, sizes, directions, and distances on the Earth's surface.

Principal Types of Projections

Projections are classified according to the method used for projecting the Earth's curved surface onto a flat map. As indicated in Figure 2.6, they can be broadly classified as cylindrical, conic, azimuthal (or planar), and miscellaneous. Several researchers have proposed several methods of projection in the broad classes, and some of them are given in the chart in the figure. Readers are advised to refer to books on map projections for further details. Widely used map projections are discussed in the following sections.

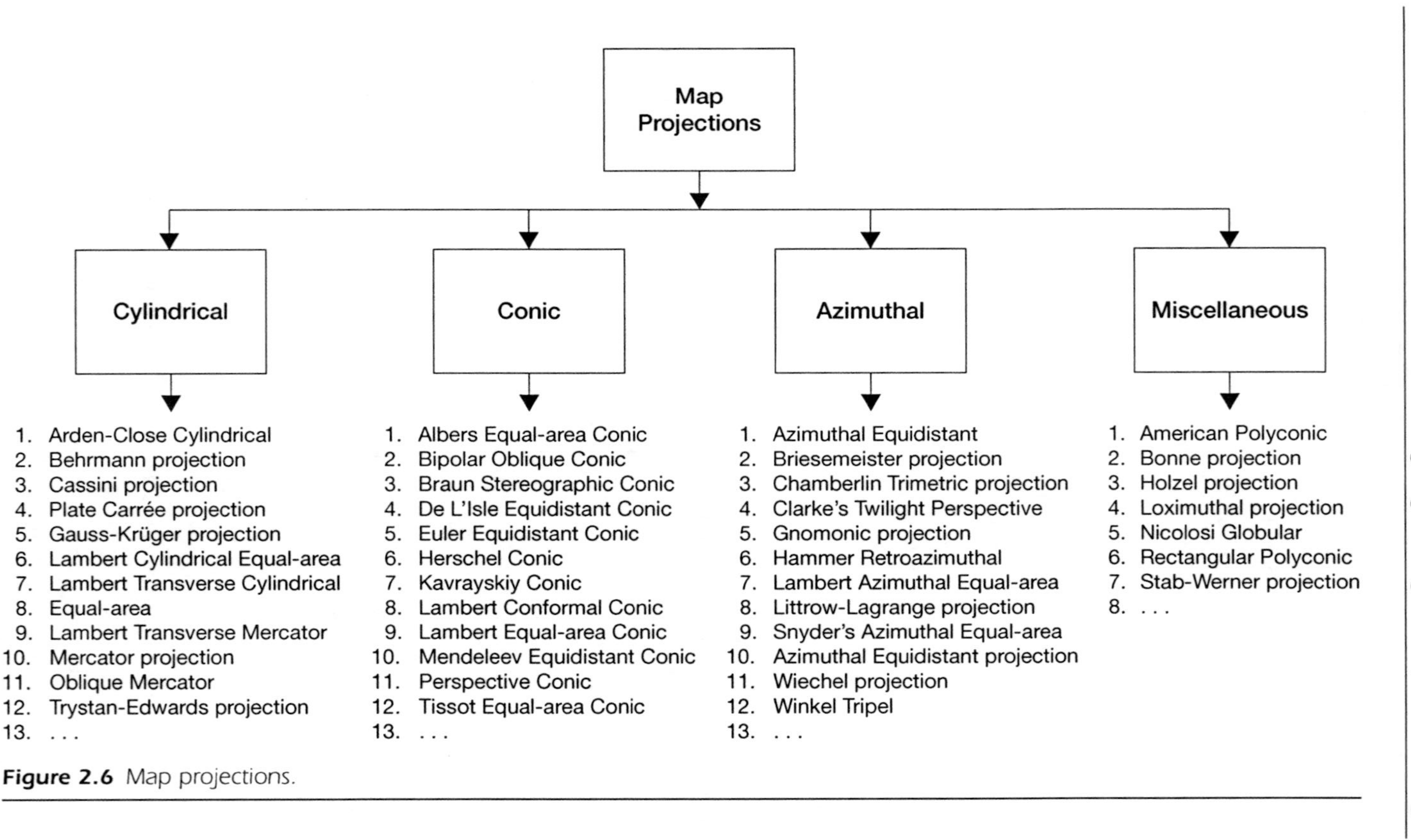

Figure 2.6 Map projections.

Cylindrical Projections

The basis of the cylindrical projection is that the points of the Earth's surface are projected onto a cylinder wrapped around the Earth and then unwrapped by cutting along any meridian. On the resulting projection, lines of longitude are equidistant, parallel straight lines, whereas lines of latitude cross the longitudinal lines at right angles and are not equally spaced. Cylindrical projection in general is more accurate for regions near the equator and less accurate, that is, with more pronounced distortion, for regions near the poles as seen in Figure 2.7.

Mercator Projection

The Mercator projection is a cylindrical projection where the cylinder touches at the equator, as shown in Figure 2.8. In this projection, longitude lines are vertical, equally spaced and parallel to each other, whereas latitude lines are horizontal and unequally spaced, crossing the longitude lines at right angles. The projection can represent a large portion of the globe except

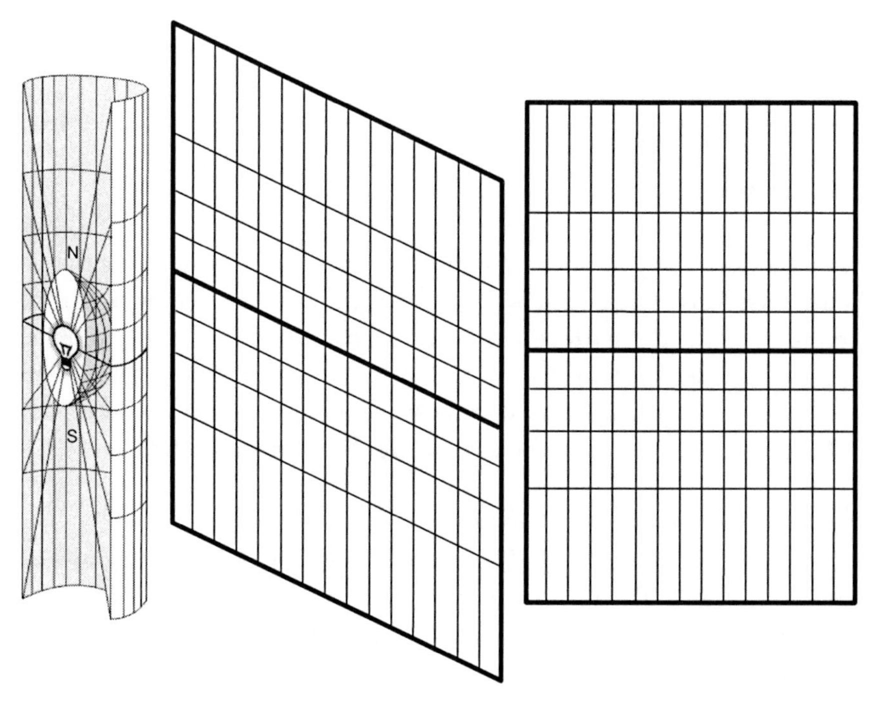

Figure 2.7 Cylindrical projections.

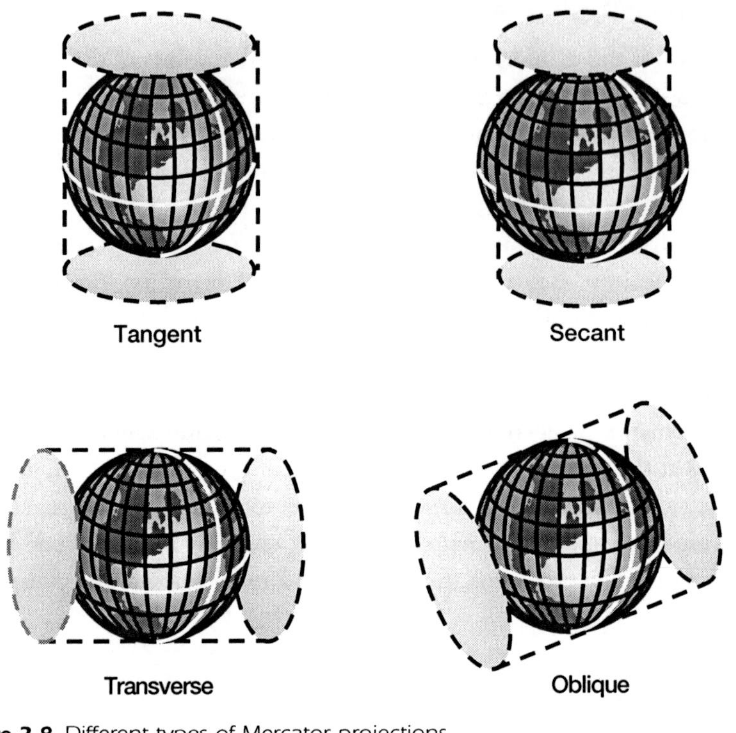

Figure 2.8 Different types of Mercator projections.

near the poles. To compensate for the distortion caused by replacing the circular meridians of longitude by parallel straight lines, the spacing of the latitude lines is increased toward the poles, so that the projection is conformal, preserving the shape of surface areas but not the size. It is used for offshore and marine navigation because straight lines are lines of constant azimuth. Different types of Mercator projections are shown in Figure 2.8.

Transverse Mercator Projection

The transverse Mercator (TM) projection, as shown in Figure 2.8, touches at a meridian. In this projection, the points are projected onto a cylinder tangent to a chosen longitude line. Latitude and longitude lines are no longer straight lines and the distortion of scale, distance, direction, and area increase away from the chosen central longitude. Calculations involving this projection are given later in this chapter.

Universal Transverse Mercator Projection

The universal transverse Mercator (UTM) projection is an ellipsoidal projection (an ellipsoid is used for projection), derived from the transverse Mercator projection using a secant cylinder. In this projection, the Earth is divided into 60 zones (each of 6° of longitude in width), which allows projection onto a map with minimum distortion. The scale variation within a zone is only 1 part in a 1000. The origin in each zone is the intersection of the equator with the central longitude, with x value of 500,000 m and y value of 0 m for the Northern Hemisphere. The values of x increases to the east and y increases to the north. The zones are numbered from west to east, starting with zero at the zero meridian. In a zone, the scale distortion is 0.9996 along the central meridian, and is 1.00158 at the edges, as shown in Figure 2.9. Calculations involving this projection are given later in this chapter.

Conic Projection

The basis of conic projection is that the points on the Earth's surface are projected radially onto a cone by aligning the axis of the cone with that of the Earth, as indicated in Figure 2.10. In this projection, the meridians become straight lines radiating from the poles, while the parallels become curved lines. On the resulting projection, parallels may be equally spaced. Common conic projections are Albers, Lambert, and polyconic. Conic projection in general is less distorted between the parallels near to the tangency of the conic surface, and more distorted away from this contact. Polyconic projections use a series of conic projections to increase the accuracy of the mapping.

Lambert Conformal Conic (LCC) Projection

Standard parallels run from east and west of the principal meridian intersecting the Earth along the parallels of latitude when a cone is inserted. In these projection directions, areas and shapes get distorted away from the standard parallels. Areas and directions are true along the standard parallels. The scale variation is shown in Figure 2.10, which depicts the compressed and elongated zones. More details about calculations using this projection are given later in this chapter.

Azimuthal (Planar) Projection

The azimuthal or zenithal projection (also called gnomonic projection) is obtained by projecting the points of the Earth's surface onto a plane that touches

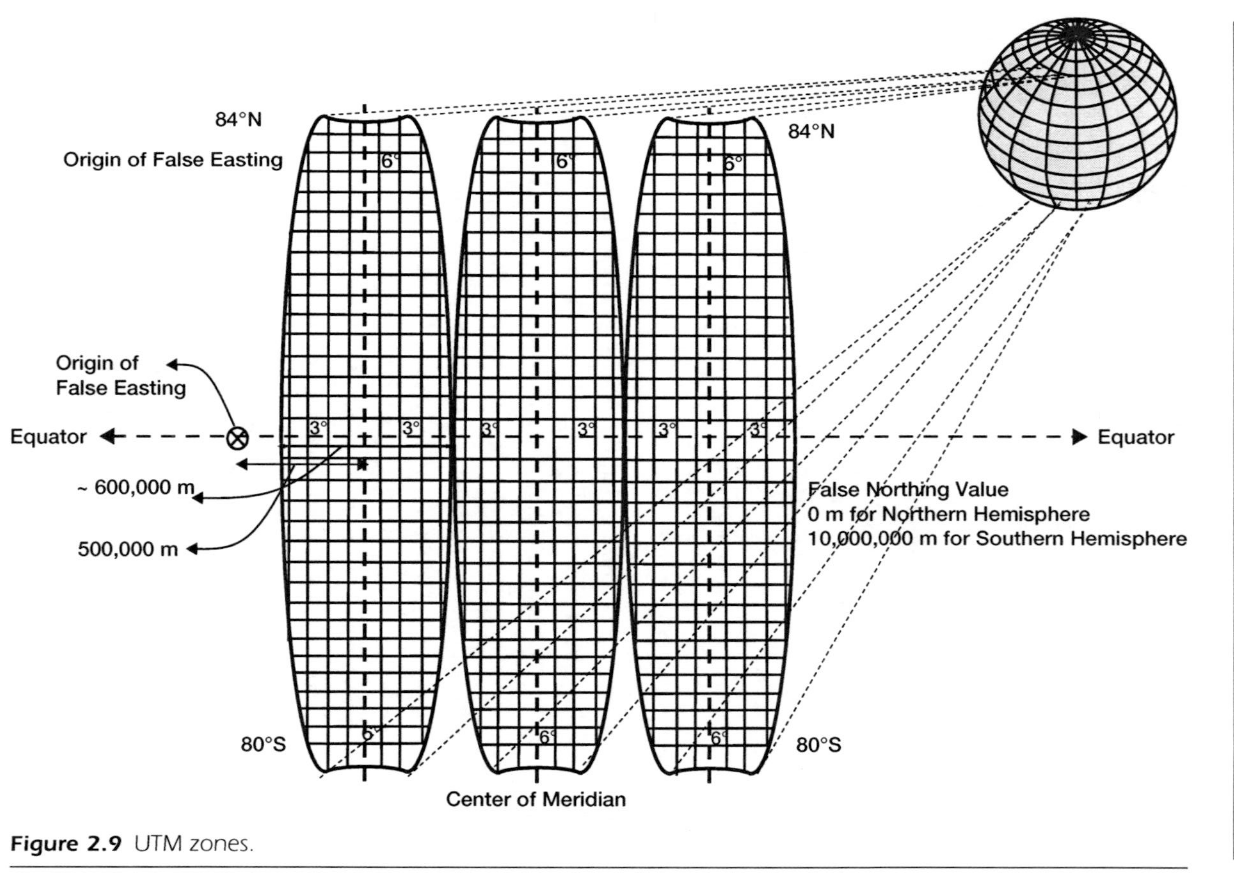

Figure 2.9 UTM zones.

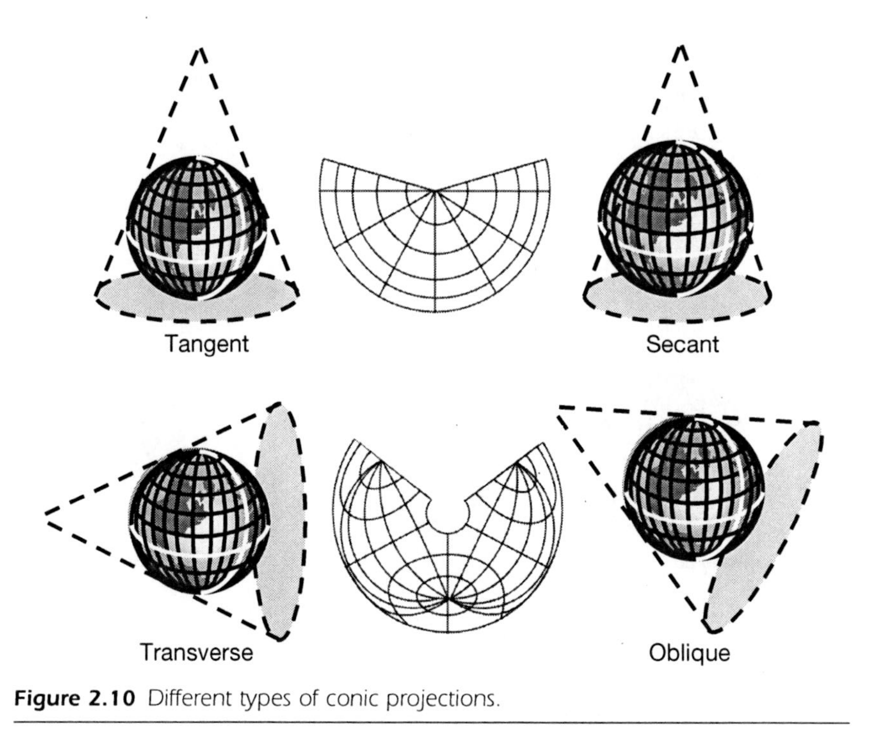

Figure 2.10 Different types of conic projections.

the Earth's surface at one point. This projection results in little distortion at the point of contact, but the distortion in distance and direction increases away from the point of contact, becoming pronounced at the edges. In the azimuthal projection, with the surface touching at a pole, the great circles of longitude become straight lines and the parallels of latitude become unevenly spaced circles. The circles are more closely spaced toward the edges of the projection, as shown in Figure 2.11. The projection can be made based on tangent, secant or oblique planes. These different types of planar projections, as well as the resulting maps, are shown in Figure 2.11.

The chart in Table 2.1 indicates the areas of applicability of the different map projections.

Datum

The datum is used to provide a reference frame or a coordinate system to calculate the location of any point on the Earth's surface, including the height (vertical component) and the distance (horizontal component). The horizontal component is based on the reference point on the horizontal surface,

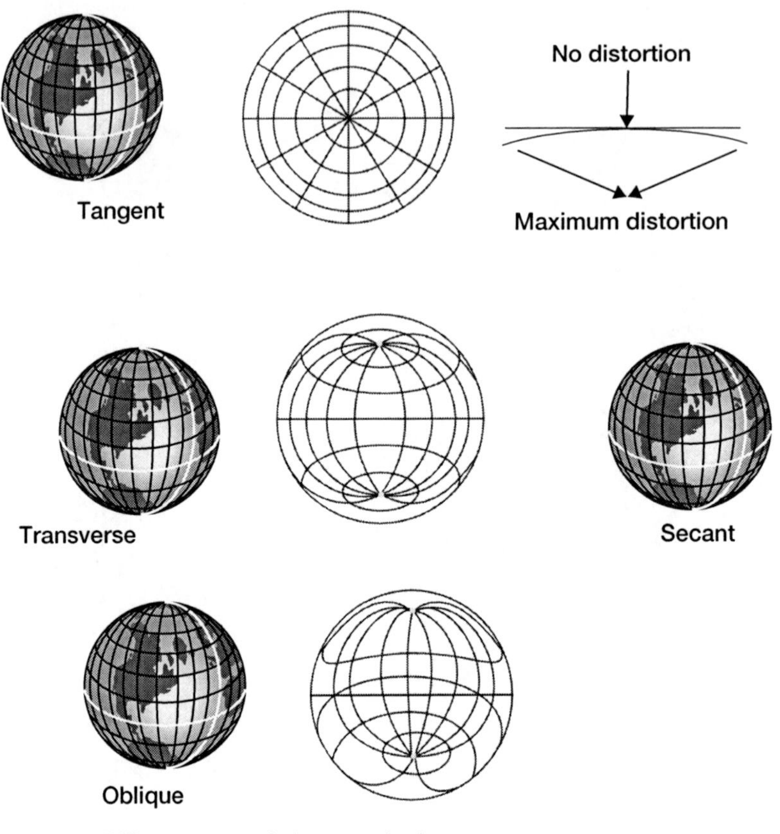

Figure 2.11 Different types of planar projections.

whereas the vertical component is based on the vertical surface, taking into account the gravity anomaly. As seen earlier the coordinate *t* can be either Cartesian (i.e., x, y, z) or geographical (i.e., ϕ, λ, H). Horizontal datums used are North American datum, 1983 (NAD83), and World Geodetic System of 1984 (WGS84). Vertical Datums used are North American Vertical Datum of 1988 (NAD88), National Geodetic Vertical datum of 1929 (NGVD29), Earth Gravity Model of 1996 (EGM96), World Geodetic System of 1984—Ellipsoid (WGS84E), and World Geodetic System of 1984—Geoid (WGS84G).

Charts using different datum are based on different ellipsoids and coordinate reference frames. Thus, the coordinates of the same point will be different. Before a chart is used, the datum on which the chart was constructed should be carefully verified. Different charts use different principles to reduce the local gravity anomaly or the local geoid-ellipsoid separation. There-

Table 2.1 Use of Map Projections

Projection	Area	Distance	Direction	Shape	Region	Topography
Transverse Mercator	✔		✘	✘	✔	✔
Miller Cylindrical					✔	
Lambert Azimuthal Equal Area			✘		✔	
Lambert Equidistant Azimuthal		✘	✘		✔	✔
Albers Equal Area Conic		✘	✘		✔	✔

Note: ✔ = yes; ✘ = partly.

fore, the latitude and longitude of a place are not the same in different charts, and have to be qualified with reference datum, as well as with the type of projection, to avoid computational errors and positional inaccuracies.

Geodetic Transformation

There are several methods for the transformation of spatial data from one datum to another datum. Commonly used methods are geocentric transformation (three parameters) and seven-parameter transformation.

Geocentric transformation involves only translations along the three axes (i.e., Δx, Δy, Δz). Seven-parameter transformation involves three translations, three rotations, and one scale distance change. The seven-parameter transformation from one Cartesian coordinate system to another can be mathematically expressed as:

$$\begin{bmatrix} x_2 \\ y_2 \\ z_2 \end{bmatrix} = \begin{bmatrix} \Delta x \\ \Delta y \\ \Delta z \end{bmatrix} + \begin{bmatrix} 1+\Delta s & \omega & -\psi \\ -\omega & 1+\Delta s & \zeta \\ \psi & -\zeta & 1+\Delta s \end{bmatrix} \begin{bmatrix} x_1 \\ y_1 \\ z_1 \end{bmatrix} \tag{2.15}$$

where

x_1, y_1, z_1 are the local datum Cartesian coordinates
x_2, y_2, z_2 are the global datum Cartesian coordinates

The Cartesian coordinates x, y, z can be expressed in terms of geographical coordinates ϕ, λ, H as follows:

$$x_i = (h_i + N_i)\cos\phi_i \cos\lambda_i \tag{2.16}$$

$$y_i = (h_i + N_i)\cos\phi_i \sin\lambda_i \tag{2.17}$$

$$z_i = \left[h_i + (1 - e^2)N_i\right]\sin\phi_i \quad \text{or} \quad z_i = \left[h_i + \frac{b^2}{a^2}N_i\right]\sin\phi_i \tag{2.18}$$

where

$$N = \frac{a^2}{\sqrt{1 - e^2\sin^2\phi}} = \frac{a^2}{\sqrt{a^2\cos^2\phi + b^2\sin^2\phi}} \tag{2.19}$$

Grid North and True North

All lines of longitude are true north lines and follow the direction of the north–south grid lines on the map. True north uses latitude and longitude coordinates of the curved Earth as the reference. Actual lines of longitude converge to the rotational pole. In the grid system, the lines of longitude are parallel to the y-axis and do not converge to a single point. Since the actual lines of longitude are not parallel, the grid north on the map is different from the true north, as shown in Figure 2.12. True north and grid north or map north is the same at the central meridian. The difference between the true north and grid north is called the grid convergence. In the Northern Hemisphere, as shown in Figure 2.12, the grid convergence is positive east of the central meridian and negative west of the central meridian, and vice versa in the Southern Hemisphere.

Magnetic Reference and Interference

The magnetic north is the direction to which the north pole of a compass would point to when there is no interference. The difference between the grid north and magnetic north is called magnetic variation. The angular difference in azimuth readings between magnetic north and true north is called magnetic declination. The magnetic declination for compass correction is positive when the magnetic north lies east of true north, and negative when the magnetic north lies west of true north (Figure 2.13). It is actually the error between the true north and magnetic north for a specific location.

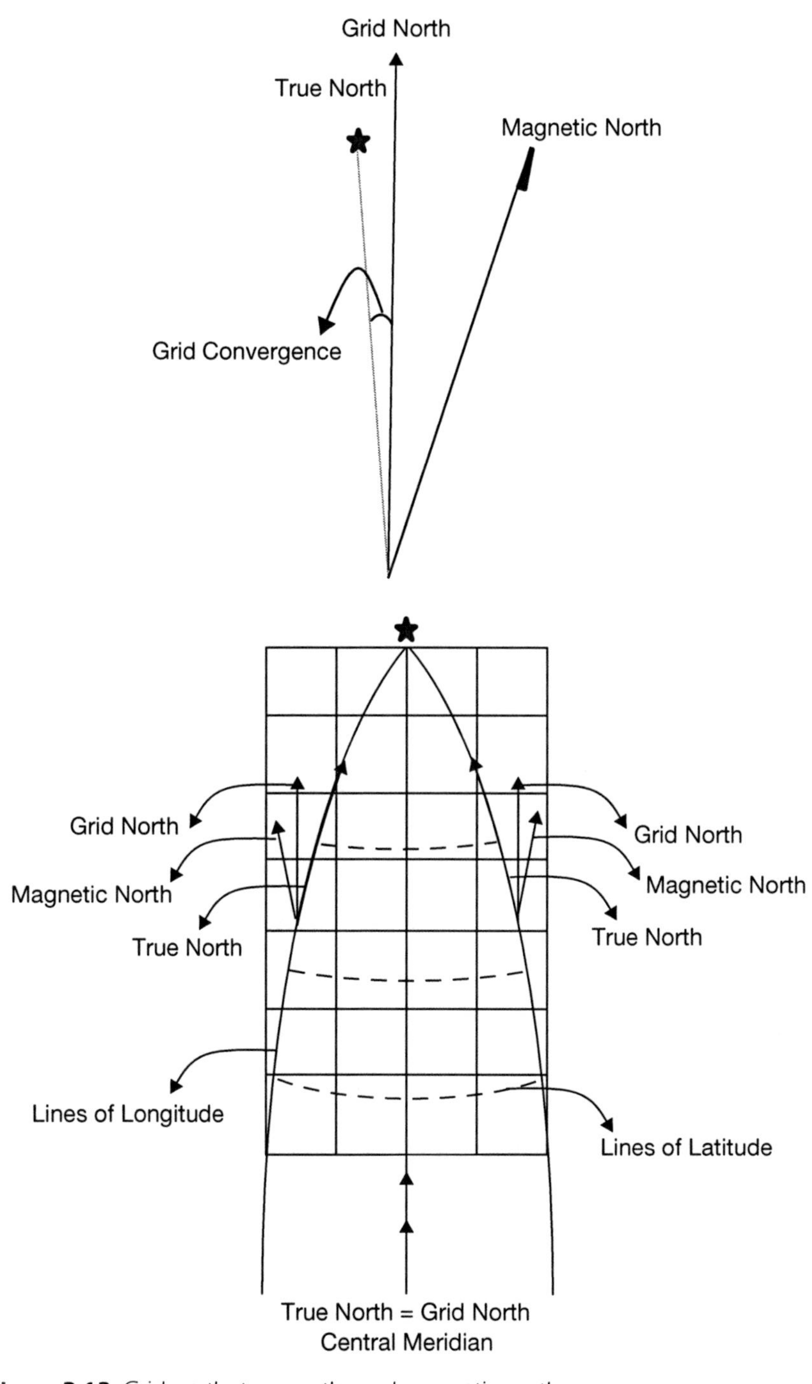

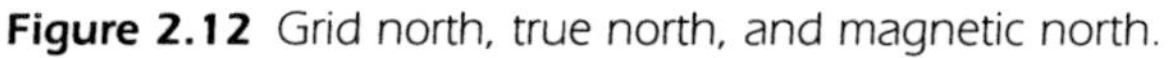

Figure 2.12 Grid north, true north, and magnetic north.

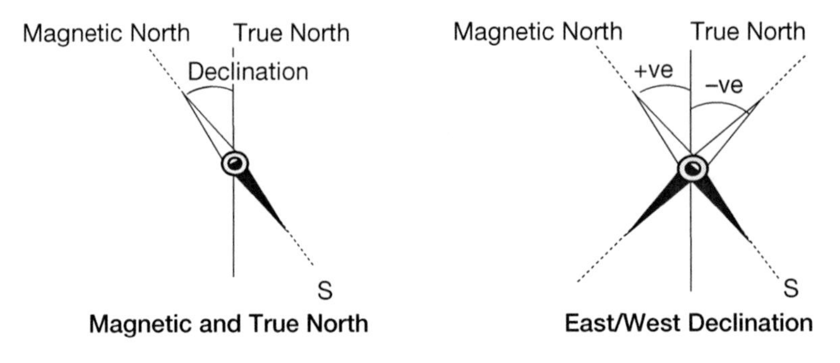

Figure 2.13 Magnetic and true north, east/west declination.

Table 2.2 Rules for Magnetic Declination

Option	Easterly	Westerly
Grid to Magnetic	Subtract	Add
Magnetic to Grid	Add	Subtract

The magnetic declination varies from place to place, and at a given place, changes slowly with time. Magnetic interference can be caused by the proximity of metal components, such as collars or adjacent casings. In map readings, grid north and magnetic north are used.

The simple rules given in Table 2.2 will help to properly obtain the corrected azimuth angle.

PROBLEM 2.3

Determine the azimuth with respect to the true north for the following wells:

N45E, Declination 5°W

N45E, Declination 5°E

Solution

Azimuth = 45 – 5 = 40° = N40E

Azimuth = 45 + 5 = 50° = N50E

The declinations are shown in Figure 2.14.

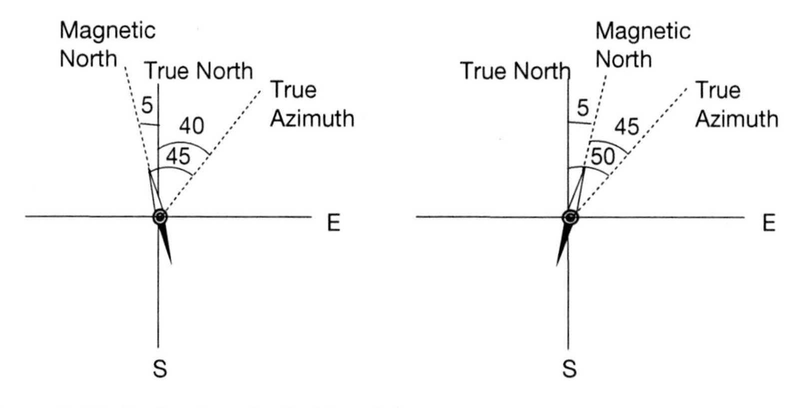

Figure 2.14 Declinations for Problem 2.3.

PROBLEM 2.4

Convert the following from grid north to magnetic north:

Grid Azimuth 111°, Declination 11°E
Grid Azimuth 199°, Declination 6°W
Grid Azimuth 55°, Declination 5°W
Grid Azimuth 79°, Declination 9°E

Convert the following from magnetic north to grid north:

Magnetic Azimuth 9°, Declination 11°E
Magnetic Azimuth 113°, Declination 11°E
Magnetic Azimuth 171°, Declination 11°W
Magnetic Azimuth 211°, Declination 21°W

Solution

Grid north to magnetic north:

Magnetic Azimuth: 111 – 11 = 100°
Magnetic Azimuth: 199 + 6 = 205°
Magnetic Azimuth: 55 + 5 = 60°
Magnetic Azimuth: 79 – 9 = 70°

Magnetic north to grid north:

Grid Azimuth: 9 + 11 = 20°
Grid Azimuth: 113 + 11 = 124°
Grid Azimuth: 171 – 11 = 160°
Grid Azimuth: 211 – 21 = 190°

Projection Calculations

This section discusses important projection methods and the associated calculations. Several map projection methods are available and the most commonly used calculations are discussed here for basic understanding. For other calculations, readers are advised to refer to the References and other cited sources.

Transverse Mercator Projection

The transverse Mercator projection is based on the Gauss-Krüger method, and uses the ellipsoid as datum. Gauss-Krüger proposed the following method of projection. The projection of the latitude and longitude are given by[3,4]:

$$\phi = \phi_f + \frac{t_f}{2N_f^2}\left(-1-\eta_f^2\right)y^2 + \frac{t_f}{24N_f^4}\left(5+3t_f^2+6\eta_f^2-6t_f^2\eta_f^2-3\eta_f^4-9t_f^2\eta_f^4\right)y^4$$

$$+\frac{t_f}{720N_f^6}\left(-61-90t_f^2-45t_f^4-107\eta_f^2+162t_f^2\eta_f^4+45t_f^4\eta_f^2\right)y^6+\ldots \quad (2.20)$$

$$\lambda = \lambda_0 + \frac{1}{N_f\cos\phi_f}y + \frac{1}{6N_f^3\cos\phi_f}\left(-1-2t_f^2-\eta_f^2\right)y^3$$

$$+\frac{1}{120N_f^5\cos\phi_f}\left(5+28t_f^2+24t_f^4+6\eta_f^2+8t_f^2\eta_f^2\right)y^5+\ldots \quad (2.21)$$

where λ_0 is the longitude at the selected origin.

The radius of the prime vertical is:

$$N_f = \frac{a^2}{b\sqrt{1-e_2^2\sin^2\phi_f}} \quad (2.22)$$

$$t_f = \tan\phi_f \quad (2.23)$$

$$\eta_f = e_2^2\cos^2\phi_f \quad (2.24)$$

ϕ_f is the footpoint latitude and is given by:

$$\phi_f = \frac{x}{\bar{\alpha}} + \left(\frac{3}{2}e_r - \frac{27}{32}e_r^3 + \frac{269}{512}e_r^5 + \ldots\right)\sin\frac{2x}{\bar{\alpha}} + \left(\frac{21}{16}e_r^2 - \frac{55}{32}e_r^4 + \ldots\right)\sin\frac{4x}{\bar{\alpha}}$$

$$+\left(\frac{151}{96}e_r^3 - \frac{417}{128}e_r^5 + \ldots\right)\sin\frac{6x}{\bar{\alpha}} + \left(\frac{1097}{512}e_r^4 + \ldots\right)\sin\frac{8x}{\bar{\alpha}} + \ldots$$

$$(2.25)$$

where the parameters $\bar{\alpha}, e_r$ are given by Eqs. 2.26 and 2.27:

$$\bar{\alpha} = \left(\frac{a+b}{2}\right)\left(1+\frac{1}{4}e_r^2+\frac{1}{64}e_r^4+\ldots\right) \quad (2.26)$$

$$e_r = \frac{1-\sqrt{1-e^2}}{1+\sqrt{1-e^2}} \tag{2.27}$$

Footpoint latitude is defined as the latitude at the central meridian that has the same coordinate as that of point P as shown in Figure 2.15. It is obtained by drawing a parallel line to the x-axis at point P.

Universal Transverse Mercator Projection

UTM is a modification of the transverse Mercator projection for global implementation. In this projection, the Earth is divided into 60 zones (each of 6° longitude width), which enables projection onto a map with minimum distortion.

The UTM coordinates can be converted from the Gauss-Krüger coordinates as follows:

$$X_{UTM} = S \times X_{GK} \tag{2.28}$$

$$Y_{UTM} = S \times Y_{GK} \tag{2.29}$$

where X_{UTM} and Y_{UTM} are UTM coordinates and X_{GK} and Y_{GK} are Gauss-Krüger coordinates as obtained in the preceding section; and S is a scale factor that is defined as the ratio between the actual distance projected on the map to the actual distance on the sphere.

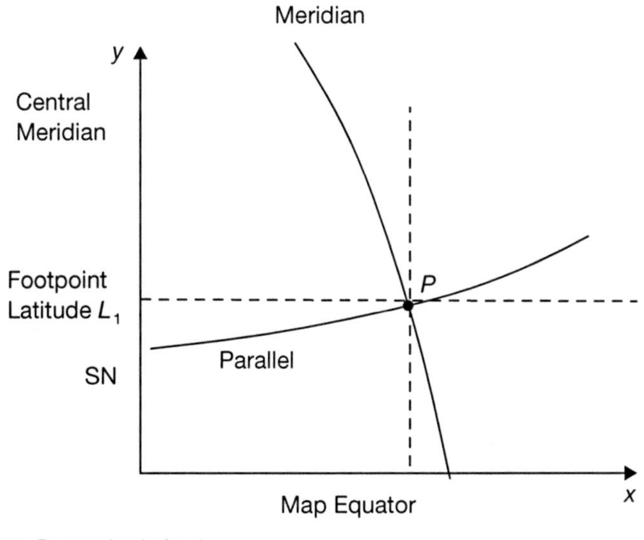

Figure 2.15 Footpoint latitude.

Lambert Conformal Conic Projection

LCC is also a widely used projection. It is a secant method using two standard parallels with angles ϕ_1 and ϕ_2, as shown in Figure 2.16. The map coordinates can be estimated with the geographical coordinates based on the ellipsoid using the following equations.[5,6]

$$X = r_0 \sin\theta;\; Y = r_1 - r_0 \cos\theta \tag{2.30}$$

Here, θ is the polar angle of the cone and is given by:

$$\theta = (\lambda - \lambda_0)\xi \tag{2.31}$$

where

λ_0 is the longitude at the origin selected
r_0 is the polar radius to the origin and is given by:

$$r_0 = \psi\left[t_0\left(\frac{m_0}{n_0}\right)^{\frac{e}{2}}\right]^{\xi} \quad \text{and} \quad r_1 = \psi\left[t_1\left(\frac{m_1}{n_1}\right)^{\frac{e}{2}}\right]^{\xi} \tag{2.32}$$

The functions ξ and ψ in Eqs. 2.31 and 2.32 are as follows:

$$\xi = \frac{\ln\left[\left(\frac{\cos\phi_1}{\cos\phi_2}\right)\left(\frac{m_2 n_2}{m_3}\right)^{\frac{1}{2}}\right]}{\ln t_1\left(\frac{m_1}{n_1}\right)^{\frac{e}{2}} - \ln t_2\left(\frac{m_2}{n_2}\right)^{\frac{e}{2}}} \quad \text{and} \quad \psi = \frac{\frac{a\cos\phi_1}{\sqrt{m_1 n_1}}}{\xi\left[t_1\left(\frac{m_1}{n_1}\right)^{\frac{e}{2}}\right]^{\xi}} \tag{2.33}$$

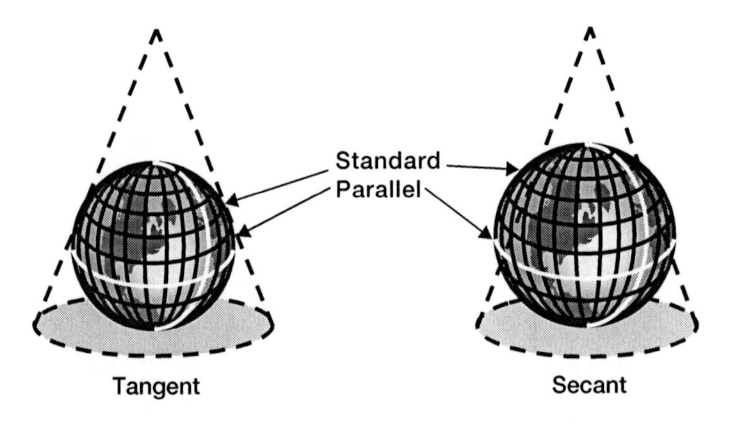

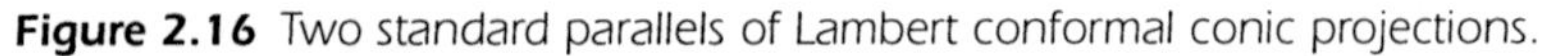
Figure 2.16 Two standard parallels of Lambert conformal conic projections.

where $m_{0,1,2,3}$, $n_{0,1,2,3}$, and $t_{0,1,2}$ are functions of the eccentricity and the latitude and are given by:

$$m_0 = 1 + e\sin\phi_0;\ n_0 = 1 - e\sin\phi_0;\ m = 1 + e\sin\phi;\ n = 1 - e\sin\phi \tag{2.34}$$

$$m_1 = 1 + e\sin\phi_1;\ n_1 = 1 - e\sin\phi_1;\ m_2 = 1 + e\sin\phi_2;\ n_2 = 1 - e\sin\phi_2 \tag{2.35}$$

$$m_3 = 1 + e^2\sin\phi_1;\ n_3 = 1 - e^2\sin\phi_1 \tag{2.36}$$

$$t_0 = \tan\left(\frac{\pi}{4} - \frac{\phi_0}{2}\right);\ t = \tan\left(\frac{\pi}{4} - \frac{\phi}{2}\right);\ t_1 = \tan\left(\frac{\pi}{4} - \frac{\phi_1}{2}\right);\ t_2 = \tan\left(\frac{\pi}{4} - \frac{\phi_2}{2}\right) \tag{2.37}$$

The above equations can be used for spherical projection by setting the eccentricity to zero and $R = a$.

PROBLEM 2.5

An oil company plans to move a rig with the following geographical coordinates:

$\phi = 29°$N latitude
$\lambda = 96°$W longitude

Using the Lambert conformal conic method, calculate the Cartesian coordinates of the location. Use the Bessel ellipsoid data. Other data are:

Standard parallels: $\phi_1 = 30°$ and $\phi_2 = 42°$
Origin data: $\phi_0 = 26°$N and $\lambda_0 = 135°$W

Solution

Data for the Bessel ellipsoid are:

$a = 6{,}377{,}397.155$ m
$b = 6{,}356{,}078.963$ m

$$\text{Eccentricity } e = \frac{\sqrt{6{,}377{,}397.155^2 - 6{,}356{,}078.963^2}}{6{,}377{,}397.155} = 0.081696831$$

$m_0 = 1 + 0.0816968 \sin 26 = 1.0358153$
$n_0 = 1 - 0.0816968 \sin 23 = 0.96418646$
$m = 1 + 0.0816968 \sin 29 = 1.03960741$
$n = 1 - 0.0816968 \sin 29 = 0.96039259$
$m_1 = 1 + 0.0816968 \sin 30 = 1.04084841$
$n_1 = 1 - 0.0816968 \sin 30 = 0.95915158$
$m_2 = 1 + 0.0816968 \sin 42 = 1.05466585$
$n_2 = 1 - 0.0816968 \sin 42 = 0.94533415$

$m_3 = 1 + 0.0066743 \sin 30 = 0.99833140$

$n_3 = 1 - 0.00667435 \sin 30 = 0.99701164$

$$t_0 = \tan\left(\frac{\pi}{4} - 13\right) = 0.62486935; \; t = \tan\left(\frac{\pi}{4} - 14.5\right) = 0.58904501$$

$$t_1 = \tan\left(\frac{\pi}{4} - 15\right) = 0.57735026; \; t_2 = \tan\left(\frac{\pi}{4} - 21\right) = 0.44522868$$

Calculating the value of ξ:

$\xi = 0.58888449$ and $\psi = 1{,}294{,}605.7$

Further radii are calculated as:

$r_0 = 9{,}497{,}499.97$ and $r_1 = 9{,}831{,}716.62$

The polar angle is:

$\theta = (-96 - (-135))0.58888449 = 22.97°$

The geographical coordinates can be obtained as:

$X = 3{,}705{,}855.90$ m; $Y = 1{,}087{,}053.23$ m

Polar Azimuthal Equidistant Projection

These map coordinates can be obtained from the geographical coordinates based on the ellipsoid using the following equations[6,7]:

$$X = r \sin(\lambda - \lambda_0); \; Y = -r \cos(\lambda - \lambda_0) \tag{2.38}$$

where

$$r = M \pm M_P \text{ (+ for south polar mapping, – for north polar mapping)} \tag{2.39}$$

$$M = a(M_1\phi + M_2 \sin 2\phi + M_3 \sin 4\phi + M_4 \sin 6\phi + \ldots) \tag{2.40}$$

$$M_1 = 1 - \frac{1}{4}e^2 - \frac{3}{64}e^4 - \frac{5}{256}e^6 \ldots \tag{2.41}$$

$$M_2 = -\frac{3}{8}e^2 - \frac{3}{32}e^4 - \frac{45}{1024}e^6 \ldots \tag{2.42}$$

$$M_3 = \frac{15}{256}e^4 + \frac{45}{1024}e^6 \ldots \tag{2.43}$$

$$M_4 = -\frac{35}{3072}e^6 \ldots \tag{2.44}$$

$$M_1 = 1 - \frac{1}{4}e^2 - \frac{3}{64}e^4 - \frac{5}{256}e^6 \ldots \tag{2.45}$$

M_P is calculated using the value $\phi = 90°$.

The scale factor is given as:

$$S = \frac{r\sqrt{1 - e^2 \sin^2 \phi}}{a \cos \phi}$$

PROBLEM 2.6

An oil company plans to move a rig with the following geographical coordinates:

ϕ = 29°N latitude
λ = 96°W longitude

Using the azimuthal equal-area projection method, calculate the Cartesian coordinates of the location 65°S and 20°E. Use the Clark 1880 data. Other data are origin data (south pole): 90°S and λ_0 = 90°E.

Solution

Data for Clark 1880 datum are:

a = 6,378,249.145 m; b = 6,356,514.869 m

Given location ϕ = 65°S and λ = 20°E

$$M = 6{,}378{,}249.145\begin{pmatrix} 0.9982969 \times 65^{\circ} - 0.00255567 \times \sin 2 \times 65 \\ +2.7260130\text{E} - 06 \times \sin 4 \times 65 - 3.5879481\text{E} - 09 \sin 6 \times 65 + \ldots \end{pmatrix}$$

$$= -7{,}211{,}066.814$$

Using the value of 90°S, M_p = −10,001,867.55

r = −17,212,934

Coordinates are calculated using Eq. 2.38:

X = 15,600,216.45 m; Y = 7,274,500.40 m

Supplementary Problems

1. With the given geographical coordinates, latitude = 43°N, longitude = 79°W, h = 215 m, estimate the Cartesian coordinates X, Y, and Z. Semi-major axis = 6,378,300 m, semi-minor axis = 6,356,860 m.
2. Using the calculated Cartesian coordinates of Supplementary Problem 1, calculate the latitude, longitude, and height.
3. What is the difference between a geoid and an ellipsoid and explain their use in geodesy.

4. What is false northing and false easting?
5. For small graphs and for less accurate calculations, the Earth's polar flattening is ignored and the Earth is considered as a sphere. Prove that the equivalent radius of the sphere is given by:

$$R_e = \sqrt{\frac{a^2}{2} + \frac{b^2}{2e}\ln\left(\frac{b}{a(1-e)}\right)}$$

6. Explain the difference between geodetic coordinate and geocentric coordinate.
7. Prove that $e_r = \dfrac{a-b}{a+b}$.
8. With the given Cartesian coordinates in meters, determine the geodetic latitude, longitude, and radial distance in degrees, arc minutes and arc seconds.

$$\begin{bmatrix} x \\ y \\ z \end{bmatrix} = \begin{bmatrix} 1,717,455.456 \\ -5,617,543.675 \\ 2,476,765.992 \end{bmatrix}$$

9. Calculate the loxodromic distance (constant direction) between Miami Beach, Florida, and London, England, with the coordinates:
   25°47'25"N, 80°7'49"W
   51°30'28"N, 00°07'41"W
10. The positions of two rigs are surveyed and the measured rig coordinates (x, y, z) are given in meters below. Determine the geodetic latitude, longitude, and altitude above the ellipsoid of the rigs.

$$\text{Rig 1} = \begin{bmatrix} x \\ y \\ z \end{bmatrix} = \begin{bmatrix} 2,640,525.534 \\ 3,146,855.792 \\ 4,862,865.302 \end{bmatrix}$$

$$\text{Rig 2} = \begin{bmatrix} x \\ y \\ z \end{bmatrix} = \begin{bmatrix} 2,585,207.208 \\ 3,192,465.03 \\ 4,862,872.962 \end{bmatrix}$$

11. Identify the universal transverse Mercator zone number for the following points:
    a. Latitude: –23°12'32" Longitude: 123°32'12"
    b. Latitude: 123°32'12" Longitude: –23°12'32"
    c. Latitude: –123°12'32" Longitude: 23°12'12"

    Also find out the meridian and false origin associated with these points.

12. Calculate the ellipsoidal radius of curvature for the normal section at the following geodetic coordinates:
    Latitude: 33°12'44.33" Longitude: 133°02'02.782"
13. Given an east or west declination angle, η, and a given azimuth angle, ϕ, determine the corrected azimuth angle, ϕ, in terms of ϕ and η for both E and W declination.
14. Derive the equation for declination correction to well directions given in the following quadrature. Use ε_e and ε_w as the east and west declinations, respectively.
    SθW, SθE, NθW, NθE
15. Sketch the map projections for the following types and discuss their characteristics and their distortion with respect to the accuracy of the projections.
    a. Cylindrical—secant
    b. Cylindrical—transverse
    c. Cylindrical—oblique
    d. Conic—secant
    e. Conic—oblique
    f. Planar—oblique
16. What are the three types of north used for geographical reference and explain the difference between them?
17. A rig is located in a field with the coordinates:
    Easting (m): 754,014.100
    Northing (m): 9,269,870.500
    Using the data given below, calculate the geographical data using the Gauss-Krüger coordinates.
    Data for the Bessel ellipsoid are:
    $a = 6,377,397.155$ m; $b = 6,356,078.963$ m
18. A well is located in Dallas County with the global coordinates:
    $Y = 11056369.52$ N; $X = 3359270.87$ E
    Calculate the geographical coordinates using the Bessel ellipsoid.
    Recalculate the geographical coordinates using the WGS84 ellipsoid.
    Compare with the results using the datum transformation method.
    WGS84 ellipsoid: $a = 6,378,137.0$ m and $f = 1/298.257223563$.
    ΔX, ΔY, ΔZ shift of origin: 587.0 m, 16.0 m, and 393.0 m, respectively.
19. Using the following data and the Lambert conformal conic projection method, calculate the geographic coordinates of the location:

$X = 37 - 5855.907$ m; $Y = 1{,}087{,}053.230$ m
Use the Bessel ellipsoid data. Other data are:
Standard parallels: $\phi_1 = 30°$ and $\phi_2 = 42°$
Origin data: $\phi_0 = 26°$N and $\lambda_0 = 135°$W

References

1. Hoffman-Wellenhof, B, H. Lichtenegger, and J. Collins. *GPS Theory and Practice* New York: Springer-Verlag, Wein, 1992.
2. Parkinson, B. W., J. Spilker, P. Axelrad, and P. Enge. *Global Positioning System: Theory and Applications.* Washington DC: American Institute of Aeronautical Astronomy, 1996.
3. Yang, Q., J. P. Snyder, and W. R. Tobler. *Map Projection Transformation, Principles and Applications.* London: Taylor & Francis, 2000.
4. Hofmann-Wellenhof, B., G. Kienast, and H. Lichtenegger. *GPS in der Praxis.* New York: Springer-Verlag, Wein, 1994.
5. Torge, W. *Geodesy*, 2nd Edition. Berlin, New York: Walter de Gruyter, 1991.
6. Snyder, John P. *Map Projections: A Working Manual.* USGS Professional Paper 1395, Washington, DC: United States Government Printing Office, 1987.
7. Frederick Pearson, H. *Map Projections: Theory and Applications.* Boca Raton, FL: CRC Press, 1990.

3

Well Path Trajectory

Planning a well path is an important aspect of the well construction process, which requires detailed well path trajectory design to reduce the overall cost of the well. Target locations, well path types, and well profiles are core components of the design. Trajectory design affects other designs such as the drill-string design; the casing design; torque and drag estimation; hole cleaning requirements; swab, surge, and wellbore pressure calculations; etc. Improper designs may lead to costly problems, such as wellbore instability, loss of circulation, drill-string failure, and so on. Proper planning is even more important for designing multiple wells from a single pad or multiple platforms. This chapter describes the different well path sections that could be embedded while designing a well trajectory or while monitoring it during drilling. This chapter examines the mathematical treatments involved in well path planning, in addition to defining the basic methodology of obtaining the well path design parameters, and acquaints the reader with the use of different well path sections at appropriate places in the well path. Representing the well path shape and obtaining parameters such as borehole curvature and torsion using different methods are also described in this chapter.

Well path design can be of two types: one type being done before drilling and the other type being done during drilling, with a continuously updated well trajectory that is to be adjusted to fit the planned well path. A predrilled well profile consists of straight lines and smooth curves with appropriate geometrical description. Whether it is a predrilled well path or an actual trajectory, both are geometrically described by continuous and smooth curves. Geometrical parameters of the curves are used to characterize the well profiles in terms of well path parameters. Some of the basic well path parameters (including brief description) are given in the following section.

Basic Definitions

Some of the basic terms used in the design of well paths are:

- Inclination angle, azimuth angle
- Measured depth (MD), true vertical depth (TVD)
- Kick-off point (KOP), turn-off point (TOP)
- Build-rate angle (BRA), drop-rate angle (DRA), turn-rate angle (TRA)
- Lead angle (LA)
- Survey station
- Closure, departure
- Closure azimuth
- Course length (CL)
- Tie-in points
- Toolface angle

Inclination Angle

Inclination angle, or the deviation angle, at any point of a wellbore is defined as the angle between the tangent line to the wellbore at that point and a vertical line passing through that point, which is parallel to the Earth's gravity at that point.

Azimuth Angle

Azimuth angle, also called the direction or bearing, is defined as the angle between the north direction and the tangent line at any point in the well-bore as measured when projected onto a horizontal plane. It is measured from 0° to 360° clockwise relative to due north with 0° azimuth being taken as exactly due north.

Measured Depth

Measured depth is the actual distance of the well path or trajectory.

True Vertical Depth

True vertical depth is the vertical distance from a point in the wellbore and the horizontal plane that intersects the well at the reference datum of the Earth's surface. Many calculations for parameters (such as pore pressure, fracture gradient, wellbore pressure, equivalent circulating density, equivalent mud weight, temperature gradient, and so on) are usually based on TVD.

Kick-Off Point

The kick-off point, also known as kick-off depth (KOD), is the depth at which the incremental building of the hole inclination or deviation angle begins.

Turn-Off Point

The turn-off point, also known as turn-off depth (TOD), is the depth at which the direction of the wellbore starts changing.

Build-Rate Angle

The build-rate angle is the incremental increase of the wellbore inclination angle from the vertical per unit distance traversed along the wellbore. This rate is specified in degrees per 100 feet or degrees per 30 meter of hole section.

Drop-Rate Angle

The drop-rate angle is the incremental decrease of the wellbore inclination angle from the vertical per unit distance traversed along the wellbore. This rate is specified in degrees per 100 feet or degrees per 30 meter of hole section.

Turn-Rate Angle

The turn-rate or walk-rate angle is the incremental change in the wellbore direction (azimuth) per unit distance traversed along the wellbore. This rate is specified in degrees per 100 feet or degrees per 30 meter of hole section. A sign convention for the walk rate is necessary to differentiate between left and right turns. A positive sign is used when the wellbore is turning to the right, whereas a negative sign signifies that the wellbore is turning to the left.

Lead Angle (LA)

Because of the clockwise rotation of the drill string, drill bits may have a tendency to walk to the right or the left in the horizontal plane. To account for this natural bit walk, it is a common practice to initiate the turn of the well path with a specified lead angle to the left or right of the target area, depending on whether the bit walks to the right or left, respectively. The magnitude of the correction angle is generally based on experience, or from past records of wells drilled in that area.

Survey Station

The survey station, or survey point, is the location along the wellbore where survey data such as inclination or azimuth are measured.

Departure

Departure is the horizontal distance between two survey depths along the wellbore. Total departure is the total horizontal distance from the starting point to the target depth.

Closure

Closure is the drift or the horizontal distance between the well origin and the target location, as shown in Figure 3.1. At the target depth, the closure is the total departure.

Closure Azimuth

Closure azimuth is the azimuth direction of the line between the well origin and the target location, as shown in Figure 3.1. It is given in degrees.

Course Length

Course length is the measured distance between two survey points.

Tie-In Points

Tie-in (or tie-on) points are used as a common reference in the survey for multiple wells. They can be from the wellhead location or from previously drilled wells for which the survey plans are available, either in the case of a single sidetrack or multiple sidetracks with different kick-off depths. Other associated terms, such as tie-in measured depth, tie-in total vertical depth, tie-in *Y* offset, and tie-in *X* offset, are defined below:

- *Tie-in measured depth:* The measured depth or distance from the elevation reference to the tie-in point of the survey, when a survey begins below the elevation reference. The tie-in point is the shallowest point of a survey that is used to link its depth to an original wellbore, previous wellbore, or from a lateral to a spoke in a horizontal well.

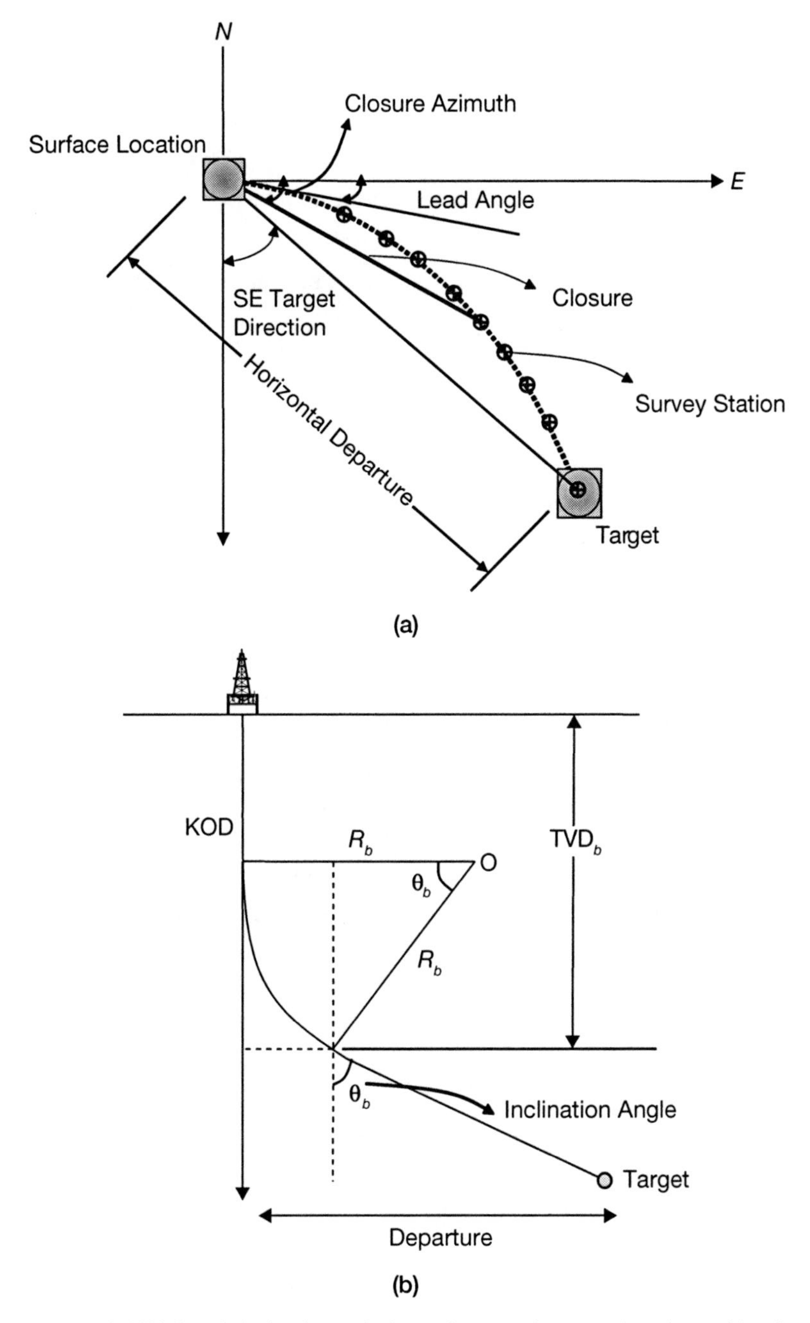

Figure 3.1 (a) Well path in horizontal plane closure, closure azimuth, and lead angle; (b) departure.

- *Tie-in total vertical depth:* The vertical depth or distance from the elevation reference point to the tie-in point of the survey.
- *Tie-in* Y *offset:* The north or south distance between a vertical axis passing through the tie-in point and a vertical axis passing through the zero vertical elevation point. A positive number denotes north, a negative number denotes south.
- *Tie-in* X *offset:* The east or west distance between a vertical axis passing through the tie-in point and a vertical axis passing through the zero vertical elevation point. A positive number denotes east, a negative number denotes west.

Toolface Angle

To drill a curved borehole, some special deflection tools are used. Several of these tools are discussed in *Downhole Drilling Tools—Theory and Practice*.[1] These tools will have a scribe line to orient the toolface. The toolface is the side of the tool on which the curved borehole will be drilled. A curvature is obtained as the result of the axial penetration rate combined with the lateral penetration rate. A deflection tool delivers a curvature measured in dog-leg severity (DLS). The DLS depends on several factors. To control the direction of the deflection, the toolface must be set in an appropriate position in the borehole. This position is called toolface angle and is measured with respect to the geographic north (azimuth of the toolface) for inclination less than 5°, and with respect to the high side (HS) of the borehole for inclinations greater than 5°. To drill a circular arc in 3D space, the toolface angle should be changed as the borehole is drilled. If the toolface angle is kept constant, a helical trajectory will be created.

Plots and Coordinate Systems

Graphical Representation of a Well Path

Specifying and using an appropriate coordinate system to define the position and orientation of a well path is very important for precise and consistent computations. Various graphical methods, coupled with space and plane coordinates, are used to describe a well path. General types of plots used are:

- Three-dimensional (3D) plot
- Cylindrical plot
- Projection plot

These three types of plots have their own characteristics and applicability based on the specific circumstances of the wells. In general, for better description of cluster wells, 3D plots are used, whereas for spatial relationships of directional wells and horizontal wells, cylindrical plots are preferred. Projection plots are useful during the actual drilling for monitoring the trajectory and comparing it with the designed well path.

3D Plot

In this plotting method, three-dimensional coordinates are used to describe the wellbore trajectory along the depth of the well, as shown in Figure 3.2. Usually, the origin of the wellbore is at the wellhead and the north (*N*), east (*E*), and vertical (*H*) directions are used to define a right-handed Cartesian coordinate system.

A three-dimensional plot has the advantage of describing a well path completely. However, when there are multiple wells, it may not provide the complete perspective of the placement and position of all the well paths. So, often some auxiliary plots are required with additional capabilities such as rotation and zooming for proper viewing.

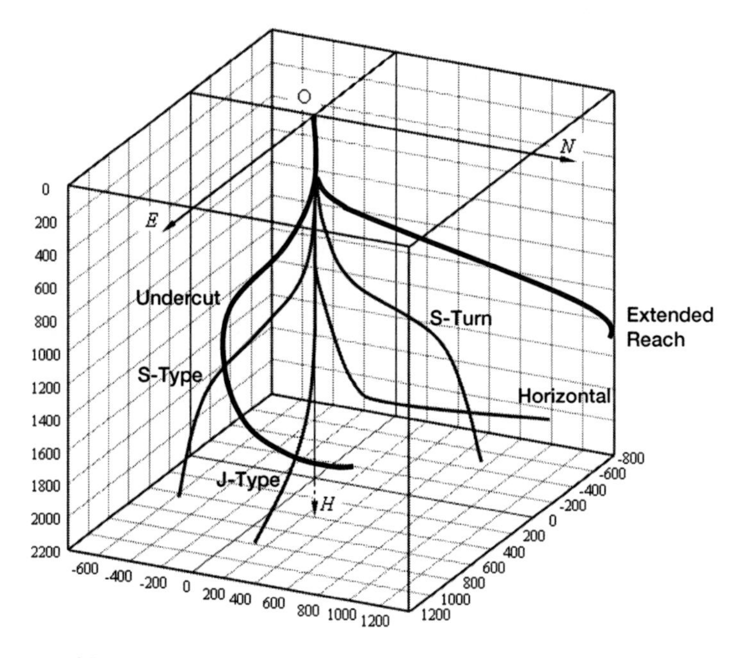

Figure 3.2 3D plot.

Cylindrical Plot

The cylindrical surface mapping method requires two 2D mappings, which are, respectively, a vertical cross-section mapping and a horizontal projection mapping. The process of developing these two projections can be understood as follows: Since the well path can be considered as a curve in space, a series of perpendicular lines can be drawn along the points on the well path and these perpendicular lines will constitute a 2D surface constituted by a series of conjoined cylindrical surfaces, as illustrated in Figure 3.3. The intersection of these cylindrical surfaces with the horizontal plane results in the horizontal projection of the well path profile. If the cylindrical surfaces are

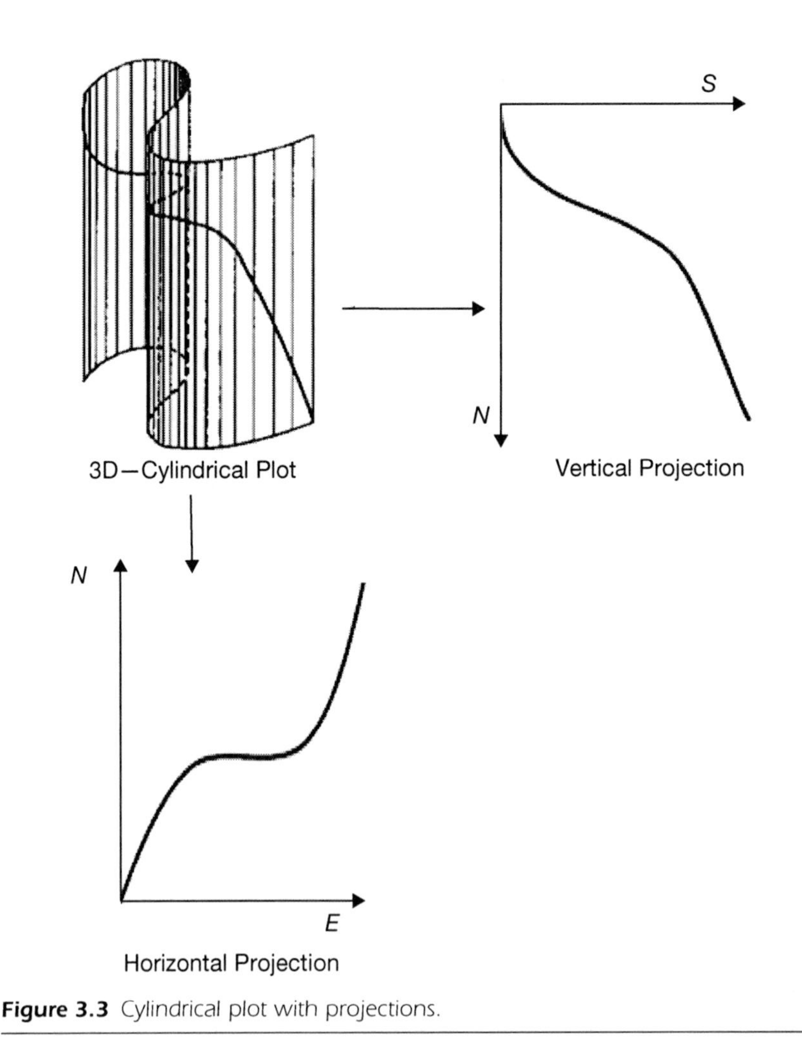

Figure 3.3 Cylindrical plot with projections.

unfolded, the space curve transforms into a vertical section (v-section) plan, as shown in the bottom illustration of Figure 3.3. Since the vertical profile is on a cylindrical surface, the developed view of the horizontal coordinate length, S, cannot be negative, and the vertical distance of points on the well path will always be the vertical coordinate axis (H) to the right of the origin of the horizontal coordinate length, S. Since the vertical and horizontal projections of the well path are derived from the initial projection of the cylindrical map, this method is called the cylindrical method of projection.

Depicting a well trajectory on a cylindrical map has the following main advantages:

- The vertical projection provides an easy means of visualizing the spatial shape of the trajectory. By bending the vertical cross-section plots using the shape of the well trajectory from the horizontal projection plots, the spatial shape of the well trajectory in 3D can be restored.
- The cylindrical map method allows direct visualization of the majority of well paths with the true value of wellbore parameters, in particular, depth, deviation, and azimuth angles.

Projection Plot

This projection method consists of vertical and horizontal projection charts. Vertical sections, or v-sections, are projected in Cartesian coordinates on vertical planes through the local origin and oriented in a particular direction. This direction is also used to calculate the well displacement or horizontal distance (departure). This view is equivalent to the elevation view in conventional mechanical-engineering drawings, as shown in Figure 3.4.

The horizontal projection is the projection of the well profile on the horizontal plane and the plot provides the direction of the well that is being drilled or planned.

The v-section coordinates are obtained from the projection of the well path onto a vertical plane using the following origin coordinates:

- Origin N is the north (+) or south (–) distance from the wellhead to the local origin.
- Origin E is the east (+) or west (–) distance from the wellhead to the local origin.

Values of 500 ft and 500 ft for origins N and E, respectively, indicate that the local origin is 500 ft north and 500 ft east of the wellhead. The values 0.00 ft and 0.00 ft indicate that the local origin is located at the wellhead.

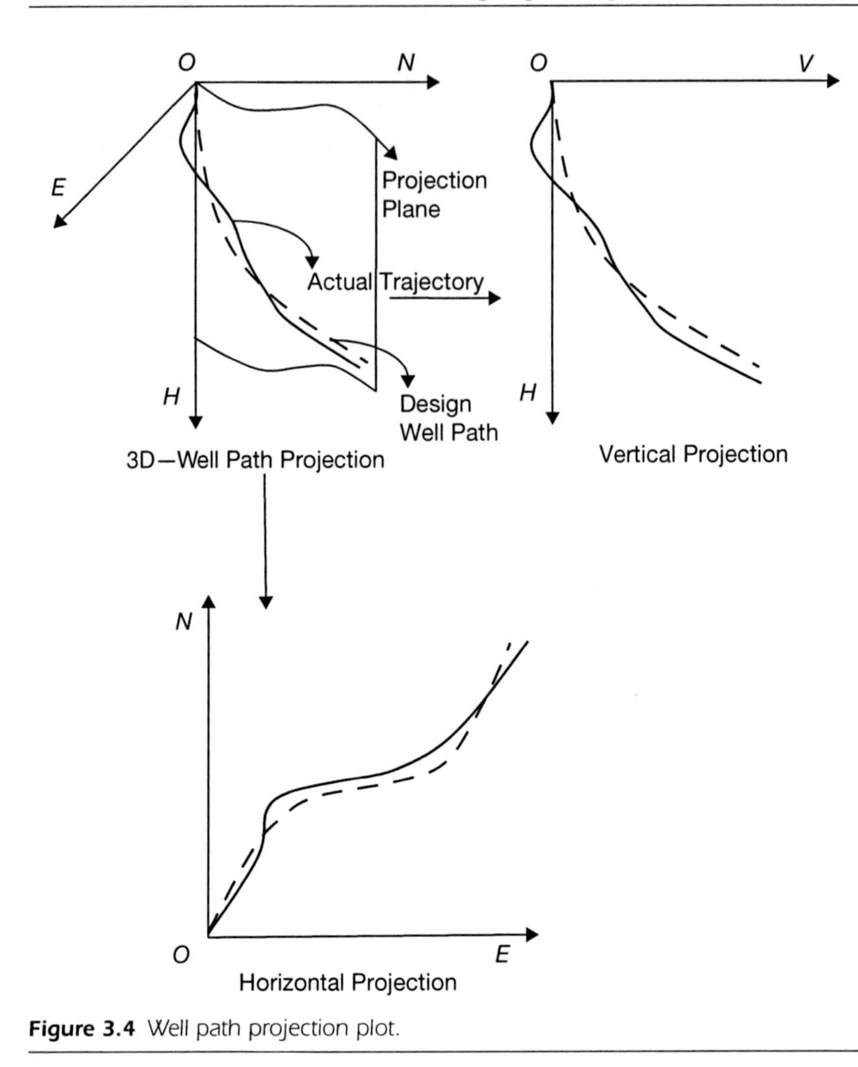

Figure 3.4 Well path projection plot.

The following calculations provide guidelines on how to calculate the vertical displacement distance between a survey station and the given v-section origin coordinates.

Horizontal distance from a NS/EW survey station from the reference v-section coordinates (2D distance) can be calculated and is given as the vertical section closure distance:

$$V_d = \sqrt{\left(NS_n - NS_{ref}\right)^2 + \left(EW_n - EW_{ref}\right)^2} \tag{3.1}$$

where n = nth survey point.

Vertical section closure azimuth, ϕ_d (also called the vertical section angle, or the polar coordinate angle), of the vertical plane onto which the well trajectory is projected to get the v-section can be calculated as follows for various conditions (several conditions apply in arriving at the final vertical displacement projected on the desired plane):

If $(EW_n - EW_{ref}) = 0$ and $(NS_n - NS_{ref}) \geq 0$, then $\phi_d = \pi$ (3.2)

If ΔNS condition exists such that $(NS_n - NS_{ref}) \geq 0$, then ϕ_d can be given as:

$$\phi_d = \frac{\pi}{2} - \arctan\left(\frac{NS_n - NS_{ref}}{EW_n - EW_{ref}}\right) \tag{3.3}$$

In addition, $EW_n - EW_{ref}$ condition should be checked. If $EW_n - EW_{ref} < 0$, then:

$$\phi_d = \frac{3\pi}{2} - \arctan\left(\frac{\left(NS_n - NS_{ref}\right)}{\Delta EW}\right) \tag{3.4}$$

Using the above equations, $\Delta\phi_c$, $\Delta\phi_c = \phi_{ref} - \phi_d$ can be calculated under various conditions as given next.

Here, ϕ_{ref} is the direction of the line from the local origin to the *n*th survey point.

If $\Delta\phi \leq \pi$, then $\Delta\phi_c = \Delta\phi$

If $\phi_d \geq \phi_{ref}$, then $\Delta\phi_c = 2\pi - \phi_d + \phi_{ref}$

If $\phi_d < \phi_{ref}$, then $\Delta\phi_c = -(2\pi - \phi_d + \phi_{ref})$

Thus, the v-section displacement distance is

$$dist \times \cos\left(\left|\Delta\phi_c\right|\right) + \text{offset} \tag{3.5}$$

Here, offset is the tie-in X offset if present.

PROBLEM 3.1

Given the following survey points, calculate the v-section distance.

Survey point N/S	–575.2 ft
Survey point E/W	363.2 ft

Reference vertical section coordinates:

Vertical section N/S	212 ft
Vertical section E/W	–108 ft
Vertical section	112° (azimuth)

Solution

Given data: $NS_{ref} = 212$ ft, $EW_{ref} = -108$ ft, $\phi_{ref} = 112°$, $NS_2 = -575.2$ ft, $EW_1 = 363.2$ ft. The given data are also diagrammatically depicted in Figure 3.5.

Step 1: Calculate the horizontal distance from the reference point to the survey point.

$\Delta NS = (-575.2 - 212) = -787.2$ ft; $\Delta EW = (363.2 + 108) = 471.2$ ft

$$V_d = \sqrt{(-787.2)^2 + (471.2)^2} = 917.42\ \text{ft}$$

Step 2: Calculate the azimuth angle from the survey point to the v-section reference coordinates.

Since $\Delta NS \geq 0$, then $\phi_{vs} = \frac{\pi}{2} - \text{arc tan}\frac{-787.2}{471.2} = 2.60$

Step 3: Calculate the difference between the two azimuth angles and express it in absolute value.

$\Delta\phi = 2.60 - 1.947 + 0.647$

If $\Delta\phi \leq \pi$, then $\Delta\phi = \Delta\phi$

Step 4: Calculate the v-section displacement distance from the local origin.

$V_d = 917.42 \times \cos(0.647) + 0 = 732$ ft

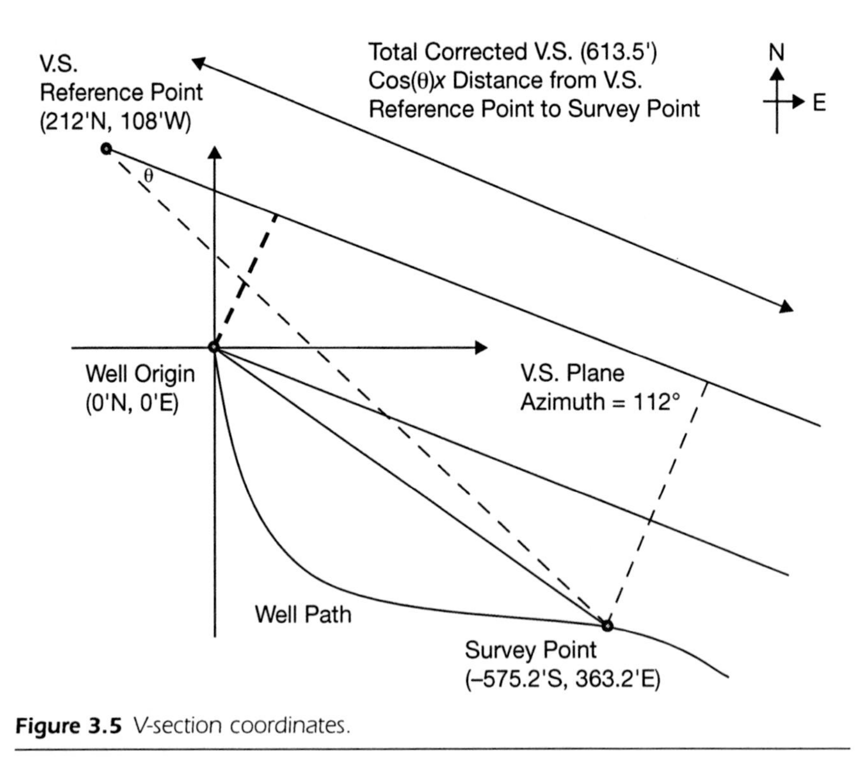

Figure 3.5 V-section coordinates.

Crossing Vertical

In the vertical projection plot, the trajectory of the well path could go to the left of the origin and cross the vertical coordinate axis (H), and in this case, the value of the vertical section (V) may become negative. Whenever there is a drastic change in the azimuth, especially by more than 90°, the well path can be interpreted in two ways. One interpretation being that the well path could be a continuous spiral around the vertical axis with a relatively small amount of drop or build between the survey stations. If this is the case, then the N/S, E/W, and TVD coordinates can be calculated using the conventional equations. On the other hand, it is also possible that the well path is a continuous curve oriented primarily in the vertical plane. It may remain on an approximately constant azimuth while dropping at an angle to the vertical point, and then it begins to build until the hole reaches the final inclination between the survey stations. In this drop and build section, the azimuth change is relatively small. In such a case the calculations need to be performed in two sections: first, from the survey station 1 to the vertical point, and then from the vertical point to the survey station 2. The results are reported as the change in N/S, E/W, and TVD coordinates between survey stations 1 and 2 that are calculated by the summation of the respective changes in each of the two wellbore sections. The following example will clarify the details of the calculations for the second case.

Consider two survey stations 1 and 2 at depths D_1 and D_2 with the inclination and azimuth values as α_1, α_2 and ϕ_1, ϕ_2, respectively. If the difference in depth $(D_2 - D_1)$ is less than 125 ft (a typical assumption) and if the difference in the azimuth is between 95° and 265°, then the walk can be described for two different conditions: one when the azimuth at survey station 2 is greater than at survey station 1, and the other when the reverse applies, as follows:

- When $\phi_2 > \phi_1$, walk $= \phi_2 - \phi_1 - 180°$ (adjusted 0–360°)
- When $\phi_1 > \phi_2$, walk $= \phi_2 - \phi_1 + 180°$ (adjusted 0–360°)

The wellbore is divided into two sections between the two survey stations. The inclination at a point that is vertical in the wellbore is 0. Thus, it can be determined how far below the upper survey station the vertical is by dividing the total inclination change from α_1 to 0 and from 0 to α_2 (and not just $\alpha_1 - \alpha_2$) by the total change in depth between the two survey stations. The result is multiplied by α_1 to calculate the distance from the top to the vertical point. Using the two sections, the two sets of coordinates are added together to get the total change between the two survey stations. The coordinates of the point above the crossing vertical point is:

$$\bar{D} = D_1 + \frac{\alpha_1}{\alpha_1 + \alpha_2}(D_2 - D_1) \tag{3.6}$$

$$\bar{\alpha} = 0$$

$$\bar{\phi} = \phi_1 \frac{\alpha_1}{\alpha_1 + \alpha_2} \times \text{walk} \quad \text{(adjusted 0–360°)}$$

The coordinates of the point below the crossing vertical point are:

$$\bar{D} = D_1 + \frac{\alpha_1}{\alpha_1 + \alpha_2}(D_2 - D_1) \tag{3.7}$$

$$\bar{\alpha} = 0$$

$$\bar{\phi} = \bar{\phi} \text{ (from above point)} + 180° \text{ (adjusted 0–360°)}$$

Well Trajectory Coordinate Systems

A coordinate system or spatial reference system is required for defining the field surveys made and for a consistent representation of the well profile. This is also required for the accurate projection of the well profile onto charts, as well as subsequent computation of the well path parameters needed for carrying out well path design. The following types of coordinate systems are generally used:

- General coordinate system (*O-NEH*)
- Borehole coordinate system (*P-xyz*)
- Local coordinate system (P-$\xi\eta\zeta$)

General Coordinate System

Well path design and the associated calculations need a global 3D coordinate system for reference, as well as for map projections. A general coordinate system is also essential to define the position of a well and the well path relative to other wells in a cluster well environment, and to avoid the explicit use of an individual coordinate system for each well. The global coordinate system uses north-east-*H* (depth) coordinates, with positive direction along the respective axes. The *H* axis is postulated to be the vertical depth with the positive direction pointing downwards as shown in Figures 3.2 to 3.5.

It is to be noted that the north coordinate could be aligned with the magnetic north or the grid north or the true north. As the three definitions of north vary relative to each other, and also with respect to time and place, the wellbore trajectory design and monitoring should consistently select one definition of north only. Once the N axis and the origin of the coordinate system in the plane have been fixed, the perpendicular to the N axis toward the east direction, and in the same plane, forms the E axis coordinate.

Borehole Coordinate System

To obtain easily the deviation angle and azimuth changes for a well path, a reference borehole coordinate system is used, which describes the well path including the reverse bending profiles. To define the wellbore borehole coordinate system, consider a point P in the well path as the origin, such that the z-axis is a line tangent to the well path and the x-axis is perpendicular to the z-axis and oriented toward the high-side of the borehole, thereby defining a standard Cartesian frame of reference, as shown in Figure 3.6. Therefore, the x-axis gives the borehole deviation, the y-axis gives the borehole direction and the z-axis gives the borehole forward direction (that is, tangent direction). Using differential geometry, the borehole P-xyz coordinates can be transformed from the general O-NEH coordinates as follows:

$$\begin{bmatrix} x \\ y \\ z \end{bmatrix} = \begin{bmatrix} \cos\alpha_p \cos\phi_P & \cos\alpha_p \sin\phi_P & -\sin\alpha_p \\ -\sin\phi_P & \cos\phi_P & 0 \\ \sin\alpha_p \cos\phi_P & \sin\alpha_p \sin\phi_P & \cos\alpha_p \end{bmatrix} \begin{bmatrix} N - N_p \\ E - E_p \\ H - H_p \end{bmatrix} \tag{3.8}$$

where

α = Inclination angle, (°)
ϕ = Azimuth angle, (°)
N = North coordinate (south is negative), m or ft
E = East coordinate (west is negative), m or ft
H = Total vertical depth, m or ft
x = High side coordinate, m or ft
y = Right-hand coordinate, m or ft
z = Onward coordinate, m or ft

Local Coordinate System

Sometimes, for some partial well design computations and for local well trajectory corrections, using a local coordinate system with coordinates ξ, η, ζ is more convenient. Usually, for the local coordinate system, the ζ axis is taken to be the same as the z-axis of the borehole coordinate system (tangent line to the well path), while the ξ, η axes are obtained by a rotation of the xy plane of the borehole coordinate system by an angle ω wellbore, as illustrated in Figure 3.6.

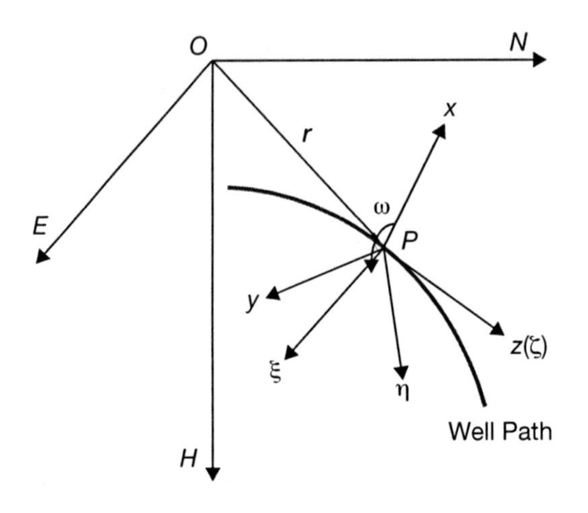

Figure 3.6 Well path coordinate system.

The transformation of the local coordinate system with P-$\xi\eta\zeta$ coordinates from the borehole coordinate system with P-xyz coordinates is as follows:

$$\begin{bmatrix} \xi \\ \eta \\ \zeta \end{bmatrix} = \begin{bmatrix} \cos\omega & \sin\omega & 0 \\ -\sin\omega & \cos\omega & 0 \\ 0 & 0 & 1 \end{bmatrix} \begin{bmatrix} x \\ y \\ z \end{bmatrix} \tag{3.9}$$

where

ω = Toolface angle (clockwise is positive), (°)
ξ = Principal normal coordinate, m or ft
η = Binormal coordinate, m or ft
ζ = Tangent coordinate, m or ft

The above transformation is a simple rotation of the coordinate system around the z-axis through the toolface angle ω.

Using Eq. 3.8 in Eq. 3.9, the relationship between the local coordinate system P-$\xi\eta\zeta$ and the general coordinate system O-NEH can be written in the form:

$$\begin{Bmatrix} \xi \\ \eta \\ \zeta \end{Bmatrix} = [T] \begin{Bmatrix} N - N_P \\ E - E_P \\ H - H_P \end{Bmatrix} \tag{3.10}$$

Here, the elements of the matrix $[T]$ are:

$$\begin{cases} T_{11} = \cos\alpha_P \cos\phi_P \cos\omega - \sin\phi_P \sin\omega \\ T_{12} = \cos\alpha_P \sin\phi_P \cos\omega + \cos\phi_P \sin\omega \\ T_{13} = -\sin\alpha_P \cos\omega \end{cases}$$

$$\begin{cases} T_{21} = -\cos\alpha_P \cos\phi_P \sin\omega - \sin\phi_P \cos\omega \\ T_{22} = -\cos\alpha_P \sin\phi_P \sin\omega + \cos\phi_P \cos\omega \\ T_{23} = \sin\alpha_P \sin\omega \end{cases}$$

$$\begin{cases} T_{31} = \sin\alpha_P \cos\phi_P \\ T_{32} = \sin\alpha_P \sin\phi_P \\ T_{33} = \cos\alpha_P \end{cases}$$

Borehole Curvature and Torsion

Wellbore curvature and torsion greatly influence the forces applied on the downhole pipe string (drill string, casing, oil production pipe and rod, etc.) and its consequent deformation. They are also the base data required for calculating the drag and torque of a pipe string and for checking its strength, and they also serve as indices to evaluate the quality of design of a well trajectory. Accordingly, curvature and torsion affect well drilling and its completion, production, and even any required work-over operations. It is necessary to calculate accurately and describe the shape of a well trajectory in order to monitor and control it effectively. A wellbore trajectory is a continuous and smooth curve in space, which bends upwards or downwards (builds or drops) and turns right or left simultaneously.

The primary parameters related to wellbore trajectory curvature are the bending angle (dogleg angle) and the borehole curvature. In order to calculate and analyze them, researchers have provided analytic methods, diagram-searching methods, graphic methods, and the Ouija board method. Although the last three methods show considerable error and are inconvenient to use, they were commonly used into the 1980s. Today, the analytical method is the one most often used to calculate the bending angle and curvature of a well trajectory.

Initially, formula were used to calculate the borehole torsion for a cylindrical-helix trajectory[3]; later on, systematical calculations were presented to calculate the curvature and torsion of wellbore trajectories in space, using differential geometry.[9,10] By establishing a moving frame on the wellbore trajectory and associating it with the trajectory's equation, a new perspective and method to describe the shape of wellbore trajectory in space was introduced.[2,12,14] From the perspective of differential geometry, this expands the definitions and geometric meanings of bending angle, borehole curvature, torsion angle, and borehole torsion.

Moving Frame

As mentioned, a wellbore trajectory is a continuous and smooth curve in space, bending and turning simultaneously to change direction. There are some intrinsic relations between the parameters that serve as the basis for describing and calculating a well trajectory.

Basic Equation

A tangent line is drawn across a given point P on a well trajectory and intercepts a unit length on the tangent line as unit tangent vector $\boldsymbol{t}$, which points to the onward direction of the well trajectory (Figure 3.7). Next, a unit vector is drawn in the bending direction of the well trajectory as unit principal normal vector $\boldsymbol{n}$, which points in the concave direction of the well trajectory. Finally, a unit vector is drawn perpendicular to both $\boldsymbol{t}$ and $\boldsymbol{n}$, called the binormal vector $\boldsymbol{b}$, viz. $\boldsymbol{b} = \boldsymbol{t} \times \boldsymbol{n}$. The three unit vectors form a right-hand system and constitute the moving frame of the well trajectory at point P, also known as Frenet's frame.[2,9,12,14]

Finding out the relation between the moving frame and well trajectory requires combining the equation of the well trajectory with the equation of the moving frame. The wellbore trajectory in vector form can be defined as:

$$\boldsymbol{r} = \boldsymbol{r}(L) = N\boldsymbol{i} + E\boldsymbol{j} + H\boldsymbol{k} \tag{3.11}$$

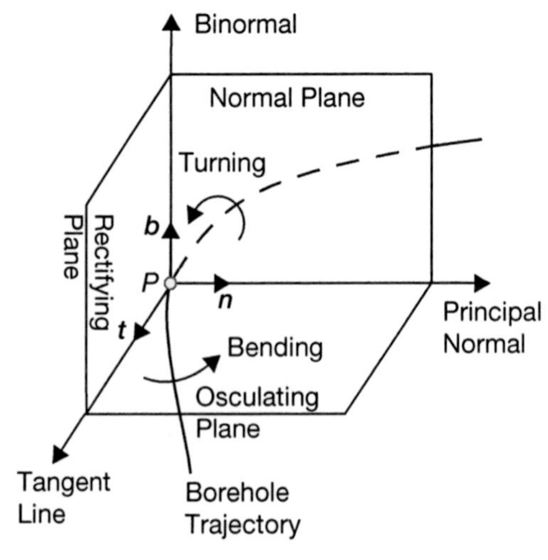

Figure 3.7 Moving frame.

where

L = Measured depth, m or ft
$\boldsymbol{r}$ = Coordinate vector of wellbore trajectory
$\boldsymbol{i}$ = Unit vector on N axis
$\boldsymbol{j}$ = Unit vector on E axis
$\boldsymbol{k}$ = Unit vector on H axis

According to differential geometry,[11,12] the relationship between the unit tangent vector $\boldsymbol{t}$, the unit principal normal vector $\boldsymbol{n}$, the unit binormal vector $\boldsymbol{b}$, and the coordinate vector $\boldsymbol{r}$ of well trajectory can be expressed as follows:

$$\begin{cases} \boldsymbol{t} = \dot{\boldsymbol{r}} = \dfrac{d\boldsymbol{r}}{dL} \\ \boldsymbol{n} = \dfrac{\dot{\boldsymbol{t}}}{|\dot{\boldsymbol{t}}|} = \dfrac{\ddot{\boldsymbol{r}}}{|\ddot{\boldsymbol{r}}|} \\ \boldsymbol{b} = \boldsymbol{t} \times \boldsymbol{n} \end{cases} \tag{3.12}$$

where

$\boldsymbol{t}$ = Unit tangent vector of wellbore trajectory
$\boldsymbol{n}$ = Unit principal normal vector of wellbore trajectory
$\boldsymbol{b}$ = Unit binormal vector of wellbore trajectory
$\dot{\boldsymbol{t}}$ = First derivative of unit tangent vector
$\dot{\boldsymbol{r}}$ = First derivative of coordinate vector
$\ddot{\boldsymbol{r}}$ = Second derivative of coordinate vector

For brevity, the derivatives have been denoted using the dot notation. Thus, the moving frame has been combined with the equation of the well trajectory, which makes it feasible to describe the curvature and torsion of the well trajectory in space using the parameters of the moving frame (Figure 3.8), as explained next.

Using a differential model of the wellbore trajectory (Figure 3.9), the relations between coordinate increments and curved-section length, inclination angle, and azimuth angle for a small interval are given by:

$$\begin{cases} \dfrac{dN}{dL} = \sin\alpha\cos\phi \\ \dfrac{dE}{dL} = \sin\alpha\sin\phi \\ \dfrac{dH}{dL} = \cos\alpha \\ \dfrac{dS}{dL} = \sin\alpha \end{cases} \tag{3.13}$$

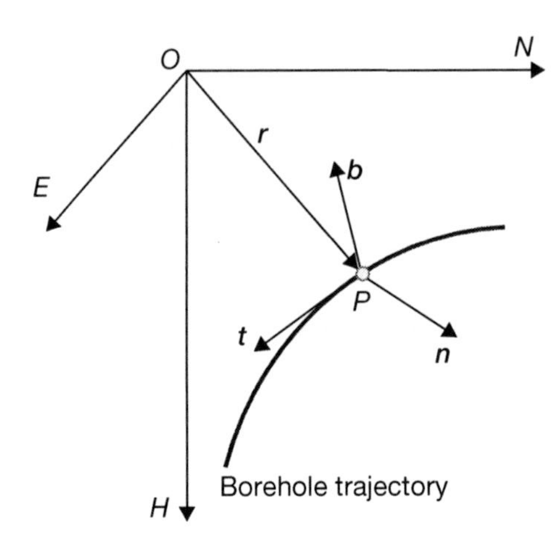

Figure 3.8 Borehole trajectory.

where

S = Horizontal curvilinear displacement, m

Substituting Eqs. 3.11 and 3.13 into Eq. 3.12 yields Eq. 3.14, which is the vector equation of the moving frame on a well trajectory:

$$\begin{cases} \boldsymbol{t} = \sin\alpha\cos\phi\,\boldsymbol{i} + \sin\alpha\sin\phi\,\boldsymbol{j} + \cos\alpha\,\boldsymbol{k} \\ \boldsymbol{n} = \left(\lambda_\alpha\cos\alpha\cos\phi - \lambda_\phi\sin\alpha\sin\phi\right)\boldsymbol{i} + \left(\lambda_\alpha\cos\alpha\sin\phi + \lambda_\phi\sin\alpha\cos\phi\right)\boldsymbol{j} \\ \quad + \left(-\lambda_\alpha\sin\alpha\right)\boldsymbol{k} \\ \boldsymbol{b} = \left(-\lambda_\alpha\sin\phi - \lambda_\phi\sin\alpha\cos\alpha\cos\phi\right)\boldsymbol{i} + \left(\lambda_\alpha\cos\phi - \lambda_\phi\sin\alpha\cos\alpha\sin\phi\right)\boldsymbol{j} \\ \quad + \left(\lambda_\alpha\sin^2\alpha\right)\boldsymbol{k} \end{cases} \tag{3.14}$$

where

$$\begin{cases} \lambda_\alpha = \dfrac{\kappa_\alpha}{\kappa} \\ \lambda_\phi = \dfrac{\kappa_\phi}{\kappa} \end{cases}$$

κ_α = Rate of inclination change, (°)/30 m or (°)/100 ft

κ_ϕ = Rate of azimuth change, (°)/30 m or (°)/100 ft

κ = Curvature of wellbore trajectory, (°)/30 m or (°)/100 ft

λ_α = Ratio of inclination change rate to wellbore curvature, dimensionless

λ_ϕ = Ratio of azimuth change rate to wellbore curvature, dimensionless

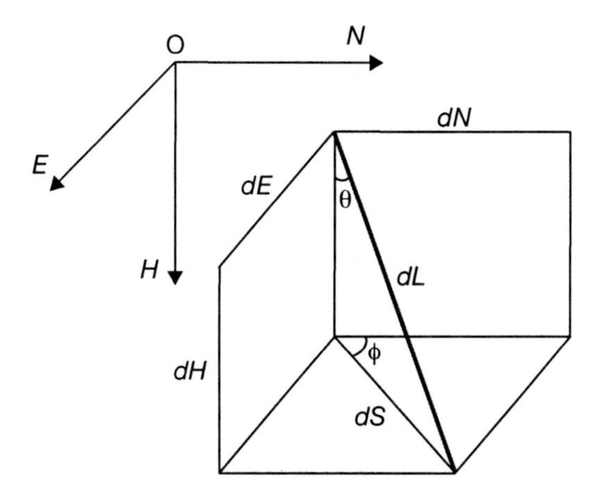

Figure 3.9 Infinitesimal section model of wellbore trajectory.

Kinematics Equation

A moving frame can be formed at any point P on a well trajectory. From the point of view of kinematics, when point P moves along the well trajectory, the moving frame will move along with it. It has been mathematically proven that Eqs. 3.15 or 3.16, namely Frenet's equation, can depict the kinematics behavior of a moving frame.[11,12]

$$\begin{cases} \dot{\boldsymbol{t}} = \dfrac{\boldsymbol{n}}{R} \\ \dot{\boldsymbol{n}} = -\dfrac{\boldsymbol{t}}{R} + \dfrac{\boldsymbol{b}}{\rho} \\ \dot{\boldsymbol{b}} = -\dfrac{\boldsymbol{n}}{\rho} \end{cases} \tag{3.15}$$

$$\begin{bmatrix} \dot{\boldsymbol{t}} \\ \dot{\boldsymbol{n}} \\ \dot{\boldsymbol{b}} \end{bmatrix} = \begin{bmatrix} 0 & \dfrac{1}{R} & 0 \\ -\dfrac{1}{R} & 0 & \dfrac{1}{\rho} \\ 0 & -\dfrac{1}{\rho} & 0 \end{bmatrix} \begin{bmatrix} \boldsymbol{t} \\ \boldsymbol{n} \\ \boldsymbol{b} \end{bmatrix} \tag{3.16}$$

where

R = Curvature radius, m or ft
ρ = Torsion radius, m or ft

$\dot{n}$ = First derivative of unit principal normal vector
$\dot{b}$ = First derivative of unit binormal vector

That is to say, the derivatives of the unit vectors $\boldsymbol{t}$, $\boldsymbol{n}$, and $\boldsymbol{b}$ with respect to measured depth L can be expressed as a linear combination of the vectors $\boldsymbol{t}$, $\boldsymbol{n}$, and $\boldsymbol{b}$. In this way, combining Eqs. 3.15 or 3.16 with Eq. 3.12 depicts the law of motion of the frame while the point P moves along a wellbore trajectory.

In oil-well drilling engineering, the common units for borehole curvature and torsion are (°)/m, (°)/30 m, or (°)/100 ft. Using a conversion constant C_κ, the value of which depends on the units used for borehole curvature and torsion, the corresponding curvature radius and torsion radius are given as follows:

$$R = \frac{180 C_\kappa}{\pi \kappa} \quad (3.17)$$

$$\rho = \frac{180 C_\kappa}{\pi \tau} \quad (3.18)$$

where

C_κ = Constant related to the units of borehole curvature and torsion. Its value is the number used in the unit. For example, if the unit for borehole curvature and torsion is (°)/30 m and (°)/100 ft, then C_κ = 30 and 100, respectively.

τ = Torsion of wellbore trajectory, (°)/30 m or (°)/100 ft

A moving frame can be regarded as a rigid body turning around point P, while it moves along a well trajectory. According to the principles of kinematics, instantaneous rotational velocity of the moving frame in vector form is given by:

$$\omega = \tau \boldsymbol{t} + \kappa \boldsymbol{b} \quad (3.19)$$

where

ω = Instantaneous rotational velocity vector of the moving frame

The vector of instantaneous rotational velocity is called Darboux's vector and lies in the rectifying plane and can be resolved into two components, $\tau\boldsymbol{t}$ and $\kappa\boldsymbol{b}$. Thus, the instantaneous rotational velocity of the moving frame can be regarded as the sum of two kinds of rotations: one rotating around the axis in the $\tau\boldsymbol{t}$ direction, and the other rotating around the axis in the $\kappa\boldsymbol{b}$ direction. Therefore, the kinematics meanings of borehole curvature and torsion become explicit: borehole curvature equals the component of the moving frame rotating around the binormal, and borehole torsion equals the component of the moving frame rotating around the tangent line. In addition, Eq. 3.20 gives Frenet's relations in another form:

$$\begin{cases} \dfrac{d\boldsymbol{t}}{dL} = \omega \times \boldsymbol{t} \\ \dfrac{d\boldsymbol{n}}{dL} = \omega \times \boldsymbol{n} \\ \dfrac{d\boldsymbol{b}}{dL} = \omega \times \boldsymbol{b} \end{cases} \tag{3.20}$$

Borehole Bending

Originally, the oil and gas industry focused on drilling vertical wells. The concept of inclination angle arose when engineers realized that a drilled trajectory need not be a plumb line. Then the concept of azimuth angle came into being when they further understood that the direction of a drilled trajectory need not change only in a vertical plane.

The change in direction of a well trajectory can be described by means of the inclination angle and the azimuth angle, but the curvature of the trajectory also needs to be taken into account as it greatly affects well drilling, completion, and even production.

Bending Angle

The oil industry defines the angle included between two tangent vectors at two different points of a wellbore trajectory as a dogleg angle.[13–16] Generally, these two tangent vectors are not in the same plane and need not meet at a point, so that the dogleg angle presents itself as a space angle (Figure 3.10). Dogleg angle is also called an overall angle,[13] which means that it includes both an inclination change and an azimuth change. In mathematics and drilling engineering, this is called the bending angle. Two ways of calculating the bending angle are described next.

A plumb line is drawn at a certain point A on a well trajectory and intersects a horizontal plane at point F (Figure 3.10). Next, a tangent line to the well trajectory at point A is drawn and intersects the horizontal plane at point C. Again, a line is drawn through point A that is parallel to the tangent line of the well trajectory at a lower point B, and intersects the horizontal plane at point D. According to the definitions of inclination angle, azimuth angle, and bending angle, the angle $\angle CAF$, designated as α_1, is equal to the inclination of the well trajectory at point A; the angle $\angle DAF$, designated as

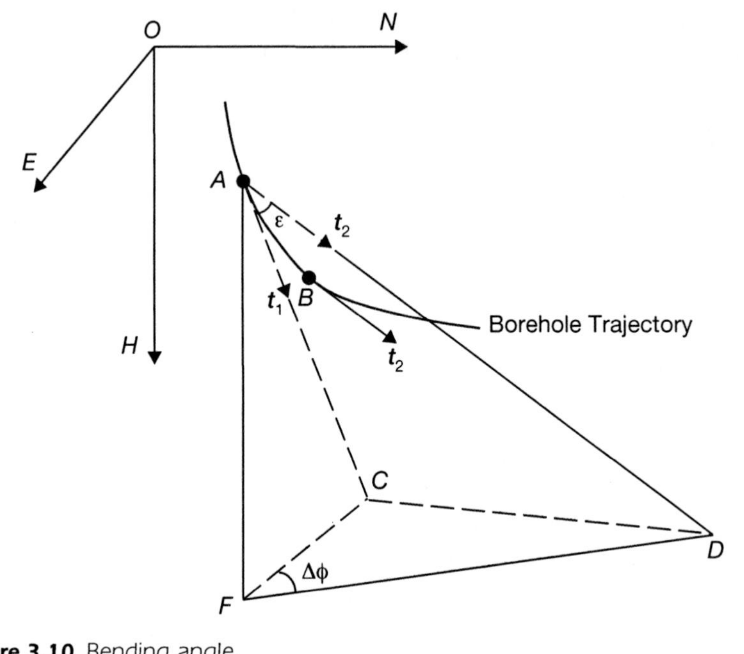

Figure 3.10 Bending angle.

α_2, is equal to the inclination at point *B*; the angle $\angle DFC$, designated as $\Delta\phi = \phi_2 - \phi_1$, is equal to the azimuth change between point *A* and point *B*; and the bending angle is, of course, equal to the angle $\angle DAC$ and is designated as ε.

The following relationships are obvious when considering the right triangles ΔAFC and ΔAFD, respectively:[2,13]

$$\left|\overline{FC}\right| = \left|\overline{AF}\right| \tan \alpha_1 \tag{3.21}$$

$$\left|\overline{FD}\right| = \left|\overline{AF}\right| \tan \alpha_2 \tag{3.22}$$

$$\left|\overline{AC}\right| = \frac{\left|\overline{AF}\right|}{\cos \alpha_1} \tag{3.23}$$

$$\left|\overline{AD}\right| = \frac{\left|\overline{AF}\right|}{\cos \alpha_2} \tag{3.24}$$

In the triangles ΔCFD and ΔCAD:

$$\left|\overline{CD}\right|^2 = \left|\overline{FC}\right|^2 + \left|\overline{FD}\right|^2 - 2\left|\overline{FC}\right| \bullet \left|\overline{FD}\right| \bullet \cos \Delta\phi \tag{3.25}$$

$$\left|\overline{CD}\right|^2 = \left|\overline{AC}\right|^2 + \left|\overline{AD}\right|^2 - 2\left|\overline{AC}\right| \bullet \left|\overline{AD}\right| \bullet \cos \varepsilon \tag{3.26}$$

Substituting Eqs. 3.21 through 3.24 into Eqs. 3.25 and 3.26 solving for $\cos\varepsilon$, and proceeding through a few transformations, gives:

$$\cos\varepsilon = \cos\alpha_1 \cos\alpha_2 + \sin\alpha_1 \sin\alpha_2 \cos\Delta\phi \tag{3.27}$$

where

$\Delta\phi = \phi_2 - \phi_1$
ε = Bending angle, (°)
$\Delta\phi$ = Section increment of azimuth angle, (°)

Eq. 3.27 would be convenient only if ε were large. For relatively small values of ε, $\cos\varepsilon$ is almost equal to unity and ε cannot be calculated accurately using mathematical tables. For this reason, Eq. 3.27 is transformed as follows:

$$\cos\varepsilon = \cos\Delta\alpha - \sin\alpha_1 \sin\alpha_2 (1 - \cos\Delta\phi) \tag{3.28}$$

where

$\Delta\alpha = \alpha_2 - \alpha_1$
$\Delta\alpha$ = Section increment of inclination angle, (°)

The following relationships are well known:

$$\begin{cases} \cos\varepsilon = 1 - 2\sin^2\dfrac{\varepsilon}{2} \\ \cos\Delta\alpha = 1 - 2\sin^2\dfrac{\Delta\alpha}{2} \\ 1 - \cos\Delta\phi = 2\sin^2\dfrac{\Delta\phi}{2} \end{cases} \tag{3.29}$$

Substituting Eq. 3.29 into Eq. 3.28 and solving for $\sin\frac{\varepsilon}{2}$ gives:

$$\sin\frac{\varepsilon}{2} = \sqrt{\sin^2\frac{\Delta\alpha}{2} + \sin^2\frac{\Delta\phi}{2}\sin\alpha_1 \sin\alpha_2} \tag{3.30}$$

In order to improve the precision in the calculation of the bending angle ε, when it is very small, $\cos\varepsilon$ has been expressed in terms of $\sin\frac{\varepsilon}{2}$. Eqs. 3.27 and 3.30 are mathematically equivalent.

Generally, the inclinations α_1 and α_2 at two successive surveying stations are close to each other. Therefore, these inclinations may each be replaced by the average value $\bar{\alpha}$, and Eq. 3.30 then becomes:

$$\sin\frac{\varepsilon}{2} = \sqrt{\sin^2\frac{\Delta\alpha}{2} + \sin^2\frac{\Delta\phi}{2}\sin^2\bar{\alpha}} \tag{3.31}$$

where

$$\bar{\alpha} = \frac{\alpha_1 + \alpha_2}{2}$$

$\bar{\alpha}$ = Average inclination angle (°)

The approximation made in Eq. 3.31 is as follows:

$$\sin^2 \bar{\alpha} = \sin \alpha_1 \sin \alpha_2 \qquad (3.32)$$

If the value of $\bar{\alpha}$ were to be calculated from α_1 and α_2 using Eq. 3.32, then Eq. 3.31 would be exact and would yield the same value of ε as calculated by Eqs. 3.27 or 3.30.

According to differential geometry,[15,16] Eq. 3.33 can be used to calculate the bending angle between two points on a wellbore trajectory.

$$\cos \varepsilon = \frac{\boldsymbol{t}_1 \bullet \boldsymbol{t}_2}{|\boldsymbol{t}_1||\boldsymbol{t}_2|} \qquad (3.33)$$

As mentioned in the case of Eq. 3.14, the unit tangent vectors $\boldsymbol{t}_1$ and $\boldsymbol{t}_2$ of a wellbore trajectory at two points can be expressed respectively by:

$$\boldsymbol{t}_1 = \sin \alpha_1 \cos \phi_1 \boldsymbol{i} + \sin \alpha_1 \sin \phi_1 \boldsymbol{j} + \cos \alpha_1 \boldsymbol{k} \qquad (3.34)$$

$$\boldsymbol{t}_2 = \sin \alpha_2 \cos \phi_2 \boldsymbol{i} + \sin \alpha_2 \sin \phi_2 \boldsymbol{j} + \cos \alpha_2 \boldsymbol{k} \qquad (3.35)$$

Substituting Eqs. 3.34 and 3.35 into Eq. 3.33 yields Eq. 3.36, which is identical to Eq. 3.27.[8,9]

$$\cos \varepsilon = \cos \alpha_1 \cos \alpha_2 + \sin \alpha_1 \sin \alpha_2 \cos \Delta\phi \qquad (3.36)$$

The bending angle of a wellbore trajectory or well path at two points is only related to the values of inclination angle and azimuth angle at those points and does not consider the rate of these changes (increase or decrease) within the trajectory interval.

Borehole Curvature

Borehole curvature, which represents the extent of a well trajectory's departure from a straight line, is the rate of rotation of the tangent vector to the well trajectory with respect to the curved length of trajectory. It depicts the extent of bending of a wellbore trajectory. Eq. 3.37 is the expression that defines borehole curvature, and shows that the rate of rotation of the tangent vector to a trajectory with respect to curved length of the trajectory gives the extent of bending of the wellbore trajectory:

$$\kappa = \left| \frac{d\boldsymbol{t}}{dL} \right| = |\dot{\boldsymbol{t}}| \qquad (3.37)$$

Besides being an important index for checking the buildup rates at the bottomhole assembly (BHA), for understanding the deflection behavior of the formation, and for evaluating the quality of a wellbore trajectory, the borehole curvature also serves as the basis for monitoring a wellbore trajectory during drilling, and for calculating the forces applied on the drill string and its deformation.

There are two ways of deducing the formula used for calculating borehole curvature. One is to substitute the expression of unit tangent vector ***t***, from Eq. 3.14 into Eq. 3.37,[4,16] and another is to substitute the formula of bending angle ε, from Eq. 3.27 into Eq. 3.28, for obtaining the defining expression of borehole curvature. By mathematical deduction, it can be shown that both methods yield the same expression, which is given in Eq. 3.39 and applies for calculating the borehole curvature at a given point:[4,5]

$$\kappa = \lim_{\Delta L \to 0} \left| \frac{\varepsilon}{\Delta L} \right| \tag{3.38}$$

or in generalized form,

$$\kappa = \sqrt{\kappa_\alpha^{\,2} + \kappa_\phi^{\,2} \sin^2 \alpha} \tag{3.39}$$

where

$\Delta L = L_2 - L_1$
ΔL = Curved section length, m or ft

A wellbore trajectory is often shown using a vertical expansion plot and a horizontal projection plot. In these plots, the curvature of the wellbore trajectory is related to the rate of change of the inclination and the rate of change of the azimuth, as follows:

$$\kappa_V = \kappa_\alpha \tag{3.40}$$

$$\kappa_V = \frac{d\alpha}{dL} = \kappa_\alpha$$

$$\kappa_H = \frac{\kappa_\phi}{\sin\alpha} \tag{3.41}$$

$$\kappa_H = \frac{d\phi}{dS} = \frac{d\phi}{dL \sin\alpha} = \frac{\kappa_\phi}{\sin\alpha}$$

where

κ_V = Curvature of the wellbore trajectory in the vertical expansion plot, (°)/30 m or (°)/100 ft
κ_H = Curvature of the wellbore trajectory in the horizontal projection plot, (°)/30 m or (°)/100 ft
S = Arc length in the azimuthal direction, m or ft

Eqs. 3.33 and 3.34 are important as they make the connection between the change of a wellbore trajectory in a cylinder plot with the actual change in space. Substituting Eqs. 3.40 and 3.41 into Eq. 3.39, the formula for calculating borehole curvature is also expressed as:

$$\kappa = \sqrt{\kappa_V^{\,2} + \kappa_H^{\,2} \sin^4 \alpha} \tag{3.42}$$

κ_V is also called the build (or drop) rate curvature in the vertical build or (drop plane), whereas κ_H is called the walk-rate curvature in the horizontal walk plane. A sign convention for the walk-rate curvature is used to differentiate between left or right turns. A positive sign indicates that the wellbore is turning to the right, whereas a negative sign indicates turning to the left. The curvature κ is also called the dogleg severity, or in short, DLS, and is typically expressed in degrees per 100 ft or degrees per 30 m.

Average Curvature

During the process of drilling, engineers usually pay attention to the average curvature of a survey interval. Generally, the borehole curvature is not constant and varies with the measured depth, making it practical to use an average curvature. Two different equations are commonly used to estimate the average curvature of a survey interval, as given below:

$$\bar{\kappa} = \sqrt{\left(\frac{\Delta\alpha}{\Delta L}\right)^2 + \left(\frac{\Delta\phi}{\Delta L}\right)^2 \sin^2 \bar{\alpha}} \tag{3.43}$$

$$\bar{\kappa} = \frac{\varepsilon}{\Delta L} \tag{3.44}$$

where

$\bar{\kappa}$ = average borehole curvature

$\varepsilon = a\cos\left(\cos\alpha_1 \cos\alpha_2 + \sin\alpha_1 \sin\alpha_2 \cos\Delta\phi\right)$

It should be noted here that:

- In most cases, Eqs. 3.43 and 3.44 yield very similar results.[2,13] Thus, it is acceptable to use either formula for calculating the average curvature of a survey interval.
- For natural-curve trajectories and cylinder-helix trajectories, the value of the average inclination angle of a survey interval equals the value of the inclination angle at the midpoint of the survey interval, as the inclination angle is a linear function of the measured depth. Thus, the calculated value of average curvature from Eq. 3.43 actually represents the value of the borehole curvature at the midpoint for natural-curve trajectories and cylinder-helix trajectories.
- Eq. 3.44 is derived from the spatial-arc model (described later).

- There always exists some error when using Eq. 3.43 for calculating the average curvature of a spatial-arc interval. For larger κ and ΔL, the error is larger. Except for $\omega = 0°$ or $\omega = 180°$, errors exist at all other toolface angles.

While Eq. 3.39 is an exact formula for calculating the borehole curvature at a given point, Eq. 3.43 can sometimes show considerable error in calculating the average curvature of a survey interval. Caution should be exercised in using Eq. 3.43. Eq. 3.44, however, is deduced from the mathematical definition of borehole curvature and is more reasonable in terms of theoretical analyses and calculated results.

PROBLEM 3.2

While drilling a directional well, a survey shows the initial inclination to be 32° and azimuth to be 112°. The inclination and directional curvatures are to be maintained at 4°/100 ft and 7°/100 ft, respectively, for a course length of 100 ft. Calculate the final inclination angle and final direction at the end of the course length. Also, determine the curvature, the vertical curvature, and the horizontal-walk curvature.

Solution

Given data:

$$\alpha_1 = 32°;\ \phi_1 = 112°;\ \kappa_\alpha = 4°/100\text{ ft};\ \kappa_\phi = 7°/100\text{ ft};\ \Delta L = 100\text{ ft}$$

The final inclination angle is:

$$\alpha_2 = \alpha_1 + ka\left(L_2 - L_1\right) = 32 + \frac{4}{100} \times 100 = 36°$$

The final direction is:

$$\phi_2 = \phi_1 + \kappa_\phi\left(L_2 - L_1\right) = 112 + \frac{7}{100} \times 100 = 119°$$

The average inclination angle is:

$$\bar{\alpha} = \frac{\alpha_1 + \alpha_2}{2} = \frac{32 + 36}{2} = 34°$$

The average curvature is:

$$\kappa = \sqrt{4^2 + 7^2 \sin^2 34} = 5.596°/100\text{ ft}$$

Using the bending or dogleg angle for calculating average curvature:

$$\cos \varepsilon = \cos \alpha_1 \cos \alpha_2 + \sin \alpha_1 \sin \alpha_2 \cos \Delta\phi$$

$$\bar{\kappa} = \frac{a\cos\left(\cos 32 \cos 36 + \sin 32 \sin 36 \cos 7\right)}{100} = 5.591°/100\text{ ft}$$

A small difference is observed between the two methods of calculation of the average curvature.

The vertical curvature and the horizontal-walk curvature are:

$$\kappa_V = \kappa_\alpha = 4°/100 \text{ ft}$$

$$\kappa_H = \frac{7}{\sin 34} = 12.518°/100 \text{ ft}$$

Borehole Turning

Borehole turning is another important parameter of the wellbore trajectory and the well path design. A wellbore trajectory is a continuous and smooth curve in space, which bends and turns simultaneously to change direction. Drilling experience shows that pipe sticking sometimes occurs during the process of a round trip of the drill string, even though the borehole curvature is small, and the weight is not effectively applied on the bit while drilling, especially when the stiffness and bent angle of a steerable motor are large. Therefore, the extensive turning of a well trajectory may twist a downhole pipe string, greatly increasing the drag and torque applied to the string and causing deformation.[6]

Torsion Angle

The torsion angle is defined as the included angle between two binormal vectors at two different points of a wellbore trajectory, as shown in Figure 3.11. According to the definition of the included angle between two vectors, Eq. 3.45 gives the torsion angle between the two given points on a wellbore

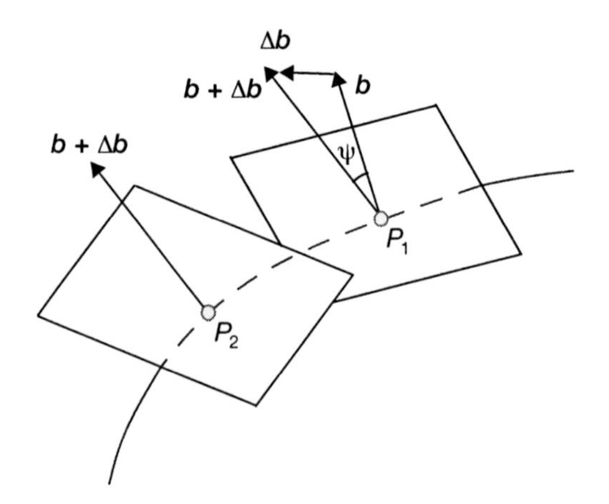

Figure 3.11 Torsion angle.

trajectory. Substituting Eq. 3.14 into Eq. 3.45 yields Eq. 3.46, which is the formula for calculating the torsion angle.[1,2,5,10]

$$\cos\theta = \frac{\boldsymbol{b}_1 \bullet \boldsymbol{b}_2}{|\boldsymbol{b}_1||\boldsymbol{b}_2|} \tag{3.45}$$

$$\cos\theta = a\cos\Delta\phi - b\sin\Delta\phi + c \tag{3.46}$$

where

$$a = \lambda_{\alpha 1}\lambda_{\alpha 2} + \frac{1}{4}\lambda_{\phi 1}\lambda_{\phi 2}\sin 2\alpha_1 \sin 2\alpha_2$$

$$b = \frac{1}{2}\left(\lambda_{\alpha 1}\lambda_{\phi 2}\sin 2\alpha_2 - \lambda_{\alpha 2}\lambda_{\phi 1}\sin 2\alpha_1\right)$$

$$c = \lambda_{\phi 1}\lambda_{\phi 2}\sin^2\alpha_1 \sin^2\alpha_2$$

θ = Torsion angle, (°)

Eq. 3.46, the universal formula for torsion angle, can be simplified for given mathematical models of the wellbore trajectory, such as spatial-arc trajectory, natural-curve trajectory, and cylinder-helix trajectory.[2,14]

It is necessary to point out that the bending angle cannot directly give the extent of bending of a wellbore trajectory. Nevertheless, as an important indirect parameter, use of the bending angle can simplify the calculation method, well path planning process, and performance of survey calculations. In the same way, torsion angle does not directly give the extent of torsion of a wellbore trajectory.

Borehole Torsion

Borehole torsion gives the extent of departure of a well trajectory from a plane curve. It is the rate of rotation of the binormal vector of the well trajectory with respect to curved length. It depicts the extent of torsion of a wellbore trajectory.[2,9,14] The expression defining borehole torsion is as follows:

$$\tau = \begin{cases} +|\dot{\boldsymbol{b}}|, & \text{if } \dot{\boldsymbol{b}} \text{ and } \boldsymbol{n} \text{ are in the opposite direction} \\ -|\dot{\boldsymbol{b}}|, & \text{if } \dot{\boldsymbol{b}} \text{ and } \boldsymbol{n} \text{ are in the same direction} \end{cases} \tag{3.47}$$

The positive and negative signs in Eq. 3.47 for borehole torsion have the following meanings: If a point moves along a wellbore trajectory in the onward direction and if the derivative of the unit binormal vector with respect to the measured depth, $\dot{\boldsymbol{b}}$, is in the direction opposite to the unit principal normal vector, $\boldsymbol{n}$, then the sign of borehole torsion, τ, is positive. Otherwise, the sign of τ is negative (Figure 3.12).

From the definitions of moving frame and borehole torsion we can derive $\dot{\boldsymbol{b}} // \boldsymbol{n}$ and get the following equation:

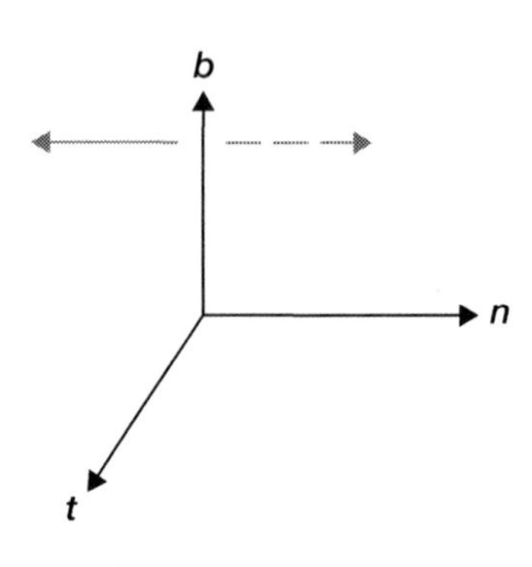

Figure 3.12 Borehole torsion sign.

$$\dot{\boldsymbol{b}} = -\tau \boldsymbol{n} \tag{3.48}$$

Multiplying Eq. 3.48 by $\boldsymbol{n}$ from the right yields:

$$\tau = -\dot{\boldsymbol{b}} \bullet \boldsymbol{n} = \boldsymbol{b} \bullet \dot{\boldsymbol{n}} \tag{3.49}$$

Proceeding through some transformations using differential geometry, the basic formula for calculating borehole torsion is:[2,5,6]

$$\tau = \frac{\left(\dot{\boldsymbol{r}}, \ddot{\boldsymbol{r}}, \dddot{\boldsymbol{r}}\right)}{\kappa^2} = \frac{1}{\kappa^2}\begin{bmatrix} \dot{N} & \dot{E} & \dot{H} \\ \ddot{N} & \ddot{E} & \ddot{H} \\ \dddot{N} & \dddot{E} & \dddot{H} \end{bmatrix} \tag{3.50}$$

The successive derivatives of Eq. 3.13 with respect to curved length are:

$$\begin{cases} \dot{N} = \sin\alpha\cos\phi \\ \dot{E} = \sin\alpha\sin\phi \\ \dot{H} = \cos\alpha \end{cases} \tag{3.51}$$

$$\begin{cases} \ddot{N} = \kappa_\alpha \cos\alpha\cos\phi - \kappa_\phi \sin\alpha\sin\phi \\ \ddot{E} = \kappa_\alpha \cos\alpha\sin\phi + \kappa_\phi \sin\alpha\cos\phi \\ \ddot{H} = -\kappa_\alpha \sin\alpha \end{cases} \tag{3.52}$$

$$\begin{cases} \dddot{N} = \dot{\kappa}_\alpha \cos\alpha\cos\phi - \dot{\kappa}_\phi \sin\alpha\sin\phi - 2\kappa_\alpha\kappa_\phi \cos\alpha\sin\phi - \left(\kappa_\alpha^2 + \kappa_\phi^2\right)\sin\alpha\cos\phi \\ \dddot{E} = \dot{\kappa}_\alpha \cos\alpha\sin\phi + \dot{\kappa}_\phi \sin\alpha\cos\phi + 2\kappa_\alpha\kappa_\phi \cos\alpha\cos\phi - \left(\kappa_\alpha^2 + \kappa_\phi^2\right)\sin\alpha\sin\phi \\ \dddot{H} = -\dot{\kappa}_\alpha \sin\alpha - \kappa_\alpha^2 \cos\alpha \end{cases} \tag{3.53}$$

where

τ = Torsion of the wellbore trajectory, (°)/30 m or (°)/100 ft

ω = Instantaneous rotational velocity vector of the moving frame
ε = Bending angle, (°)
$\Delta\phi$ = Section increment of azimuth angle, (°)
$\Delta\alpha$ = Section increment of inclination angle, (°)
$\bar{\alpha}$ = Average inclination angle, (°)
ΔL = Curved section length, m or ft
κ_V = Curvature of wellbore trajectory in the vertical expansion plot, (°)/30 m or (°)/100 ft
κ_H = Curvature of wellbore trajectory in horizontal projection plot, (°)/30 m or (°)/100 ft
$\bar{\kappa}$ = Average borehole curvature
θ = Torsion angle, (°)
$\dot{\kappa}_\alpha$ = First derivative of inclination change rate, viz. second derivative of inclination angle
$\dot{\kappa}_\phi$ = First derivative of azimuth change rate, viz. second derivative of azimuth angle

Substituting Eqs. 3.51 through 3.53 into Eq. 3.50 yields Eq. 3.54, a formula to calculate the borehole torsion at any point:[2,5,6,8,10]

$$\tau = \frac{\kappa_\alpha \dot{\kappa}_\phi - \kappa_\phi \dot{\kappa}_\alpha}{\kappa^2} \sin\alpha + \kappa_\phi \left(1 + \frac{\kappa_\alpha^2}{\kappa^2}\right) \cos\alpha \tag{3.54}$$

The value of borehole curvature is always positive, but that of borehole torsion can be positive or negative. A zero borehole curvature, $\kappa = 0$, will depict a straight section, and a zero borehole torsion, $\tau = 0$, will depict a plane curve, and vice versa.

Average Torsion

Often it is necessary to calculate the average borehole torsion over a survey interval of a drilled well trajectory, or a section of a planned well path. Using the definition of borehole torsion and the calculation method for average borehole curvature as a reference, the formula for calculating average borehole torsion will be similar to Eq. 3.44. Eq. 3.55 gives the formula to calculate average borehole torsion, taking into account the fact that the value of the torsion angle is always positive, while that of borehole torsion can be either positive or negative.

$$\bar{\tau} = \text{sgn}(\Delta\phi)\frac{\theta}{\Delta L} \tag{3.55}$$

where

$$\operatorname{sgn}(x) = \begin{cases} +1, & \text{if } x > 0 \\ 0, & \text{if } x = 0 \\ -1, & \text{if } x < 0 \end{cases}$$

where

$\bar{\tau}$ = Average borehole torsion

Needless to say, the sign of the torsion is important when presenting the values. Calculating borehole curvature and torsion provides a basis for effectively monitoring and controlling wellbore trajectories, analyzing the forces applied on the drill string and its deformation, calculating the drag and torque on the pipe string, and checking its strength.[17,18]

Models for Well Path Trajectories

In this section, both 2D and 3D models for the shapes of well paths using standard curves will be covered. The vast majority of directional wells and horizontal wells are designed as two-dimensional well paths. Commonly used sections of the well path profile are based on straight-line and circular-arc methods. In engineering design, the catenary shape formed by suspending a flexible cable loosely held at both fixed ends is extensively used. The shape thus formed may look like a parabola, but it is not. Extended-reach drilling (ERD) and ultra-extended-reach drilling (u-ERD) technology have been rapidly developed during the past decades, and the horizontal departure nowadays can be over 10 km. Wellbore friction is an important issue for ultra-long wells, and optimizing the shape of the well path is an effective means of reducing the torque and drag. One of the shapes that can effectively be used to achieve this is the catenary shape. In order to reduce the torque and drag in the drill string, catenary and parabolic profiles[19-26] are the methods of choice for extended-reach and ultra-extended-reach wells.

2D Mechanical Models

While estimating drag forces, two types of models are considered: soft-string and stiff-string models. In the soft-string model, the stiffness of the drill string or the work string is neglected and it is assumed that the entire string behaves as a flexible cable or catenary. The string axis aligns with the wellbore axis, if the well path profile is also of the catenary type, which can greatly reduce the friction in the drill string or the work string.[2,5]

Some of the other notable uses of the catenary curve in the petroleum industry are for the design of riser shape and of anchor-cable modeling.

Drilling in deepwater presents many challenges, not only to the operators, but also to the well planners. The riser shape plays an important role in the torque and drag analysis, which may result in additional costs because of fatigue damage and wear of the pipes. The increased water depth and severe water currents place severe strain on the drilling risers. A simple, flexible steel, catenary riser, also known as SCR, is usually considered in arriving at the profile of the riser. An SCR is a steel pipe suspended from the platform to the seabed in the near-catenary shape. If the bending stiffness of the pipe is assumed to be negligible, the riser will take the shape of a simple catenary when suspended between a vessel and the seafloor. Catenary shape is also called alysoid or chainette.

Catenary Model

Before studying the details of the use of the catenary curve in well path planning, it is important to study the basic characteristics of the catenary shape. It is assumed that the drill string behaves as a flexible cable, with only gravity loading along its length and no stiffness. The drill string is fixed at the two ends A and B and the shape formed is as shown in Figure 3.13(a), using xy coordinates with the origin at the point A. Resolution of the forces on a differential element Δl of the drill string is shown in Figure 3.13(b).

The tension, T, along the drill string is in the tangent direction and will continuously change along the length. The differential element, Δl, has a weight, $q(l)\Delta l$, acting downward at a distance of $\delta(\Delta x)$ from the endpoint O of $\Delta l (0 < \delta < 1)$. Here,

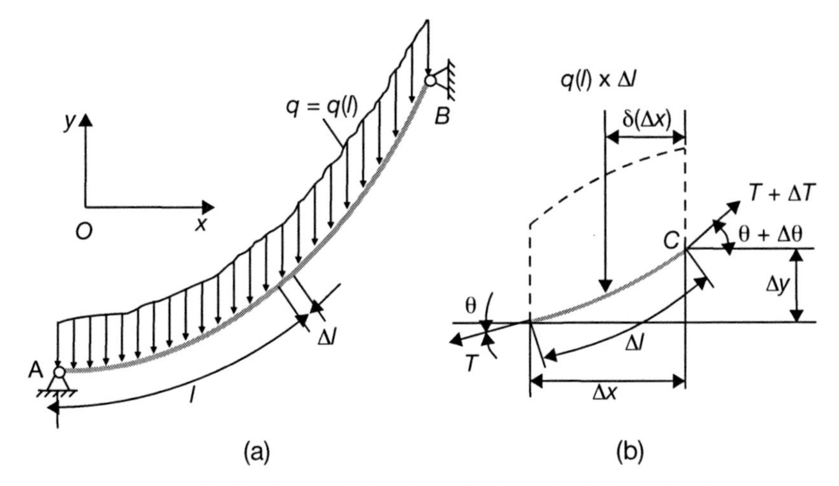

Figure 3.13 Load distribution analysis along the length of the drill string.

$q(l)$ = weight per unit length along the drill-string length l

The conditions for equilibrium can be written as:[2,5,27]

$$-T\cos\theta + (T + \Delta T)\cos(\theta + \Delta\theta) = 0 \quad (3.56)$$
$$-T\sin\theta - q(l)\Delta l + (T + \Delta T)\sin(\theta + \Delta\theta) = 0 \quad (3.57)$$
$$q(l)\Delta l \times \delta(\Delta x) - T\cos\theta \times \Delta y + T\sin\theta \times \Delta x = 0 \quad (3.58)$$

where θ, $\theta + \Delta\theta$ are angles made with the horizontal.

It can be noted that $\Delta l \to 0$ when $\Delta x \to 0$, $\Delta y \to 0$, $\Delta\theta \to 0$, $\Delta T \to 0$. Therefore, applying the limits the following relationships can be obtained:

$$\frac{d(T\cos\theta)}{dl} = 0 \quad (3.59)$$
$$\frac{d(T\sin\theta)}{dl} - q(l) = 0 \quad (3.60)$$
$$\frac{dy}{dx} = \tan\theta \quad (3.61)$$

Using Eqs. 3.59 and 3.60:

$$T\cos\theta = F_H = \text{constant} \quad (3.62)$$
$$T\sin\theta = \int q(l)dl \quad (3.63)$$

where

F_H = Horizontal component of the pulling force or the tension T

Using Eqs. 3.61 through Eq. 3.63 results in:

$$\frac{dy}{dx} = \frac{1}{F_H}\int q(l)dl \quad (3.64)$$

Using differential geometry, the following can be written:

$$\frac{dy}{dx} = \sqrt{\left(\frac{dl}{dx}\right)^2 - 1} \quad (3.65)$$

Consequently, Eq. 3.64 becomes:

$$\frac{dl}{dx} = \sqrt{1 + \left[\frac{1}{F_H}\int q(l)dl\right]^2} \quad (3.66)$$

Separation of variables and integrating with boundary condition yields:

$$x = \int \frac{dl}{\sqrt{1 + \left[\frac{1}{F_H}\int q(l)dl\right]^2}} \quad (3.67)$$

This is the universal deflection equation of a flexible cable.

In the special case of uniform loading of the cable $q(l)$ = constant = q_m, and with this condition the right side of Eq. 3.64 can be integrated as:

$$\frac{dy}{dx}=\frac{1}{F_H}\left(q_m l+C_1'\right) \tag{3.68}$$

Eq. 3.67 can be rewritten as:

$$x=\int\frac{dl}{\sqrt{1+\frac{1}{F_H{}^2}\left(q_m l+C_1'\right)^2}} \tag{3.69}$$

Integrating Eq. 3.69 yields:

$$x=\frac{F_H}{q_m}\left[\sinh^{-1}\left(\frac{q_m l+C_1'}{F_H}\right)+C_2'\right] \tag{3.70}$$

Combining Eqs. 3.70 and 3.68, Eq. 3.71 is obtained:

$$\frac{dy}{dx}=\sinh\left(\frac{q_m}{F_H}x+C_1\right) \tag{3.71}$$

Eq. 3.71 can be integrated to obtain:

$$y=a\cosh\left(\frac{x}{a}+C_1\right)+C_2 \tag{3.72}$$

where

$$a=\frac{F_H}{q_m}$$

and where C_1, C_2, C_1', C_2' are constants of integration.

The catenary constant a is called the control parameter of the catenary shape, which controls the length of the curve to the width between the fixed ends. It has the dimension of length.

Choosing an appropriate coordinate system can simplify the equation. As shown in Figure 3.14, the lowest point F of the curve in the xy coordinate system has a zero slope with the distance from F to O providing the value of the catenary constant.

In the O-xy coordinate system, it is evident that the boundary conditions are:

$$\begin{cases} y\big|_{x=0}=a \\ \left.\dfrac{dy}{dx}\right|_{x=0}=0 \end{cases} \tag{3.73}$$

Applying the boundary conditions $C_1 = C_2 = 0$, Eq. 3.72 becomes:

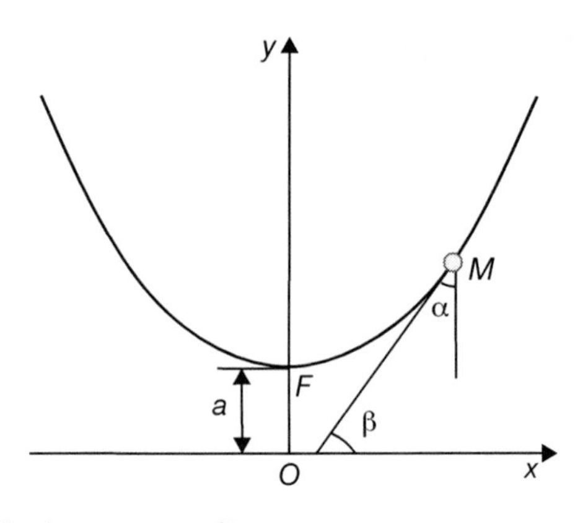

Figure 3.14 Standard catenary profile.

$$y = a\cosh\frac{x}{a} \tag{3.74}$$

This is the classic catenary line equation that is described by a hyperbolic function.

Parabolic Model

The previous section presented the catenary analysis with the load distribution considered as a function of length l of the drill string. The load distribution can also be assumed to be a function of x, the horizontal projection on the x-axis, with the shape as shown in Figure 3.15(a). The free-body diagram of the force distribution in a differential element is shown in Figure 3.15(b).

If F_H is the horizontal tension at the lowest point A in the xy coordinate system the origin of which is also at the lowest point A, and if T and $T + \Delta T$ are the tensions in the differential element, as shown in Figure 3.15, then from the force balance analysis the following equations are obtained:

$$-T\cos\theta + (T + \Delta T)\cos(\theta + \Delta\theta) = 0 \tag{3.75}$$

$$-T\sin\theta - q(x)\Delta x + (T + \Delta T)\sin(\theta + \Delta\theta) = 0 \tag{3.76}$$

$$q(x)\Delta x \times \delta(\Delta x) - T\cos\theta \times \Delta y + T\sin\theta \times \Delta x = 0 \tag{3.77}$$

Applying the limits, as done earlier for the catenary model, gives:

$$\frac{d(T\cos\theta)}{dx} = 0 \tag{3.78}$$

$$\frac{d(T\sin\theta)}{dx} - q(x) = 0 \tag{3.79}$$

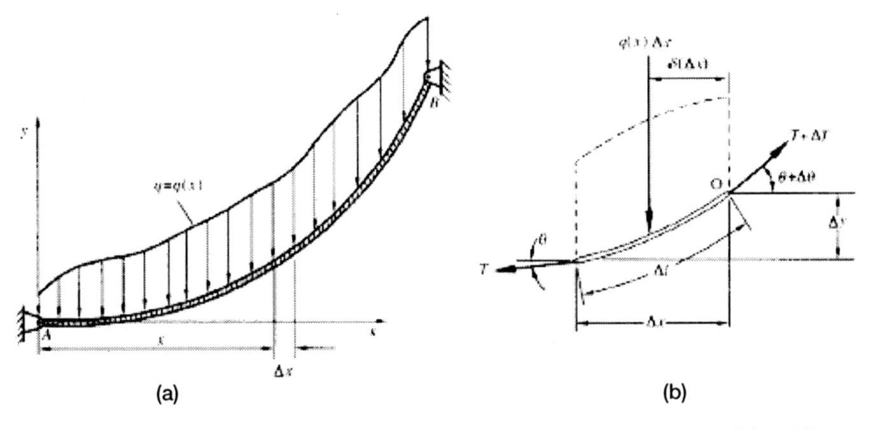

Figure 3.15 Load distribution analysis along the horizontal direction of the drill string.

$$\frac{dy}{dx} = \tan\theta \tag{3.80}$$

Using Eqs. 3.78 and 3.79:

$$T\cos\theta = F_H = \text{constant} \tag{3.81}$$

$$T\sin\theta = \int q(x)dx \tag{3.82}$$

Using Eqs. 3.80 through 3.82:

$$y = \frac{1}{F_H}\int\left[\int q(x)dx\right]dx \tag{3.83}$$

If $q(x)$ = constant = q_x, then (the development of the equation is much simpler in the present case than in the earlier case of the catenary model) integrating Eq. 3.83 yields:

$$y = \frac{1}{F_H}\left(\frac{q_x}{2}x^2 + C_1x + C_2\right) \tag{3.84}$$

This is the deflection equation of a flexible cable with uniform load distribution, q_x. The integration constants C_1 and C_2 are determined by applying the boundary conditions as explained below. Choosing an appropriate coordinate system, as indicated in Figure 3.16, simplifies the equations for the boundary conditions as follows:

$$\begin{cases} y\big|_{x=0} = 0 \\ \dfrac{dy}{dx}\bigg|_{x=0} = 0 \end{cases} \tag{3.85}$$

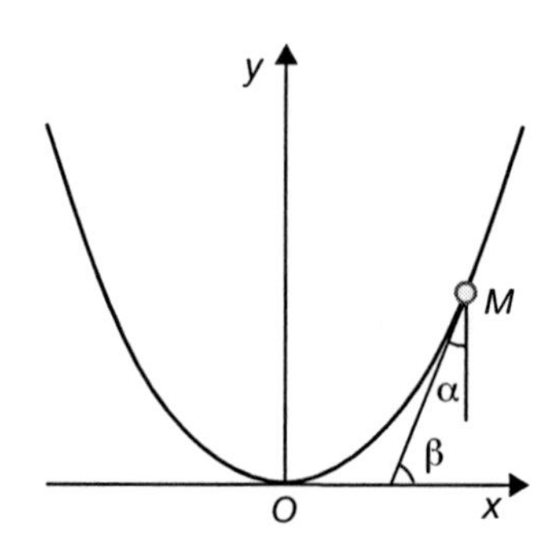

Figure 3.16 Standard parabolic profile.

Consequently, with these boundary conditions, $C_1 = C_2 = 0$; Eq. 3.84 becomes:

$$y = \frac{1}{2P}x^2 \tag{3.86}$$

where

$$P = \frac{F_H}{q_x}$$

Eq. 3.86 is the equation of a standard parabola with the vertical axis as y-axis and with the vertex coinciding with the origin of the coordinate system. The constant P is called the control parameter of the parabola, which controls the length of the curve to the width of the fixed ends. Thus, when the drill string is loaded uniformly along the horizontal axis, the curve takes the shape of a parabola.

Slant-Line or Circular-Arc Model

There are many different ways of treating the load distribution along the drill string. In the earlier paragraphs, it was seen that the resulting string shape is a catenary if the load distribution is considered to be uniform along the drill-string length, and the shape is parabolic if the load distribution of the string is considered to be uniform along the horizontal direction. Another option is to neglect the load distribution of the drill string if it is small, which makes the model very simple (i.e., $q(l) = q(x) = 0$). This condition results in:

$$T \cos \theta = F_H = \text{constant} \tag{3.87}$$

$$T \sin \theta = F_V = \text{constant} \tag{3.88}$$

$$\frac{dy}{dx} = \tan \theta \tag{3.89}$$

where

F_V = Vertical component of the tension

Combining Eqs. 3.87 and 3.88:

$$\tan\theta = \frac{F_V}{F_H} \tag{3.90}$$

If the ratio,

$$\frac{F_V}{F_H}$$

is indeed a constant, then the following equation can be arrived at using Eqs. 3.89 and 3.90:

$$y = ax + b \tag{3.91}$$

where a and b are constants.

However, in fact, the approximation that there is no load distribution will result in $T = 0$, and then $F_H = F_V = 0$, which makes the ratio

$$\frac{F_V}{F_H}$$

undetermined. So Eq. 3.91 is not the general solution, but only a special case. The general solution is derived below.

When the coordinate system with origin O and axes as shown in the schematic in Figure 3.17, and point M on the drill string, where the tangent line makes an angle θ with the x-axis and where the radius of curvature is R, the following geometrical relationship can be written from Figure 3.17:

$$\tan\theta = \frac{x}{R - y} \tag{3.92}$$

Using Eqs. 3.89 and 3.92 gives:

$$xdx = (R - y)dy \tag{3.93}$$

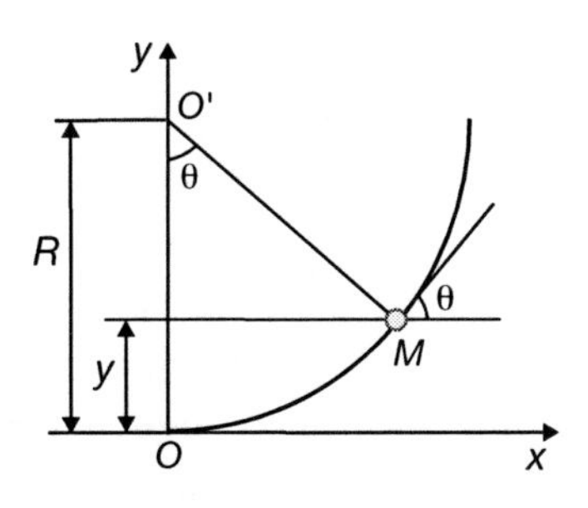

Figure 3.17 Profile of a no-load drill string.

Integration of Eq. 3.93 yields:

$$x^2 + (y - R)^2 = C \tag{3.94}$$

where C is a constant of integration.

Using the boundary condition $y|_{x=0} = 0$ gives $C = R^2$. Therefore, Eq. 3.94 takes the form:

$$x^2 + (y - R)^2 = R^2 \tag{3.95}$$

Eq. 3.95 provides the general solution for the shape of the drill string with zero load distribution. It can be noted that Eq. 3.95 is the equation of a circular arc that in the limit of $R \to \infty$ is equivalent to the straight-line case. From this it can be deduced that a flexible cable with a zero load distribution has the shape of a circular arc that in the limit becomes a straight line when $R \to \infty$.

2D Mathematical Models

There are more than ten parameters describing a well trajectory. For a 2D well trajectory: azimuth change rate is $\kappa_\phi = 0$, the curvature of the well trajectory on the horizontal projection plot is $\kappa_H = 0$, and the reference plane angle is $\omega = 0°$ (build) or 180° (drop); well torsion $\tau = 0$; the values of well inclination change rate, the curvature of well trajectory on the vertical projection plot, and well curvature are equal, or $\kappa_\alpha = \kappa_V = \kappa$; the horizontal length S and horizontal displacement A are equal, or $S = A$; the north coordinate, east coordinate, horizontal displacement, and azimuth change have a certain relationship; and the displacement azimuth and azimuth of each point of the well trajectory are equal to the designed azimuth. Thus, given the designed azimuth, the 2D well trajectory model requires the changes of well inclination angle α, true vertical depth H, horizontal displacement A, and well curvature κ (or curvature radii R, r), depending on well depth L.

Straight-Line Model

The straight-line linear model is the simplest well path model and is used for vertical wells, or vertical and horizontal sections in deviated wells. In a certain sense, vertical well design, or a vertical section and a horizontal section, are particular cases of the straight inclined section.

As shown in the Figure 3.18 the line AB depicts the inclined straight section of a well shown in the AOH coordinate system. When M is a point on the AB section, the required wellbore parameters such as inclination, vertical depth, horizontal displacement, and curvature can be written as:

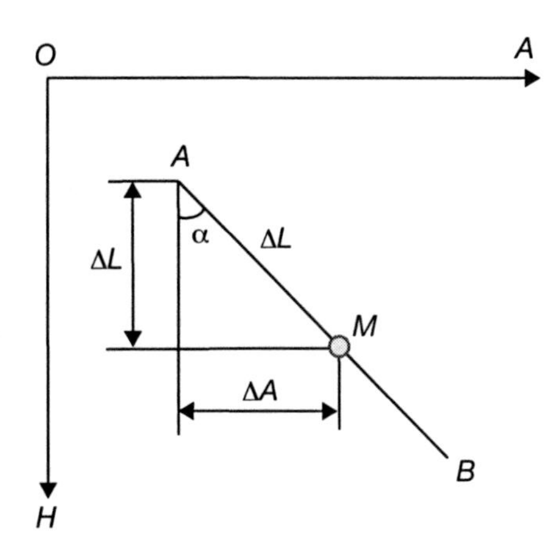

Figure 3.18 Straight linear profile.

$$\alpha = \alpha_0 \tag{3.96}$$
$$H = H_0 + \Delta L \cos \alpha_0 \tag{3.97}$$
$$A = A_0 + \Delta L \sin \alpha_0 \tag{3.98}$$
$$\kappa = 0 \tag{3.99}$$

where

$$\Delta L = L - L_o$$

Clearly, in the straight-line model of a wellbore trajectory, the curvature κ is constant and 0. Also, the vertical distance, as well as the horizontal displacement, varies linearly with depth changes.

Circular-Arc Model

Circular arcs used to describe the build and drop sections of a conventional well path are shown in Figure 3.19.

If, at a point M of the circular arc, the radius of the curvature is R and the curvature is κ, then the borehole parameters for a build section can be written as:

$$\alpha = \alpha_0 + \frac{\kappa}{C_\kappa} \Delta L = \alpha_0 + \frac{180}{\pi} \times \frac{\Delta L}{R} \tag{3.100}$$

$$H = H_o + R(\sin \alpha - \sin \alpha_0) \tag{3.101}$$
$$A = A_o + R(\cos \alpha_0 - \cos \alpha) \tag{3.102}$$

Similarly, for a drop section:

$$\alpha = \alpha_0 - \frac{\kappa}{C_\kappa} \Delta L = \alpha_0 - \frac{180}{\pi} \times \frac{\Delta L}{R} \tag{3.103}$$

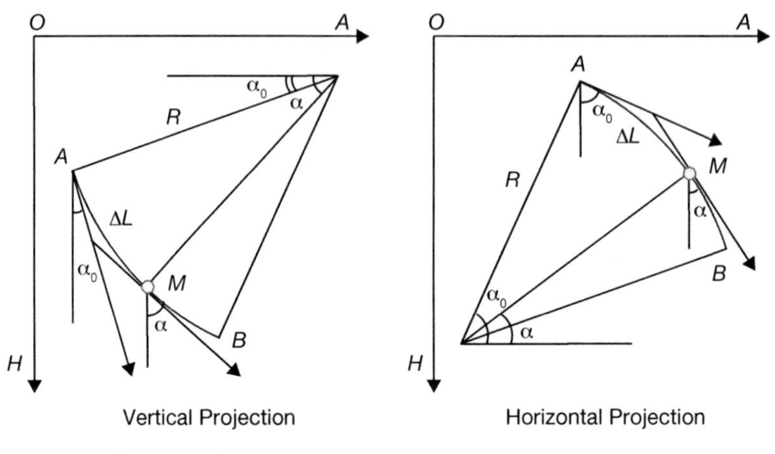

Figure 3.19 Well path circular-arc model.

$$H = H_0 + R(\sin \alpha_0 - \sin \alpha) \quad (3.104)$$

$$A = A_0 + R(\cos \alpha - \cos \alpha_0) \quad (3.105)$$

Conventionally, the radius of curvature and the curvature are both positive. However, if the radius of curvature R is defined as negative for a drop section, then Eqs. 3.100 through 3.102 could be used as common equations for both the build and the drop sections.

Catenary Model

The mathematical model using the catenary line shape for a wellbore is shown in Figure 3.20. The properties and details of the catenary shape have been already described under the section treating the catenary mechanical model.

Using Eq. 3.74, the slope of a catenary line can be written as:

$$\frac{dy}{dx} = \sinh \frac{x}{a} \quad (3.106)$$

It is apparent from Figure 3.20 that the slope can also be written as:

$$\frac{dy}{dx} = \tan \beta = \tan(90° - \alpha) \quad (3.107)$$

Hence,

$$\tan \alpha = \frac{1}{\sinh \frac{x}{a}} \quad (3.108)$$

The above equations need to be converted to usable forms so that the borehole parameters can be calculated easily from measured lengths along the well path.

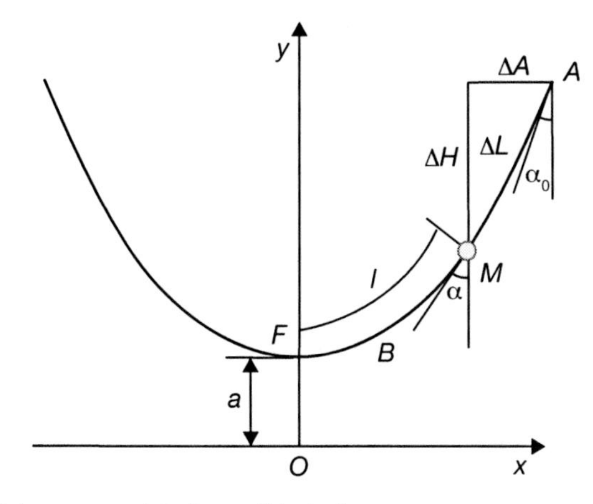

Figure 3.20 Catenary model of a well trajectory.

The length of a catenary segment is:

$$l = \int \sqrt{1 + \left(\frac{dy}{dx}\right)^2}\, dx \tag{3.109}$$

Eq. 3.109 can be integrated to yield Eq. 3.110 after applying the boundary conditions:

$$l = a \sinh \frac{x}{a} \tag{3.110}$$

The catenary shape is often incorporated in a well path profile to achieve a smooth well path, so the calculation of the length of the catenary is important. The length included in any catenary path between two points A and M can be written as:

$$\Delta L = l_A - l_M = a\left(\sinh \frac{x_0}{a} - \sinh \frac{x}{a}\right) \tag{3.111}$$

Using Eqs. 3.108 and 3.111, α can be expressed in convenient borehole parameters[2,9,23] as follows:

$$\Delta L = a\left(\frac{1}{\tan \alpha_0} - \frac{1}{\tan \alpha}\right) \tag{3.112}$$

From which it follows that:

$$\tan \alpha = \frac{1}{\dfrac{1}{\tan \alpha_0} - \dfrac{\Delta L}{a}} \tag{3.113}$$

It is evident that Eq. 3.113 enables the calculation of the inclination at any point along a catenary well path.

Furthermore, vertical depth or incremental vertical distance ΔA, and horizontal displacement ΔH, as well as the curvature along the wellbore in the catenary path, can be calculated as follows:

$$\begin{cases} \Delta A = x_o - x \\ \Delta H = y_o - y \end{cases} \tag{3.114}$$

Combining Eqs. 3.108 and 3.114 and rearranging gives:

$$A = A_0 + a(X_0 - X) \tag{3.115}$$

where

$$X = \sinh^{-1}\left(\frac{1}{\tan\alpha}\right)$$

Using Eqs. 3.74 and 3.108 with Eq. 3.115, gives the value of H as:

$$H = H_0 + a(\cosh X_0 - \cosh X) \tag{3.116}$$

In the plane Cartesian coordinate system, the curvature of the curve is calculated using the conventional formula:

$$\kappa = \frac{|y''|}{\left(1 + y'^2\right)^{\frac{3}{2}}} \tag{3.117}$$

The curvature of a point M on the catenary of the wellbore can therefore be expressed as[2,23]:

$$\kappa = \frac{180C_\kappa}{\pi a \cosh^2 X} \tag{3.118}$$

To avoid using the inconvenient hyperbolic functions for estimating the course coordinate position parameters, dimensionless parameters are defined. This avoids the use of hyperbolic functions, while still using the exact mathematical solutions. The method of explicit solution, avoiding a trial-and-error procedure and iterative calculations, provides excellent maneuverability for the requirements of well planning, and is convenient for oilfield applications. The results show that the essential elements of planning a catenary profile are to determine the shape and position of the catenary section, the characteristic parameter of the catenary well path describing its shape, and the parameters, such as starting inclination, ending inclination, length of the succeeding hold-up section, and position of the catenary section in the well profile.

It is possible to define the dimensionless X coordinate parameter, Y coordinate parameter, and length parameter as follows:

$$\begin{cases} X = \dfrac{x}{a} \\ Y = \dfrac{y}{a} \\ \Gamma = \dfrac{l}{a} \end{cases} \quad (3.119)$$

The dimensionless parameters can be expressed in various forms:

$$Y = \cosh X \quad (3.120)$$

$$\tan\alpha = \frac{1}{\sinh X} \quad (3.121)$$

$$\Gamma = \sinh X \quad (3.122)$$

Furthermore, mathematical manipulation of Eq. 3.121 yields the following relationships:

$$X = -\ln\left(\tan\frac{\alpha}{2}\right) \quad (3.123)$$

$$Y = \frac{1}{\sin\alpha} \quad (3.124)$$

$$\cosh X = \frac{1}{\sin\alpha} \quad (3.125)$$

Using Eqs. 3.123 through 3.125, the vertical depth, horizontal displacement, and curvature can be obtained as:

$$A = A_o + a(X_o - X) \quad (3.126)$$

$$H = H_o + a(Y_o - Y) \quad (3.127)$$

$$\kappa = \frac{180C_\kappa}{\pi a}\sin^2\alpha \quad (3.128)$$

This dimensionless approach transforms the basic catenary equation shown in Figure 3.20 into a dimensional form, as shown in Figure 3.21. The figure depicts the relevance of the dimensionless constants used.

Parabolic Model

The basics, analytical formulation, and approach for the parabolic model are the same as for the catenary model. As before, it has the single low point F and a vertical line of symmetry. As explained earlier in the mechanical model the load distribution is different in the parabolic model as compared to the catenary model. A parabolic trajectory model is shown in Figure 3.22, and

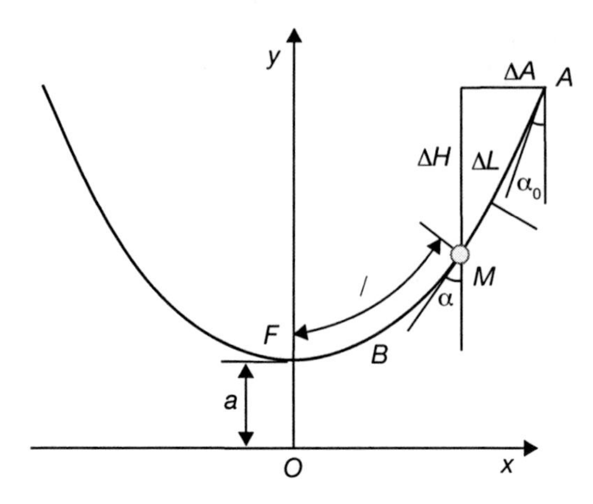

Figure 3.21 Catenary model with dimensionless parameters.

for the mathematical derivations, the geometric relationships are obtained by referring to Figure 3.16:

$$\frac{dy}{dx} = \frac{x}{P} \tag{3.129}$$

$$\tan\alpha = \frac{P}{x} \tag{3.130}$$

Arc length between point M and the lowest point F can be written as:

$$l = \int \sqrt{1 + \left(\frac{dy}{dx}\right)^2}\, dx \tag{3.131}$$

Using Eqs. 3.129 and 3.131, the length can be written as:

$$l = \frac{P}{2}\left[\frac{1}{\sin\alpha\tan\alpha} - \ln\left(\tan\frac{\alpha}{2}\right)\right. \tag{3.132}$$

Hence, the length between two points on the parabolic well profile can be written as:

$$\Delta L = l_A - l_M = \frac{P}{2}\left[f(\alpha_0) - f(\alpha)\right] \tag{3.133}$$

The right side of the expression is abbreviated by defining an angle function as:

$$f(\alpha) = \frac{1}{\sin\alpha\tan\alpha} - \ln\left(\tan\frac{\alpha}{2}\right) \tag{3.134}$$

The inclination angle along the wellbore can be written as:

$$f(\alpha) = f(\alpha_0) - \frac{2\Delta L}{P} \tag{3.135}$$

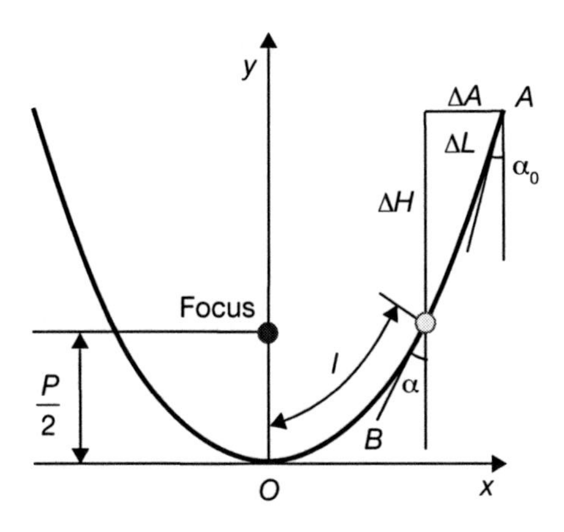

Figure 3.22 Parabolic model of a well trajectory.

Practically, the calculation has to be carried out using an iterative process to satisfy the defined algebraic conditions.

As before, ΔA and ΔH can be written as:

$$\begin{cases} \Delta A = x_o - x \\ \Delta H = y_o - y \end{cases} \tag{3.136}$$

Substituting Eq. 3.130 into Eq. 3.136, the following is obtained[24,25]:

$$A = A_o + P\left(\frac{1}{\tan\alpha_0} - \frac{1}{\tan\alpha}\right) \tag{3.137}$$

Using Eqs. 3.130 and 3.86 in Eq. 3.136 gives the displacement relationship:

$$H = H_o + \frac{P}{2}\left(\frac{1}{\tan^2\alpha_0} - \frac{1}{\tan^2\alpha}\right) \tag{3.138}$$

Starting from the fundamental equation, the curvature of the parabolic profile is given by:

$$\kappa = \frac{180 C_\kappa}{\pi P}\sin^3\alpha \tag{3.139}$$

3D Well Path Models

Although the majority of the directional wells and horizontal wells are designed using 2D profiles, in some cases 3D wellbore path planning becomes essential. For example, while drilling lateral wells or multiple well completions, the targets are not normally in the same plane, which necessitates 3D well path design to reach the targets successively, as well as successfully. In addition, describing and calculating drilled wellbore trajectory parameters are fundamentally required even for vertical and horizontal wells. Survey measurement at any survey point provides the data at the survey points, and not all along the drilled depth, and hence, cannot provide the real profile of the trajectory. So a more precise model is required to describe the well path more closely to the actually drilled well trajectories.

A 3D model of a wellbore trajectory description includes:

- Spatial-arc model
- Cylindrical-helix model
- Natural-curve model

Spatial-Arc Model

Initial studies using a 3D well profile description assumed the well path to be an oblique plane arc, and the corresponding calculations used the minimum curvature method. An oblique plane is a plane neither parallel nor perpendicular to the two projections. However, the proposed spatial-arc model formulas were very complicated and also were not accurate enough for trajectory design. Later, more practical and concise calculations[28–32] for the 3D directional wells were developed. For complex wells, a more accurate model[9,33] is required to calculate and describe the well path, and this is described in the following section.

The well path described by the spatial-arc model is in an oblique plane. The notable feature of this model is that the hole-curvature radius R and the curvature κ are constant. As shown in Figure 3.23, in the spatial-arc model, the toolface angle changes as the depth changes, and the curvature κ defines the shape of the well path. If the initial angle α_0, azimuth ϕ_0, toolface angle ω_0, and curvature κ are known at any starting point A, other well path parameters can be calculated.[9,33,34]

Well Inclination and Direction The deviation angle and the azimuth determine the wellbore profile in the tangent direction. As shown in Figures 3.23 and 3.24, in the A-$\xi\eta\zeta$ coordinate system, the unit tangent vector $\boldsymbol{t}$ of the circular arc can be expressed as:

$$\boldsymbol{t} = \sin\varepsilon\, \boldsymbol{e}_\xi + \cos\varepsilon\, \boldsymbol{e}_\zeta \tag{3.140}$$

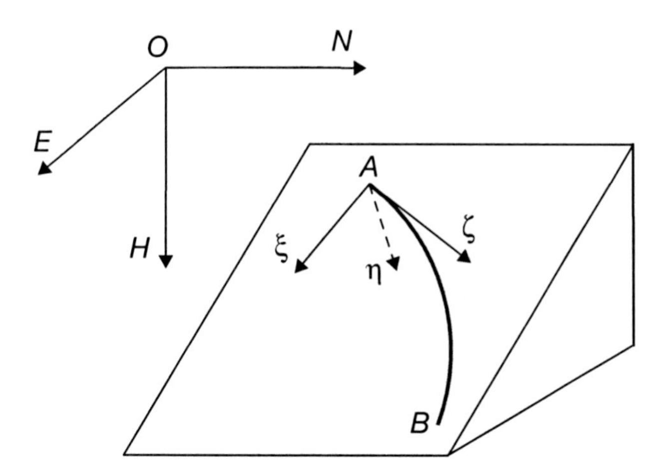

Figure 3.23 Spatial-arc model.

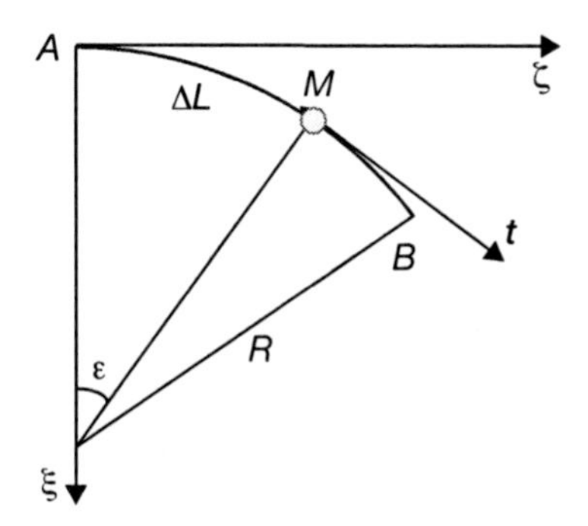

Figure 3.24 Inclined-plane circular arc.

where

$$\varepsilon = \frac{\kappa}{C_\kappa}\left(L - L_0\right) = \frac{180}{\pi} \times \frac{\Delta L}{R}$$

where

ε = Bending angle between the starting point A (at depth L_o) to the point M (at depth L) (°)

e_ξ = ξ axis coordinates of unit vector

R = Corresponding radius of curvature to the hole κ curvature, m or ft

By the unit vector coordinates conversion formula, it can be written:

$$\begin{cases} e_\xi = \dfrac{\dfrac{\partial \xi}{\partial N} i + \dfrac{\partial \xi}{\partial E} j + \dfrac{\partial \xi}{\partial H} k}{\sqrt{\left(\dfrac{\partial \xi}{\partial N}\right)^2 + \left(\dfrac{\partial \xi}{\partial E}\right)^2 + \left(\dfrac{\partial \xi}{\partial H}\right)^2}} \\ e_\zeta = \dfrac{\dfrac{\partial \zeta}{\partial N} i + \dfrac{\partial \zeta}{\partial E} j + \dfrac{\partial \zeta}{\partial H} k}{\sqrt{\left(\dfrac{\partial \zeta}{\partial N}\right)^2 + \left(\dfrac{\partial \zeta}{\partial E}\right)^2 + \left(\dfrac{\partial \zeta}{\partial H}\right)^2}} \end{cases} \tag{3.141}$$

According to Eq. 3.10, it is known that:

$$\begin{cases} \dfrac{\partial \xi}{\partial N} = T_{11}, & \dfrac{\partial \xi}{\partial E} = T_{12}, & \dfrac{\partial \xi}{\partial H} = T_{13} \\ \dfrac{\partial \zeta}{\partial N} = T_{31}, & \dfrac{\partial \zeta}{\partial E} = T_{32}, & \dfrac{\partial \zeta}{\partial H} = T_{33} \end{cases} \tag{3.142}$$

so

$$\begin{cases} e_\xi = T_{11} i + T_{12} j + T_{13} k \\ e_\zeta = T_{31} i + T_{32} j + T_{33} k \end{cases} \tag{3.143}$$

Similarly, substituting Eq. 3.143 into Eq. 3.140 results in:

$$t = (T_{11} \sin\varepsilon + T_{31} \cos\varepsilon) i + (T_{12} \sin\varepsilon + T_{32} \cos\varepsilon) j + (T_{13} \sin\varepsilon + T_{33} \cos\varepsilon) k \tag{3.144}$$

Therefore, the deviation angle and azimuth, respectively, are:

$$\cos\alpha = T_{13} \sin\varepsilon + T_{33} \cos\varepsilon \tag{3.145}$$

$$\tan\alpha = \frac{\sqrt{\left(T_{11} \sin\varepsilon + T_{31} \cos\varepsilon\right)^2 + \left(T_{12} \sin\varepsilon + T_{32} \cos\varepsilon\right)^2}}{T_{13} \sin\varepsilon + T_{33} \cos\varepsilon} \tag{3.146}$$

$$\tan\phi = \frac{T_{32} + T_{12} \tan\varepsilon}{T_{31} + T_{11} \tan\varepsilon} \tag{3.147}$$

Similar to Eq. 3.10, substituting T_{ij} into the above formula and rearranging results in[9,33]:

$$\cos\alpha = \cos\alpha_0 \cos\varepsilon - \sin\alpha_0 \cos\omega_0 \sin\varepsilon \tag{3.148}$$

$$\tan\alpha = \frac{\sqrt{\sin^2\omega_0 \sin^2\varepsilon + \left(\cos\alpha_0 \cos\omega_0 \sin\varepsilon + \sin\alpha_0 \cos\varepsilon\right)^2}}{\cos\alpha_0 \cos\varepsilon - \sin\alpha_0 \cos\omega_0 \sin\varepsilon} \tag{3.149}$$

$$\tan\phi = \frac{\sin\alpha_0 \sin\phi_0 + \left(\cos\alpha_0 \sin\phi_0 \cos\omega_0 + \cos\phi_0 \sin\omega_0\right) \tan\varepsilon}{\sin\alpha_0 \cos\phi_0 + \left(\cos\alpha_0 \cos\phi_0 \cos\omega_0 - \sin\phi_0 \sin\omega_0\right) \tan\varepsilon} \tag{3.150}$$

$$\tan\Delta\phi = \frac{\sin\varepsilon \sin\omega_0}{\sin\alpha_0 \cos\varepsilon + \cos\alpha_0 \sin\varepsilon \cos\omega_0} \tag{3.151}$$

These are the spatial circular-arc model equations, and they allow calculation of the inclination and azimuth angles interchangeably.

Deflection Parameters In the spatial circular-arc model, the calculations are difficult due to the fact that the spatial arcs in the vertical and horizontal projections are not circular arcs in the direction of the inclination plane and the position plane.

Taking the derivatives on both sides of Eq. 3.148 and using the definitions of well inclination angle change rate and well curvature gives[33,34]:

$$\kappa_\alpha = \frac{\kappa}{\sin\alpha}\left(\cos\alpha_0 \sin\varepsilon + \sin\alpha_0 \cos\omega_0 \cos\varepsilon\right) \tag{3.152}$$

Similarly, taking the derivatives on both sides of Eq. 3.147, and using the definition of the azimuth change rate results in:

$$\frac{1}{\cos^2\phi}\kappa_\phi = \frac{\kappa}{\cos^2\varepsilon}\frac{T_{12}\left(T_{31} + T_{11}\tan\varepsilon\right) - T_{11}\left(T_{32} + T_{12}\tan\varepsilon\right)}{\left(T_{31} + T_{11}\tan\varepsilon\right)^2}$$

That is,

$$\kappa_\phi = \kappa\left(\frac{\cos\phi}{\cos\varepsilon}\right)^2 \frac{T_{12}T_{31} - T_{11}T_{32}}{\left(T_{31} + T_{11}\tan\varepsilon\right)^2} \tag{3.153}$$

And further,

$$\kappa_\phi = \kappa\left(\frac{\cos\phi}{\cos\varepsilon}\right)^2 \left(\frac{T_{12} - T_{11}\tan\phi}{T_{31} + T_{11}\tan\varepsilon}\right) \tag{3.154}$$

Using results of Eq. 3.10 and substituting T_{ij} into the above formula and rearranging results in[9,33]:

$$\kappa_\phi = \kappa\frac{\left(\sin\omega_0 \cos\Delta\phi - \cos\alpha_0 \cos\omega_0 \sin\Delta\phi\right)^2}{\sin\alpha_0 \sin\omega_0 \cos^2\varepsilon} \tag{3.155}$$

It can be proven that κ, κ_α, and κ_ϕ can be correlated by Eq. 3.39.

For the well trajectory in the vertical profile and horizontal-walk plane projections, using the curvature definition as given in Eqs. 3.40 and 3.41 results in:

$$\kappa_V = \kappa_\alpha \tag{3.156}$$

$$\kappa_H = \frac{\kappa_\phi}{\sin\alpha} \tag{3.157}$$

It is clearly seen that for a spatial circular-arc well path, the curvature κ is maintained constant and the borehole torsion is $\tau = 0$.

Course Coordinates As shown in Figure 3.24 for a spatial plane oblique circular-arc well path, the A-$\xi\eta\zeta$ coordinates can be written as:

$$\begin{cases} \xi = R(1 - \cos\varepsilon) \\ \eta = 0 \\ \zeta = R\sin\varepsilon \end{cases} \tag{3.158}$$

where

R = radius of curvature

ε = bending angle

Thus, using Eq. 3.10, the well path space coordinate parameters can be calculated for a spatial circular arc at a depth of L to be:

$$\begin{cases} \Delta N = T_{11}\xi + T_{31}\zeta \\ \Delta E = T_{12}\xi + T_{32}\zeta \\ \Delta H = T_{13}\xi + T_{33}\zeta \end{cases} \tag{3.159}$$

That is,

$$\begin{cases} \Delta N = R\left[(\cos\alpha_0 \cos\phi_0 \cos\omega_0 - \sin\phi_0 \sin\omega_0)(1 - \cos\varepsilon) + \sin\alpha_0 \cos\phi_0 \sin\varepsilon\right] \\ \Delta E = R\left[(\cos\alpha_0 \sin\phi_0 \cos\omega_0 + \cos\phi_0 \sin\omega_0)(1 - \cos\varepsilon) + \sin\alpha_0 \sin\phi_0 \sin\varepsilon\right] \\ \Delta H = R\left[-\sin\alpha_0 \cos\omega_0 (1 - \cos\varepsilon) + \cos\alpha_0 \sin\varepsilon\right] \end{cases} \tag{3.160}$$

Because the spatial-arc projection in the horizontal plane is elliptical, it is impossible to get an explicit expression for the length of the horizontal projection and numerical integration is required. That is,

$$\Delta S = \int_{L_0}^{L} \sin\alpha \, dL \tag{3.161}$$

The deviation angle α is calculated from Eq. 3.148.

If the horizontal projection of the circular arc is elliptic, the approximate solution is[2,33]:

$$\Delta S = \lambda \times R(\sin\alpha_0 + \sin\alpha) \tag{3.162}$$

where

$$\lambda = \frac{\pi}{180} \times \tan\frac{\varepsilon}{2} \times \frac{\frac{\Delta\phi}{2}}{\tan\frac{\Delta\phi}{2}}$$

The spatial-arc model describes the well trajectory as an arc on a reference 2D surface. In a well trajectory, the inclinations, azimuths, and coordinates change with the well depth, and hence, calculations in 3D space are preferred; however, in order to study well path deviation and turn, the trajectory is treated as a 2D curve, with zero torsion. The curvature of the spatial-arc well trajectory determines the shape of the trajectory, inclination, azimuth, and kick-off point.

Cylindrical-Helix Model

Several methods of describing well paths using cylindrical-helical models have been proposed through the years.[6,7,35-39] Well trajectory design based on the classical cylindrical-helix model uses geometrical relationships to obtain the formulas for calculating the coordinates. To make the formulas comprehensive and to calculate the well trajectory parameters, the characteristics of the cylindrical-helix model are coupled with principles of differential geometry. This aids in establishing the expressions for the deviation angle and the azimuth angle, the coordinate position parameters, as well as the formulas for calculating the deflection parameters, such as curvature and torsion, as a function of depth.[2] The formulation and mathematical treatment involved in this method are given in this section.

When using well profiles based on the cylindrical-helix model, the vertical projection curvature κ_V, as well as the horizontal-walk plane curvature κ_H, remain constant. The notable feature of the model is that the vertical profile and horizontal projection plots describe circular arcs, as shown in Figure 3.25.

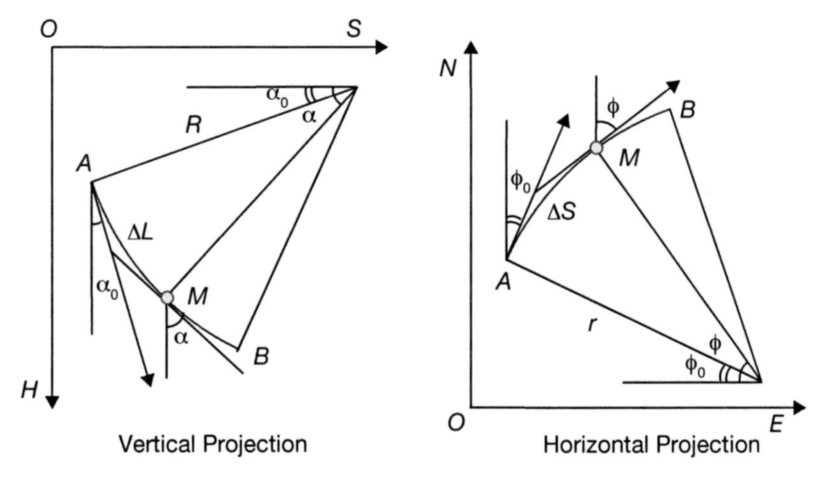

Figure 3.25 Cylindrical-helix model of a well bore trajectory.

Well Inclination and Direction Using the vertical projection of the hole with the curvature path as seen in Figure 3.25, the well path inclination angle is given as:

$$\alpha = \alpha_0 + \frac{\kappa_V}{C_\kappa}\left(L - L_0\right) \tag{3.163}$$

where

κ_V = Curvature of the wellbore trajectory in the vertical projection plot, (°)/30 m or (°)/100 ft

Obviously when $\kappa_V = 0$, $\alpha = \alpha_o$; that is, in the vertical projection, the well path is a straight line. In order to obtain the azimuth, Eq. 3.41 can be used as follows:

$$\frac{d\phi}{dL} = \kappa_H \sin\alpha \tag{3.164}$$

where

κ_H = Curvature of the wellbore trajectory in the horizontal projection plot, (°)/30 m or (°)/100 ft

The well path azimuth angle is given as:

$$\phi = \phi_0 + \frac{\kappa_H}{C_\kappa}\sin\alpha_0\left(L - L_0\right) \tag{3.165}$$

When $\kappa_V \neq 0$, using Eq. 3.163 in Eq. 3.164 after integration, yields:

$$\phi = \phi_0 + \frac{180}{\pi}\frac{\kappa_H}{\kappa_V}\left(\cos\alpha_0 - \cos\alpha\right) \tag{3.166}$$

When $\kappa_H = 0$, $\phi = \phi_o$; that is, in the horizontal projection, the well path is a straight line.

Deflection Parameters Applying Eqs. 3.40 and 3.41 to a cylindrical helical path, and calculating the rate of deviation change and the rate of direction change, results in:

$$\kappa_\alpha = \kappa_V \tag{3.167}$$

$$\kappa_\phi = \kappa_H \sin\alpha \tag{3.168}$$

Also, applying Eq. 3.42 for the cylindrical-helix model gives the wellbore curvature as:

$$\kappa = \sqrt{\kappa_V^{\,2} + \kappa_H^{\,2}\sin^4\alpha} \tag{3.169}$$

When the wellbore curvature is $\kappa = 0$, the well path is a straight line, which will result in zero torsion. When the hole curvature is $\kappa \neq 0$, using

Eqs. 3.167 and 3.168 and substituting in Eq. 3.54 gives the torsion equation for the cylindrical-helix model:

$$\tau = \kappa_H \left(1 + \frac{2\kappa_V^{\,2}}{\kappa^2}\right) \sin\alpha \cos\alpha \tag{3.170}$$

PROBLEM 3.3

The original inclination of a hole is 22°. To reach the target, it is desired to build an angle of 26° over a course length of 100 ft. The directional change is –5°. What is the resulting curvature and torsion that will achieve the desired objective?

Solution

Given data:

$\alpha_1 = 22°$; $\alpha_2 = 26°$; $\Delta\phi = -5°$; $\kappa_\alpha = 4°/100$ ft; $\kappa_\phi = 7°/100$ ft;
$\Delta L = 100$ ft

The vertical curvature is:

$$\kappa_\alpha = \frac{\alpha_2 - \alpha_1}{(L_2 - L_1)} = \frac{4}{100} \times 100 = 4°/100 \text{ ft}$$

The directional curvature is:

$$\kappa_\phi = \frac{\phi_2 - \phi_1}{(L_2 - L_1)} = \frac{-10}{100} \times 100 = -10°/100 \text{ ft}$$

The average inclination angle is:

$$\bar{\alpha} = \frac{\alpha_1 + \alpha_2}{2} = \frac{22 + 26}{2} = 24°$$

$$\kappa_\alpha = \kappa_V = 4°/100 \text{ ft}$$

$$\kappa_H = \frac{\kappa_\varphi}{\sin\alpha} = \frac{-10}{\sin 24} = -12.294°/100 \text{ ft}$$

Using Eq. 3.162 the wellbore curvature can be calculated as:

$$\kappa = \sqrt{\kappa_V^{\,2} + \kappa_H^{\,2} \sin^4\alpha} = 4.487°/100 \text{ ft}$$

Using Eq. 3.132 to calculate the torsion:

$$\tau = -12.294\left(1 + \frac{2 \times 4^2}{4.487^2}\right) \sin 24 \cos 24 = -11.827°/100 \text{ ft}$$

Course Coordinates The course coordinates are presented using the differential geometry approach.[2] According to the differential model of wellbore trajectory (Figure 3.8), the relation between coordinate increments and curved-section length, inclination angle, and azimuth angle for a small interval length is given by:

$$\begin{cases} \Delta N = \int_{L_0}^{L} \sin\alpha \cos\phi \, dL \\ \Delta E = \int_{L_0}^{L} \sin\alpha \sin\phi \, dL \\ \Delta H = \int_{L_0}^{L} \cos\alpha \, dL \\ \Delta S = \int_{L_0}^{L} \sin\alpha \, dL \end{cases} \tag{3.171}$$

Equation 3.171 is instructive as it includes different solutions for calculating the course coordinates under different conditions. Depending on the particular case, different conditions are imposed for estimating the coordinate parameters. As an illustrative example, the incremental north (ΔN) equation defined by the differential model of wellbore trajectory given in Eq. 3.171 is used with the different conditions to obtain explicit solutions.

Taking note of Eqs. 3.167 and 3.168, if $\kappa_V = 0$, $\kappa_H = 0$, then $\alpha = \alpha_0$ and $\phi = \phi_0$. Hence,

$$\Delta N = \Delta L \sin \alpha_0 \cos \phi_0 \tag{3.172}$$

If $\kappa_V = 0$ and $\kappa_H \neq 0$, then $\alpha = \alpha_0$. Hence,

$$\Delta N = \left[\sin\alpha_0 \frac{1}{\kappa_\phi} \sin\phi \right]_{L_0}^{L} = \frac{1}{\kappa_H}\left(\sin\phi - \sin\phi_0\right) \tag{3.173}$$

If $\kappa_V \neq 0$ and $\kappa_H = 0$, then $\phi = \phi_0$. Hence,

$$\Delta N = \left[\frac{1}{\kappa_\alpha}\left(-\cos\alpha\right)\cos\phi_0 \right]_{L_0}^{L} = \frac{1}{\kappa_V}\left(\cos\alpha_0 - \cos\alpha\right)\cos\phi_0 \tag{3.174}$$

If $\kappa_V \neq 0$ and $\kappa_H \neq 0$, then,

$$\Delta N = \left(\frac{1}{\kappa_\phi} \sin\alpha \sin\phi \right)\Bigg|_{L_0}^{L} - \int_{L_0}^{L} \left(\frac{\sin\alpha}{\kappa_\phi} \right)' \sin\phi \, dL \tag{3.175}$$

Substituting Eq. 3.168 into Eq. 3.175 yields:

$$\Delta N = \left(\frac{1}{\kappa_H} \sin\phi \right)\Bigg|_{L_0}^{L} - \int_{L_0}^{L} \left(\frac{1}{\kappa_H} \right)' \sin\phi \, dL = \frac{1}{\kappa_H}\left(\sin\phi - \sin\phi_0\right) \tag{3.176}$$

The results for the north course coordinate for the different conditions are summarized as follows:

$$\Delta N = \begin{cases} \Delta L \sin\alpha_0 \cos\phi_0, & \text{if } \kappa_V = 0, \kappa_H = 0 \\ \dfrac{1}{\kappa_H}\left(\sin\phi - \sin\phi_0\right), & \text{if } \kappa_V = 0, \kappa_H \neq 0 \\ \dfrac{1}{\kappa_V}\left(\cos\alpha_0 - \cos\alpha\right)\cos\phi_0, & \text{if } \kappa_V \neq 0, \kappa_H = 0 \\ \dfrac{1}{\kappa_H}\left(\sin\phi - \sin\phi_0\right), & \text{if } \kappa_V \neq 0, \kappa_H \neq 0 \end{cases} \quad (3.177)$$

It should be noted that in Eq. 3.177, the curvatures are in radian per feet or radian per meter. The expressions can be converted to the more common unit degrees as follows:

$$F_S\left(\theta,\kappa,\lambda\right) = \begin{cases} \lambda\cos\theta, & \text{if } \kappa = 0 \\ \dfrac{180C_\kappa}{\pi\kappa}\left[\sin\left(\theta + \dfrac{\kappa}{C_\kappa}\lambda\right) - \sin\theta\right], & \text{if } \kappa \neq 0 \end{cases} \quad (3.178)$$

$$F_C\left(\theta,\kappa,\lambda\right) = \begin{cases} \lambda\sin\theta, & \text{if } \kappa = 0 \\ \dfrac{180C_\kappa}{\pi\kappa}\left[\cos\theta - \cos\left(\theta + \dfrac{\kappa}{C_\kappa}\lambda\right)\right], & \text{if } \kappa \neq 0 \end{cases} \quad (3.179)$$

Eq. 3.178 can be further expressed as:

$$\Delta N = \begin{cases} F_C\left(\alpha_0,\kappa_V,\Delta L\right)\cos\phi_0, & \text{if } \kappa_H = 0 \\ \dfrac{180C_\kappa}{\pi\kappa_H}\left(\sin\phi - \sin\phi_0\right), & \text{if } \kappa_H \neq 0 \end{cases} \quad (3.180)$$

Similarly, the other incremental coordinates can be calculated as:

$$\Delta E = \begin{cases} F_C\left(\alpha_0,\kappa_V,\Delta L\right)\sin\phi_0, & \text{if } \kappa_H = 0 \\ \dfrac{180C_\kappa}{\pi\kappa_H}\left(\cos\phi_0 - \cos\phi\right), & \text{if } \kappa_H \neq 0 \end{cases} \quad (3.181)$$

$$\Delta H = F_S(\alpha_0, \kappa_v, \Delta L) \quad (3.182)$$

$$\Delta S = F_C(\alpha_0, \kappa_v, \Delta L) \quad (3.183)$$

The cylindrical-helix model gives the various well path parameters as a function of wellbore depth. Therefore, if the well path profiles are known from the vertical and the horizontal projection charts, the vertical curvature κ_V, and the horizontal-walk plane curvature κ_H, can be calculated.

Curvature κ and radius of curvature R are related by Eq. 3.17, and if one is known, the other parameter can be calculated. Thus, by knowing the radii of the curvature of the well trajectory projections, R and r, on the vertical surface and the horizontal surface respectively, their corresponding curvatures,

κ_V and κ_H, can be found, and then the above equations can be applied for further calculations.

Natural-Curve Model

While the radius of curvature method is more suitable for describing a wellbore trajectory drilled in the rotary mode, the minimum curvature method is more suitable for slide drilling. On the other hand the natural-curve method is suitable for both rotary and slide drilling.[40-48] The most convenient way to calculate the inclination and azimuth at any depth is by using interpolation and extrapolation assuming the rates of inclination and azimuth change in each course to be independently constant. This forms the basis of the natural-parameter or natural-curve method. In differential geometry, the parameterization is done using the length of the curve. The derivatives are taken with respect to this parameter instead of any other parameter. Thus, this method of calculating trajectory is termed as the natural parameter method. This method uses the inclinations and azimuths at two survey stations on the course length. Survey instruments measure the inclination and direction at various depths, and so this method of calculation is simple and relatively easy to use, in addition to having the advantage of a mathematically stable solution.

Well Inclination and Direction According to the natural-curve model, with the underlying assumption that the rate of inclination change, as well as the rate of azimuth change, are constants, the inclination and azimuth can be expressed in terms of measured depth as:

$$\alpha = \alpha_0 + \frac{\kappa_\alpha}{C_\kappa}\left(L - L_o\right) \tag{3.184}$$

$$\phi = \phi_0 + \frac{\kappa_\phi}{C_\kappa}\left(L - L_o\right) \tag{3.185}$$

It can be seen that for the natural-curve model, inclination angle and azimuth angle are a linear function of depth.

Deflection Parameters As seen earlier it is convenient to use vertical and horizontal section plots to show the wellbore trajectories. A zero vertical build rate ($\kappa_\alpha = 0$) will depict a linear section, and a nonzero vertical build rate ($\kappa_\alpha \neq 0$) will depict an arc section on the vertical section plot. If the well path is a holding section (constant inclination) with no inclination build ($\kappa_\alpha \neq 0$), then a zero azimuth build rate ($\kappa_\phi \neq 0$) will depict a linear section and a nonzero azimuth build rate ($\kappa_\phi \neq 0$) will depict an arc section on the horizontal section plot. In other words, the well path on the vertical section plot can be either a linear section or an arc section, but will be a linear section or an arc section on the horizontal plot only if the well path is a holding section. The curvatures of a wellbore trajectory are related to the rate of change of the inclination and the rate of change of the azimuth, as explained earlier. For the wellbore trajectory described by the natural-parameter model, the vertical and horizontal curvatures can be determined by Eqs. 3.40 and 3.41 and the borehole curvature by using Eq. 3.37 (these equations are repeated here for convenience):

$$\kappa_V = \kappa_\alpha \tag{3.186}$$

$$\kappa_H = \frac{\kappa_\phi}{\sin\alpha} \tag{3.187}$$

$$\kappa = \sqrt{\kappa_\alpha{}^2 + \kappa_\phi{}^2 \sin^2\alpha} \tag{3.188}$$

Assumptions of the model results in:

$$\dot{\kappa}_\alpha = \dot{\kappa}_\phi = 0$$

Therefore, Eq. 3.54 simplifies to:

$$\tau = \kappa_\phi \left(1 + \frac{\kappa_\alpha{}^2}{\kappa^2}\right)\cos\alpha \tag{3.189}$$

Obviously when the hole curvature is $\kappa = 0$, the wellbore trajectory is a straight line resulting in zero borehole torsion.

PROBLEM 3.4

The original inclination of a hole is 22°. To reach the target, it is desired to build an angle of 26° over a course length of 100 ft. The directional change is −5°. What is the resulting curvature and torsion that will achieve the desired objective?

Solution

Given data:

$$\alpha_1 = 22°;\ \alpha_2 = 26°;\ \Delta\phi = -5°;\ \kappa_\alpha = 4°/100\text{ ft};\ \kappa_\phi = 7°/100\text{ ft};\ \Delta L = 100\text{ ft}$$

The vertical curvature is:

$$\kappa_\alpha = \frac{\alpha_2 - \alpha_1}{(L_2 - L_1)} = \frac{4}{100} \times 100 = 4°/100\,\text{ft}$$

The horizontal curvature is:

$$\kappa_\varphi = \frac{\phi_2 - \phi_1}{(L_2 - L_1)} = \frac{-10}{100} \times 100 = -10°/100\,\text{ft}$$

The average inclination angle is:

$$\bar{\alpha} = \frac{\alpha_1 + \alpha_2}{2} = \frac{22 + 26}{2} = 24°$$

The average curvature is:

$$\kappa = \sqrt{4^2 + (-10)^2 \sin^2 24} = 4.487°/100 \text{ ft}$$

Using Eq. 3.132 to calculate the torsion:

$$\tau = -10\left(1 + \frac{4^2}{4.487^2}\right)\cos 24 = -8.197°/100\,\text{ft}$$

Course Coordinates As an illustration, the calculations for the incremental north coordinate, ΔN, under different conditions are derived as follows:

$$\begin{cases} A_P = \alpha_0 + \phi_0 \\ A_Q = \alpha_0 - \phi_0 \end{cases} \tag{3.190}$$

$$\begin{cases} \kappa_P = \kappa_\alpha + \kappa_\phi \\ \kappa_Q = \kappa_\alpha - \kappa_\phi \end{cases} \tag{3.191}$$

where

A_P, A_Q = Intermediate angle variables, (°)
κ_P, κ_Q = Intermediate curvature variables, (°)/30 m or (°)/100 ft
α_o = Initial inclination angle, (°)
ϕ_o = Initial azimuth angle, (°)

Using Eq. 3.171 and applying the trigonometric product and difference relationship,

$$\Delta N = \frac{1}{2}\int_{L_0}^{L} \sin(\alpha + \phi)\,dL + \frac{1}{2}\int_{L_0}^{L} \sin(\alpha - \phi)\,dL \tag{3.192}$$

Using the relationships provided in Eqs. 3.190 and 3.191, Eq. 3.192 can be expressed as:

$$\Delta N = \frac{1}{2}\int_{L_0}^{L} \sin\left[A_P + \kappa_P(L - L_0)\right]dL + \frac{1}{2}\int_{L_0}^{L} \sin\left[A_Q + \kappa_Q(L - L_0)\right]dL \tag{3.193}$$

Integration yields:

$$\Delta N = -\frac{1}{2}\left[\frac{1}{\kappa_P}\cos\left[A_P + \kappa_P(L - L_o)\right] + \frac{1}{\kappa_Q}\cos\left[A_Q + \kappa_Q(L - L_o)\right]\right]_{L_o}^{L} \tag{3.194}$$

Hence, applying limits:

$$\Delta N = -\frac{1}{2\kappa_P}\left[\cos\left(A_P + \kappa_P \Delta L\right) - \cos A_P\right] - \frac{1}{2\kappa_Q}\left[\cos\left(A_Q + \kappa_Q \Delta L\right) - \cos A_Q\right] \tag{3.195}$$

It can be observed that Eq. 3.195 is valid for $\kappa_P \neq 0$ and $\kappa_Q \neq 0$ that is $|\kappa_\alpha| = |\kappa_\phi|$. Therefore, separate analysis is needed for the special cases of $\kappa_P = 0$ or $\kappa_Q = 0$ or when both are zero, as indicated below.

Using Eqs. 3.171 or 3.193 when $\kappa_P = \kappa_Q = 0$, that is $\kappa_\alpha = \kappa_\phi = 0$, it can be easily obtained that:

$$\Delta N = \Delta L \sin \alpha_0 \cos \phi_0 \tag{3.196}$$

When $\kappa_P \neq 0$, $\kappa_Q = 0$, it can be observed from Eq. 3.191 that $\kappa_\alpha = \kappa_\phi \neq 0$, then:

$$\Delta N = \frac{1}{2}\int_{L_o}^{L} \sin\left[A_P + \kappa_P\left(L - L_o\right)\right] dL + \frac{1}{2}\int_{L_o}^{L} \sin A_Q \, dL \tag{3.197}$$

After integration,

$$\Delta N = -\frac{1}{2\kappa_P}\left[\cos\left(A_P + \kappa_P \Delta L\right) - \cos A_P\right] + \frac{1}{2}\Delta L \sin A_Q \tag{3.198}$$

Similarly, when $\kappa_P = 0$, $\kappa_Q \neq 0$, that is $\kappa_\alpha = -\kappa_\phi = 0$, then:

$$\Delta N = \frac{1}{2}\int_{L_o}^{L} \sin A_P \, dL + \frac{1}{2}\int_{L_o}^{L} \sin\left[A_Q + \kappa_Q\left(L - L_o\right)\right] dL \tag{3.199}$$

And after integration,

$$\Delta N = \frac{1}{2}\Delta L \sin A_P - \frac{1}{2\kappa_Q}\left[\cos\left(A_Q + \kappa_Q \Delta L\right) - \cos A_Q\right] \tag{3.200}$$

In the above equations, the unit used for the rate of change of the inclination and the rate of change of the azimuth is radians/feet. If the inclination and azimuth are in radians and the curvature in radians per feet, the coordinates' parameters take the form as in Eqs. 3.201 through 3.204, which are similar to Eqs. 3.178 and 3.179:

$$\Delta N = \frac{1}{2}\left[F_C\left(A_P, \kappa_P, \Delta L\right) + F_C\left(A_Q, \kappa_Q, \Delta L\right)\right] \tag{3.201}$$

$$\Delta E = \frac{1}{2}\left[F_S\left(A_P, \kappa_P, \Delta L\right) - F_S\left(A_Q, \kappa_Q, \Delta L\right)\right] \tag{3.202}$$

$$\Delta H = F_S(\alpha_0, \kappa_\alpha, \Delta L) \tag{3.203}$$

$$\Delta S = F_C(\alpha_0, \kappa_\alpha, \Delta L) \tag{3.204}$$

This model and its corresponding formulas show that the inclination, azimuth, and course coordinates are all functions of measured depth, the natural parameter of a wellbore trajectory. Thus, in this model, interpolation is a very convenient means for calculating the course coordinate at any point of the wellbore trajectory.

PROBLEM 3.5

Calculate the course coordinates with the following data: Initial direction and azimuth are 35° and 182°, respectively. $\kappa_\alpha = 4°/100$ ft and $\kappa = 0°/100$ ft. The tangent length = 200 ft. Calculate the course coordinates.

Solution

The new angle and azimuth are

$$\alpha = 35 + \frac{4}{100} \times 200 = 43° \text{ and } \phi = 182 + \frac{0}{100} \times 200 = 182°$$

$$\kappa_V = 4°/100 \text{ ft and } \kappa_H = \frac{0}{\sin 43} = 0°/100\,\text{ft}$$

$$\kappa = \sqrt{4^2 + 0} = 4°/100\,\text{ft}$$

Calculating the intermediate variables:

$$\begin{cases} A_P = 35 + 182 = 217° \\ A_Q = 35 - 182 = -147° \end{cases} \text{and} \begin{cases} \kappa_P = 4 + 0 = 4°/100 \text{ ft} \\ \kappa_Q = 4 - 0 = 4°/100 \text{ ft} \end{cases}$$

$$F_S(217, 4, 200) = \frac{180 \times 100}{\pi \times 4}\left[\sin\left(217 + \frac{4}{100} \times 200\right) - \sin 217\right] = -150.8 \text{ ft}$$

$$F_C(217, 4, 200) = \frac{180 \times 100}{\pi \times 4}\left[\cos 217 - \sin\left(217 + \frac{4}{100} \times 200\right)\right] = -131 \text{ ft}$$

$$F_S(-147, 4, 200) = \frac{180 \times 100}{\pi \times 4}\left[\sin\left(-147 + \frac{4}{100} \times 200\right) - \sin(-147)\right] = -159.6 \text{ ft}$$

$$F_C(-147, 4, 200) = \frac{180 \times 100}{\pi \times 4}\left[\cos(-147) - \sin\left(-147 + \frac{4}{100} \times 200\right)\right] = -120.26 \text{ ft}$$

$$F_S(35, 4, 200) = \frac{180 \times 100}{\pi \times 4}\left[\sin\left(35 + \frac{4}{100} \times 200\right) - \sin(35)\right] = 155.3 \text{ ft}$$

$$F_C(35, 4, 200) = \frac{180 \times 100}{\pi \times 4}\left[\cos(35) - \sin\left(35 + \frac{4}{100} \times 200\right)\right] = 125.76 \text{ ft}$$

The course coordinates are:

$$\Delta N = \frac{1}{2}\left[F_C(217, 4, 200) + F_C(-147, 4, 200)\right] = -125.68 \text{ ft}$$

$$\Delta E = \frac{1}{2}\left[F_S(-147, 4, 200) - F_S(-147, 4, 200)\right] = -4.4 \text{ ft}$$

$$\Delta H = F_S(50, 4, 200) = 155 \text{ ft}$$

$$\Delta S = F_C(50, 4, 200) = 125.76 \text{ ft}$$

Using SI units, the same problem is worked out as follows. The new angle and azimuth are

$$\alpha = 35 + \frac{4}{30.48} \times 60.96 = 43° \quad \text{and } \phi = 182 + \frac{0}{100} \times 200 = 182°$$

And so $\kappa = \sqrt{4^2 + 0} = 4°/30.48$ m.

Calculating the intermediate variables:

$$\begin{cases} A_P = 35 + 182 = 217° \\ A_Q = 35 - 182 = -147° \end{cases} \text{and} \begin{cases} \kappa_P = 4 + 0 = 4°/30.48 \text{ m} \\ \kappa_Q = 4 - 0 = 4°/30.48 \text{ m} \end{cases}$$

$$F_S(217, 4, 60.96) = \frac{180 \times 30.48}{\pi \times 4}\left[\sin\left(217 + \frac{4}{30.48} \times 60.96\right) - \sin 217\right]$$
$$= -45.96 \text{ m}$$

$$F_C(217, 4, 60.96) = \frac{180 \times 30.48}{\pi \times 4}\left[\cos 217 - \sin\left(217 + \frac{4}{30.48} \times 60.96\right)\right]$$
$$= -39.96 \text{ m}$$

$$F_S(-147, 4, 60.96) = \frac{180 \times 30.48}{\pi \times 4}\left[\sin\left(-147 + \frac{4}{30.48} \times 60.96\right) - \sin(-147)\right.$$
$$= -48.64 \text{ m}$$

$$F_C(-147, 4, 60.96) = \frac{180 \times 30.48}{\pi \times 4}\left[\cos(-147) - \sin\left(-147 + \frac{4}{30.48} \times 60.96\right)\right.$$
$$= -36.65 \text{ m}$$

$$F_S(35, 4, 60.96) = \frac{180 \times 30.48}{\pi \times 4}\left[\sin\left(35 + \frac{4}{30.48} \times 60.96\right) - \sin(35)\right]$$
$$= 47.33 \text{ m}$$

$$F_C(35, 4, 60.96) = \frac{180 \times 30.48}{\pi \times 4}\left[\cos(35) - \sin\left(35 + \frac{4}{30.48} \times 60.96\right)\right]$$
$$= 38.33 \text{ m}$$

The course coordinates are:

$$\Delta N = \frac{1}{2}\left[F_C(217, 4, 60.96) + F_C(-147, 4, 60.96)\right] = -38.30 \text{ m}$$

$$\Delta E = \frac{1}{2}\left[F_S(-147, 4, 60.96) - F_S(-147, 4, 200)60.96\right] = -1.33 \text{ m}$$

$$\Delta H = F_S(35,4,60.96) = 47.33 \text{ m}$$

$$\Delta S = F_C(35,4,60.96) = 38.33 \text{ m}$$

Supplementary Problems

1. Show that the v-section displacement distance does not change if the v-section angle is zero with the constant north and east vertical section reference coordinates.
2. Show that the v-section distance does not change even if the reference east coordinate is changed with the constant north reference coordinate and zero vertical section angles.
3. Using the information given in Table 3.1, compute the vertical section displacement relative to the local origin for both survey stations.
   V-section definition:
   Origin – N = 120 ft
   Origin – E = –65 ft
   Azimuth = 45°
4. Using the data given in Figure 3.26, calculate the v-section orientation azimuth.
5. Determine the v-section closure distance with the following data:
   a. N/S point: 200 ft south = –200 ft
      E/W point: 500 ft east = 500 ft
      Vertical section N/S reference = 0 ft
      Vertical section E/W reference = 0 ft
      Vertical section azimuth = 165°
   b. N/S point: 200 ft south = 333 ft
      E/W point: 500 ft east = 333 ft
      Vertical section N/S reference = 0 ft
      Vertical section E/W reference = 0 ft
      Vertical section azimuth = 330°

Table 3.1 Data for Problem 3

Measured Depth	Inclination	Azimuth
0 ft	0°	0°
1000 ft	12°	30°
2000 ft	15°	45°

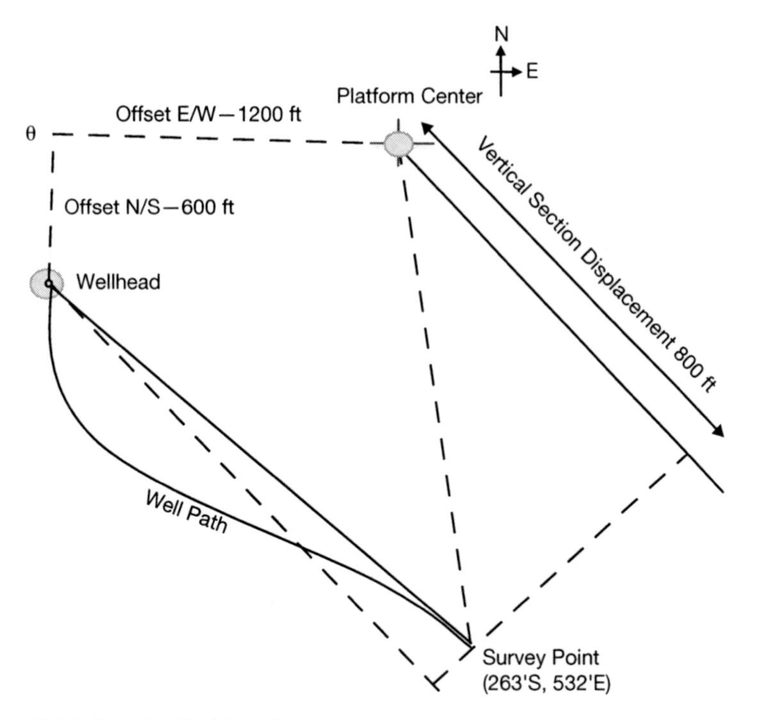

Figure 3.26 Data for Problem 4.

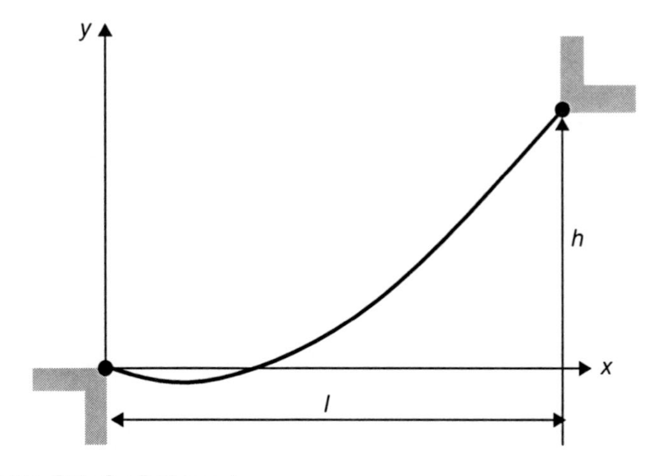

Figure 3.27 Data for Problem 6.

6. Figure 3.27 shows a uniform riser with constant weight w, per unit length. The offset of the riser is l, distance between the two supports is h. Derive the equation to calculate the length of the riser L, and the deflection of the riser y.
   (Hint: You may start from the equation $\frac{dy}{dx} = \frac{1}{H}\int w ds + c_1$.)
7. Calculate the wellbore torsion and torsion angle for the spatial-arc, natural-curve, helical, and constant toolface trajectories for the following data:
   $\Delta MD = 100$ ft; $\theta_1 = 22°$; $\theta_2 = 28°$; and $\Delta\phi = -20°$
8. Prove that the toolface angle can be given by the following relation:
   $$\omega = a\cos\left(\frac{\cos\alpha_1 \cos\varepsilon - \cos\alpha_2}{\sin\alpha_1 \sin\varepsilon}\right)$$
9. Original hole inclination is 38.2°. To reach the limits of the target, it is desired to drop an angle of 26° in a course length of 180 ft. The directional change is –10°. What is the resulting curvature and torsion that will achieve the desired objective? Use both natural-curve and circular-helix models. Comment on the results.
10. A survey shows 10° inclination with the direction far off the course. A deflection assembly for 2.4° dogleg is run and oriented for maximum turn. Determine the magnitude of the tool rotation angle required to obtain maximum turn, hole direction change, and new inclination.
11. With the following data:
    Riser weight: 150 lb/ft
    Wellhead height from mud line: 50 ft
    Total vertical depth of the riser: 1000 ft
    Horizontal offset: 50 ft
    a. Find the equation of the shape of the riser using the catenary model.
    b. Calculate the length of the riser.
    c. Calculate the minimum and maximum values of tension.
    d. Repeat the calculation from 150 to 200 lb/ft, plot riser weight versus riser length, and analyze the results.
12. A survey taken at the course of drilling a directional path shows an inclination of 22° and an azimuth of N38E. A directional driller would like to turn 6° to the left and build an angle. Determine the new hole inclination, toolface angle from the original course direction, and tool direction if a deflection tool with a 3° bend angle is used.

13. Calculate the deflection rate of a single bent assembly with the following given input data using both methods and compare the results.
    Bent angle: $\beta = 1°, 2°, 3°$
    Distance of bottom stabilizer (near bit stabilizer) to bit segment = 5 ft
    Distance of bent angle and bottom stabilizer segment = 4.5 ft
    Distance of top stabilizer to bent angle segment = 15 ft
    Plot the influencing parameter versus bent angle.
14. Calculate the change in the azimuth and inclination for the following data:
    Inclination at the present location = 5°
    Expected dogleg = 4°
    Well is planned to be turned to the right using a whipstock.
    Also, calculate the toolface setting that will give maximum turn to the right.
15. Given the following data:
    Deflection tool: whipstock
    Initial inclination = 10°
    Acceptable dogleg severity = 4°/100 ft
    Toolface setting = 40° to the right
    Calculate the final inclination if the wellbore is turned 20° to the right.
16. Given the following data:
    Deflection tool: whipstock
    Present inclination = 4°
    Present direction = S40W
    Acceptable dogleg severity = 4°/100 ft
    Determine the whipstock setting to achieve a N40E turn.
17. A survey shows 10° inclination with direction far off course. A deflection assembly for 2.4° dogleg is run and oriented for maximum turn. Determine the magnitude of tool rotation required to obtain maximum turn, hole direction change, and new inclination.
18. A survey shows 21.3° inclination and N40E direction. It is desired to turn 5° to the left and build an angle. Determine the new hole inclination, tool rotation from original course direction, and tool direction if a deflection tool with 2.5° dogleg is used.
19. A directional survey shows that the original hole direction and inclination are S10W and 5°, respectively. To obtain new hole direction of S20E, plan to use a deflection tool with the deflection angle of 3.4°.

Determine the required deflection tool direction if the anticipated walk to the right is 15°.

20. Original hole direction is N20E and hole inclination is 10°. It is desired to deviate the borehole so that the new inclination angle and direction will be N30E and 13°, respectively. What is the tool orientation, tool deflection, and tool deflection angle necessary to achieve this turn and build?
21. It is desired to build an angle from 28° to 32° in the next 200 ft. The directional change is to be maintained between ±5°. What will be the expected curvature and torsion?
22. Derive Eqs. 3.195 through 3.197 with all the steps involved.
23. Calculate the trajectory from 10,000 ft to 10,500 ft for a build section with the following data. The inclination and azimuth at 10,000 ft are 8° and 280°, respectively. The build and walk curvatures are 6°/100 ft and 3°/100 ft, respectively. Calculate the course coordinates between 10,000 ft and 10,500 ft.
24. Calculate the trajectory from 5000 m to 5150 m for a build section with the following data. The inclination and azimuth at 5000 m are 8° and 280°, respectively. The build and walk curvatures are 6°/31 m and 3°/31 m, respectively. Calculate the course coordinates between 5000 m and 5150 m.
25. A survey taken at the course of drilling a directional path shows an inclination of 22° and an azimuth of N38E. A directional driller would like to turn 6° to the left and build an angle. Determine the new hole inclination and the course coordinates after drilling 200 m. The build rate curvature is 4°/30 m.
26. A survey taken at the course of drilling a directional path shows an inclination of 22° and an azimuth of N38E. A directional driller would like to turn 36° to the right and build an angle. Determine the new hole inclination and the course coordinates after drilling 600 ft. The build rate curvature is 6°/100 ft. Estimate the course coordinates if the same turn is achieved to the left.
27. The original hole direction is N20E and the hole inclination is 10°. It is desired to deviate the borehole so that the new inclination angle and direction will be N30E and 10°, respectively. What are the course coordinates for a course length of 300 ft?

28. If α and ϕ are hole inclination and azimuth angles, respectively, at a given measured depth, then position vector is given as $\boldsymbol{r}(L) = N\boldsymbol{i} + E\boldsymbol{j} + H\boldsymbol{k}$ where $\boldsymbol{i}, \boldsymbol{j}, \boldsymbol{k}$ are unit vectors and N, E, H are north, east, vertical coordinates, respectively, and

$$\frac{dN}{dL} = \sin\alpha\cos\phi; \frac{dE}{dL} = \sin\alpha\sin\phi; \frac{dH}{dL} = \cos\alpha$$

Prove that curvature can be given as:

$$k = \pm\sqrt{\left(\frac{d\alpha}{dL}\right)^2 + \sin^2\alpha\left(\frac{d\phi}{dL}\right)^2}$$

and torsion as:

$$\tau = \frac{1}{k^2}\left\{\sin\alpha\left(\frac{d\alpha}{dL}\frac{d^2\phi}{dL^2} - \frac{d^2\alpha}{dL^2}\frac{d\phi}{dL}\right) + \cos\alpha\left[2\left(\frac{d\alpha}{dL}\right)^2\left(\frac{d\phi}{dL}\right) + \sin^2\alpha\left(\frac{d\phi}{dL}\right)^3\right]\right.$$

29. Using the equations for curvature and torsion in problem 28, calculate the dogleg severity and torsion using the data between the survey stations for a distance of $dL = 100$ ft:

$\alpha_1 = 46.31°$; $\phi_1 = 65.5°$

$\alpha_2 = 46.31°$; $\phi_2 = 73.78°$

30. If a tangent vector is given as $\boldsymbol{t}(L) = (\sin\alpha\cos\phi)\boldsymbol{i} + (\sin\alpha\sin\phi)\boldsymbol{j} + (\cos\alpha)\boldsymbol{k}$, calculate the dogleg angle using the data given in problem 28.

31. Show that the angle between the principal normals at two neighboring survey points in a well path is

$$S\sqrt{\frac{1}{R^2} + \frac{1}{\rho^2}}$$

where S is the arcual distance between the survey points.

References

1. Samuel, R. G. *Downhole Drilling Tools—Theory and Practice for Students and Engineers*. Houston: Gulf Publishing, 2007.
2. Xiushan, L. *Geometry of Wellbore Trajectory*. Beijing, China: Petroleum Industry Press, 2006.
3. Fitchard, E. E., and S. A. Fitchard. "The Effect of Torsion on Borehole Curvature." *Oil & Gas Journal* (January 17, 1983): 121–124.
4. Xiushan, L., Z. Daqian, and Q. Lin. "Accurate Description of Curved Shape in Space for an Actual Borehole Trajectory." Dezhou, China. *Petroleum Drilling Techniques* 20, no. 2 (June 1992): 18–20.

5. Xiushan, L., W. Shan, J. Zhongxuan, et al. *Designing Theory and Describing Method for Wellbore Trajectory*. Harbin, China: Heilongjiang Science and Technology Press, 1993.
6. Shan, W., L. Xiushan, Z. Daqian, et al. "The Shape of the Space Curve of Borehole Trajectory." *Journal of Daqing Petroleum Institute* 17, no. 3 (September 1993): 32–36.
7. Xiushan, L., S. Zaihong, and Z. Daqian. "The Curve Structure Method of Borehole Trajectory Calculation." Beijing, China. *Acta Petrolei Sinica* 15, no. 3 (July 1994): 126–33.
8. Xiushan, L., and S. Zaihong. "Numerical Approximation Improves Well Survey Calculation." *Oil & Gas Journal* (April 9, 2001): 50–54.
9. Xiushan, L. "Average Borehole Curvature Calculation of Hole Trajectory." Renqiu, China. *Oil Drilling & Production Technology* 27, no. 5 (October 2005): 11–15.
10. Xiushan, L. "New Technique Calculates Borehole Curvature, Torsion." *Oil & Gas Journal* (October 23, 2006): 41–49.
11. Oprea, J. *Differential Geometry and Its Applications*, 2nd ed. Englewood Cliffs, NJ: Prentice-Hall, 2003.
12. Xiangming, M., and H. Jingzhi. *Differential Geometry*, 3rd ed. Beijing, China: Higher Education Press, 2003.
13. Lubinski, A. "How to Determine Hole Curvature." *The Petroleum Engineer* (February 1957): 42–47.
14. Lubinski, A. "Maximum Permissible Dog-Legs in Rotary Boreholes." *Journal of Petroleum Technology* (February 1961): 175–194.
15. Wilson, G. J. "Dog-Leg Control in Directional Drilled Wells." *Journal of Petroleum Technology* (January 1967): 107–112.
16. Wilson, G. J. "An Improved Method for Computing Directional Surveys." *Journal of Petroleum Technology* (August 1968): 871–876.
17. Walker, B. H., and M. B. Friedman. "Three-Dimensional Force and Deflection Analysis of a Variable Cross Section Drill String." *Journal of Pressure Vessel Technology* (May 1977): 367–373.
18. Ho, H.-S. "An Improved Modeling Program for Computing the Torque and Drag in Directional and Deep Wells." SPE Paper No. 18047, SPE Annual Technical Conference and Exhibition, Houston, Texas, October 2–5, 1988.
19. Anders, E. O. "Method and Apparatus for Drilling a Well Bore." United States Patent 4440241, April 3, 1984.
20. McClendon, R. T. "Directional Drilling Using the Catenary Method." SPE Paper No. 13478, SPE/IADC Drilling Conference, New Orleans, Louisiana, March 5–8, 1985.
21. Zhiyong, H. "Method of Non-Dimensional Design of Catenary Profile of Directional Well." Renqiu, China. *Oil Driling & Production Technology* 19, no. 4 (August 1997): 13–16.
22. Aadnoy, B. S., V. T. Fabiri, and J. Djurhuus. "Construction of Ultralong Wells Using a Catenary Well Profile." SPE Paper No. 98890, IADC/SPE Drilling Conference, Miami, Florida, February 21–23, 2006.
23. Xiushan, L. "Catenary Line on the Track Design Method." Chengdu, China. *Natural Gas Industry* 27, no. 7 (July 2007): 73–75.

24. Xiushan, L., Z. Daqian, L. Shibin, et al. "Designing Principles and Methods for Parabolic Sections of Directional Wells." Daqing, China. *Journal of Daqing Petroleum Institute* 13, no. 4 (December 1989): 29–37.
25. Xiushan, L. "Mathematical Model and Design Methods of Parabola Well Paths." Renqiu, China. *Oil Driling & Production Technology* 28, no. 4 (August 2006): 7–9, 13.
26. Sheppard M. C., C. Wick, and T. Burgess. "Designing Well Paths to Reduce Drag and Torque." *SPE Drilling Engineering* 2, no. 6 (December 1987): 344–350. SPE Paper No. 15463.
27. Hibbeler, R. C. *Engineering Mechanics—Statics,* 10th ed. Englewood Cliffs, NJ: Prentice Hall, 2003.
28. Taylor, H. L., and C. M. Mason. "A Systematic Approach to Well Surveying Calculations." SPE Paper No. 3362, SPE 46th Annual Fall Meeting, New Orleans, Louisiana, October 3–6, 1971.
29. Taylor, H. L., and C. M. Mason. "A Systematic Approach to Well Surveying Calculations." *SPE Journal* (December 1972): 474–488.
30. Zaremba, W. A. "Directional Survey by the Circular Arc Method." Standard Oil Co. of California Report, La Habra, California, 1970.
31. Zaremba, W. A. "Directional Survey by the Circular Arc Method." *SPE Journal* (February 1973): 5–11.
32. Blythe, Jr. E. J. "Computing Accurate Directional Surveys." *World Oil* (August 1975): 25–28.
33. Xiushan, L., and G. Jun. "Description and Calculation of the Well Path with Spatial Arc Model." Chendu, China. *Natural Gas Industry* 20, no. 5 (2000): 44–47.
34. Xiushan, L. S., and S. Zaihong. "Improved Method Makes a Soft Landing of Well Path." *Oil & Gas Journal* 99, no. 43 (2001): 47–51.
35. Selcull Bayin, S. *Mathematical Methods in Science and Engineering.* New Jersey: Wiley Interscience, 2006.
36. Wilson, G. J. "Radius of Curvature Method for Computing Directional Surveys." SPWLA Ninth Annual Logging Symposium, Oklahoma City, Oklahoma, June 23–26, 1968.
37. Zhiyong, H. *Directional Well Design and Computation.* Beijing, China: Petroleum Industry Press, 1989.
38. Craig, J. T., and B. V. Randall. "Directional Survey Calculation." *Petroleum Engineer* (March 1976): 38–45.
39. Callas, N. P. "Computing Directional Surveys with a Helical Method." *SPE Journal* (December 1976): 327–336.
40. Planeix, M. Y., and R. C. Fox. "Use of an Exact Mathematical Formulation to Plan Three Dimensional Directional Wells." SPE Paper No. 8338, SPE Annual Technical Conference and Exhibition, Las Vegas, Nevada, September 23–26, 1979.
41. Xiushan, L., and H. Shushan. "Design of Bit Walk Path for Slant Well." *Petroleum Drilling Techniques* 29, no. 3 (2001): 15–17.
42. Xiushan, L., and S. Zaihong. "Technique Yields Exact Solution for Planning Bit-Walk Paths." *Oil & Gas Journal* 100, no. 5 (2002): 45–50.
43. Xiushan, L., Q. Tongci, S. Zhongguo, et al. "Design of a 3D Drift Wellpath." Beijing, China. *Acta Petrolei Sinica* 16, no. 4 (1995): 118–124.

44. Xiushan, L., P. Guosheng, and Z. Xiaoxiang. "An Advanced Method for Planning and Surveying Well Trajectories of Rotary Steering Drilling." *Acta Petrolei Sinica* 24, no. 4 (2003): 81–85.
45. Xiushan, L., L. Rushan, and S. Mingxin. "New Techniques Improve Well Planning and Survey Calculation for Rotary-Steerable Drilling." IADC/SPE Paper No. 87976, IADC/SPE Asia Pacific Drilling Technology Conference and Exhibition, Kuala Lumpur, Malaysia, September 13–15, 2004.
46. Xiushan, L., S. Zaihong, and F. Sen. "Natural Parameter Method Accurately Calculates Well Bore Trajectory," *Oil & Gas Journal* 95, no. 4 (January 27, 1997): 90–92.
47. Xiushan, L., S. Zaihong, and W. Xingguo. "A New Method of Survey Calculation for Well Bore Trajectory." Second International Symposium on Measuring Technique of Multiphase Flow, Beijing, China, August 30–September 1, 1998.
48. Xiushan, L., and R. Samuel. "Actual 3D Shape of Wellbore Trajectory: An Objective Description for Complex Steered Wells." SPE Paper 115714, ATCE, Denver, Colorado, 2008.

4

Well Path Planning—2D

A well path is designed with a desired profile so that the maximum recovery of hydrocarbons is achieved from the reservoir, and the type of well profile selected meets the overall plan and requirements of the exploration and field development program, taking into account the structural characteristics of the reservoir. Judicious planning is required for reducing the overall well cost, especially in the case of directional, horizontal, extended-reach, and ultra-extended-reach wells. The type of design selected, such as vertical, near-vertical, directional, horizontal, near-horizontal, extended-reach, and ultra-extended-reach, should be optimized to achieve a high-quality well bore with reduced well construction and completion cost.[1-6] According to established well path design principles, in the absence of special requirements, the hole is to be designed using a two-dimensional (2D) profile.

This chapter focuses on different types of standard curves and well profiles. It introduces different profiles used for the planning of 2D paths of directional, horizontal, extended, and ultra-extended-reach wells. Although 2D well paths use a simplified planar approach, they are important and useful even for more complex geometries. It is particularly advantageous in determining the influence of design parameters on well path profiles. Special types of well profiles using different transition curves, such as catenary, parabola, and clothoid curves, are also discussed. The selection of the transition curves and determination of the well path coordinates, while embedding them at the appropriate depth, should result in an acceptable solution. The basic theory of well path profiles and the coupling of the mechanical model with the geometrical model, that were presented in Chapter 3, are applied in this chapter. To illustrate the practical design aspects and to provide broader understanding, a few examples are provided.

Directional Wells

First, the definitions of three terms that will be often used in what follows are recalled. As one progresses along a well path, a "build" section has an increasing inclination to the vertical, a "drop" section has a decreasing inclination to the vertical, and a "hold" section has a constant inclination to the vertical. Theoretically, there are more than ten different types of directional and horizontal well path designs. Of these, the commonly used profiles can be identified as:

- Straight vertical/inclined-build-hold profile, which can be used for moderate to large displacement wells (also called J-shaped wells).
- Straight vertical/inclined-build-hold (tangent)-drop-hold profile, which allows the placement of wells (also called S-shape wells) in a multiple pay area.
- Continuous-build profile, which is used to reach the target depth starting from a kick-off at a deeper depth.

In addition, drilling a well may involve a vertical or slant rig, and the starting straight section may be vertical or slant. In order to accommodate these types of rigs and to make the discussion below valid for a design with a more general profile it is assumed that the first section has $\alpha_1 \neq 0$, as indicated in Figure 4.1. So, depending on the case the straight section may be straight vertical with $\alpha_1 = 0$ or straight inclined with $\alpha_1 \neq 0$.

A general directional well path profile may be taken to consist of five sections—straight vertical/inclined-build-hold-drop-hold—dependent on the characteristic parameters, such as the tangent length in the hold section, build and drop rates, and radii of curvature of the build and hold curves. In the mathematical development that follows, the parameters in each section are identified by a symbol for the parameter, followed by a subscript denoting the section number. For proper completion of a well at the desired depth, an additional depth called "rat hole" may be needed for efficient casing landing as well as for good cementing at the shoe. In this case, the well may be extended to the final depth f, with the same hold angle resulting in an extended final vertical depth H_f and a horizontal displacement A_f, as

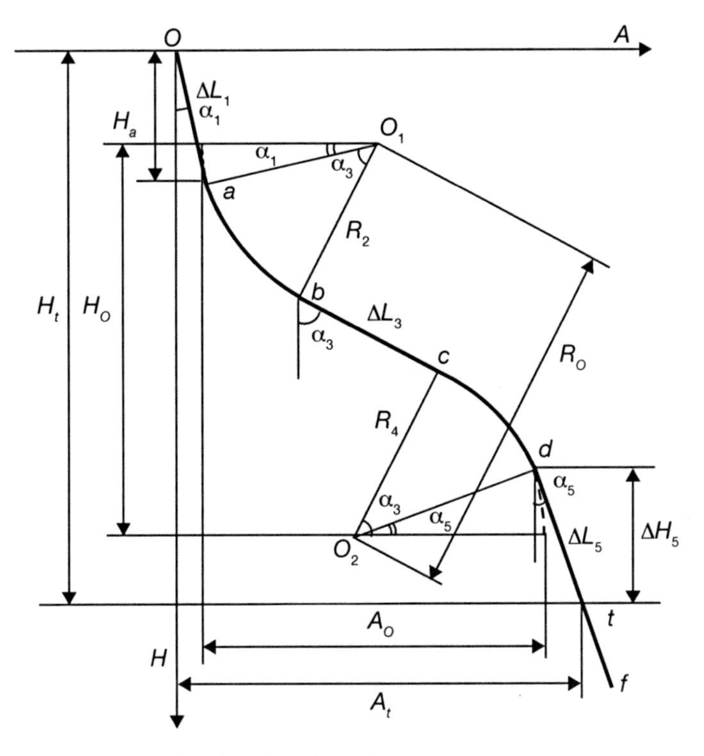

Figure 4.1 Typical directional well path profile.

shown in Figure 4.1, where the target vertical depth H_t and horizontal displacement A_t will be at the initial target point t.

From Figure 4.1, the following geometrical relationships can be obtained:

$$\begin{cases} \sum\limits_{i=1,3,5} \Delta L_i \cos\alpha_i + R_2(\sin\alpha_3 - \sin\alpha_1) + R_4(\sin\alpha_3 - \sin\alpha_5) = H_t \\ \sum\limits_{i=1,3,5} \Delta L_i \sin\alpha_i + R_2(\cos\alpha_1 - \cos\alpha_3) + R_4(\cos\alpha_5 - \cos\alpha_3) = A_t \end{cases} \quad (4.1)$$

where subscripts 1–5 correspond to the respective sections, and

ΔL = Measured incremental distance or course length between the sections
α = Inclination angle
H = Vertical depth
A = Horizontal distance

The parameters used in Eq. 4.1 are illustrated in Figure 4.1.

Substituting the respective parameters for all the sections, the foregoing expression takes the form:

$$\begin{cases} \Delta L_3 \cos\alpha_3 + (R_2 + R_4)\sin\alpha_3 \\ \quad = H_t - \Delta L_1 \cos\alpha_1 - \Delta L_5 \cos\alpha_5 + R_2 \sin\alpha_1 + R_4 \sin\alpha_5 \\ \Delta L_3 \sin\alpha_3 + (R_2 + R_4)(1 - \cos\alpha_3) \\ \quad = A_t - \Delta L_1 \sin\alpha_1 - \Delta L_5 \sin\alpha_5 + R_2(1 - \cos\alpha_1) + R_4(1 - \cos\alpha_5) \end{cases} \tag{4.2}$$

Introducing the parameters H_o, A_o, R_o, defined as:

$$\begin{cases} H_o = H_t - \Delta L_1 \cos\alpha_1 - \Delta L_5 \cos\alpha_5 + R_2 \sin\alpha_1 + R_4 \sin\alpha_5 \\ A_o = A_t - \Delta L_1 \sin\alpha_1 - \Delta L_5 \sin\alpha_5 + R_2(1 - \cos\alpha_1) + R_4(1 - \cos\alpha_5) \\ R_o = R_2 + R_4 \end{cases} \tag{4.3}$$

Then,

$$\begin{cases} \Delta L_3 \cos\alpha_3 + R_o \sin\alpha_3 = H_o \\ \Delta L_3 \sin\alpha_3 + R_o(1 - \cos\alpha_3) = A_o \end{cases} \tag{4.4}$$

where for the sake of convenience and for repeated use in the mathematical formulation in the rest of the book, H_o, R_o, and A_o are called the *design center parameters.*

The geometrical significance of the center parameters is as follows:

- H_o is the vertical distance between the centers of the build and drop sections.
- A_o is the horizontal distance between the points where the circular arcs of the build and drop sections intersect the horizontal line from their respective centers.
- R_o is the distance between lines through the center of curvatures of the build and drop sections parallel to the direction of the third hold (tangent) section.

Elimination of ΔL_3 from Eq. 4.4 results in:

$$H_o \sin\alpha_3 + (R_o - A_o)\cos\alpha_3 = R_o \tag{4.5}$$

The sine and cosine terms in Eq. 4.5 can be replaced by tangents using the trigonometric relationships:

$$\sin\alpha_3 = \frac{2\tan\frac{\alpha_3}{2}}{1 + \tan^2\frac{\alpha_3}{2}} \quad \text{and} \quad \cos\alpha_3 = \frac{1 - \tan^2\frac{\alpha_3}{2}}{1 + \tan^2\frac{\alpha_3}{2}}$$

Substituting these in Eq. 4.5 and rearranging gives:

$$(2R_o - A_o)\tan^2\frac{\alpha_3}{2} - 2H_o\tan\frac{\alpha_3}{2} + A_o = 0 \tag{4.6}$$

Therefore,

$$\tan\frac{\alpha_3}{2} = \frac{H_o \pm \sqrt{H_o^2 + A_o^2 - 2R_oA_o}}{2R_o - A_o} \tag{4.7}$$

It can be seen from Eq. 4.7 that there are two solutions. It has been proven that the two solutions of Eq. 4.7 are the well inclinations of the common tangent to the two curvature circles of the building section and the dropping section. For a directional well with the geometry indicated in Figure 4.1, the radical should take the negative sign and therefore the equation takes the form:

$$\tan\frac{\alpha_3}{2} = \frac{H_o - \sqrt{H_o^2 + A_o^2 - 2R_oA_o}}{2R_o - A_o} \tag{4.8}$$

In order to obtain a relationship between ΔL_3 and the other center parameters, taking squares on both equations in Eq. 4.4 gives:

$$\begin{cases} \Delta L_3^2\cos^2\alpha_3 + 2R_o\Delta L_3\sin\alpha_3\cos\alpha_3 + R_o^2\sin^2\alpha_3 = H_o^2 \\ \Delta L_3^2\sin^2\alpha_3 + 2R_o\Delta L_3\sin\alpha_3(1-\cos\alpha_3) + R_o^2(1-\cos\alpha_3)^2 = A_o^2 \end{cases} \tag{4.9}$$

Adding both the equations in Eq. 4.9 results in:

$$\Delta L_3^2 + 2R_o\Delta L_3\sin\alpha_3 + 2R_o^2(1-\cos\alpha_3) = H_o^2 + A_o^2 \tag{4.10}$$

Using Eq. 4.4 in Eq. 4.10 results in the ΔL_3 relationship:

$$\Delta L_3 = \sqrt{H_o^2 + A_o^2 - 2R_oA_o} \tag{4.11}$$

Eqs. 4.8 and 4.11 provide the key equations for the directional well.

Eq. 4.8 is valid for $2R_o - A_o \neq 0$, and when $2R_o - A_o = 0$, Eq. 4.6 reduces to:

$$\tan\frac{\alpha_3}{2} = \frac{A_o}{2H_o} \tag{4.12}$$

In addition, Eqs. 4.8 and 4.11 have the following condition:

$$H_o^2 + A_o^2 - 2R_oA_o \geq 0$$

In particular, when

$$H_o^2 + A_o^2 - 2R_oA_o \geq 0$$

Eq. 4.8 can be simplified as:

$$\tan\frac{\alpha_3}{2} = \frac{A_o}{H_o} \tag{4.13}$$

The foregoing derivations are for the type straight vertical/inclined-build-hold-drop-hold profile. For other profiles, equations for the design center parameters H_o, A_o, and R_o can be modified to include the correct sections. For example, for a straight vertical/inclined-build-hold profile, in the absence of sections 4 and 5, the parameters related to these sections have to be removed from the H_o, R_o, and A_o equations, which results in the following equations:

$$\begin{cases} H_o = H_t - \Delta L_1 \cos\alpha_1 + R_2 \sin\alpha_1 \\ A_o = A_t - \Delta L_1 \sin\alpha_1 + R_2 (1 - \cos\alpha_1) \\ R_o = R_2 \end{cases} \tag{4.14}$$

Eqs. 4.3 and 4.4 lead to the traditional equations for the design of directional well paths.[4-6] It is also possible to obtain alternate design equations as explained below.

If the alternate design center parameters are defined as follows:

$$\begin{cases} H_o = H_t - \Delta L_1 \cos\alpha_1 - \Delta L_5 \cos\alpha_5 + R_2 \sin\alpha_1 + R_4 \sin\alpha_5 \\ A_o = A_t - \Delta L_1 \sin\alpha_1 - \Delta L_5 \sin\alpha_5 - R_2 \cos\alpha_1 - R_4 \cos\alpha_5 \\ R_o = R_2 + R_4 \end{cases} \tag{4.15}$$

then Eq. 4.2 will become:

$$\begin{cases} \Delta L_3 \cos\alpha_3 + R_o \sin\alpha_3 = H_o \\ \Delta L_3 \sin\alpha_3 - R_o \cos\alpha_3 = A_o \end{cases} \tag{4.16}$$

These equations provide concise and clear geometric meaning of the design center parameters A_o, H_o, and R_o, including the curvature relationship. This alternate approach also helps in the generalization of the design equations, making them similarly applicable for horizontal well paths. And eliminating ΔL_3 from Eq. 4.16 gives the relationship between the alternate design center parameters:

$$H_o \sin\alpha_3 - A_o \cos\alpha_3 = R_o \tag{4.17}$$

As before, replacing the sine and cosine terms in Eq. 4.17 by tangent terms using the trigonometric relationships gives:

$$(R_o - A_o)\tan^2\frac{\alpha_3}{2} - 2H_o \tan\frac{\alpha_3}{2} + (R_o + A_o) = 0 \tag{4.18}$$

Hence,

$$\tan\frac{\alpha_3}{2} = \frac{H_o - \sqrt{H_o^2 + A_o^2 - R_o^2}}{R_o - A_o} \tag{4.19}$$

To obtain an expression for ΔL_3, the two equations in Eq. 4.16 are squared and added to give:

$$\Delta L_3 = \sqrt{H_o^2 + A_o^2 - R_o^2} \tag{4.20}$$

Eq. 4.19 is valid for $R_o - A_o \neq 0$. When $R_o - A_o = 0$, Eq. 4.18 results in:

$$\tan\frac{\alpha_3}{2} = \frac{A_o}{H_o} \tag{4.21}$$

In general, Eqs. 4.19 and 4.20 are valid for $H_o^2 + A_o^2 - R_o^2 \geq 0$. In the particular case when $H_o^2 + A_o^2 - R_o^2 = 0$ and $\Delta L_3 = 0$, Eqs. 4.16 and 4.19 will result in:

$$\tan\alpha_3 = -\frac{H_o}{A_o} \tag{4.22}$$

and in:

$$\tan\frac{\alpha_3}{2} = \frac{H_o}{R_o - A_o} \tag{4.23}$$

For other well path profiles, for example of the type straight vertical/inclined-build-hold, the general Eq. 4.15 for the alternate *design center parameters* can be readily modified as:

$$\begin{cases} H_o = H_t - \Delta L_1 \cos\alpha_1 + R_2 \sin\alpha_1 \\ A_o = A_t - \Delta L_1 \sin\alpha_1 - R_2 \cos\alpha_1 \\ R_o = R_2 \end{cases} \tag{4.24}$$

PROBLEM 4.1

Design a directional well with a target vertical depth H_t of 2500 m, with a horizontal departure at the target depth of 900 m and with a displacement angle of 300°. Use a straight vertical/build-tangent-drop-hold profile, with the vertical depth to the kick-off point of 500 m.

The first build rate is 6°/100 m, and the drop rate is 4°/100 m. At the entrance point of the hold section in the oil layer, the inclination angle is 12°, TVD increment is 300 m, and extra TVD for "rat hole" is 50 m.

Solution

Given data:

$H_a = \Delta L_1 = 500$ m; $\Delta H_{tf} = 50$ m; $\Delta H_5 = 300$ m

Curvatures:

$\kappa_2 = 6°/100$ m; $\kappa_4 = 4°/100$ m; $\alpha_5 = 12°$

Target data:

$H_t = 2500$ m; $A_t = 900$ m

Radii of curvature of the build and drop sections are:

$$R_2 = \frac{180 \times 100}{\pi \times 6} = 954.930 \text{ m}; \; R_4 = \frac{180 \times 100}{\pi \times 4} = 1432.395 \text{ m}$$

Incremental measured length inside the reservoir is:

$$\Delta L_5 = \frac{300}{\cos 12°} = 306.702 \text{ m}$$

Using the traditional design equations, the center parameters are:

$$H_o = 2500 - 500 \times \cos 0° - 306.702 \times \cos 12° + 954.930 \times \sin 0° + 1432.395 \times \sin 12° = 1997.812 \text{ m}$$

$$A_o = 900 - 500 \times \sin 0° - 306.702 \times \sin 12° + 954.930 \times (1 - \cos 0°) + 1432.395 \times (1 - \cos 12°) = 867.534 \text{ m}$$

$R_o = 954.930 + 1432.395 = 2387.325$ m

The tangent length and the tangent angle are calculated as:

$$\Delta L_3 = \sqrt{1997.812^2 + 867.534^2 - 2 \times 2387.325 \times 867.534} = 775.691 \text{ m}$$

$$\alpha_3 = 2 \times \tan^{-1}\left(\frac{1997.812 - 775.691}{2 \times 2387.325 - 867.534}\right) = 34.739°$$

Using the alternate design equations, the alternate center parameters are:

$$H_o = 2500 - 500 \times \cos 0° - 306.702 \times \cos 12° + 954.930 \times \sin 0° + 1432.395 \times \sin 12° = 1997.812 \text{ m}$$

$$A_o = 900 - 500 \times \sin 0° - 306.702 \times \sin 12° - 954.930 \times \cos 0° - 1432.395 \times \cos 12° = -1519.791 \text{ m}$$

$R_o = 954.930 + 1432.395 = 2387.325$ m

Similarly, the tangent length and the tangent angle are calculated as:

$$\Delta L_3 = \sqrt{1997.812^2 + (-1519.791)^2 - 2387.325^2} = 775.691 \text{ m}$$

$$\alpha_3 = 2 \times \tan^{-1}\left[\frac{1997.812 - 775.691}{2387.325 - (-1519.791)}\right] = 34.739°$$

The directional well design and related well path data are summarized in Table 4.1.

Table 4.1 Calculated Well Path Data for Problem 4.1

Well Depth (m)	Deviation (°)	Azimuth (°)	Vertical Depth (m)	North Coordinate (m)	East Coordinate (m)	Horizontal Displacement (m)	Curvature (°/30 m)	Remarks
0.00	0.00	—	0.00	0.00	0.00	0.00	0.00	Wellhead
500.00	0.00	300.00	500.00	0.00	0.00	0.00	0.00	Kick-off
1078.98	34.74	300.00	1044.15	85.10	–147.40	170.21	6.00	
1854.67	34.74	300.00	1681.58	306.11	–530.20	612.22	0.00	
2423.14	12.00	300.00	2200.00	418.12	–724.20	836.23	4.00	
2729.84	12.00	300.00	2500.00	450.00	–779.42	900.00	0.00	Target
2780.96	12.00	300.00	2550.00	455.31	–788.63	910.63	0.00	Bottom

Horizontal Wells

One of the main advantages of a horizontal well is that it provides extended contact with the reservoir. Horizontal drilling is also carried out for the purpose of enhancing the productivity of hydrocarbons from one or more thinly stacked layers of potential formations. This reduces the number of wells by simultaneously developing multiple reservoirs. It also helps to develop tight reservoirs, consisting of naturally fractured prospective channel sands that intersect reservoirs with a high-formation dip angle. The design of horizontal wells is characterized by the radius of curvatures as follows (see Table 4.2):

- Long radius design
- Medium radius design
- Short radius design
- Ultra-short radius design

Drilling ultra-short radius wells needs special downhole tools, such as articulated (or wiggly-type) mud motors, downhole measurement tools, and flexible drill collars. An articulate multishot surveying assembly is used for accurate surveying.

Being given the wellhead and the target coordinates (such as altitude, target vertical depth, etc.), the following steps are involved in the design of 2D well paths of horizontal wells:

- Choose the type of profile
- Determine the kick-off points (lateral drilling, branch)

Table 4.2 Comparative Characteristics of Horizontal Drilling Methods (SPE/IADC)[3]

	Long Radius	*Medium Radius*	*Short Radius*
Build Rate	Up to 6° per 100'	6° to 35° per 100'	1.5° to 3° per 1'
Build Radius	1000' to 3000'	955' to 286'	20' to 40'
Hole Size	No limitation	4¾", 6⅛", 8½", 9⅞"	4¾" 6½"
Drilling Method	Rotary or steerable motor systems for curved and horizontal sections	Special design motors for angle—build section; rotary or steerable motor system for horizontal sections	Special design deflection tool or articulated motor for angle—build section; rotary tools and special drill pipe for horizontal sections

- Determine the build and drop rates
- Calculate the well parameters for the profile
- Check the deviation angle and the hole curvature parameters
- Calculate the well path parameters at various depths
- Generate the required plots

The choice of well profile type, determination of the kick-off depth, maximum deviation angle, build or drop rate, and allowed hole curvature are important aspects of the well path design.

Different horizontal well path designs can be implemented using the methods and designs discussed in the previous sections of this chapter. Different well path profiles, such as long radius, medium radius, short radius, and ultra-short radius, can be developed using a combination of sections, such that a near-horizontal or 90° inclination well profile is achieved. Apart from the conventional horizontal well path designs, the following complex type of profiles can be also designed:

- Horizontal arch design
- Horizontal ladder (staircase) design

Horizontal Arch Design

The horizontal arch design, or inverted design profile, uses a single circular arc as the transition section and is used to develop two separate formation layers. The horizontal length may be sloped to intersect many stratified sands, and allows gravity drainage to act over the life of the well. As shown in Figure 4.2, the horizontal arch well path design method is similar to the

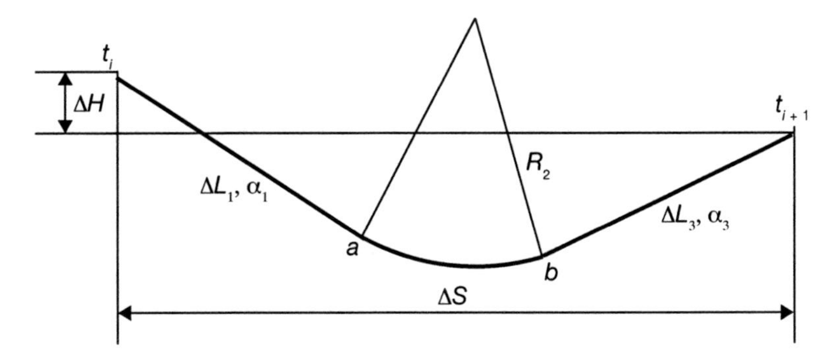

Figure 4.2 Profile of a horizontal arch section.

conventional J-shaped well profile design method. Since the profile is similar to an inverted arch, it is called the arch design.

The horizontal arch design consists of two tangent sections connected by an arch or a circular-arc section. In the discussion that follows, the well path parameters in each section are tagged with the corresponding section numbers. An arch-shaped horizontal section is composed of two hold sections and one build or drop section. When the positions of the two adjacent targets t_i and t_{i+1} are given, the true vertical depth difference ΔH and horizontal displacement difference ΔA are fixed. Using the method mentioned previously, based on the reservoir layer tilt angle, direction (declination angle), and the azimuth and inclination angles of the horizontal section, the two inclination angles of the two hold sections are known, and so α_1 and α_3 are usually known parameters. Moreover, the difference between α_1 and α_3 is often very small, so that the build or drop section is very short and can be considered to be a transition section or adjusted section. Thus, for the design of an arch-shaped horizontal well path, it is important to accurately find the lengths ΔL_1 and ΔL_2 of the two hold sections.

According to the geometric relations, the following equation can be easily obtained[7]:

$$\begin{cases} \Delta L_1 = \dfrac{\Delta H \sin\alpha_3 - \Delta A \cos\alpha_3}{\sin(\alpha_3 - \alpha_1)} - R_2 \tan\dfrac{\alpha_3 - \alpha_1}{2} \\ \Delta L_3 = \dfrac{\Delta A \cos\alpha_1 - \Delta H \sin\alpha_1}{\sin(\alpha_3 - \alpha_1)} - R_2 \tan\dfrac{\alpha_3 - \alpha_1}{2} \end{cases} \tag{4.25}$$

It can be seen that Eq. 4.25 requires $\alpha_1 \neq \alpha_3$ or, in other words, the inclination angles of the two hold sections should not be same. In fact, this is a required condition, as otherwise it would be unnecessary to design an arch-shaped well path.

PROBLEM 4.2

A horizontal well with azimuth 60° is planned for exploiting two thin reservoir layers at different depths, separated by a shale layer. Based on the dip angle and the direction of the reservoir layer, the inclination angles of two well sections at these two layers are estimated to be $\alpha_1 = 87°$ and $\alpha_4 = 92°$. The true vertical depth difference between the targets is $\Delta H = 2$ m, and the horizontal displacement is $\Delta A = 420$ m. If the build rate of the shale section is $\kappa_2 = 4°/30$ m, design the horizontal well sections.

Solution

According to the requirements, a horizontal arch-shaped trajectory should be used. From Eq. 4.25, $\Delta L_1 = 204.38$ m and $\Delta L_3 = 258.57$ m are obtained. If the origin of the coordinate system is set as the starting point, the results for the horizontal well sections are as given in Table 4.3.

It is necessary to also take into account the associated problems in these types of horizontal arch designs. Proper hole cleaning in the arched section and the axial force transfer while drilling different sections of the well need to be ensured. Effective transfer of weight to the bit is required, so that the desired curvature and the effective reservoir length are achieved for suitable completion. The design can be extended to multiple arch sections for drilling anisotropic formations with different vertical and horizontal permeabilities. Formations with low vertical permeability are exploited by drilling undulated horizontal wells, so that maximum recovery is achieved. This provides more drainage, but adds complexity to the drilling and the completion of the zone, due to humps and arches with downhill and uphill sections. Single staircase construction design can be extended for multiple staircase wells.

Horizontal Ladder (Staircase) Design

Another type of complex horizontal well design is called the ladder (or staircase) design. A horizontal or near-horizontal well path with one or more ladder-shaped profiles allows development of two or more pay-zones at the same time. For such a condition, the well path has to be optimized so that the torque and drag are reduced, especially while drilling the convoluted shape and, in addition, the well path should be conducive to effective hole cleaning.

A ladder horizontal well path is sketched in Figure 4.3. Two neighboring depths are designated as t_i and t_{i+1}, with circular arcs as transition curves.

In a horizontal well section, the rotary drilling method is often the choice. A low build rate not only improves the penetration rate and hole cleaning, but also reduces friction and improves the downhole safety. A large build rate can cause higher well bore friction while drilling, and may even result in key seating and eventually pipe sticking. As discussed earlier, the vertical and horizontal displacements of the two target points are often fixed, and the inclination angles α_1 and α_4 of the two formation sections (hold sections) are usually known design parameters. Moreover, the lengths of the two formation layer sections and the build rate of the drilling tool are easy to estimate, and so they also can be considered as known design parameters. However, the inclination at the inflection point b is an unknown,

Table 4.3 Design Results for Horizontal Arch Section for Problem 4.2

Points	Depth (m)	Inclination (°)	Azimuth (°)	TVD (m)	North (m)	East (m)	Departure (m)	Displacement (°)	Horizontal Length (m)
t_1	0.00	87.00	60.00	0.00	0.00	0.00	0.00	—	0.00
a	204.39	87.00	60.00	10.70	102.05	176.76	204.10	60.00	204.10
b	241.89	92.00	60.00	11.02	120.80	209.22	241.59	60.00	241.59
t_2	500.45	92.00	60.00	2.00	250.00	433.01	500.00	60.00	500.00

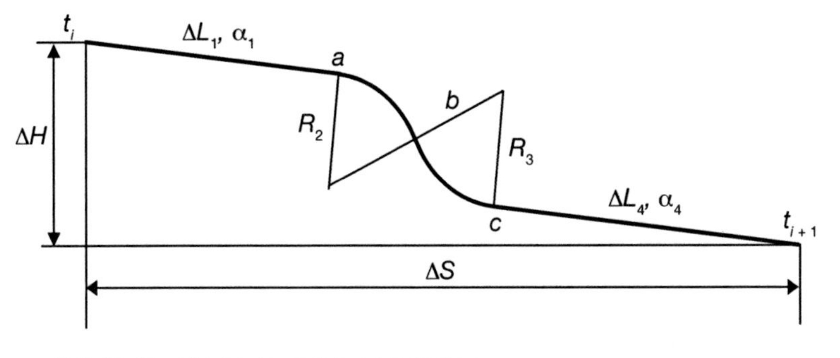

Figure 4.3 Profile of a horizontal ladder section.

and an iterative method has to be carried out to get satisfactory design results.

Theoretically, two characteristic parameters can be selected for the horizontal ladder design. For planning horizontal ladder design, the common three solution combinations are as follows[8]:

- Length ΔL_1 and the inclination angle α_b
- Length ΔL_4 and the inclination angle α_b
- Build or drop rate of the transition and the inclination angle α_b

The solutions for these possible combinations are as follows:

1. Solving for the length ΔL_1 of the first reservoir layer and the angle α_b at the inflection point[8]:

$$\tan\frac{\alpha_b}{2} = \begin{cases} \dfrac{B-\sqrt{A^2+B^2-C^2}}{C+A}, & \text{if } C+A \neq 0 \\ \dfrac{C-A}{2B}, & \text{if } C+A=0 \end{cases} \tag{4.26}$$

$$\Delta L_1 = D - \sqrt{D^2 - E} \tag{4.27}$$

where

$$A = (R_2 - R_3)\cos\alpha_1$$
$$B = (R_2 - R_3)\sin\alpha_1$$
$$C = \Delta H \sin\alpha_1 - \Delta A \cos\alpha_1 + R_2 - R_3 \cos(\alpha_4 - \alpha_1) + \Delta L_4 \sin(\alpha_4 - \alpha_1)$$
$$D = U\cos\alpha_1 + V\sin\alpha_1$$
$$E = U^2 + V^2 - (R_2 - R_3)^2$$

$$U = \Delta H + R_2 \sin\alpha_1 - R_3 \sin\alpha_4 - \Delta L_4 \cos\alpha_4$$
$$V = \Delta A - R_2 \cos\alpha_1 + R_3 \cos\alpha_4 - \Delta L_4 \sin\alpha_4$$

2. Solving for the length ΔL_4 of the second reservoir layer and the angle α_b at the inflection point:

$$\tan\frac{\alpha_b}{2} = \begin{cases} \dfrac{B - \sqrt{A^2 + B^2 - C^2}}{C + A}, & \text{if } C + A \neq 0 \\ \dfrac{C - A}{2B}, & \text{if } C + A = 0 \end{cases} \tag{4.28}$$

$$\Delta L_4 = D - \sqrt{D^2 - E} \tag{4.29}$$

where

$$A = (R_2 - R_3)\cos\alpha_4$$
$$B = (R_2 - R_3)\sin\alpha_4$$
$$C = \Delta H \sin\alpha_4 - \Delta A \cos\alpha_4 - R_3 + R_2 \cos(\alpha_4 - \alpha_1) - \Delta L_1 \sin(\alpha_4 - \alpha_1)$$
$$D = U\cos\alpha_4 + V\sin\alpha_4$$
$$E = U^2 + V^2 - (R_2 - R_3)^2$$
$$U = \Delta H - \Delta L_1 \cos\alpha_1 + R_2 \sin\alpha_1 - R_3 \sin\alpha_4$$
$$V = \Delta A - \Delta L_1 \sin\alpha_1 - R_2 \cos\alpha_1 + R_3 \cos\alpha_4$$

3. Solving for the build or drop rate of the transition section (when $\kappa_2 = -\kappa_3$) and the inclination angle α_b at the inflection point:

$$\tan\frac{\alpha_b}{2} = \begin{cases} \dfrac{B + \sqrt{A^2 + B^2 - C^2}}{C + A}, & \text{if } C + A \neq 0 \\ \dfrac{C - A}{2B}, & \text{if } C + A = 0 \end{cases} \tag{4.30}$$

$$R_3 = -R_2 = D + \sqrt{D^2 + E} \tag{4.31}$$

where

$$A = \Delta H - \Delta L_1 \cos\alpha_1 - \Delta L_4 \cos\alpha_4$$
$$B = \Delta A - \Delta L_1 \sin\alpha_1 - \Delta L_4 \sin\alpha_4$$
$$C = -0.5A(\cos\alpha_1 + \cos\alpha_4) - 0.5B(\sin\alpha_1 + \sin\alpha_4)$$
$$D = \frac{B(\cos\alpha_1 + \cos\alpha_4) - A(\sin\alpha_1 + \sin\alpha_4)}{1 - \cos(\alpha_4 - \alpha_1)}$$

$$E = \frac{A^2 + B^2}{1 - \cos(\alpha_4 - \alpha_1)}$$

For Eq. 4.31 to be valid, $\alpha_1 \neq \alpha_4$ (i.e., the two oil layer angles are not equal). In case $\alpha_1 = \alpha_4$, then:

$$R_3 = -R_2 = \frac{A^2 + B^2}{2[A(\sin\alpha_1 + \sin\alpha_4) - B(\cos\alpha_1 + \cos\alpha_4)]} \tag{4.32}$$

PROBLEM 4.3

A horizontal well at an azimuth angle of 30° is to be drilled to exploit two different thin reservoir layers at different depths. According to the dip angles and the directions of the reservoir layers, the inclination angles of the two well sections at these two layers are $\alpha_1 = 88°$ and $\alpha_4 = 90°$, respectively. The required target's vertical displacement is $\Delta H = 10$ m, and the horizontal displacement is $\Delta A = 420$ m. Design the well profile.

Solution

Based on the requirements, a ladder well path design is planned. If the build rates of the two arc sections are $\kappa_2 = -3°/30$ m and $\kappa_3 = 4.5°/30$ m, and the well length in the second reservoir layer is $\Delta L_4 = 150$ m, then by using Eqs. 4.26 and 4.27, the inclination angle at inflection is obtained as $\alpha_b = 85.65°$, and the well length in the first reservoir layer is $\Delta L_1 = 217.68$ m. The design results for this horizontal well path are shown in Table 4.4.

To design a ladder horizontal well section, the build rate of the first arc section is negative (drop rate), and the build rate of the second arc section is positive; and for a reverse ladder horizontal well section the opposite is true. There is no substantial difference between these two cases.

The problem is worked out in API units for convenience as follows:
The given data are:

$\alpha_1 = 88°$; $\alpha_4 = 90°$; $\kappa_2 = -3°/30$ m; $\kappa_3 = 4.5°/30$ m

Target data are:

$\Delta H = 32.808$ ft and $\Delta A = 1378$ ft

$\Delta L_4 = 492.146$ ft

Radii of curvatures are:

$$R_2 = -\frac{180 \times 30 \times 3.2804}{\pi \times 3} = -1879.53 \text{ ft}$$

$$R_3 = -\frac{180 \times 30 \times 3.2804}{\pi \times 4.5} = 1253.19 \text{ ft}$$

Table 4.4 Design Results for the Horizontal Ladder Well Path for Problem 4.3

Points	*Depth (m)*	*Inclination (°)*	*Azimuth (°)*	*TVD (m)*	*North (m)*	*East (m)*	*Departure (m)*	*Displacement (°)*	*Horizontal Length (m)*
t_i	0.00	88.00	30.00	0.00	0.00	0.00	0.00	—	0.00
a	217.68	88.00	30.00	7.60	188.40	108.78	217.55	30.00	217.55
b	241.19	85.65	30.00	8.90	208.73	120.51	241.02	30.00	241.02
c	270.20	90.00	30.00	10.00	233.83	135.00	270.00	30.00	270.00
t_{i+1}	420.20	90.00	30.00	10.00	363.73	210.00	420.00	30.00	420.00

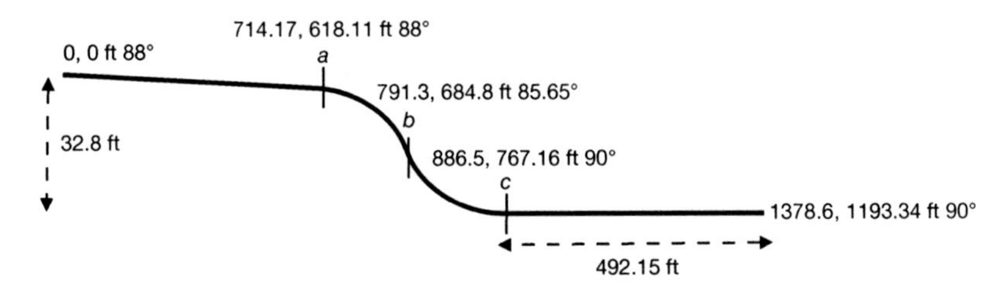

Figure 4.4 Results for the well path profile of Problem 4.3.

Calculating the values of A, B, and C:

$$A = (-1879.53 - 1253.19) \cos 88 = -109.34$$

$$B = (-1879.53 - 1253.19) \sin 88 = -3133$$

$$C = 32.808 \sin 88 - 1378 \cos 88 - 7879.53 - 1253.19 \cos (90 - 88) + 492.15 \sin (90 - 88) = -3130.34$$

Further calculating the values of U and V:

$$U = 32.808 + R_2 \sin 88 - R_3 \sin 90 - 492.146 \cos 90 = -3099$$

$$V = 1378 - R_2 \cos 88 + R_3 \cos 90 - 492.146 \sin 90 = 951.43$$

Finally, calculating the values of D and E to estimate the length of the first reservoir layer:

$$D = -944.58 \cos 90 + 290 \sin 90 = 842.7$$

$$E = -944.58^2 + 290^2 - (-1879.53 - 1253.19)^2 = 693{,}621.6$$

Substituting the calculated values in Eq. 4.27:

$$\Delta L_1 = 842.7 - \sqrt{842.7^2 - 693{,}621.6} = 714.18 \text{ ft}$$

Similarly, the other parameters at the required points can be calculated and are shown in Figure 4.4.

Planning the Well Path Just Before Targets

At present, long radius horizontal well designs commonly use the straight vertical/inclined-build-tangent-build-hold sections[3,7-10] and this profile is also known as the "double-build" curve design. The tangent section between the two curved sections is used to compensate for the uncertainty in well path build rates due to the tool tolerances and their build rates. The tangent section angle depends on the target depth and is affected by the uncertainties in the tool errors, build rates, and target formation thickness to be drilled. The well path is shown in Figure 4.5.

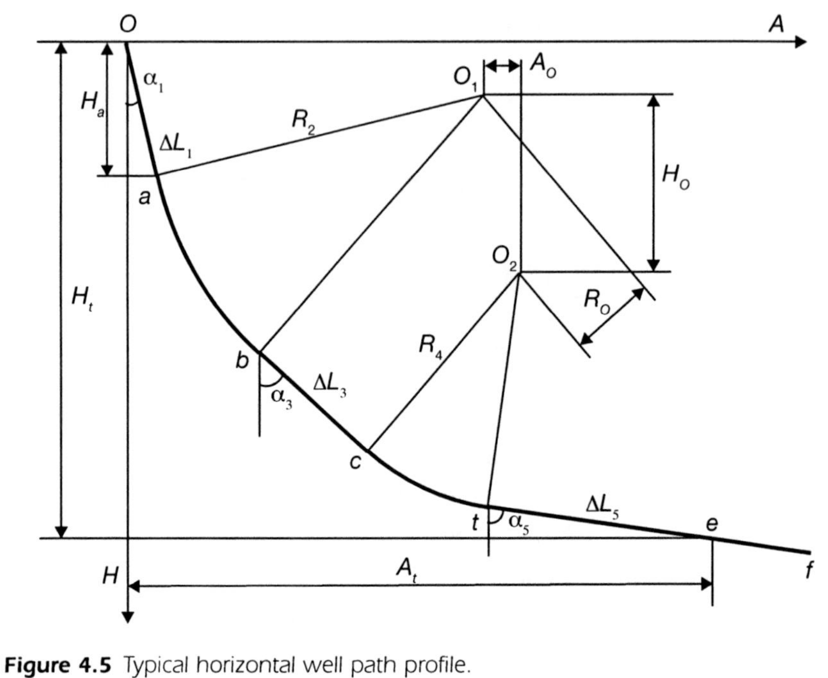

Figure 4.5 Typical horizontal well path profile.

In the double-build curve well path design, it is necessary to find the tangent angle α_3 and the tangent length ΔL_3 of the tangent section, so that the second build rate and desired landing into the formation are achieved.

Referring to Figure 4.5 and using the geometry of the sections, it can be written:

$$\begin{cases} \sum\limits_{i=1,3,5} \Delta L_i \cos\alpha_i + R_2(\sin\alpha_3 - \sin\alpha_1) + R_4(\sin\alpha_5 - \sin\alpha_3) = H_e \\ \sum\limits_{i=1,3,5} \Delta L_i \sin\alpha_i + R_2(\cos\alpha_1 - \cos\alpha_3) + R_4(\cos\alpha_3 - \cos\alpha_5) = A_e \end{cases} \tag{4.33}$$

Using the respective subscripts for the straight sections and rearranging, the resulting equations are:

$$\begin{cases} \Delta L_3 \cos\alpha_3 + (R_2 - R_4)\sin\alpha_3 \\ \quad = H_e - \Delta L_1 \cos\alpha_1 - \Delta L_5 \cos\alpha_5 + R_2 \sin\alpha_1 - R_4 \sin\alpha_5 \\ \Delta L_3 \sin\alpha_3 - (R_2 - R_4)\cos\alpha_3 \\ \quad = A_e - \Delta L_1 \sin\alpha_1 - \Delta L_5 \sin\alpha_5 - R_2 \cos\alpha_1 + R_4 \cos\alpha_5 \end{cases} \tag{4.34}$$

It is not difficult to express the design center parameters as:

$$\begin{cases} H_o = H_e - \Delta L_1 \cos\alpha_1 - \Delta L_5 \cos\alpha_5 + R_2 \sin\alpha_1 - R_4 \sin\alpha_5 \\ A_o = A_e - \Delta L_1 \sin\alpha_1 - \Delta L_5 \sin\alpha_5 - R_2 \cos\alpha_1 + R_4 \cos\alpha_5 \\ R_o = R_2 - R_4 \end{cases} \tag{4.35}$$

Eq. 4.34 can be simplified using Eq. 4.35, and the resulting equations are:

$$\begin{cases} \Delta L_3 \cos\alpha_3 + R_o \sin\alpha_3 = H_o \\ \Delta L_3 \sin\alpha_3 - R_o \cos\alpha_3 = A_o \end{cases} \tag{4.36}$$

Eq. 4.36 can be solved for the tangent angle α_3 and the tangent length ΔL_3:

$$\begin{cases} \tan\dfrac{\alpha_3}{2} = \dfrac{H_o - \sqrt{H_o^2 + A_o^2 - R_o^2}}{R_o - A_o} \\ \text{or} \\ \alpha_3 = 2 \times \tan^{-1}\left(\dfrac{H_o - \sqrt{H_o^2 + A_o^2 - R_o^2}}{R_o - A_o} \right) \\ \Delta L_3 = \sqrt{H_o^2 + A_o^2 - R_o^2} \end{cases} \tag{4.37}$$

It can be clearly seen that H_o, A_o, and R_o are, respectively, the vertical distance, the horizontal departure, and the distance between the centers of the curvature of the two curved sections, as shown in Figure 4.5.

The S-shaped directional well and the double-build horizontal well profile differ in the second circular arc, which is a drop section in one case and a build section in the other case. However, the design formulas have the same form. Comparing Eqs. 4.15 and 4.35, it is easy to see that if the inclined to drop radius of curvature were not negative, then the two H_o, A_o, and R_o values are the same. Thus, typical directional and horizontal well profile design methods have the same form, resulting in simplified well path design equations. In addition, if the hole is not considered beyond the target depth, Eq. 4.37 is still valid and the equations for the center parameters H_o, A_o, and R_o are:

$$\begin{cases} H_o = H_t - \Delta L_1 \cos\alpha_1 + R_2 \sin\alpha_1 - R_4 \sin\alpha_5 \\ A_o = A_t - \Delta L_1 \sin\alpha_1 - R_2 \cos\alpha_1 + R_4 \cos\alpha_5 \\ R_o = R_2 - R_4 \end{cases} \tag{4.38}$$

PROBLEM 4.4

Design a well path with the given data.

Wellhead coordinates:

$X_o = 4{,}557{,}543$ m; $Y_o = 21{,}402{,}647.4$ m (Gaussian projection plane coordinates are used with the north direction for the X coordinate and the east direction for the Y coordinate)

The first target coordinates are:

$X_t = 4{,}557{,}493.511$ m; $Y_t = 21{,}402{,}366.731$ m

The final target coordinates are:

$X_e = 4{,}557{,}406.716$ m; $Y_e = 21{,}401{,}874.493$ m

The respective vertical depths are:

$H_t = 2000$ m; $H_e = 2013$ m

The profile selected is a straight vertical/inclined-build-tangent-build-hold section and kick-off point data are:

$H_a = \Delta L_1 = 1700$ m

The curvatures of the two build sections are:

$\kappa_2 = 8°/30$ m; $\kappa_4 = 10°/30$ m

Solution

Using the Gauss plane coordinate system, the wellhead and the target real Y coordinates are:

$$Y_o = -97{,}352.6 \text{ m}; \; Y_t = -97{,}633.269 \text{ m}; \; Y_e = -98{,}125.507 \text{ m}$$

Therefore,

$$\Delta N_5 = 4{,}557{,}406.716 - 4{,}557{,}493.511 = -86.795 \text{ m}$$
$$\Delta E_5 = (-98{,}125.507) - (-97{,}633.269) = -492.238 \text{ m}$$
$$\Delta Z_5 = 2013 - 2000 = 13 \text{ m}$$
$$\Delta L_5 = \sqrt{(-86.795)^2 + (-492.238)^2 + 13^2} = 500 \text{ m}$$
$$\alpha_5 = \cos^{-1}\left(\frac{13}{500}\right) = 88.510°$$

Additionally,

$$\Delta N_{ot} = 4{,}557{,}493.511 - 4{,}557{,}543 = -49.489 \text{ m}$$
$$\Delta E_{ot} = (-97{,}633.269) - (-97{,}352.6) = -280.669 \text{ m}$$

and,

$$\frac{\Delta E_{ot}}{\Delta N_{ot}} = \frac{-280.669}{-49.489} = \frac{-492.238}{-86.795} = \frac{\Delta E_5}{\Delta N_5}$$

Therefore, if the wellhead and the two targets lie in the same azimuth line, it becomes a two-dimensional design. Also, $\phi_d = 260°$.

This well path design has the following calculation steps:

$$\alpha_1 = 0°$$

$$R_2 = \frac{180 \times 30}{\pi \times 8} = 214.859 \text{ m}$$

$$R_4 = \frac{180 \times 30}{\pi \times 10} = 171.887 \text{ m}$$

$$A_t = \sqrt{(-49.489)^2 + (-280.669)^2} = 285 \text{ m}$$

Appropriate substitution yields:

$$H_o = 2000 - 1700 \times \cos 0° + 214.859 \times \sin 0° - 171.887 \times \sin 88.510°$$
$$= 128.171 \text{ m}$$

$$A_o = 285 - 1700 \times \sin 0° - 214.859 \times \cos 0° + 171.887 \times \cos 88.510°$$
$$= 74.610 \text{ m}$$

$$R_o = 214.859 - 171.887 = 42.972 \text{ m}$$

Hence,

$$\alpha_3 = 2 \times \tan^{-1}\left(\frac{128.171 - \sqrt{128.171^2 + 74.610^2 - 42.972^2}}{42.972 - 74.610}\right) = 47.047°$$

$$\Delta L_3 = \sqrt{128.171^2 + 74.610^2 - 42.972^2} = 141.942 \text{ m}$$

For a typical horizontal well path, the horizontal section of which is the hold section, the well path can also be designed by using the target parameters:

$$\Delta N_{oe} = 4{,}557{,}406.716 - 4{,}557{,}543 = -136.284 \text{ m}$$

$$\Delta E_{oe} = (-98{,}125.507) - (-97{,}352.6) = -772.907 \text{ m}$$

$$A_e = \sqrt{(-136.284)^2 + (-772.907)^2} = 784.830 \text{ m}$$

In this case, the tangent length and tangent angle of the tangent section are to be calculated using the center parameters:

$$H_o = 2013 - 1700 \times \cos 0° - 500 \times \cos 88.510° + 214.859 \times \sin 0°$$
$$- 171.887 \times \sin 88.510° = 128.170 \text{ m}$$

$$A_o = 784.830 - 1700 \times \sin 0° - 500 \times \sin 88.510° - 214.859 \times \cos 0°$$
$$+ 171.887 \times \cos 88.510° = 74.610 \text{ m}$$

$$R_o = 214.859 - 171.887 = 42.972 \text{ m}$$

$$\alpha_3 = 2 \times \tan^{-1}\left(\frac{128.170 - \sqrt{128.170^2 + 74.610^2 - 42.972^2}}{42.972 - 74.610}\right) = 47.047°$$

$$\Delta L_3 = \sqrt{128.170^2 + 74.610^2 - 42.972^2} = 141.942 \text{ m}$$

Thus, the results of the design using these two methods are the same. Design results for this horizontal well path and the course parameters are shown in Table 4.5.

The condensed solution in API units is as follows:

$H_a = \Delta L_1 = 5577.42$ ft

The first target coordinates are:

$X_t = 14{,}952{,}407$ ft; $Y_t = 70{,}217{,}739$ ft

The final target coordinates are:

$X_e = 14{,}952{,}122$ ft; $Y_e = 70{,}216{,}124$ ft

Under the Gauss plane coordinate system, the wellhead and the target real Y coordinates are:

$Y_o = -319{,}398.3$ ft; $Y_t = -320{,}319.1$ ft; $Y_e = -321{,}934.1$ ft

Therefore:

$\Delta N_5 = 14{,}952{,}407 - 14{,}952{,}122 = -284.76$ ft
$\Delta E_5 = (-321{,}934.1) - (-320{,}319.1) = -1615$ ft
$\Delta Z_5 = 42.65$ ft

$$\Delta L_5 = \sqrt{(-284.76)^2 + (-1615)^2 + 42.65^2} = 1640.42 \text{ ft}$$

The solution for the well path data in API units is schematically shown in Figure 4.6.

Slight-Build Transition Well Path

During the process of drilling a horizontal well, the drilling tool at a hold section could have a high stiffness, which, however, may not be safe for a succeeding build section with a relatively higher build rate. Hence, a lower-stiffness drilling tool needs to be used for the hold section. To have a compromise the well profile is modified, such that the hold section of the original double-build design is also made into an intermediate build section and the resulting design is called slight-build transition well path,[4,11] as shown in Figure 4.7.

The main advantages of the slight- or low-build transition well path include: reducing the number of trips of the drill string by using the same set of bottomhole assembly tools to complete the first build section, as well as to continue the original hold section; permitting a less-stiff bottomhole assembly and thereby increased annular clearance in the build section; allowing

Table 4.5 Design Results for the Horizontal Double-Build Well Path for Problem 4.4

Depth (m)	*Inclination (°)*	*Azimuth (°)*	*Displacement (m)*	*TVD (m)*	*North (m)*	*East (m)*	*Curvature (°/30 m)*	*Location*
0.00	0.00	260.00	0.00	0.00	0.00	0.00	0.00	Wellhead
1700.00	0.00	260.00	1700.00	0.00	0.00	0.00	0.00	Kick-off point
1876.43	47.05	260.00	1857.26	−11.89	−67.42	68.45	8.00	
2018.37	47.05	260.00	1953.98	−29.93	−169.73	172.34	0.00	
2142.76	88.51	260.00	2000.00	−49.49	−280.67	285.00	10.00	First target
2642.76	88.51	260.00	2013.00	−136.28	−772.91	784.83	0.00	Final

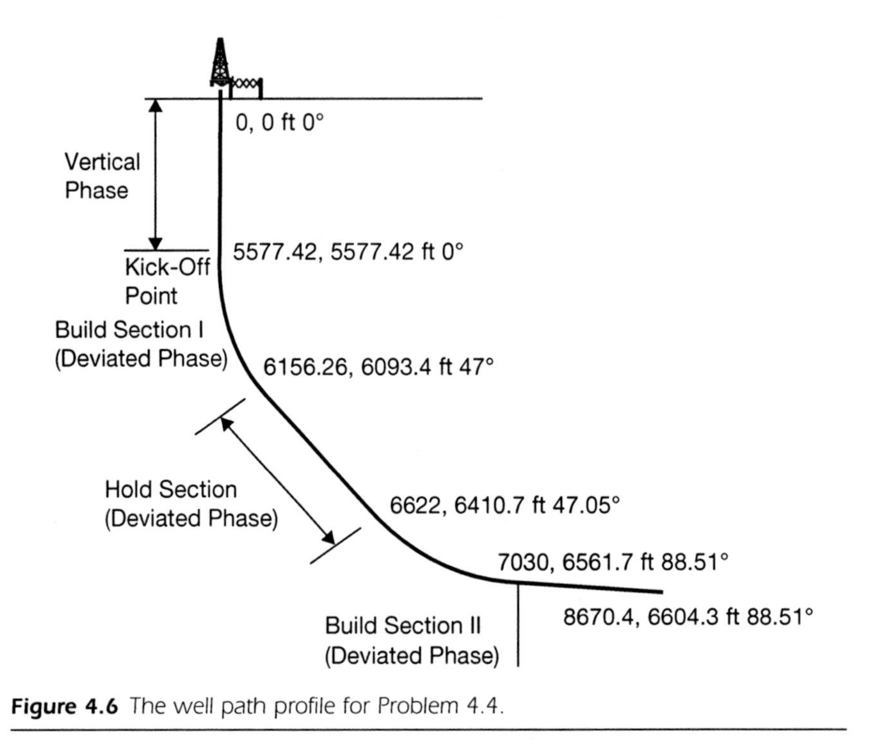

Figure 4.6 The well path profile for Problem 4.4.

adjustment for any deficient build rate angle in the previous section; and helping regain control of the well path trajectory.

The slight- or low-build type of the well profile is normally based on a three-arc design, and the parameters to be discovered are the deviation angles α_b and α_c. These build angles are important for achieving a smooth landing of the well into the target reservoir. Based on the geometrical parameters as shown in Figure 4.7, the vertical distance and target horizontal departure should satisfy the following conditions:

$$\begin{cases} \Delta L_1 \cos\alpha_1 + R_2(\sin\alpha_b - \sin\alpha_1) + R_3(\sin\alpha_c - \sin\alpha_b) + R_4(\sin\alpha_5 - \sin\alpha_c) = H_t \\ \Delta L_1 \sin\alpha_1 + R_2(\cos\alpha_1 - \cos\alpha_b) + R_3(\cos\alpha_b - \cos\alpha_c) + R_4(\cos\alpha_c - \cos\alpha_5) = A_t \end{cases} \tag{4.39}$$

If

$$\begin{cases} H_o = H_t - \Delta L_1 \cos\alpha_1 + R_2 \sin\alpha_1 - R_4 \sin\alpha_5 \\ A_o = A_t - \Delta L_1 \sin\alpha_1 - R_2 \cos\alpha_1 + R_4 \cos\alpha_5 \\ R = R_2 - R_3 \\ r = R_3 - R_4 \end{cases} \tag{4.40}$$

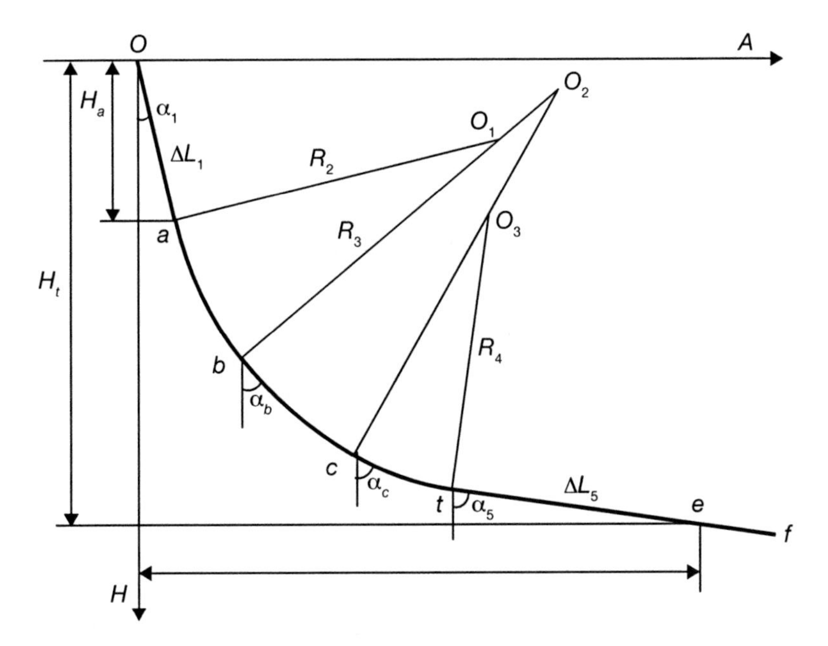

Figure 4.7 Horizontal well path design using slight-build transition profile.

then

$$\begin{cases} R\sin\alpha_b + r\sin\alpha_c = H_o \\ -R\cos\alpha_b - r\cos\alpha_c = A_o \end{cases} \tag{4.41}$$

Solving Eq. 4.41 yields:

$$\tan\frac{\alpha_b}{2} = \begin{cases} \dfrac{A+\sqrt{A^2+B^2-C^2}}{C-B}, \text{if } B \neq C \\ \dfrac{C+B}{2A}, \qquad \text{if } B = C \end{cases} \tag{4.42}$$

$$\tan\frac{\alpha_c}{2} = \begin{cases} \dfrac{a-\sqrt{a^2+b^2-c^2}}{c-b}, \text{if } b \neq c \\ \dfrac{c+b}{2a}, \qquad \text{if } b = c \end{cases} \tag{4.43}$$

where the expressions for A, B, C and a, b, c are,

$$\begin{cases} A = 2RH_o \\ B = 2RA_o \\ C = H_o^{\,2} + A_o^{\,2} + R^2 - r^2 \end{cases}$$

$$\begin{cases} a = 2rH_o \\ b = 2rA_o \\ c = H_o^2 + A_o^2 + r^2 - R^2 \end{cases}$$

For a horizontal well with the hole section beyond the target as a straight inclined/vertical-build-build-build-hold section, Eqs. 4.42 and 4.43 still apply, but Eq. 4.40 becomes:

$$\begin{cases} H_o = H_e - \Delta L_1 \cos\alpha_1 - \Delta L_5 \cos\alpha_5 + R_2 \sin\alpha_1 - R_4 \sin\alpha_5 \\ A_o = A_e - \Delta L_1 \sin\alpha_1 - \Delta L_5 \sin\alpha_5 - R_2 \cos\alpha_1 + R_4 \cos\alpha_5 \\ R = R_2 - R_3 \\ r = R_3 - R_4 \end{cases} \tag{4.44}$$

PROBLEM 4.5

If the horizontal well given in Problem 4.4 uses the slight-build rate design with the following data, design the well path κ_2 = 7°/30 m; κ_3 = 3°/30 m; κ_4 = 10°/30 m with other conditions unchanged.

Solution

The respective radii of the curvature can be calculated from the curvatures given:

$$R_2 = \frac{180\times 30}{\pi\times 7} = 245.533 \text{ m}$$

$$R_3 = \frac{180\times 30}{\pi\times 3} = 572.958 \text{ m}$$

$$R_4 = \frac{180\times 30}{\pi\times 10} = 171.887 \text{ m}$$

Angles α_b and α_c are calculated as:

$$H_o = 2000 - 1700\times\cos 0° + 245.553\times\sin 0° - 171.887\times\sin 88.510°$$
$$= 128.171 \text{ m}$$

$$A_o = 285 - 1700\times\sin 0° - 245.553\times\cos 0° + 171.887\times\cos 88.510°$$
$$= 43.915 \text{ m}$$

R = 245.553 – 572.958 = –327.405 m

r = 572.958 – 171.887 = 401.071 m

Calculating the values of A, B, and C to determine the angle α_b:

$$A = 2\times(-327.405)\times 128.171 = -83{,}926.998$$

$$B = 2\times(-327.405)\times 43.915 = -28{,}755.981$$

$$C = 128.171^2 + 43.915^2 + (-327.405)^2 - 401.071^2 = -35{,}307.837$$

$$\alpha_b = 2\times\tan^{-1}\frac{(-83{,}926.998)+\sqrt{(-83{,}926.998)^2+(-28{,}755.981)^2-(-35{,}307.837)^2}}{(-35{,}307.837)-(-28{,}755.981)}$$

$$= 42.364°$$

Similarly, calculating the values of a, b, and c to determine the angle α_c:

$$a = 2 \times 401.071 \times 128.171 = 102{,}810.540$$

$$b = 2 \times 401.071 \times 43.915 = 35{,}226.066$$

$$c = 128.171^2 + 43.915^2 + 401.071^2 - (-327.405)^2 = 72{,}019.989$$

$$\alpha_c = 2\times\tan^{-1}\frac{102{,}810.540-\sqrt{102{,}810.540^2+35{,}226.066^2-72{,}019.989^2}}{72{,}019.989-35{,}226.066}$$

$$= 60.418°$$

The horizontal well path design and node data are shown in Table 4.6.

Case of Uncertainty in the Position of the Reservoir Top

If the top position of the reservoir layer is not certain, then the current build section can be designed as a build-hold-build section, and the hold section can be used as the landing section on the top of the reservoir. This method is very useful, especially in the case of a thin reservoir layer with the top position uncertain. In this case, the two-build well path profile will change to a vertical-build-hold-build-hold-build-horizontal well path, and the slight- or low-build transition profile will become a vertical-build-build-build-hold-build-horizontal well path. The key to this method of solving the uncertainty in the position of the reservoir top is to design reasonable values for the inclination angle of the hold section, inclination angle of the horizontal section, and build rate of the tool.

Extended-Reach Wells

Extended-reach drilling (ERD) technology has rapidly developed over the past two decades. Extended-reach wells are usually characterized using ERD ratio or step-out ratio, which is defined as the ratio of the total horizontal departure to the total vertical depth. The present and future ERD envelop is shown in Figure 4.8.[12] If the ERD ratio is greater than two, then the well is classified

Table 4.6 Design Results for Horizontal Wells with Slight- or Low-Build Transition Profile for Problem 4.5

Depth (m)	*Inclination (°)*	*Azimuth (°)*	*Azimuth (°)*	*TVD (m)*	*North (m)*	*East (m)*	*Curvature (°/30 m)*	*Location*
0.00	0.00	260.00	0.00	0.00	0.00	0.00	0.00	Wellhead
1700.00	0.00	260.00	1700.00	0.00	0.00	0.00	0.00	Kick-off point
1881.56	42.36	260.00	1865.47	–11.13	–63.15	64.12	7.00	
2062.10	60.42	260.00	1977.65	–35.53	–201.51	204.61	3.00	
2146.37	88.51	260.00	2000.00	–49.49	–280.67	285.00	10.00	First target
2646.37	88.51	260.00	2013.00	–136.28	–772.91	784.83	0.00	Final

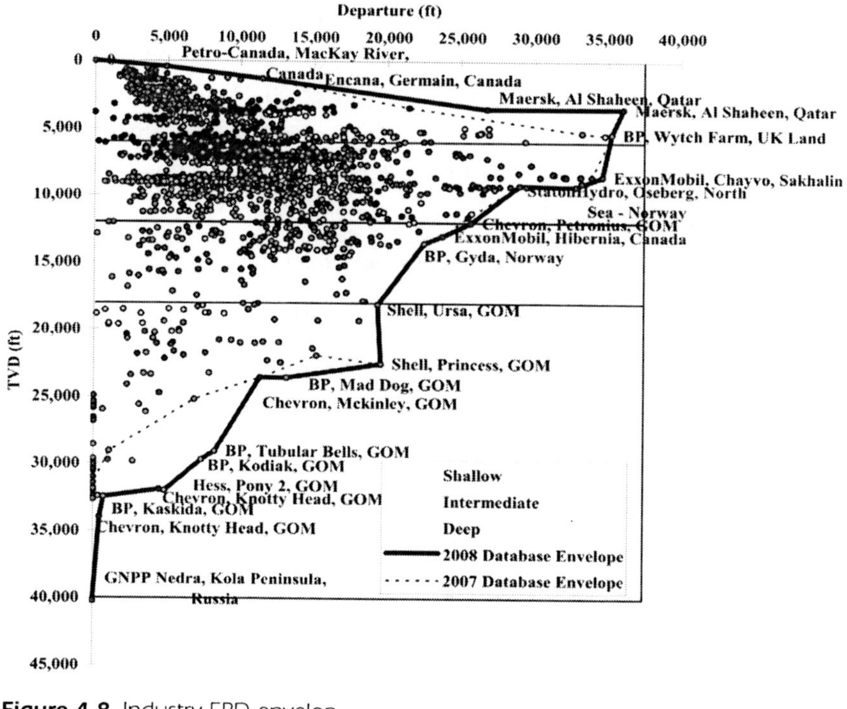

Figure 4.8 Industry ERD envelop.

as an extended-reach well. However, the ERD ratio does not necessarily characterize the difficulty of drilling and completing such wells. Extending the reach to a greater depth requires both improved models and comprehensive analysis. Wellbore friction is an important issue for ultra-long wells, and optimizing the design of the well path is an effective means to reduce torque and drag. In order to reduce torque and drag and effectively run the casing to the desired depth, modified optimal well path designs are required for both conventional or buoyancy-assisted design methods. For the conventional design, circular arcs and straight sections discussed in previous chapters can be used and will not be repeated here. The primary focus here will be on designs with catenary, parabola, or clothoid curves. Although it is not possible to have the whole well path with catenary, parabolic, or clothoid profiles, embedding these shapes in the design as transition sections adds value to the design for extended-reach wells.

Advantages are:

- The designed well profile will result in no contact and hence in low friction between the drill string and wellbore wall in the catenary section of the well path for the specified drilling operations. This results in reduced torque and drag in the drill string and the downhole tools, and further mitigates drill string and casing wear.

- These transition curves allow gradual increase of weight-on-bit (WOB) in the build section, which helps to improve the penetration rate.
- Results in reduced fatigue in the drill string.
- Reduces the risk of key seating and consequent pipe sticking.
- Allows to achieve better cementing and zonal isolation.

Typical catenary, parabolic, or clothoid profiles are embedded in the four-section well path designs and these profiles are generally identified as follows, as shown in Figure 4.9:

- Straight vertical/inclined-build-catenary-hold
- Straight vertical/inclined-build-parabola-hold
- Straight vertical/inclined-build-clothoid-hold

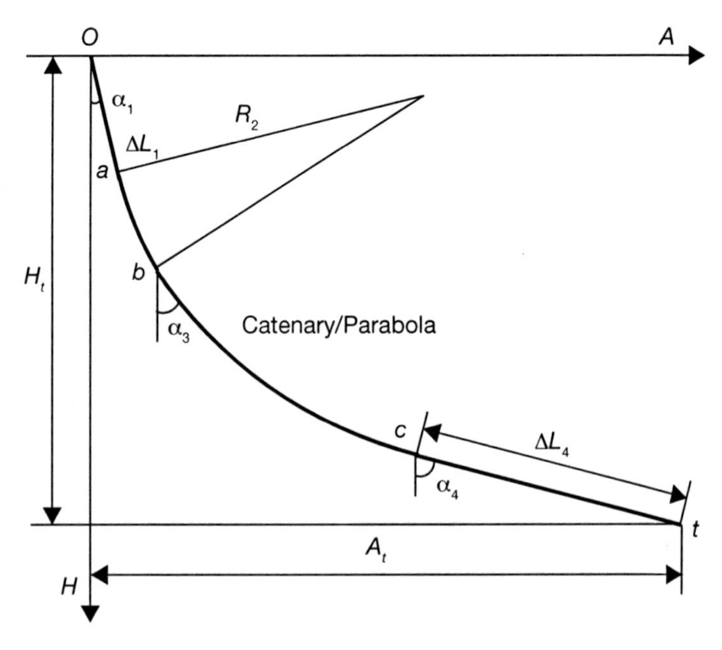

Figure 4.9 Catenary/parabolic transition section.

In what follows, the third section is taken to be either a catenary, a parabola curve, or a clothoid curve, while the case of a clothoid third section is taken up for discussion later. It should be noted that the start build angle of the catenary/parabolic shape is $\alpha_b = \alpha_3$.

When designing a well path, the true vertical depth of the target H_t and total horizontal displacement A_t are known, based on the given target zone. The characteristic parameters include: ΔL_1, α_1, R_2, α_b, a or P, ΔL_4, α_4. Section *oa* is the straight vertical/inclined section, and the end is the kick-off point (KOP) if the section is straight vertical; section *ab* is an arc transition section for which the radius of curvature R_2 is known; section *ct* is the hold section close to the target, of which the inclination α_4 can be determined based on the drilling method and values of the friction factor, fluid pressure, WOB, and other factors, so as to ensure achieving the WOB effectively. Design of the catenary well path or parabolic well path requires determination of any two of the three parameters, α_b, a or P, ΔL_4.

The catenary parameter a and the parabolic parameter P, respectively, determine the shape of the catenary or parabola; section length ΔL_4; and b, the starting depth of the section. The design should calculate these parameters in such a way that they are placed at the appropriate depth in the overall well path design.

Catenary Profile

Referring to Eqs. 3.126 and 3.127 in Chapter 3, the incremental coordinates of the depth and departure of a catenary profile third section are given in terms of dimensional parameters as follows:

$$\begin{cases} \Delta H_3 = a(Y_a - Y_b) \\ \Delta A_3 = a(X_a - X_b) \end{cases} \tag{4.45}$$

Using Eq. 4.45 with the catenary build section, the vertical depth and horizontal departure can be written as:

$$\begin{cases} \Delta L_1 \cos\alpha_1 + R_2(\sin\alpha_b - \sin\alpha_1) + a(Y_a - Y_b) + \Delta L_4 \cos\alpha_4 = H_t \\ \Delta L_1 \sin\alpha_1 + R_2(\cos\alpha_1 - \cos\alpha_b) + a(X_a - X_b) + \Delta L_4 \sin\alpha_4 = A_t \end{cases} \tag{4.46}$$

Replacing the dimensionless variables X and Y using Eqs. 3.123 and 3.124, Eq. 4.46 becomes:

$$\begin{cases} \Delta L_1 \cos\alpha_1 + R_2(\sin\alpha_b - \sin\alpha_1) + a\left(\dfrac{1}{\sin\alpha_b} - \dfrac{1}{\sin\alpha_4}\right) + \Delta L_4 \cos\alpha_4 = H_t \\ \Delta L_1 \sin\alpha_1 + R_2(\cos\alpha_1 - \cos\alpha_b) + a\ln\left(\dfrac{\tan\dfrac{\alpha_4}{2}}{\tan\dfrac{\alpha_b}{2}}\right) + \Delta L_4 \sin\alpha_4 = A_t \end{cases} \tag{4.47}$$

From Eq. 4.47, the relation involving α_b and the catenary parameter a can be obtained as[13,14]:

$$\frac{1}{\sin\alpha_b} - \frac{1}{\sin\alpha_4} - \frac{H_o - R_2\sin\alpha_b}{A_o + R_2\cos\alpha_b}\ln\left(\frac{\tan\dfrac{\alpha_4}{2}}{\tan\dfrac{\alpha_b}{2}}\right) = 0 \tag{4.48}$$

$$a = \frac{H_o - R_2\sin\alpha_b}{\dfrac{1}{\sin\alpha_b} - \dfrac{1}{\sin\alpha_4}} \tag{4.49}$$

where

$$\begin{cases} H_o = H_t - \Delta L_1\cos\alpha_1 - \Delta L_4\cos\alpha_4 + R_2\sin\alpha_1 \\ A_o = A_t - \Delta L_1\sin\alpha_1 - \Delta L_4\sin\alpha_4 - R_2\cos\alpha_1 \end{cases}$$

Using Eq. 4.48, α_b can be obtained by an iterative trial-and-error method. Since α_1 and α_4 are known, and α_b is situated in between these angles, it is easy to determine the initial starting value for the iteration. Then, using Eq. 4.49, the catenary parameter a can be calculated. It is noted from Eqs. 4.48 and 4.49 that the catenary line satisfies the conditions $\alpha_b \neq 0$, $\alpha_4 \neq 0$, and $\alpha_b \neq \alpha_4$.

If α and ΔL_4 are the parameters to be calculated, then using Eq. 4.47[13]:

$$a = \frac{H_o\sin\alpha_4 - A_o\cos\alpha_4}{b\sin\alpha_4 - c\cos\alpha_4} \tag{4.50}$$

$$\Delta L_4 = \frac{bA_o - cH_o}{b\sin\alpha_4 - c\cos\alpha_4} \tag{4.51}$$

where H_o, A_o, b, and c are related by:

$$\begin{cases} H_o = H_t - \Delta L_1\cos\alpha_1 - R_2(\sin\alpha_b - \sin\alpha_1) \\ A_o = A_t - \Delta L_1\sin\alpha_1 - R_2(\cos\alpha_1 - \cos\alpha_b) \end{cases}$$

$$b = \frac{1}{\sin\alpha_b} - \frac{1}{\sin\alpha_4}$$

$$c = \ln\left(\frac{\tan\frac{\alpha_4}{2}}{\tan\frac{\alpha_b}{2}}\right)$$

Finally, the catenary line of the wells is calculated as the length:

$$\Delta L_3 = a\left(\frac{1}{\tan\alpha_b} - \frac{1}{\tan\alpha_4}\right) \tag{4.52}$$

PROBLEM 4.6

It is desired to drill a directional well using straight vertical/inclined-build-catenary-hold sections to reach a TVD of 2800 m and a horizontal departure of 6000 m. The planned kick-off depth is at 300 m, with a starting angle of 0°. The curvature of the build section is 10°/30 m.

Solution

Given data are:

$\alpha_1 = 0$; $\kappa_2 = 10°/30$ m; $H_t = 2800$ m; $A_t = 600$ m; $\Delta L_1 = 300$ m

If the length of the hold section $\Delta L_4 = 3000$ m is assumed to be given, then a and α_b are the unknowns. The starting angle α_b of the catenary section is calculated to be 45.17°. The catenary characteristic parameter a and the section length ΔL_3 are estimated to be 4525.99 m and 3536.69 m, respectively. Details of the well path sections are given in Table 4.7.

If the inclination angle α_b at the end of the build section (or at the start of the catenary section) is assumed to be given, then a and ΔL_4 are the unknowns. Calculation yields $\Delta L_4 = 3364.55$ m and the catenary line characteristic

Table 4.7 Design Results for the Catenary Section for Problem 4.6 (α and α_b are the unknowns)

Point	*Well Depth (m)*	*Inclination (°)*	*TVD (m)*	*Horizontal Displacement (m)*	*Well Path Curvature (°/30 m)*
O	0.00	0.00	0.00	0.00	0.00
a	300.00	0.00	300.00	0.00	0.00
b	435.52	45.17	421.91	50.71	10.00
c	3972.21	78.00	2176.26	3065.56	0.36
t	6972.21	78.00	2800.00	6000.00	0.00

Table 4.8 Design Results for the Catenary Section for Problem 4.6 (α and ΔL_4 are the unknowns)

Point	Well Depth (m)	Inclination (°)	TVD (m)	Horizontal Displacement (m)	Well Path Curvature (°/30 m)
O	0.00	0.00	0.00	0.00	0.00
a	300.00	0.00	300.00	0.00	0.00
b	426.00	42.00	415.02	44.15	10.00
c	3631.93	78.00	2100.47	2708.97	0.46
t	6996.48	78.00	2800.00	6000.00	0.00

Table 4.9 Additional Design Results for the Catenary Section for Problem 4.6

Well Depth (m)	Inclination (°)	TVD (m)	Horizontal Displacement (m)	Well Path Curvature (°/30 m)
426.00	42.00	415.02	44.15	0.22
500.00	42.54	469.78	93.92	0.22
900.00	45.64	757.18	372.06	0.25
1300.00	49.11	1028.17	666.19	0.28
1700.00	52.99	1279.75	977.08	0.31
2100.00	57.31	1508.47	1305.12	0.34
2500.00	62.09	1710.44	1650.25	0.38
2900.00	67.34	1881.47	2011.68	0.41
3600.00	77.51	2093.70	2677.77	0.46
3631.93	78.00	2100.47	2708.97	0.46

parameter a = 3569.85 m. The section length ΔL_3 = 3205.93 m. The well path node data are given in Table 4.8.

As a further example of the calculation for a catenary section, the design results for the catenary section in the second case, using 100 m as the incremental depth, are listed in Table 4.9.

Parabolic Profile

For the case of insertion of a parabolic profile section as the transition section in a well path, the section vertical depth and the section horizontal departure can be respectively obtained using Eqs. 3.137 and 3.138, as follows:

$$\begin{cases} \Delta H_3 = \dfrac{P}{2}\left(\dfrac{1}{\tan^2 \alpha_b} - \dfrac{1}{\tan^2 \alpha_4}\right) \\ \Delta A_3 = P\left(\dfrac{1}{\tan \alpha_b} - \dfrac{1}{\tan \alpha_4}\right) \end{cases} \tag{4.53}$$

Using Eq. 4.53 in the general well path design equation results in:

$$\begin{cases} \Delta L_1 \cos\alpha_1 + R_2(\sin\alpha_b - \sin\alpha_1) + \dfrac{P}{2}\left(\dfrac{1}{\tan^2 \alpha_b} - \dfrac{1}{\tan^2 \alpha_4}\right) + \Delta L_4 \cos\alpha_4 = H_t \\ \Delta L_1 \sin\alpha_1 + R_2(\cos\alpha_1 - \cos\alpha_b) + P\left(\dfrac{1}{\tan \alpha_b} - \dfrac{1}{\tan \alpha_4}\right) + \Delta L_4 \sin\alpha_4 = A_t \end{cases} \tag{4.54}$$

If P and α_b are the parameters to be found, then

$$2H_o\left(\frac{1}{\tan\alpha_b} - \frac{1}{\tan\alpha_4}\right) - A_o\left(\frac{1}{\tan^2\alpha_b} - \frac{1}{\tan^2\alpha_4}\right) = 0 \tag{4.55}$$

$$P = \frac{A_o}{\dfrac{1}{\tan\alpha_b} - \dfrac{1}{\tan\alpha_4}} \tag{4.56}$$

where

$$H_o = H_t - \Delta L_1 \cos\alpha_1 - \Delta L_4 \cos\alpha_4 - R_2(\sin\alpha_b - \sin\alpha_1)$$
$$A_o = A_t - \Delta L_1 \sin\alpha_1 - \Delta L_4 \sin\alpha_4 - R_2(\cos\alpha_1 - \cos\alpha_b)$$

First of all, using Eq. 4.55, α_b can be obtained by an iterative method. Since α_1 and α_4 are known, and α_b is situated in between these angles, it is easy to determine the initial starting value for the iteration. Then, using Eq. 4.56, the parabola parameter P can be calculated.

On the other hand, if P and ΔL_4 are the parameters to be found, then Eq. 4.54 can be written in a form more conducive to solution, as shown in Eqs. 4.57 and 4.58[15,16]:

$$P = \frac{H_o \sin\alpha_4 - A_o \cos\alpha_4}{b \sin\alpha_4 - c \cos\alpha_4} \tag{4.57}$$

$$\Delta L_4 = \frac{bA_o - cH_o}{b \sin\alpha_4 - c \cos\alpha_4} \tag{4.58}$$

Appropriate substitutions are made for H_o, A_o, b, and c using the following relations:

$$H_o = H_t - \Delta L_1 \cos\alpha_1 - R_2(\sin\alpha_b - \sin\alpha_1)$$

$$A_o = A_t - \Delta L_1 \sin\alpha_1 - R_2(\cos\alpha_1 - \cos\alpha_b)$$

$$b = \frac{1}{2}\left(\frac{1}{\tan^2\alpha_b} - \frac{1}{\tan^2\alpha_4}\right)$$

$$c = \frac{1}{\tan\alpha_b} - \frac{1}{\tan\alpha_4}$$

Finally, the length of the parabolic section can be obtained as (more details can be seen in Chapter 3):

$$\Delta L_3 = \frac{P}{2}\left[f(\alpha_b) - f(\alpha_4)\right] \tag{4.59}$$

Where the function $f(\alpha)$ used in Eq. 4.59 is as follows:

$$f(\alpha) = \frac{1}{\sin\alpha\tan\alpha} - \ln\left(\tan\frac{\alpha}{2}\right)$$

PROBLEM 4.7

Plan an extended-reach well with a parabolic transition build section using the same data as given in Problem 4.6.

Solution

If the length of the hold section $\Delta L_4 = 3000$ m is assumed to be given, then P and α_b are the unknowns. After calculation, the starting angle of the parabolic section $\alpha_b = 46.45°$, the parabola characteristic parameter (the distance from the vertex to the focus) $P = 4081.51$ m, and the section length $\Delta L_3 = 3528.71$ m. The details of each section of the well path are given in Table 4.10.

If the inclination angle α_b at the end of the build section (or at the start of the parabolic section) is assumed to be given, then P and ΔL_4 are the

Table 4.10 Design Results for the Parabolic Section for Problem 4.7 (P and α_b are the unknowns)

Point	*Well Depth (m)*	*Inclination (°)*	*TVD (m)*	*Horizontal Displacement (m)*	*Well Path Curvature (°/30 m)*
O	0.00	0.00	0.00	0.00	0.00
a	300.00	0.00	300.00	0.00	0.00
b	439.36	46.45	424.58	53.46	10.00
c	3968.07	78.00	2176.26	3065.56	0.39
t	6968.07	78.00	2800.00	6000.00	0.00

Table 4.11 Design Results for the Parabolic Section for Problem 4.7 (P and ΔL_4 are the unknowns)

Point	Well Depth (m)	Inclination (°)	TVD (m)	Horizontal Displacement (m)	Well Path Curvature (°/30 m)
O	0.00	0.00	0.00	0.00	0.00
a	300.00	0.00	300.00	0.00	0.00
b	426.00	42.00	415.02	44.15	10.00
c	3463.19	78.00	2063.76	2536.27	0.58
t	7004.31	78.00	2800.00	6000.00	0.00

Table 4.12 Additional Design Results for the Parabolic Section for Problem 4.7

Well Depth (m)	Inclination (°)	TVD (m)	Horizontal Displacement (m)	Well Path Curvature (°/30 m)
426.00	42.00	415.02	44.15	0.19
500.00	42.46	469.81	93.89	0.19
900.00	45.20	758.49	370.72	0.22
1300.00	48.40	1032.48	662.06	0.26
1700.00	52.15	1288.37	969.41	0.30
2100.00	56.57	1521.75	1294.15	0.36
2500.00	61.79	1727.06	1637.28	0.42
2900.00	67.89	1897.52	1998.93	0.49
3000.00	69.56	1933.81	2092.11	0.51
3400.00	76.79	2049.97	2474.60	0.57
3463.19	78.00	2063.76	2536.27	0.58

unknowns. Calculation yields $\Delta L_4 = 3541.12$ m and the parabolic characteristic parameter $P = 2725.01$ m. The section length $\Delta L_3 = 3037.19$ m. The well path node data are given in Table 4.11.

As a further example of a calculation for a parabolic section the design results of the parabolic section in the second case, using 100 m as the incremental depth, are listed in Table 4.12.

Clothoid Profile

Over the years, mathematicians and physicists have studied the properties of various curves. Of these, the clothoid profile (also called Cornu's spiral,

Fresnel spiral, Euler spiral, or Linarc) is of great relevance to the topic of this book due to the nature and special properties of the curve. Euler has described several properties of the curve, including its quadrature. Insertion of clothoid sections between the tangent and build or drop section of a well path results in curvature continuity. This curvature bridge[17] alleviates drag problems, thereby enabling well design engineers to extend the reach of wells within the given mechanical limitations. Accuracy of the clothoid spiral matches the cubic spiral to more than eight digits.

A clothoid curve is defined as a spiral curve the curvature of which changes linearly from zero to a desired curvature proportional to the arc length. In other words, the radius of curvature at any point of the curve varies as the inverse of the arc length from the starting point of the curve.

$$R \propto \frac{1}{L} \text{ or } L_1 \times R_1 = L_2 \times R_2 = \ldots = L_n \times R_n = \sigma \tag{4.60}$$

$$\kappa(s) = \kappa(0) + \sigma s \tag{4.61}$$

where

R = Radius of curvature
L = Length of the curve
κ = Curvature
σ = Sharpness of the curve
s = Arc length of the curve

The clothoid curves can be parametrically given as below:

$$\begin{aligned} f(\ell) &= (C_f(\ell), S_f(\ell)) \\ C_f(\ell) &= \xi \int_0^{\ell} \cos\left(\frac{\pi u^2}{2}\right) du \\ S_f(\ell) &= \xi \int_0^{\ell} \sin\left(\frac{\pi u^2}{2}\right) du \end{aligned} \tag{4.62}$$

where ξ is the characteristic parameter.

The following are identified as Fresnel sine and cosine integrals:

$$\begin{aligned} \text{Fresnel } C_f(\ell) &= \int_0^{\ell} \cos\left(\frac{\pi u^2}{2}\right) du \\ \text{Fresnel } C_f(\ell) &== \int_0^{\ell} \sin\left(\frac{\pi u^2}{2}\right) du \end{aligned} \tag{4.63}$$

Since it is not possible to obtain closed-form solutions to the above equations, several approximate numerical computations are presented in the literature using Taylor series, Power series, and Maclaurin expansions. A simple expression using Maclaurin expansion and the coordinates can be expressed in terms of the length of the spiral arc as below[18]:

$$y = \frac{L^3}{6\sigma^2} - \frac{L^7}{336\sigma^6} + \frac{L^{11}}{42,240\sigma^{10}} + \ldots$$
$$x = L - \frac{L^5}{40\sigma^4} + \frac{L^9}{3456\sigma^{10}} + \ldots \tag{4.64}$$

Omitting the higher-order terms, it can be written as follows,[19] which is a cubic parabola:

$$y = \left(6\sigma^2 x\right)^{\frac{1}{3}} \tag{4.65}$$

Based on the properties of the clothoid curve, the relationship between the curvature and the scale parameter is given as:

$$L_1 \times R_1 = L_2 \times R_2 = \ldots = L_n \times R_n = \sigma^2 \tag{4.66}$$

Also it can be noted that the tangents at the connection point between the clothoid spiral and the straight segment are the same. It has been found that using a Cornu's spiral path reduces the lateral stresses on the tubulars that pass through the section.

Figure 4.10 illustrates a well path with a clothoid spiral in a curved section. It can be seen that the well path consists of the following sections:

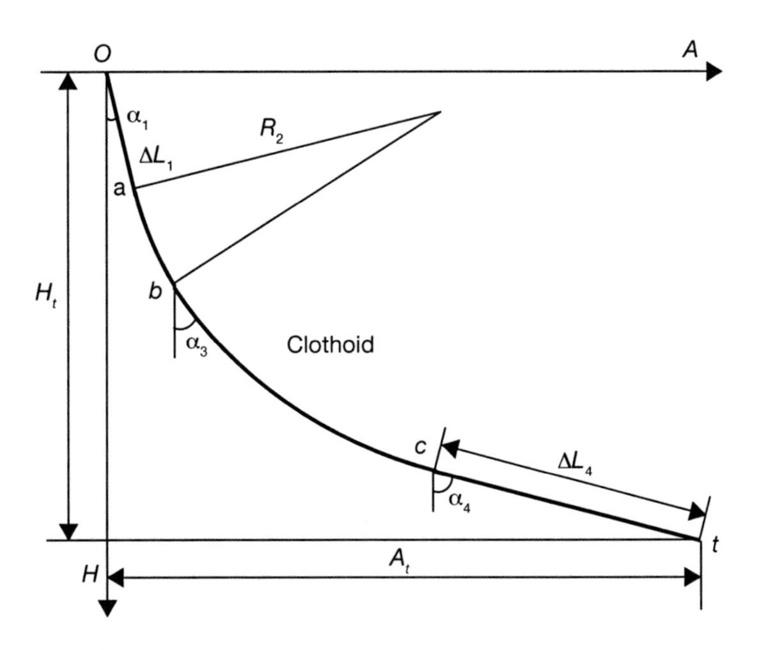

Figure 4.10 Clothoid transition section.

- Circular arc with maximum curvature κ_{max}
- Fresnel spiral arc from the circular arc
- Hold horizontal section

It should be noted that the start build angle of the clothoid shape is $\alpha_b = \alpha_3$.

Curvature

Curvature is given as below:

$$\kappa(s) = \frac{\pi a}{\sigma^2 + b^2} u \qquad (4.67)$$

It can be inferred from Eq. 4.67 that it is a continuous function with no steps.

Supplementary Problems

1. Construct a plan for a straight vertical-build-hold–type well having the following well data:
   Build rate = 3°/100 ft
   Kick-off depth = 2000 ft and angle of 0°
   True vertical depth to target = 9500 ft
   Target coordinates = 5500 ft N30°E of rig
2. Assume that the maximum angle of inclination of a straight vertical-build-hold–type well is to be between 40° and 45°. The maximum build rate angle is 3.5°/100 ft. The kick-off depth is to be between 1500 ft and 2500 ft.
   Rig coordinates = 2500'N75°W
   Target coordinates = 4500'N45°E
   Target true vertical depth = 11,000 ft
3. The following data are given for a straight-build-hold–type well:
   Target true vertical depth = 9600 ft
   Total horizontal displacement = 2000 ft
   Build rate = 2.5°/100 ft
   Maximum angle at total depth = 45°
   Determine the kick-off depth required.
4. A certain field has a thin reservoir with low vertical permeability, and it was decided to drill an arch-shaped horizontal well with three arches (i.e., one hump and two inverted arches). Develop the equations for designing the well path.

5. A horizontal well at an azimuth angle of 30° is to be drilled to exploit two different thin reservoir layers at different depths. According to the dip angles and directions of the reservoir layer, the inclination angles of the two well sections at these two layers are $\alpha_1 = 88°$ and $\alpha_4 = 90°$, respectively. The required target's vertical displacement is $\Delta H = 30$ ft, and the horizontal displacement is $\Delta A = 1260$ ft. Using a ladder well path design, calculate the length of the second reservoir and the inclination angle at the inflection point, if the build rates of the two arc sections are $\kappa_2 = -3°/100$ ft and $\kappa_3 = 4.5°/100$ ft, and the well length in the first reservoir layer is $\Delta L_1 = 600$ ft.
6. A horizontal well at an azimuth angle of 230° is to be drilled to exploit two different thin reservoir layers at different depths.

   The angles of the reservoir layers are $\alpha_1 = 88°$ and $\alpha_4 = 90°$. The target coordinates are: vertical displacement is 30 ft; horizontal displacement is 500 ft.

   Using a ladder well path design, calculate the inclination angle at the inflection point and the build/drop rates of the two arc sections ($\kappa_2 = -\kappa_3 = 4.5°/100$ ft).

   The reservoir layer lengths are 600 ft and 400 ft. Is it possible to construct the well with the obtained build rates? If not, suggest suitable alternative parameters to develop these two reservoirs with a ladder well path design.
7. For a ladder design, develop a few equations to calculate the inclination angle at the inflection point and the length of the reservoirs with the condition that $\alpha_1 = \alpha_4 = 90°$ and $\kappa_2 \neq \kappa_3$.
8. Calculate the first reservoir length and inclination angle at the inflection point using a staircase design with the following data:
   Inclinations of the reservoirs are 90° and 88°, respectively
   First and second arc curvatures are –4.5 and 3°/100 ft
   Length of the second reservoir is 500 ft
   The target depth data are: vertical depth is 100 ft; horizontal departure is 1500 ft.
9. Determine the well path of a straight vertical-build-hold-drop–type well with the following given data:
   Horizontal displacement = 6000 ft
   Target TVD = 12,000 ft
   TVD at KOP depth = 1500 ft

Build rate = 20°/100 ft
Drop rate = 1.50°/100 ft

10. Estimate the kick-off depth for a straight build-hold-build–type well with the following given data:
    Target TVD = 12,000 ft
    Total horizontal displacement = 4500 ft
    Build rate = 3.5°/100 ft
    Maximum angle at target = 25°
    Starting angle at surface = 5°
    Determine the kick-off depth required.
11. Design the well path with the following data:
    Wellhead coordinates X_o = 639,543 ft, Y_o = 2,402,647.4 ft (Gaussian projection plane coordinates with the north direction for the X coordinate and the east direction for the Y coordinate).
    The first target coordinates are: X_t = 6,377,493.511 ft,
    Y_t = 13,602,366.731 ft
    The final target coordinates are: X_e = 639,406.716 ft,
    Y_e = 13,601,874.493 ft
    The respective vertical depths are: H_t = 10,000 ft, H_e = 10,113 ft
    The profile selected is of the straight vertical/inclined-build-tangent-build-hold type. Kick-off point data are: $H_a = \Delta L_1$ = 1700 m.
    Curvatures of the two build sections are κ_2 = 8°/30 m, κ_4 = 10°/30 m.
12. Estimate the kick-off depth for a continuous-build–type well with the given data:
    Target true vertical depth = 13,000 ft
    Total horizontal displacement = 5500 ft
    Build rate = 2.5°/100 ft
    Maximum angle at target = 20°
    Starting angle at surface = 3°
    Determine the kick-off depth required.
13. It would be instructive to study various transition curves used in the design of well path profiles. Explore the possibility of using other transition curves that can be embedded in the straight-build-new-curve-hold–type well profiles. Develop related well path design equations.
14. Derive Eqs. 4.55 and 4.56.
15. Due to the uncertainty of the tools and build rates in the double-build horizontal profiles, prove that the maximum tangent angle can be given as:

$$\alpha_{tang} = \sin^{-1}\left(\sin\alpha_f - \frac{\kappa_{max} \times \kappa_{min}\left(H_t - H_{tang}\right)}{C_k\left(\kappa_{max} - \kappa_{min}\right)}\right)$$

where

α_{tang} = Tangent angle

α_f = Final well inclination

κ_{max}, κ_{min} = Maximum and minimum build rates

C_k = Conversion factor, 5790 for (°)/100 ft and 1719 for (°)/30 m

H_t = Maximum allowable target vertical depth

H_{tang} = Optimum vertical target depth

16. Using the following data and the relationships derived in the text, calculate the angle α_3.
    H_f = 8000 ft; final departure at depth point *f* is A_f = 2700 ft
    $H_b = \Delta L_1$ = 1500 ft
    Curvatures: κ_2 = 2°/100 ft; κ_2 = 1.6477°/100 ft
    Inclination angles are:
    α_1 = 0°; α_5 = 15°; α_5 = 25°
    ΔH_5 = 900 ft, ΔH_6 = 100 ft
    Also calculate the total measured depth of the well and draw the wellbore profile.

17. A catenary profile is embedded in the profile shown in Figure 4.11. Calculate the measured depth, inclination, true vertical depth, and dogleg in increments of 30 m with the following given data:
    H_d = 2800 m; D_d = 6000 m; ΔL_1 = 300 m where point *d* is the target point
    κ_2 = 10°/30 m; α_0 = 0°; α_c = 78°
    a. Find out whether a well path profile can be constructed with the above data. If not, justify with reasons.
    b. Calculate and draw the inclination, dogleg versus measured depth with ΔL_4 = 3745 m; H_d = 2800 m; D_d = 6000 m; ΔL_1 = 300 m, where point *d* is the target point. κ_2 = 10°/30 m; α_o = 0°.

18. Verify Eq. 4.32, which is reproduced below:

$$R_3 = -R_2 = \frac{A^2 + B^2}{2\left[A\left(\sin\alpha_1 + \sin\alpha_4\right) - B\left(\cos\alpha_1 + \cos\alpha_4\right)\right]}$$

19. Design a ladder well path with α_1 = 87°; α_4 = 92°; length ΔL_1 = 250 m and ΔL_2 = 250 m. The total vertical height = 15 m. Use Figure 4.12.

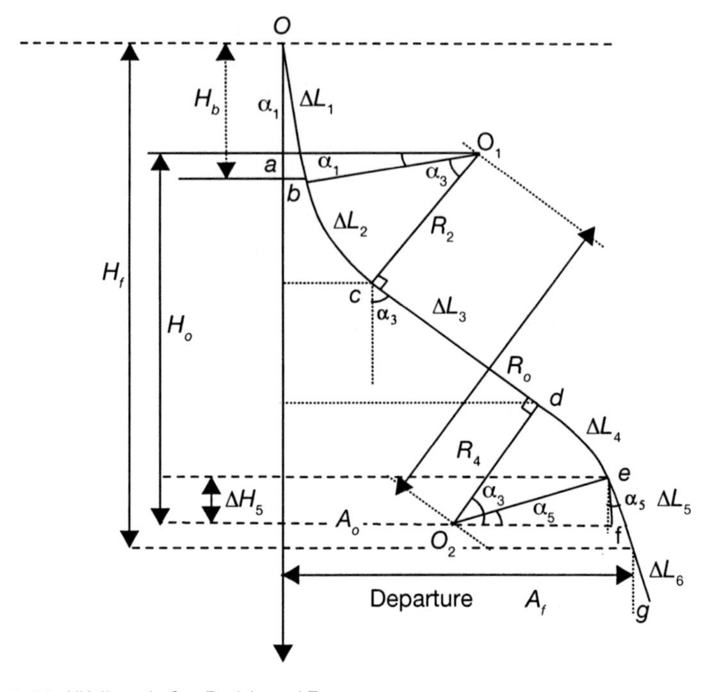

Figure 4.11 Well path for Problem 17.

20. A reversed double-build profile, as shown in Figure 4.12, is to drill horizontal wells in which the horizontal departure of the target is small compared to the vertical depth. In Figure 4.12, the distances *a*, *b*, and *d* are auxiliary elements used to derive the equations and are not design parameters. It is required that the hole reach the target using the above profile with a tangent section 23. Derive the equation to calculate inclination angles α_1 and α_2 in terms of H_o, b and R_o only, where $R_o = R_1 + R_2$.
21. Using the profile in Problem 20, determine the inclination angles α_1 and α_2. The KOP is at 1600 ft, the TVD of the target is 12,050 ft and the horizontal departure is A_t = 600 ft. The expected first build rate is 2°/100 ft and the second build rate is 1.5°/100 ft. The horizontal distance A_2 is 200 ft.
22. The well-plan engineer later on decides to incorporate a downhill slope in the final hold section, as shown in the Figure 4.13, to assist with gravity drainage during production, while maintaining the rest of the profile and data. TVD at the end of the hold section H_f is 20 ft. Calculate the angle α_2.

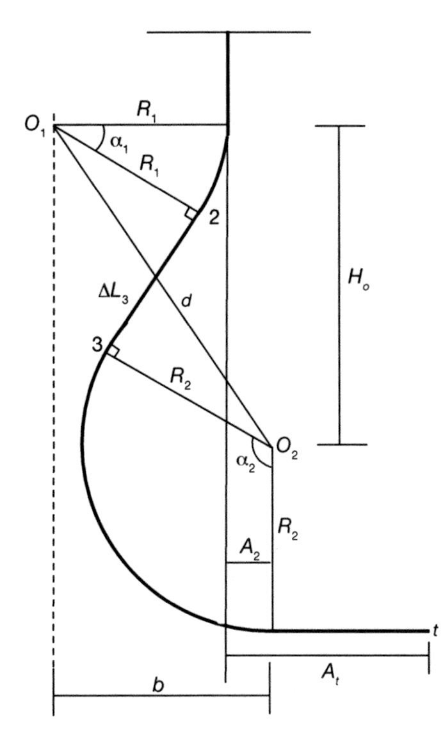

Figure 4.12 Well path for Problem 20.

23. Using data from Problem 20, the relationship between the inclination angle α_1 and the length of the initial tangent length can be given as:

$$\tan\frac{\alpha_1}{2} = \frac{H_o - \sqrt{\Delta L_3}}{R_o + b}$$

24. Show that a necessary sufficient condition that a curve be a helix is that:

$$\left[r'', r''', r^{iv}\right] = -k^5 \frac{d}{ds}\left(\frac{\tau}{\kappa}\right) = 0$$

where τ = torsion and κ = curvature.

25. Design a directional well with a target vertical depth H_t of 2500 m, with a horizontal departure at the target depth of 900 m, and with a displacement angle of 300°. Use a straight vertical-build-tangent-drop-hold profile, with the vertical depth to the kick-off point of 500 m.

The first build rate is 6°/100 m, and the drop rate is 4°/100 m. At the entrance point of the hold section in the oil layer, the inclination angle is 12°, the TVD increment is 300 m, and the extra TVD "rat hole" is

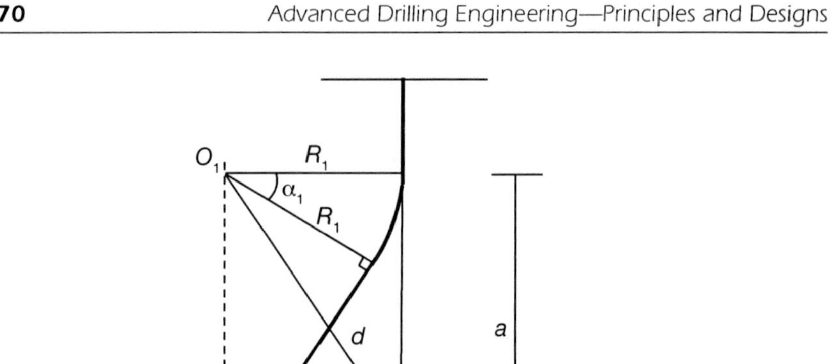

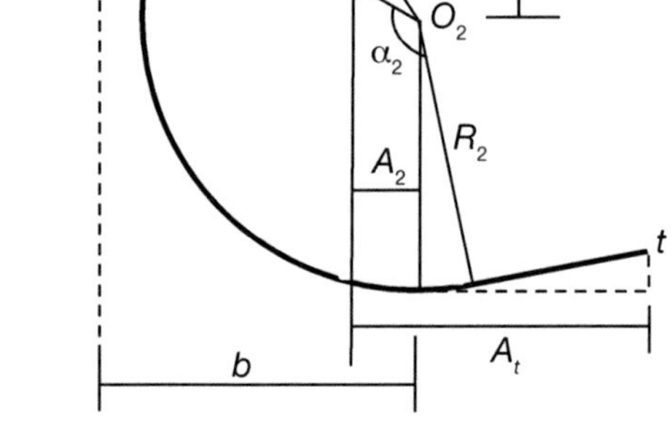

Figure 4.13 Well path for Problem 22.

50 m. Calculate in both API and SI units. Also, calculate the total measured depth of the well and draw the wellbore profile.

References

1. Bourgoyne, A. T., M. E. Chenevert, K. K. Millheim, and F. S. Young. “Applied Drilling Engineering.” Society of Petroleum Engineers, Richardson, TX, 1986.
2. Samuel G. Robello. *Downhole Drilling Tools—Theory and Practice for Students and Engineers*. Houston, TX: Gulf Publishing, 2007.
3. Azar, J. J., and S. G. Robello. *Drilling Engineering*. Tulsa, OK: Penwell Publishers, 2007.
4. Xiushan, L. *Geometry of Wellbore Trajectory*. Beijing, China: Petroleum Industry Press, 2006.
5. Xiushan, L., W. Shan, and J. Zhongxuang. *Designing Theory and Describing Method for Wellbore Trajectory*. Harbin, China: Heilongjiang Science and Technology Press, 1993.
6. Inglis, T. A. *Directional Drilling*. London: Graham & Trotman, 1987.
7. Xiushan, L., and S. Zaihong. “Practical Method for Design of Horizontal Well Trajectory.” *Oil Drilling & Production Technology* 16, no. 1 (1994): 5–8, 23.

8. Xiushan, L. "Study on Arched Horizontal Well Design." *Natural Gas Industry* 26, no. 6 (2006): 64–65.
9. Xiushan, L. "Well Trajectory Design of Staircase Horizontal Well." *Petroleum Drilling Techniques* 33, no. 3 (2005): 1–5.
10. Karisson, H., R. Cobbley, and G. E. Jaques. "New Developments in Short-, Medium-, and Long-Radius Lateral Drilling." SPE/IADC Paper No. 18706, 1989.
11. Zhenguang, W., Z. Jun, and W. Wenzhong. "Design and Application of the Third Segment Deflection Trail of Middle-Radium Horizontal Well." *West-China Exploration Engineering* 6, no. 2 (1994): 12–15.
12. Data provided by Nicky Roberts.
13. Xiushan, L. "Catenary Profile Design Method Study." *Natural Gas Industry* 27, no. 7 (2007): 74–75.
14. Xiushan, L., and R. Samuel. "Applied Models and Methods for Planning a Catenary Well Profile." Submitted for JSPE, 2009.
15. Xiushan, L., Z. Daqian, and L. Shibin. "Designing Principles and Methods for Parabolic Profiles of Directional Wells." *Journal of Daqing Petroleum Institute* 13, no. 4 (1989): 29–37.
16. Xiushan, L. "Mathematical Model and Design Methods of Parabola Well Paths." *Oil Drilling & Production Technology* 28, no. 4 (2006): 7–9, 13.
17. Samuel, R. "Ultra-Extended-Reach Drilling (u-ERD: Tunnel in the Earth)—A New Wellpath Design." SPE-119459, SPE/IADC Drilling Conference, March 17–19, 2009, Amsterdam, The Netherlands.
18. Brandse, J., M. Mulder, and M. M. van Paassen. "Clothoid-Augmented Trajectories for Perspective Flight-Path Displays." *International Journal of Aviation Psychology* 17, no. 1 (January 2007): 1–29.
19. Heald, M. A. "Rational Approximations for the Fresnel Integrals." *Mathematics of Computation* 44, no. 170 (1985): 459–461.

5

Well Path Planning—3D

Accurate well path planning is important for drilling directional or horizontal wells successfully. Well path planning in a 2D plane can satisfy the requirements only when there are no target or surface restrictions, and the well path can be designed using a combination of simple arc and straight sections. However, increasingly complex reservoirs are being exploited that require complex well construction and completion methods, needing designs to be done in 3D. In addition, complex well designs are needed for bypassing obstructions, such as geological faults, salt domes, fish (irretrievable drilling equipment), existing wells, etc. Well twinning is another complex design method for horizontally drilling a well close to and parallel to an existing well. Drilling these types of complex well profiles, especially in the case of extended-reach and ultra-extended-reach wells, requires meticulous planning. This chapter covers 3D well path planning for conventional and complex wells, such as sidetrack wells, wells for bypassing obstructions, wells that need position drift, etc. In short, 3D well designs are required for[1]:

- Multilateral wells to access multiple zones.
- Multiple target wells from a single location when the target and wellhead are not in the same vertical plane.
- Relief wells to control blowouts when conventional methods of blowout control fail.
- Sidetracking to avoid irretrievable fish left in the wellbore.
- Bypassing obstructions or potential problem zones such as salt domes, faults, previously drilled wells, etc.

- Azimuth drift due to natural bit walk.

Also nowadays, due to the use of improved and advanced downhole tools, such as rotary steerable downhole motors, bent subs, etc., continuous monitoring and effective adjustment of the drilled well path using 3D methods are required.

Multilateral Wells

For reservoirs with several potential discontinuous zones, a multilateral profile may be an option when conventional wells do not serve the purpose. Multilateral wells are complex well profiles with a main wellbore or mother bore connected to several sidetracks, and also may have laterals branching off from the sidetracks. There are many different ways to construct and drill a multilateral well, depending on the reservoir characteristics, completion type, economics of well construction, the need to avoid collision with obstructions, other risks and uncertainties, etc. The design of the well path profile also depends on several other geometrical parameters. Some of the types of multilateral well profiles are (Figure 5.1)[2]:

- Horizontal fishbone
- Crow's-foot triple lateral
- Dual/triple lateral
- Stacked dual
- Dual opposing
- Pitchfork dual

The 3D designs discussed below can be classified as:

- Simple profile with azimuth or azimuth/inclination correction
- Complex profile for multitarget drilling
- Complex design path planning

Conventional Model

Azimuth or Azimuth/Inclination Correction

This method of design is the simplest involving 3D planning.[2-6] A typical situation occurs when a correction is required while drilling a simple 2D well path (for example, a build and hold profile). This correction usually involves changes

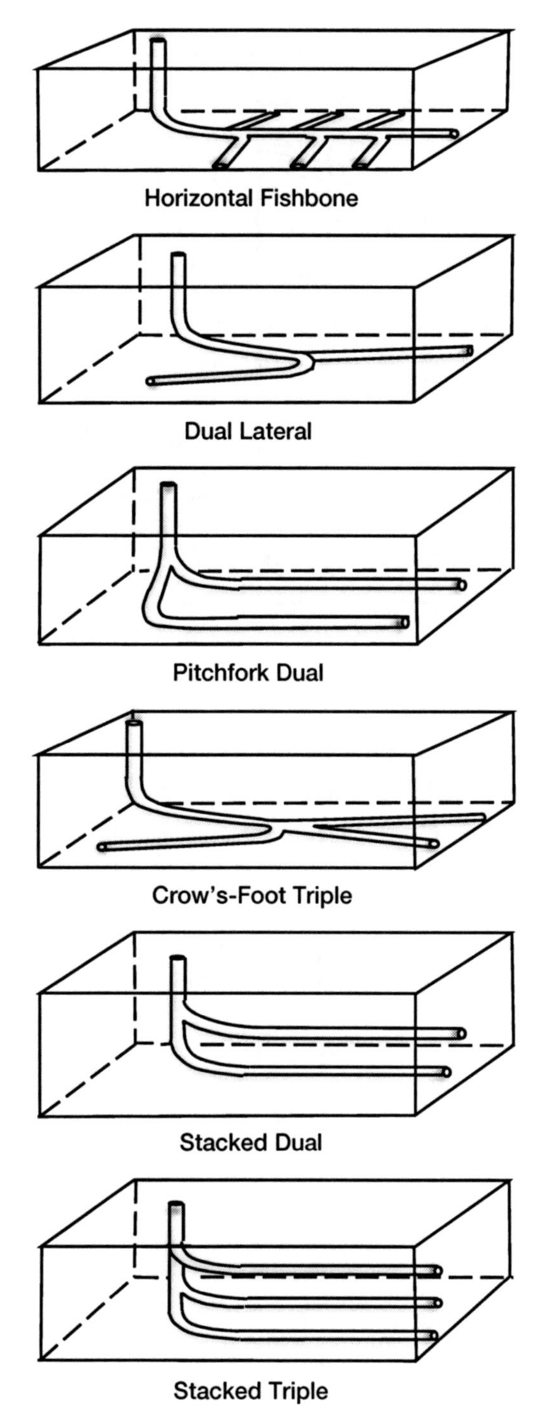

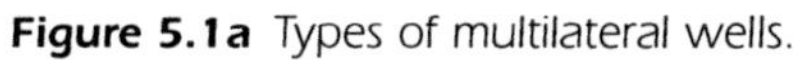

Figure 5.1a Types of multilateral wells.

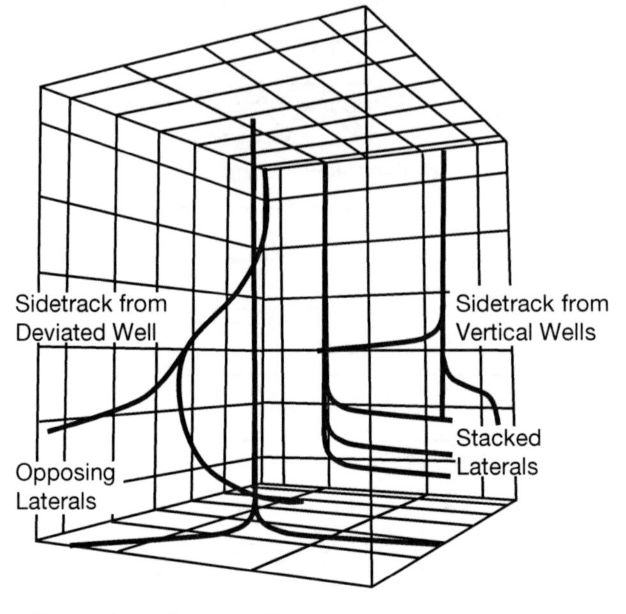

Figure 5.1b Types of multilateral wells.

in both the inclination and the azimuth. Such situations also occur in abandoned well reentry. In this case, an abandoned directional well is used to reach another target from some suitable point in the old well path, normally involving changes in inclination and azimuth. Several other well drilling situations also exist, necessitating such corrections.

Let s be the current position of a point in the hole with coordinates (N_s, E_s, H_s) and measured depth m_s. Let the inclination and azimuth at the point s be α_s and ϕ_s, respectively. It is required to reach the target t with coordinates (N_t, E_t, H_t). It is expected that the deviation tool can deliver a dogleg severity c, which corresponds to a radius of curvature:

$$R = \frac{1}{c}$$

The parameters to be calculated are[7]:

- Spatial angle of turn
- Length of the curved segment
- Toolface for the correction
- Toolface, inclination, and azimuth of any point in the curved segment (in terms of the measured depth of the point)

- Length, inclination, and azimuth of the straight segment
- Minimum dogleg severity (DLS) required to reach the target

The unit vector $c_s = (c_N, c_E, c_H)$ represents the direction of the borehole at s. According to the theory of differential geometry, the direction cosines of the vector c in the O-NEH coordinate frame are:

$$\begin{cases} c_N = \sin\alpha_s \cos\phi_s \\ c_E = \sin\alpha_s \sin\phi_s \\ c_H = \cos\alpha_s \end{cases} \tag{5.1}$$

As long as the target t is not on the line passing through s with direction $\boldsymbol{c}_s$, the points s and t and the vector $\boldsymbol{c}_s$ define a plane Ω that passes through s and t and contains $\boldsymbol{c}_s$. This is shown in Figure 5.2. The dashed lines in the diagram represent a portion of the plane Ω.

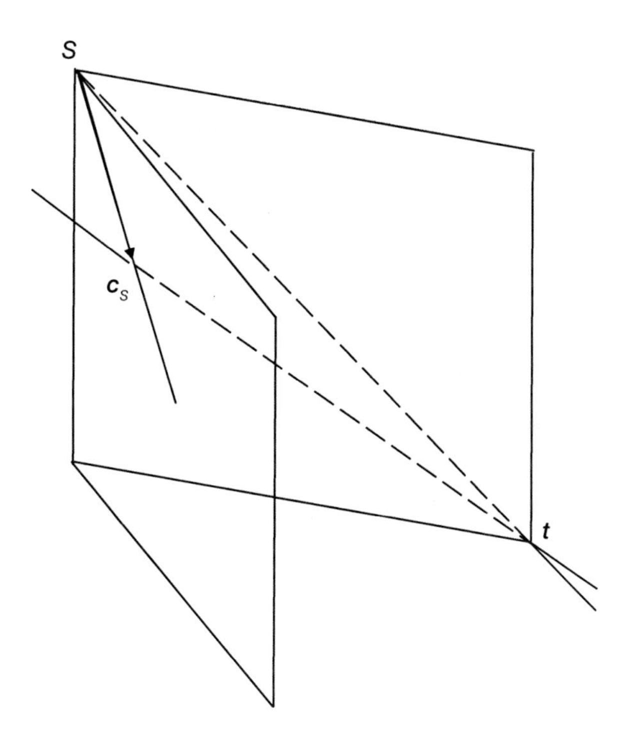

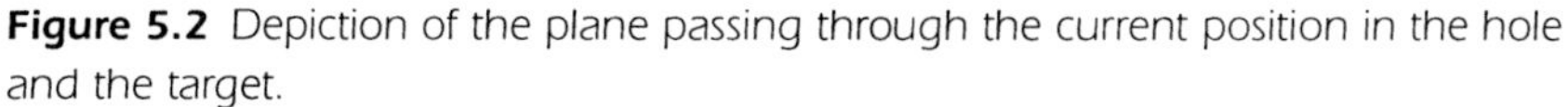

Figure 5.2 Depiction of the plane passing through the current position in the hole and the target.

The plane Ω is a generic plane in a 3D space on which the 3D well path will be planned. Figure 5.3(a) shows the Ω plane with the points s and t, and the unit vector $\boldsymbol{c}_s$.

Starting at s, a curved well path with radius R, corresponding to the dog-leg severity c, develops from s and tangent to $\boldsymbol{c}_s$ until it points to t. For this, the position of the center of curvature is denoted by C.

Consider the unit vector $\boldsymbol{t}$ with components (n_H, e_H, t_H) at s pointing to the target t. Since the coordinates of s and t are known, the unit vector $\boldsymbol{t}$ is given by the following relation (see Figure 5.3b):

$$\boldsymbol{t} = \frac{\boldsymbol{t}-\boldsymbol{s}}{\|\boldsymbol{t}-\boldsymbol{d}\|} = \frac{\boldsymbol{t}-\boldsymbol{s}}{d} \tag{5.2}$$

where d is the distance between points s and t. The actual components of the unit vector t are:

$$\begin{aligned} d &= \sqrt{(N_t - N_s)^2 + (E_t - E_s)^2 + (H_t - H_s)^2} \\ \boldsymbol{t} &= \frac{1}{d}(N_t - N_s, E_t - E_s, H_t - H_{se}) \end{aligned} \tag{5.3}$$

where

N_s, E_s, H_s = North, east, and vertical coordinates of the current point
N_t, E_t, H_t = North, east, and vertical coordinates of the target

The angle θ between the unit vectors $\boldsymbol{c}_s$ and $\boldsymbol{t}$, as shown in Figure 5.3(c), is given by:

$$\theta = \arccos(\boldsymbol{c} \bullet \boldsymbol{t}) \tag{5.4}$$

where

$$\boldsymbol{c}_s \bullet \boldsymbol{t} = c_N t_N + c_E t_E + c_H t_H$$

The position vector $\boldsymbol{C}$ of the center of curvature C is given by:

$$\boldsymbol{C} = s + \frac{R}{\sin\theta}\boldsymbol{t} - \frac{R}{\tan\theta}\boldsymbol{t}_s \tag{5.5}$$

The radial unit vector $\boldsymbol{r}_P$, as shown in Figure 5.4, points from the center of curvature C to point P and is given by:

$$\boldsymbol{r}_P = \frac{s-C}{R} = \frac{1}{\tan\alpha}\boldsymbol{c}_s - \frac{1}{\sin\alpha}\boldsymbol{t} \tag{5.6}$$

The distance e from the center of curvature C to the target t is calculated by $e = \|t - C\|$.

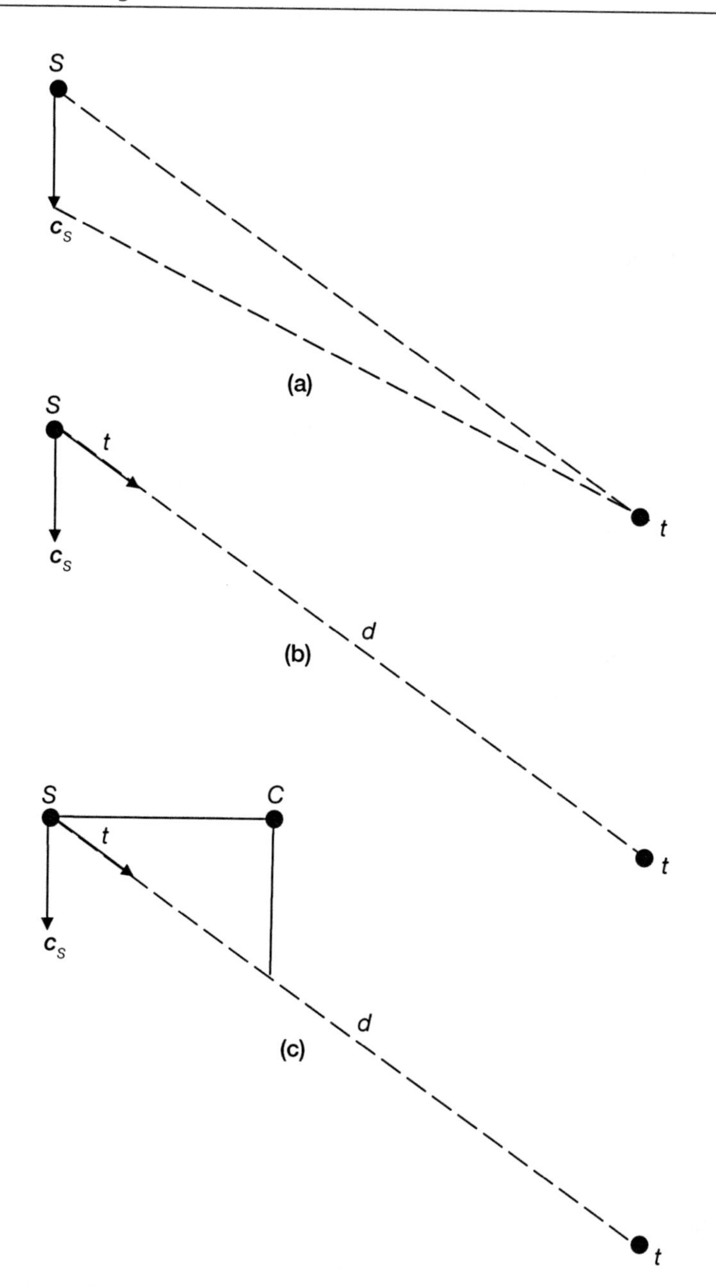

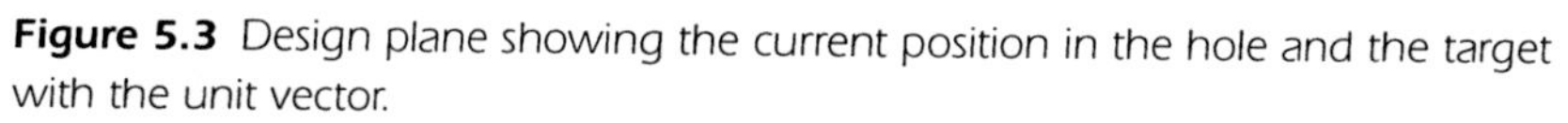
Figure 5.3 Design plane showing the current position in the hole and the target with the unit vector.

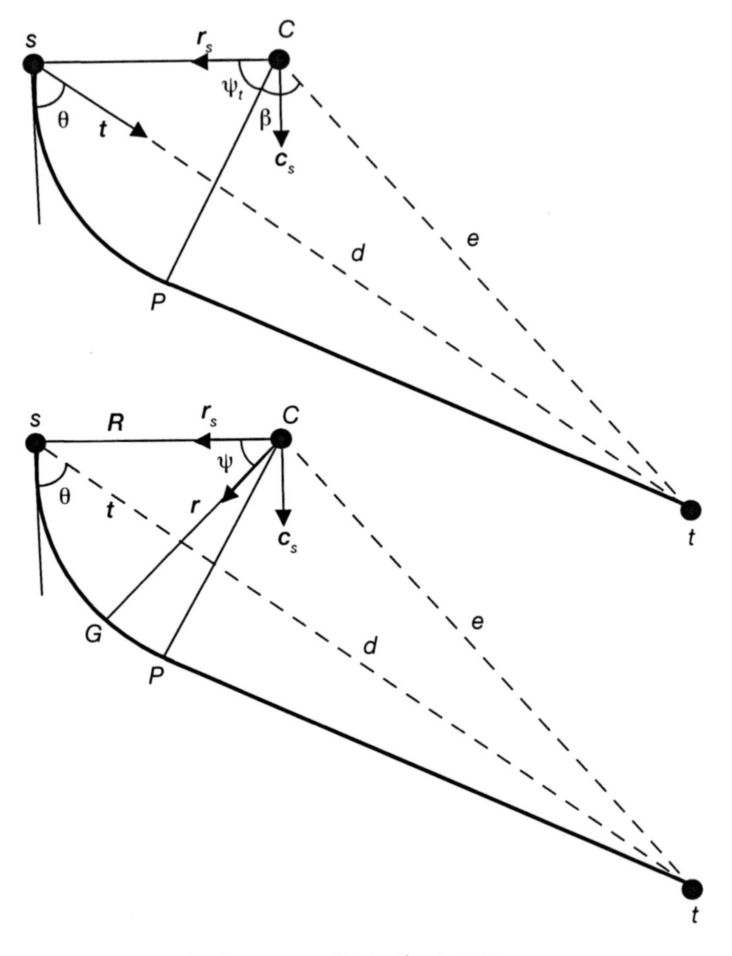

Figure 5.4 Arc of a circle drawn to point to the target.

With the center at C, an arc of a circle is drawn with radius R and angle ϕ_t, starting at s and ending at point P, such that the tangent at P passes through the target position t, as shown in Figure 5.4.

The angle Φ_T is the total angle of turn and can be calculated as follows. Considering the right triangle ΔPCt, the angle β is given by:

$$\beta = \arccos \frac{R}{e} \tag{5.7}$$

The angle β is always less than 90°. Considering the triangle ΔsCt and using the cosine law:

$$d^2 = e^2 + R^2 - 2eR\cos(\psi_t + \beta) \tag{5.8}$$

Solving for ψ_t and using the expression for β in Eq. 5.7 yields:

$$\psi_T = \arccos\frac{e^2 + R^2 - d^2}{2eR} - \arccos\frac{R}{e} \tag{5.9}$$

This expression is only valid for $\theta \leq 90°$. For $\theta > 90°$, the following expression should be used:

$$\psi_T = 360° - \arccos\frac{e^2 + R^2 - d^2}{2eR} - \arccos\frac{R}{e} \tag{5.10}$$

The coordinates of P are obtained from:

$$\boldsymbol{P} = \boldsymbol{C} + \left(R\cos\psi_t\right)\boldsymbol{r}_s + \left(R\sin\psi_t\right)\boldsymbol{t}_s$$

This expression can be rewritten as:

$$\boldsymbol{P} = \boldsymbol{C} + R\left(\cos\psi_t\boldsymbol{r}_s + \sin\psi_t\boldsymbol{c}_s\right) = \boldsymbol{C} + R\boldsymbol{r}_P \tag{5.11}$$

where $\boldsymbol{r}_P$ is the unit vector pointing from the center C to the end of the curved segment P. This expression can be generalized to obtain the coordinates of any point G in the curved segment between s and P, corresponding to a partial angle of turn ψ:

$$G = C + Rr(\psi) \tag{5.12}$$

where

$$\boldsymbol{r}(\psi) = \cos\psi\boldsymbol{r}_s + \sin\psi\xi_s$$

These elements are shown in Figure 5.4.

The measured depth at P (end of the curved section) and t (target) are given respectively by:

$$m_E = m_s + R\psi_t m_t = m_E + \sqrt{e^2 - R^2} \tag{5.13}$$

A generic point G in the curved segment is determined by its measured depth m_G. Given m_G, the partial angle of turn ψ is calculated using the expression:

$$\psi = \frac{m_G - m_s}{R} \tag{5.14}$$

To calculate the inclination and azimuth at the general point G the unit vector $\boldsymbol{r}$ pointing from C to G is considered. Since the radius of curvature R is constant, the tangent vector ξ at G is given by:

$$\omega = \frac{d\boldsymbol{r}}{d\psi} = -\sin\psi\boldsymbol{r}_P + \cos\psi\omega_P \tag{5.15}$$

If the components of ξ are (ξ_N, ξ_E, ξ_H), the inclination and azimuth at G are obtained from:

$$\alpha_G = \arccos \xi_H \quad \phi_G = \arctan \frac{\xi_E}{\xi_N} \tag{5.16}$$

Making $\psi = \psi_T$ the unit tangent vector, ξ_P is obtained at P and the inclination and azimuth of the slant section Pt. If S is a point in the slant segment with measured depth m_s, the coordinates of S can be obtained from the following expression:

$$S = E + (m_s - m_P)\xi_P \tag{5.17}$$

PROBLEM 5.1

The current coordinates of a location are:

Measured depth: 2045 m; vertical depth: 1600 m

Inclination: 47.75°; azimuth: 25.25°

North/south coordinate: 667 m; east/west coordinate: 230 m

The coordinates of the target to be reached are:

Vertical depth: 2600 m

North/south coordinate: 800 m; east/west coordinate: 600 m

The expected DLS is 3°/30 m. Plan the well path to reach the target and determine (a) the length of the curved section, (b) the measured depth at the target, and (c) the inclination and azimuth at the target.

Solution

The following are the steps in the calculation:

1. The radius of curvature of the arc is:

$$R = \frac{180}{\pi \times 3/30} = 572.958 \text{ m}$$

2. The components of the unit vector tangent to a point in the well ξ_s are:

$$\omega_P = (\cos 47.75, \sin 47.75 \times \cos 26.25, \sin 47.75 \times \sin 26.25)$$
$$\omega_P = (0.67236681, 0.66388146, 0.32739011)$$

3. The components of unit vector t are:

$$d = \sqrt{(2600-1600)^2 + (800-667)^2 + (600-230)^2}$$
$$= 1074.518mt = \frac{1}{1074.518}(2600-1,600, 800-667, 600-230)t$$
$$= (0.9306498181, 0.12377642, 0.34434043)$$

4. The angle θ is:

$\theta = \arccos(0.67236681 \times 0.93064981 + 0.66388146 \times 0.12377642$
$+ 0.32739011 \times 0.34434043)$

$\theta = 34.851°$

5. The coordinates of the center of curvature are:

$$C = (1,600,667,230) + \frac{572.958}{\sin 34.851}(0.93061981, 0.12377612, 0.34434043)$$
$$- \frac{572.958}{\tan 34.851}(0.67236681, 0.66388146, 0.32739011)$$

$C = (1979.883\text{ m}, 244.847\text{ m}, 305.871\text{ m})$

6. The radial unit vector r_s is:

$$r_s = \frac{1}{\tan 34.851}(0.67236681, 0.66388146, 0.32739011)$$
$$- \frac{1}{\sin 34.851}(0.93064981, 0.12377642, 0.34434043)$$

$r_s = (-0.66302031, 0.73679659, -0.13241925)$

7. The distance from the center to the target is:

$$e = \sqrt{(2600 - 1979.883)^2 + (800 - 244.847)^2 + (600 - 305.871)^2}$$
$$= 882.753\text{ m}$$

8. The total angle of turn ψ_t ($\theta = 25.726 < 90$) is:

$$\psi_t = \arccos\frac{882.753^2 + 572.958^2 - 1074.518^2}{2 \times 882.753 \times 572.958} - \arccos\frac{572.958}{882.753}$$
$$= 43.137°$$

9. The coordinates of the end of curved section E are:

$r_P = \cos 43.137(-0.66302031, 0.7367959, -0.13241925)$
$+ \sin 43.137(0.67236681, 0.66388146, 0.32739011)$

$r_P = (-0.02409483, 0.99158164, 0.12722140)$

$$P = (1979.883, 244.847, 305.871)$$
$$+ 572.958(-0.02409483, 0.99158164, 0.12722140)E$$
$$= (1966.077\text{ m}, 812.981\text{ m}, 378.763\text{ m})$$

10. The coordinates of the well unit vector at the end of the curved section are:

$$\xi_P = -\sin 43.137(-0.66302031, 0.73679659, -0.13241925) + \cos 43.137(0.6726681, 0.663881, 0.32739011)$$

$$\xi_P = (0.94397695, -0.01933018, 0.32944477)$$

11. The final calculations are:
    a. Length of the curved section:

$$\Delta m = R\psi_t = 572.958 \times 43.137 \times \left(\frac{\pi}{180}\right) = 431.37 \text{ m}$$

    b. Measured depth at the target:

$$m_t = m_s + \Delta m + \sqrt{e^2 + R} = 2045 + 431.37 + \sqrt{882.753^2 - 572.958^2} = 3147.913 \text{ m}$$

    c. Inclination and azimuth at the target:

$$\alpha_G = \arccos \xi_H = \arccos 0.94397695 = 19.269°$$

Toolface Angle Calculation

To determine the required toolface at any point[7] G of the curved section, consider the unit radial vector $\boldsymbol{r}$ and the unit tangent vector ξ. Consider also the unit vertical vector $v = (1,0,0)$ pointing downwards.

It should be realized that the unit vectors $\boldsymbol{r}$ and ξ are in the very same plane Ω of vectors $\boldsymbol{r}_P$ and ξ_s The vector ξ points to the direction of axial penetration, and vector $\boldsymbol{r}$ points in the direction opposite to the direction of lateral penetration. Therefore, it is to the direction of $-\boldsymbol{r}$ that the toolface of the deviation tool should be oriented.

Figure 5.5 depicts the cross section of the borehole at a generic point G of the curved well path.

The toolface angle ω can be measured as the angle between $-\boldsymbol{r}$ and the direction of the high-side point HS, or between $\boldsymbol{r}$ and the direction of the low-side point LS. Using the latter, the unit vector $\boldsymbol{v}^*$ pointing from G toward the low-side point LS must be determined first. For that, the unit vector ξ makes an angle θ with the vertical vector $\boldsymbol{v}$, and that $\boldsymbol{v}^*$ is perpendicular to ξ. Figure 5.5(b) shows the vector $\boldsymbol{v}^*$ as defined.

Since these three vectors are unit vectors, the following relationship is valid:

$$\boldsymbol{v} = \cos\theta\xi + \sin\theta\boldsymbol{v}^* \quad (5.18)$$

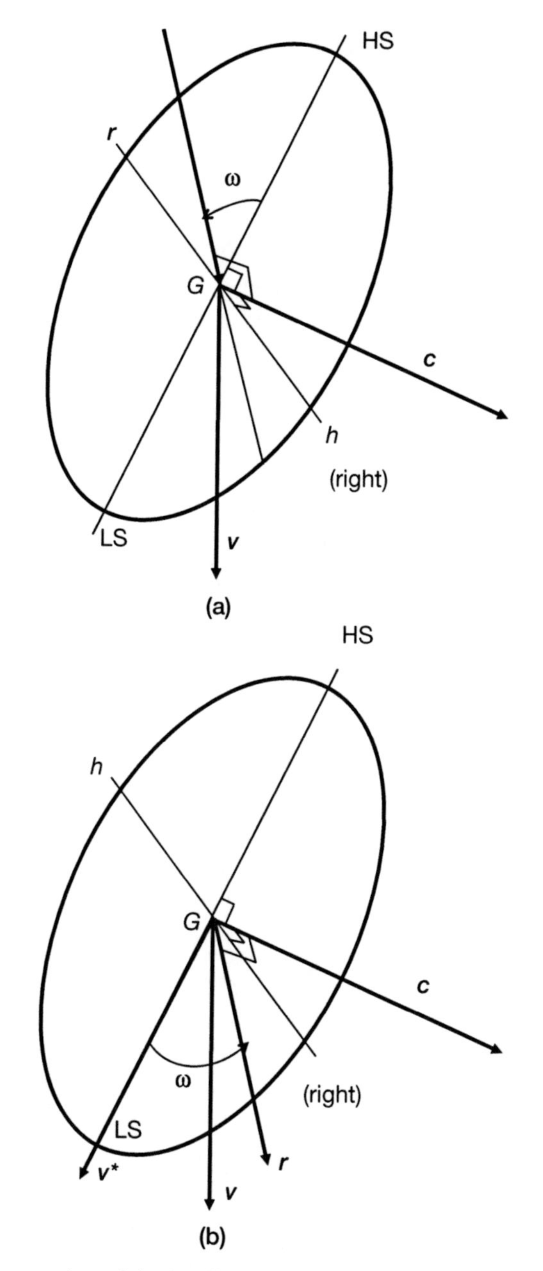

Figure 5.5 Cross section of the borehole with the unit vectors.

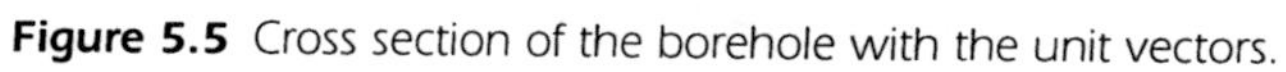

Solving for v^* the following is obtained:

$$v^* = \left(\sin\theta, -\cos\theta\cos\varphi, -\cos\theta\sin\varphi\right) \quad (5.19)$$

But the toolface angle ω is given by $\cos\gamma = r \times v^*$, so that:

$$\omega = \arccos(r \times v^*) \quad (5.20)$$

To determine whether the toolface is to the right or to the left of the high side, the cross product of r and ξ is calculated, which results in another vector f:

$$f = r \times \xi$$

Since f is perpendicular to ξ, it lies on the cross section of the borehole. Vector f is also perpendicular to r.

In Figure 5.6(a), the position of vector f is shown when the toolface is at the left of the high side. It should be noted that f is always in the upper side of the cross section. On the other hand, if the toolface is at the right of the high side, as shown in Figure 5.6(b), f is in the lower side of the cross section.

The discriminant between the two cases is, therefore, the vertical component of vector f. The component is positive when f is in the lower side of the cross section, and therefore at the right of the high side, and negative when f is in the upper side, and therefore at the right of the high side. The vertical component of f is given by:

$$f_v = r_N \xi_E - r_E \xi_N \quad (5.21)$$

If $f_H = 0$, it means that γ is either 0° or 180°. In both cases the direction is irrelevant.

PROBLEM 5.2

Calculate the position of the toolface at point P in Problem 5.1 at the start of the curved section.

Solution

The unit vector pointing to the lower side is:

$$v^* = \left(\sin 47.75, -\cos 47.75\cos 25.25, -\cos 47.75\sin 25.25\right)$$

$$v^* = \left(0.74021813, -0.60302746, -0.29738023\right)$$

The angle of the toolface is:

$$\begin{aligned}\omega &= \arccos\left(r_P \times v^*\right)\\ &= \arccos\left[\left(-0.66302, 0.73680, -0.13242\right)\times\left(-0.74022, -0.60303, -0.29738\right)\right]\\ &= 153.6°\end{aligned}$$

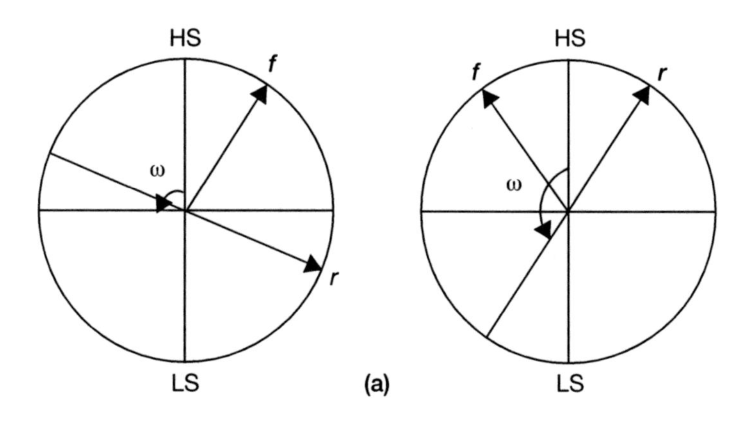

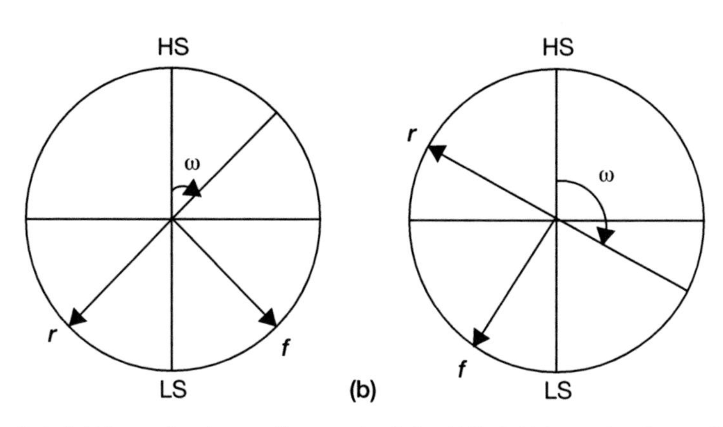

Figure 5.6 (a) Vector for the toolface to the left of HS. (b) Vector for the toolface to the right of HS.

To determine the direction, the vertical component of vector $\boldsymbol{r}$ may be calculated:

$$f_v = r_N \xi_E - r_E \xi_N = 0.73679659 \times 0.32739011 - (-0.13241925) \times 0.66388146 = 0.329 > 0$$

which indicates that the toolface must be positioned at 153.6° to the right of the high side.

Complex Models

Cubic-Type Wells[7]

There are basically two general ways to reach a subsurface target with a 3D well path[8]:

- Reaching the target with no restrictions on the inclination or the azimuth.
- Reaching the target with either a given inclination or a given azimuth, or both.

The first type (free-end) is commonly used for well path corrections and for wells originating from reentry wells. The second type (set-end) is required for some relief wells (when it is desired to reach the target with a well path parallel to the blowing well) or when one wants to take advantage of preferential permeability of the reservoir to increase productivity, as in the case of naturally fractured formations. There are also situations where only one parameter, either the inclination or the azimuth, at the final end is set and the other parameter is free.

Well Path Parameterization

For 3D well paths, generalized well path parameters can be obtained using the following generic method. A curve in a 3D geographical space system can be expressed in the form:

$$M(u)=[N(u),E(u),H(u)] \tag{5.22}$$

The three coordinates are coupled by the parameter $u = [0,1]$. Here,

N = North coordinate (south is negative), m or ft
E = East coordinate (west is negative), m or ft
H = Total vertical depth, m or ft

The parameter $u = 0$ corresponds to the starting point of the well path, whereas $u = 1$ corresponds to the final point of the well path. If N_0, E_0, H_0 and N_1, E_1, H_1 are the coordinates of the initial and final points of the well path, respectively, then:

$$\begin{aligned} N(0)=N_0, E(0)=E_0, H(0)=H_0 \\ N(1)=N_1, E(1)=E_1, H(1)=H_1 \end{aligned} \tag{5.23}$$

An infinitesimal length ds of the well path can be expressed by:

$$ds=\sqrt{\dot{N}^2+\dot{E}^2+\dot{H}^2}=\sqrt{\dot{M}\bullet\dot{M}}\,du \tag{5.24}$$

Therefore, the length $s(u)$ of the well path between a starting point and any point $M(u)$ of the well path is given by the integral:

$$s(u)=\int_0^u \sqrt{[\dot{N}(\varsigma)]^2+[\dot{E}(\varsigma)]^2+[\dot{H}(\varsigma)]^2}\,d\varsigma \tag{5.25}$$

where ς is a dummy argument, and the total length of the well path is given by $S = s(1)$.

From the definition of inclination α and azimuth ϕ of any point of the well path, the following relations exist:

$$\begin{cases} \sin\alpha\cos\phi = \dfrac{dN}{ds} = \dfrac{du}{ds}\dot{N} \\ \sin\alpha\sin\phi = \dfrac{dE}{ds} = \dfrac{du}{ds}\dot{E} \\ \cos\alpha = \dfrac{dH}{ds} = \dfrac{du}{ds}\dot{H} \end{cases} \tag{5.26}$$

with

$$\frac{du}{ds} = \frac{1}{\sqrt{\dot{N}^2 + \dot{E}^2 + \dot{H}^2}}$$

In particular, if the inclination α_0 and the azimuth ϕ_0 are known, the following relations exist:

$$\cos\phi_0 = \frac{1}{L_0}\dot{N}(0);\ \sin\alpha_0\sin\phi_0 = \frac{1}{L_0}\dot{E}(0);\ \cos\alpha_0 == \frac{1}{L_0}\dot{H}(0) \tag{5.27}$$

with

$$L_0 = \sqrt{\dot{N}(0)^2 + \dot{E}(0)^2 + \dot{H}(0)^2}$$

The final shape of the functions depends on the magnitude of L_0, which can be considered as an independent parameter. Rearranging Eq. 5.27 yields:

$$\begin{cases} \dot{N}(0) = L_0\sin\theta_0\cos\phi_0 \\ E(0) = L_0\sin\theta_0\sin\phi_0 \\ \dot{H}(0) = L_0\cos\theta_0 \end{cases} \tag{5.28}$$

If the inclination θ_1 and the azimuth ϕ_1 at the final point of the well path are known, then as before:

$$\begin{cases} \dot{N}(1) = L_1\sin\theta_1\cos\phi_1 \\ E(1) = L_1\sin\theta_1\sin\phi_1 \\ \dot{H}(1) = L_1\cos\theta_1 \end{cases} \tag{5.29}$$

with

$$L_1 = \sqrt{\dot{N}(1)^2 + \dot{E}(1)^2 + \dot{H}(1)^2}$$

The specification of L_0 and L_1 will determine the final shape of the 3D well path. The subscript 0 denotes the starting location, whereas 1 represents the final location.

Cubic Functions

In this section, cubic functions are used to generate the three coordinate functions[8] that describe a well path. The advantage of this method is that it provides a continuous and smooth mathematical 3D curve connecting the current position of the hole (possibly at the surface) to a target, in general with a given inclination and azimuth, with a degree of freedom that allows a great deal of control on the geometric characteristics of the well path.

Cubic functions can be used to design 3D well trajectories. The general expressions for a cubic function along with its first and second derivatives are:

$$\begin{cases} y(u) = C_0 + C_1 u + C_2 u^2 + C_3 u^3 \\ \dot{y}(u) = C_1 + 2C_2 u + 3C_3 u^2 \\ \ddot{y}(1) = 2C_2 + 6C_3 u \end{cases} \tag{5.30}$$

where u is an independent variable, y is related to the position of the well, and C_0, C_1, C_2, and C_3 are coefficients.

Two important cases need to be considered: (1) free slope at the final end of the interval, and (2) set slope at the final end of the interval. It is assumed that the abscissas at both ends and the slope at the initial end are known.

Free Slope Condition For this case, the slope at $u = 1$ is not specified. To implement a free slope condition, the following relation should be satisfied:

$$y(0) = Y_0;\ y(1) = Y_1;\ \dot{y}(0) = \dot{Y}_0;\ \ddot{y}(0) = 0 \tag{5.31}$$

where $Y_0, Y_1, \dot{Y}_0$ are known and the curvature at the end $\ddot{y}(0)$ is zero.

Using Eqs. 5.30 and 5.31 it is possible to write:

$$\begin{bmatrix} 1 & 0 & 0 & 0 \\ 1 & 1 & 1 & 1 \\ 0 & 1 & 0 & 0 \\ 0 & 0 & 2 & 6 \end{bmatrix} \begin{bmatrix} C_0 \\ C_1 \\ C_2 \\ C_3 \end{bmatrix} = \begin{bmatrix} Y_0 \\ Y_1 \\ \dot{Y}_0 \\ 0 \end{bmatrix} \tag{5.32}$$

The coefficients are obtained by solving Eq. 5.32 to get:

$$\begin{cases} C_0 = Y_0 \\ C_1 = \dot{Y}_0 \\ C_2 = -\dfrac{3}{2}\left(Y_0 - Y_1 + \dot{Y}_0\right) \\ C_3 = \dfrac{1}{2}\left(Y_0 - Y_1 + \dot{Y}_0\right) \end{cases} \tag{5.33}$$

Set Slope Condition In this case, the slope at $u = 1$ is specified. To implement a set slope condition, the following relations must be satisfied at the ends:

$$y(0) = Y_0;\ y(1) = Y_1;\ \dot{y}(0) = \dot{Y}_0;\ \ddot{y}(0) = \dot{Y}_1 \tag{5.34}$$

As in the earlier case, the following system of equations are formed:

$$\begin{bmatrix} 1 & 0 & 0 & 0 \\ 1 & 1 & 1 & 1 \\ 0 & 1 & 0 & 0 \\ 0 & 1 & 2 & 3 \end{bmatrix} \begin{bmatrix} C_0 \\ C_1 \\ C_2 \\ C_3 \end{bmatrix} = \begin{bmatrix} Y_0 \\ Y_1 \\ \dot{Y}_0 \\ \dot{Y}_1 \end{bmatrix} \tag{5.35}$$

Solving the equations for the coefficients gives:

$$\begin{cases} C_0 = Y_0 \\ C_2 = \dot{Y}_0 \\ C_3 = \left[-3Y_0 + 3Y_1 - 2\dot{Y}_0 - \dot{Y}_1\right] \\ C_4 = \left[2Y_0 - 2Y_1 + \dot{Y}_0 + \dot{Y}_1\right] \end{cases} \tag{5.36}$$

3D Cubic Well Paths

To construct a 3D cubic well path, the expressions derived for cubic functions are used to represent the three parametric coordinate functions.[8] It is assumed that the coordinates, inclination, and azimuth of the initial end of the well path are known. Concerning the final end of the well path there are four cases to be analyzed:

- Free inclination and free azimuth
- Set inclination and set azimuth
- Free inclination and set azimuth
- Set inclination and free azimuth

In addition, the coordinates of the final end of the well path are also known. The procedures are basically the same for the first three cases. They differ only in the manner in which the functions are selected to represent each coordinate, and how the final end conditions are imposed.

In the last case, however, due to its nature, an iterative procedure is required to obtain the azimuth of the final end.

Free Inclination and Free Azimuth In this case, the three coordinate functions are free slope cubic functions. There is only one degree of freedom (the parameter L_0). L_0 affects the shape of the 3D well path, in particular its total length and the varying curvature along the well path. The final path will be a 2D curve. Using Eq. 5.27, additional boundary conditions can be specified as:

$$\begin{cases} \dot{N}_0 = \dot{N}(0) = L_0 \sin\theta_0 \cos\phi_0 \\ \dot{E}_0 = \dot{E}(0) = L_0 \sin\theta_0 \sin\phi_0 \\ \dot{H}_0 = \dot{H}(0) = L_0 \cos\theta_0 \end{cases} \tag{5.37}$$

The coefficients for the north coordinate are calculated using Eq. 5.33:

$$\begin{cases} C_{0,N} = N_0 \\ C_{1,N} = \dot{N}_0 \\ C_{2,N} = -\dfrac{3}{2}\left(N_0 - N_1 + \dot{N}_0\right) \\ C_{3,N} = \dfrac{1}{2}\left(N_0 - N_1 + \dot{N}_0\right) \end{cases} \tag{5.38}$$

Proceeding similarly, the other coordinate coefficients can be calculated and the coordinate functions can be expressed as:

$$\begin{cases} N(u) = C_{0,N} + C_{1,N}u + C_{2,N}u^2 + C_{3,N}u^3 \\ E(u) = C_{0,E} + C_{1,E}u + C_{2,E}u^2 + C_{3,E}u^3 \\ H(u) = C_{0,H} + C_{1,H}u + C_{2,H}u^2 + C_{3,H}u^3 \end{cases} \tag{5.39}$$

The integral in Eq. 5.25 can be calculated numerically, or approximated by a summation over a reasonably fine-mesh partition in the interval [0,1]. (A mesh of constant length 0.01 will suffice for most cases.) The expressions for the derivatives are calculated as:

$$\begin{aligned} \dot{N}(u) &= C_{1,N}u + 2C_{2,N}u + 3C_{3,N}u^2 \\ \dot{E}(u) &= C_{1,E}u + 2C_{2,E}u + 3C_{3,E}u^2 \\ \dot{H}(u) &= C_{1,H}u + 2C_{2,H}u + 3C_{3,H}u^2 \end{aligned} \tag{5.40}$$

The measured depth is obtained by adding the arc length to the target depth as:

$$D(u) = D_0 + s(u) \quad (5.41)$$

PROBLEM 5.3

The coordinates and survey data of the initial point at a measured depth of 508 m and vertical depth of 500 m are:

N = 5 m; E = 18 m; inclination = 30°; azimuth = 10°

The target point is at a vertical depth of 1600 m with the coordinates N = 800 m and E = 1000 m. Use the parameter value L_0 = 3000 to design the well path to reach the target.

Solution

The sequence of calculations is identical to the previous examples. Table 5.1 presents the summary of the results of the computation of the 3D cubic well path for the data above. The arc length of the well path is 1805.26 m, and the measured depth, inclination, and azimuth of the final point are, respectively, 2313.26 m, 75.09°, and 71.32°.

Set Inclination and Set Azimuth In this case, the three coordinate functions are set slope cubic functions. There are two degrees of freedom (the parameters L_0 and L_1). They affect the shape of the 3D well path, in particular its total length and the varying curvature along the well path. L_0 has a major impact on the initial portion of the well path, while L_1 has a major impact on the final portion of the well path.

Table 5.1 Summary of the Results for the 3D Cubic Well Path Computation (Free-end) for Problem 5.3

u	*H*	*N*	*E*	*s*	*MD*	α	ϕ
0.00	500.00	5.00	18.00	0.00	508.00	30.00	10.00
0.20	935.72	262.24	110.50	515.36	1023.36	35.69	30.38
0.40	1227.63	453.98	272.27	901.67	1409.67	47.34	49.22
0.60	1411.68	596.61	485.98	1219.02	1727.02	61.76	62.16
0.80	1523.82	706.49	734.33	1513.42	2021.42	72.41	69.15
1.00	1600.00	800.00	1000.00	1805.26	2313.26	76.09	71.32

Using data for the initial end and the final end of the well path, the additional boundary conditions are calculated as:

$$\begin{cases} \dot{N}_1 = \dot{N}(1) = L_1 \sin\theta_1 \cos\phi_1 \\ \dot{E}_1 = \dot{E}(1) = L_1 \sin\theta_1 \sin\phi_1 \\ \dot{H}_1 = \dot{H}(1) = L_1 \cos\theta_1 \end{cases} \tag{5.42}$$

$$\begin{cases} \dot{N}_0 = \dot{N}(0) = L_0 \sin\theta_0 \cos\phi_0 \\ \dot{E}_0 = \dot{E}(0) = L_0 \sin\theta_0 \sin\phi_0 \\ \dot{H}_0 = \dot{H}(0) = L_0 \cos\theta_0 \end{cases} \tag{5.43}$$

The coefficients of the north coordinate functions are calculated using Eq. 5.41:

$$\begin{cases} C_{0,N} = N_0 \\ C_{1,N} = \dot{N} \\ C_{2,N} = -3N_0 + 3N_1 - 2\dot{N}_0 - \dot{N} \\ C_{3,N} = 2N_0 - 2N_1 + \dot{N}_0 + \dot{N} \end{cases} \tag{5.44}$$

The coordinate functions are the same as in Eq. 5.39.

PROBLEM 5.4

The coordinates and survey data of the initial point at a measured depth of 508 m and vertical depth of 500 m are:

N = 5 m; E = 18 m; inclination = 30°; azimuth = 10°

The target point is at a vertical depth of 1600 m with the coordinates N = 800 m, E = 1000 m, inclination = 90°, and azimuth = 0°. Use the parameter value L_0 = 3000 and L_1 = 2500 to design the well path to reach the target.

Solution

Table 5.2 presents the summary of the results of the computation of the 3D cubic well path for the data above. The arc length of the well path is 1933.57 m and the final measured depth is 2441.57 m.

Free Inclination and Set Azimuth In this case, the vertical coordinate function is a free slope cubic function and both the horizontal coordinate functions are set slope cubic functions. The two degrees of freedom are L_0 and L_1.

Since the inclination α_1 is unknown, a second free parameter $L_T = L_1 \sin \alpha_1$ is used to avoid an iterative process. The boundary conditions for the

Table 5.2 Summary of the Results for the 3D Cubic Well Path Computation (Set-end) for Problem 5.4

u	*H*	*N*	*E*	*s*	*MD*	α	φ
0.00	500.00	5.00	18.00	0.00	508.00	30.00	10.00
0.20	946.95	196.76	153.47	509.36	1017.36	31.52	62.42
0.40	1261.32	257.56	401.17	916.43	1424.43	47.59	83.09
0.60	1462.22	301.97	679.34	1263.49	1771.49	61.42	74.38
0.80	1568.74	444.59	906.21	1557.47	2065.47	77.16	37.11
1.00	1600.00	800.00	1000.00	1933.57	2441.57	90.00	0.00

horizontal coordinate functions are calculated using the appropriate expressions.

$$\dot{N}_1 = \dot{N}(1) = L_T \cos\phi_1 \tag{5.45}$$

$$\dot{E}_1 = \dot{E}(1) = L_T \sin\phi_1 \tag{5.46}$$

The coefficients for the north and east coordinate functions are calculated using Eq. 5.44. The coordinate functions are the same as in Eq. 5.39.

PROBLEM 5.5

The coordinates and survey data of the initial point at a measured depth of 508 m and vertical depth of 500 m are:

N = 5 m; E = 18 m; inclination = 30°; azimuth = 10°

The target point is at a vertical depth of 1600 m, with the coordinates N = 800 m, E = 1000 m, and azimuth = 0°. Use the parameter values $L_0 = 3000$ and $L_1 = 2500$ to design the well path to reach the target.

Solution

Proceeding as in the previous example, Table 5.3 presents the summary of the results of the computation of the 3D cubic well path for the data above. The final inclination is 82.01°. The arc length of the well path is 1904.67 m and the final point measured depth is 2412.67 m. Using the final inclination and the value of L_T, L_1 is calculated from:

$$L_1 = \frac{L_T}{\sin\alpha_1} = \frac{2500}{\sin 82.01^{\circ}} = 2524.51$$

Table 5.3 Summary of the Results for the 3D Cubic Well Path Computation (Free Inclination and Set Azimuth) for Problem 5.5

u	*H*	*N*	*E*	*s*	*MD*	α	ϕ
0.00	500.00	5.00	18.00	0.00	508.00	30.00	10.00
0.20	935.72	195.76	153.47	499.57	1007.57	32.90	62.42
0.40	1227.63	257.56	401.17	889.65	1397.65	50.22	83.09
0.60	1411.68	301.97	679.34	1227.06	1735.06	62.79	74.38
0.80	1523.82	444.59	905.21	1522.58	2030.58	73.00	37.11
1.00	1600.00	800.00	1000.00	1904.67	2412.67	82.01	0.00

Set Inclination and Free Azimuth In this case, the three coordinate functions are set slope cubic functions. Again, there are two degrees of freedom (L_0 and L_1). Since the final azimuth is not known, it is determined by setting the curvature of the horizontal projection at the final end to zero. The projection is represented parametrically by the horizontal coordinate functions $N(u)$ and $E(u)$. The condition above is satisfied when

$$\dot{N}(1)\ddot{E}(1)-\ddot{N}(1)\dot{E}(1)=0 \tag{5.47}$$

A Newton–Raphson Scheme[9] of iteration can be used to find the final azimuth, which is the root of a test function, defined as:

$$g(\alpha)=\dot{N}(1)\ddot{E}(1)-\ddot{N}(1)\dot{E}(1)=0 \tag{5.48}$$

where

$$\begin{aligned}
\dot{N}(1)&=C_{1,N}+2C_{2,N}+3C_{3,N}\dot{E}(1)\\
\dot{E}(1)&=C_{1,E}+2C_{2,E}+3C_{3,E}\ddot{N}(1)\\
\ddot{N}(1)&=2C_{2,N}+6C_{3,N}\ddot{E}(1)\\
\ddot{E}(1)&=2C_{2,E}+6C_{3,E}\ddot{N}(1)
\end{aligned}$$

Here, $C_{1,N}$, $C_{2,N}$, $C_{3,N}$ are given by Eq. 5.44 while $C_{1,E}$, $C_{2,E}$, $C_{3,E}$ are obtained from equations similar to Eq. 5.44.

A recommended initial guess is the relative azimuth of the final end horizontal coordinates to the initial end coordinates, which is given by:

$$\phi_0=\arctan\frac{E_1-E_0}{N_1-N_0} \tag{5.49}$$

Care must be exercised to find the correct quadrant. The process is stable and with a few iterations the correct angle can be obtained. With the final end azimuth, the well path is obtained using the same steps as used in the case of set inclination and set azimuth above.

PROBLEM 5.6

The coordinates and survey data of the initial point at a measured depth of 508 m and vertical depth of 500 m are:

N = 5 m; E = 18 m; inclination = 30°; azimuth = 10°

The target point is at a vertical depth of 1600 m with the coordinates N = 800 m, E = 1000 m, and inclination = 90°. Use the parameter values L_0 = 3000 and L_1 = 2500 to design the well path to reach the target.

Solution

The initial guess for the end azimuth is 51.01° and the Newton–Raphson process results in a final end azimuth of 71.32°. Table 5.4 presents the summary of the results of the computation of the 3D cubic well path for the data above. The arc length of the well path is 1907.59 m and the end point measured depth is 2415.59 m.

Inclination, Azimuth, Curvature, and Toolface Angle

After the 3D well path has been determined, several additional parameters can be determined, like inclination, azimuth, curvature, etc., at any point along the 3D curve. To determine these additional parameters, the following general procedure can be used with any suitable coordinate system. The fundamental mathematical concepts and operations can be found in texts on differential geometry.[10]

To calculate the inclination and azimuth at any point of the well path, a unit vector ***t*** tangent to the well path must be determined. Denoting $t(u)$ as

Table 5.4 Summary of the Results for the 3D Cubic Well Path Computation (Set Inclination and Free Azimuth) for Problem 5.6

u	*H*	*N*	*E*	*s*	*MD*	α	φ
0.00	500.00	5.00	18.00	0.00	508.00	30.00	10.00
0.20	945.95	251.14	77.68	513.88	1021.88	29.66	19.73
0.40	1261.32	420.70	173.81	884.62	1392.62	35.66	41.30
0.60	1462.22	545.69	338.30	1175.69	1683.69	57.79	61.17
0.80	1568.74	662.12	603.06	1485.61	1993.61	79.29	69.59
1.00	1600.00	800.00	1000.00	1907.59	2415.59	90.00	71.32

the unit tangent vector, it can be expressed at any given point $M(u)$ (refer to Eq. 5.22) of the well path as:

$$\boldsymbol{t} = \frac{dM}{ds} = \frac{dM}{du}\frac{du}{ds} = \frac{\dot{M}}{\|\dot{M}\|} = \frac{\dot{M}}{\sqrt{\dot{M} \bullet \dot{M}}} \tag{5.50}$$

where

$$\dot{M} = \left(\dot{N}, \dot{E}, \dot{H}\right)$$

$$\|\dot{M}\| = \sqrt{\dot{M} \bullet \dot{M}} = \sqrt{\dot{N}^2 + \dot{E}^2 + \dot{H}^2} \tag{5.51}$$

If the components of the unit tangent vector are t_N, t_E, t_H, then the inclination and azimuth are given respectively by:

$$\alpha = \arccos t_H; \quad \phi = \arccos \frac{t_E}{t_N} \tag{5.52}$$

The principal normal $\boldsymbol{K}$ to the well path at a point $M(u)$ is given by:

$$\boldsymbol{K} = \frac{dt}{ds} = \frac{dt}{du}\frac{du}{ds} \tag{5.53}$$

$\boldsymbol{K}$ is called the curvature vector at the point. Its modulus $\kappa = \|K\|$ is the magnitude of the curvature and is equal to the reciprocal of the radius of curvature R.

Differentiating Eq. 5.50 and using

$$\frac{du}{ds} = \frac{1}{\sqrt{\dot{M} \bullet \dot{M}}}$$

results in:

$$\boldsymbol{K} = \frac{1}{\dot{M} \bullet \dot{M}}\ddot{M} - \left[\frac{\dot{M} \bullet \ddot{M}}{\left(\dot{M} \bullet \dot{M}\right)^2}\right]\dot{M} \tag{5.54}$$

The unit principal normal vector $\boldsymbol{n}$ is defined by:

$$\boldsymbol{n} = \frac{K}{\kappa} = \left(n_N, n_E, n_H\right) \tag{5.55}$$

In terms of components, the various elements in Eq. 5.54 are given by:

$$\ddot{M} = \left(\ddot{N}, \ddot{E}, \ddot{H}\right) \tag{5.56}$$

$$\begin{aligned} \dot{M} \bullet \dot{M} &= \dot{N}^2 + \dot{E}^2 + \dot{H}^2 \\ \dot{M} \bullet \ddot{M} &= \dot{N}\ddot{N} + \dot{E}\ddot{E} + \dot{H}\ddot{H} \end{aligned} \tag{5.57}$$

The DLS expressed in (°)/30 m or (°)/100 ft is given by:

$$DLS = \left(\frac{5400}{\pi}\right)\kappa(°)\,/\,30\text{ m or } DLS = \left(\frac{1800}{\pi}\right)\kappa(°)\,/\,100\text{ ft} \quad (5.58)$$

Proceeding through some transformations using differential geometry, the basic formula for calculating the borehole torsion is as given in Eq. 3.50:

$$\tau = \frac{\left(\dot{M},\ddot{M},\dddot{M}\right)}{\kappa^2} = \frac{1}{\kappa^2}\begin{bmatrix} \dot{N} & \dot{E} & \dot{H} \\ \ddot{N} & \ddot{E} & \ddot{H} \\ \dddot{N} & \dddot{E} & \dddot{H} \end{bmatrix} \quad (5.59)$$

The toolface angle is the angle that the principal normal to the well path makes with the intersection of the vertical plane, containing the tangent to the well path and the plane perpendicular to this tangent. This intersection defines a unit vector $\boldsymbol{h}$, called the high-side vector, given by:

$$\boldsymbol{h} = \frac{\cos\alpha}{\sin\alpha}\boldsymbol{t} - \frac{1}{\sin\alpha}\boldsymbol{H} \quad (5.60)$$

where $\boldsymbol{h} = (1,0,0)$ is the vertical unit vector.

The toolface angle ω can be easily calculated from:

$$\omega = \arccos(\boldsymbol{h} \bullet \boldsymbol{n}) \quad (5.61)$$

This expression gives only the angle of the toolface, but cannot provide the direction with respect to the high side. The direction can be determined using the following expression:

$$(\boldsymbol{h} \bullet \boldsymbol{n})_H = (h_N n_E - h_E n_N) \quad (5.62)$$

The toolface angle is at the right of the high side if $(\boldsymbol{h} \bullet \boldsymbol{n})_H$ is positive, and at the left of the high side if $(\boldsymbol{h} \bullet \boldsymbol{n})_H$ is negative. This expression cannot be used when $\alpha = 90°$, and in this case, the following more general expression can be used:

$$\begin{aligned}(\boldsymbol{h}\bullet\boldsymbol{n})\bullet \boldsymbol{t} &= \left(h_N n_E - h_E n_N\right)t_H + \left(h_E n_H - h_H n_E\right)t_N \\ &\quad + \left(h_H n_N - h_N n_H\right)t_E\end{aligned} \quad (5.63)$$

Again, a positive value corresponds to an angle at the right of the high-side. From Eq. 5.60 it can clearly be seen that the high side vector is not defined for $\alpha = 0$, and in this case, the toolface angle is given by:

$$\omega = \arctan\frac{n_E}{n_n} \quad (5.64)$$

PROBLEM 5.7

After the shape of the 3D curve has been determined, it is important to calculate the additional parameters of the well path. Table 5.5 presents the continuation of Table 5.2 (set-end 3D cubic well path) showing the components of the vectors, the curvature, the dogleg severity, and the toolface angle. Similarly, the additional parameters can be calculated for the other worked-out problems given above.

Spline-in-Tension–Type Wells

Spline-in-tension (SiT) functions differ from cubic functions[11-13] as an additional parameter representing the "tension" of the curve that is used. A totally "relaxed" curve is identical to a cubic curve, and as the tension increases, a shorter curve length is obtained, affecting the profile and the curvature along the curve. As the tension increases to infinity, the SiT curve approaches a straight line. The tension offers an additional degree of freedom, which can be used to further optimize the final well path.

SiT Functions

Spline-in-tension functions are continuous functions, resulting from the solutions of a differential equation obtained by induction (see Chapter 1). They are relatively simple and general enough to be used as coordinate functions to design 3D well paths. Figure 5.7 illustrates typical cubic and SiT well paths.

The general expression of an SiT function, along with its first and second derivatives, are[13]:

$$\begin{cases} y(u) = C_0 + C_1 u + C_2 \sinh \lambda u + C_3 \cosh \lambda u \\ \dot{y}(u) = C_1 + C_2 \cosh \lambda u + C_3 \sinh \lambda u \\ \ddot{y}(u) = C_2 \lambda^2 \sinh \lambda u + C_3 \lambda^2 \cosh \lambda u \end{cases} \tag{5.65}$$

where

u = An independent variable in the interval [0,1]

λ = Tension parameter

C_0, C_1, C_2, and C_3 are coefficients that assume specific values to satisfy the required boundary conditions

Two cases to be considered are:

- Free slope at the final end of the interval.
- Set slope at the final end of the interval.

Table 5.5 Summary of the Results for the Computation of Tangent, Curvature, Normal, and High-Side Vectors for the Set-end Case for Problem 5.7

u	t_H	t_N	t_E	κ_0	*DLS*	n_H	n_n	n_E	h_H	h_n	h_E	*TF*	*AZM*
0.00	2598.08	1477.21	260.47	0.00	1.22	0.204	–0.506	0.838	–0.500	0.853	1.50E-01	114.07	0.79
0.20	1887.38	535.91	1025.07	0.00	1.73	–0.272	–0.551	0.789	–0.523	0.395	7.56E-01	58.63	0.73
0.40	1272.23	167.53	1382.82	0.00	1.27	–0.725	–0.109	0.580	–0.738	0.811	5.69E-01	10.96	0.13
0.60	752.62	372.09	1330.73	0.00	2.26	–0.485	0.874	0.302	–0.878	0.129	4.61E-01	55.45	–0.40
0.80	328.54	1149.58	869.79	0.00	4.96	–0.360	0.62	–0.691	–0.975	0.177	1.34E-01	68.32	–0.21
1.00	0.00	2500.00	0.00	0.00	1.53	–0.253	0	–0.967	1.0	0.612	0.00E+00	75.35	0.00

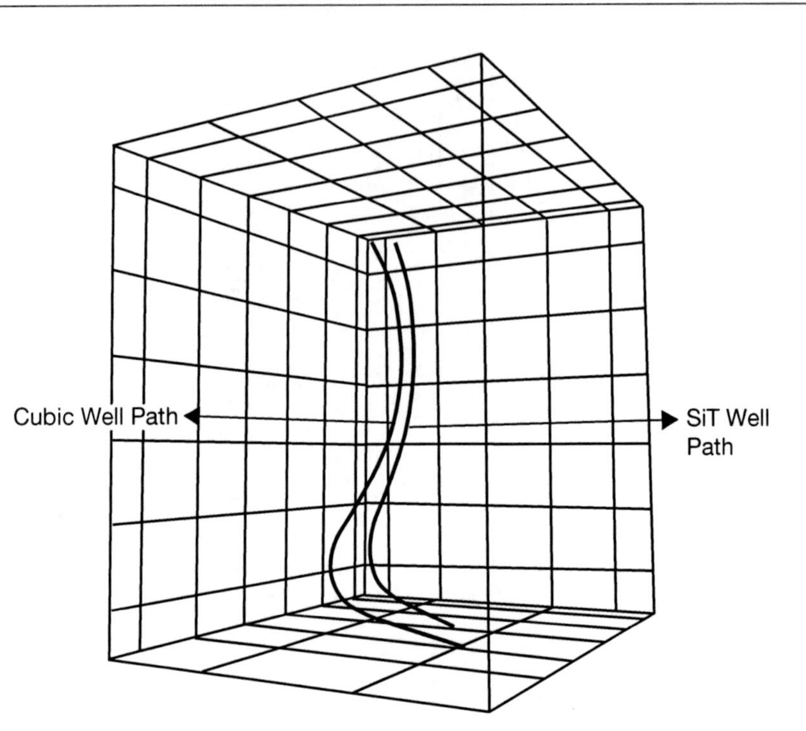

Figure 5.7 Cubic and SiT well paths.

Free Slope Condition In this case, the slope at u = 1 is not specified. To implement a free slope condition, the following should be satisfied:

$$y(0) = Y_0;\ y(1) = Y_1;\ \dot{y}(0) = \dot{Y}_0;\ \ddot{y}(0) = 0 \tag{5.66}$$

where $Y_0, Y_1, \dot{Y}_0$ are known and the curvature at the end $\ddot{y}(0)$ is zero.

Using Eqs. 5.65 and 5.66, the equation for the coefficients can be written in matrix form:

$$\begin{bmatrix} 1 & 0 & 0 & 1 \\ 1 & 1 & S & C \\ 0 & 1 & \lambda & 0 \\ 0 & 0 & \lambda^2 S & \lambda^2 C \end{bmatrix} \begin{bmatrix} C_0 \\ C_1 \\ C_2 \\ C_3 \end{bmatrix} = \begin{bmatrix} Y_0 \\ Y_1 \\ \dot{Y}_0 \\ 0 \end{bmatrix} \tag{5.67}$$

where C = cosh λ and S = sinh λ.

The coefficients are obtained by solving Eq. 5.67 as:

$$\begin{cases} C_0 = \dfrac{1}{\Delta f}\left(\lambda C Y_0 - S Y_1 + S\dot{Y}_0\right) \\ C_2 = \dfrac{1}{\Delta f}\left(-\lambda C Y_0 + \lambda C Y_1 - S\dot{Y}_0\right) \\ C_3 = \dfrac{C}{\Delta f}\left(Y_0 - Y_1 + \dot{Y}_0\right) \\ C_4 = \dfrac{1}{\Delta f}\left(-Y_0 + Y_1 - \dot{Y}_0\right) \end{cases} \tag{5.68}$$

where $\Delta f = \lambda C - S$.

Set Slope Condition In this case, the slope at u = 1 is specified. To implement a set slope condition, the following boundary conditions are defined:

$$y(0) = Y_0;\ y(1) = Y_1;\ \dot{y}(0) = \dot{Y}_0;\ \ddot{y}(0) = \dot{Y}_1 \tag{5.69}$$

As before, the system of equations are written in matrix form as:

$$\begin{bmatrix} 1 & 0 & 0 & 0 \\ 1 & 1 & S & C \\ 0 & 1 & \lambda & 0 \\ 0 & 0 & \lambda C & \lambda S \end{bmatrix} \begin{bmatrix} C_0 \\ C_1 \\ C_2 \\ C_3 \end{bmatrix} = \begin{bmatrix} Y_0 \\ Y_1 \\ \dot{Y}_0 \\ \dot{Y}_1 \end{bmatrix} \tag{5.70}$$

Solving the system for the coefficients gives:

$$\begin{cases} C_0 = \dfrac{1}{\Delta f}\left[(\lambda S + 1 - C)Y_0 + (1-C)Y_1 - \left(\dfrac{s}{\lambda} - C\right)\dot{Y}_0 + \left(\dfrac{s}{\lambda} - 1\right)\dot{Y}_1\right] \\ C_2 = \dfrac{1}{\Delta f}\left[-\lambda S Y_0 + \lambda S Y_1 + (1-C)\dot{Y}_0 + (1-c)\dot{Y}_1\right] \\ C_3 = \dfrac{1}{\Delta f}\left[S Y_0 + S Y_1 + \left(S + \dfrac{1-C}{\lambda}\right)\dot{Y}_0 - \dfrac{1-C}{\lambda}\dot{Y}_1\right] \\ C_4 = \dfrac{1}{\Delta f}\left[(1-C)Y_0 - (1-C)Y_1 + \left(\dfrac{s}{\lambda} - C\right)\dot{Y}_0 - \left(\dfrac{s}{\lambda} - 1\right)\dot{Y}_1\right] \end{cases}$$

$$\tag{5.71}$$

where $\Delta f = \lambda S - 2C + 2$.

3D Spline-in-Tension Well Paths

For a 3D SiT well path the expressions derived for SiT functions can be used to calculate the three well path parametric coordinate functions. It is assumed that the coordinates, inclination, and azimuth of the starting location of the well path are known. The procedure used is exactly the same as that discussed for the case of the cubic well path design. Following are the four possible cases for the final end conditions:

- Free inclination and free azimuth
- Set inclination and set azimuth
- Free inclination and set azimuth
- Set inclination and free azimuth

In addition, the coordinates of the final end of the well path are also known. The details of the calculations are similar to that for cubic well paths.

Free Inclination and Free Azimuth In this case, the three coordinate functions are free slope SiT functions. There are four degrees of freedom (the parameter L_0, and one tension for each coordinate function). To simplify, the same tension for all coordinate functions can be used, which will then reduce the degrees of freedom to two. The remaining two parameters affect the shape of the 3D well path; in particular, its total length and the varying curvature along the well path and the final path will be a 2D curve.

Using Eq. 5.28 and the data for the initial end of the well path, the additional boundary conditions are calculated as follows:

$$\begin{cases} \dot{N}_0 = \dot{N}(0) = L_0 \sin\theta_0 \cos\theta_0 \\ \dot{E}_0 = \dot{E}(0) = L_0 \sin\theta_0 \sin\theta_0 \\ \dot{H}_0 = \dot{H}(0) = L_0 \cos\theta_0 \end{cases} \tag{5.72}$$

The subscript 0 denotes the starting location, whereas 1 represents the ending location.

The coefficients for north coordinate are calculated using Eq. 5.68:

$$\begin{cases} C_{0,N} = \dfrac{1}{\Delta f}\left(\lambda C N_0 - S N_1 + S\dot{N}_0\right) \\ C_{1,N} = \dfrac{1}{\Delta f}\left(-\lambda C N_0 + \lambda C N_1 - S\dot{N}_0\right) \\ C_{2,N} = \dfrac{C}{\Delta f}\left(N_0 - N_1 + \dot{N}_0\right) \\ C_{3,N} = \dfrac{1}{\Delta f}\left(-N_0 + N_1 - \dot{N}_0\right) \end{cases} \tag{5.73}$$

where $\Delta f = \lambda C - S$.

Similarly, the coefficients can be obtained for the east and the depth coordinates. Further the three coordinates can be expressed as:

$$\begin{cases} N(u) = C_{0,N} + C_{1,N}u + C_{2,N}\sinh(\lambda u) + C_{3,N}\cosh(\lambda u) \\ E(u) = C_{0,E} + C_{1,E}u + C_{2,E}\sinh(\lambda u) + C_{3,E}\cosh(\lambda u) \\ H(u) = C_{0,H} + C_{1,H}u + C_{2,H}\sinh(\lambda u) + C_{3,H}\cosh(\lambda u) \end{cases} \tag{5.74}$$

The integral in Eq. 5.25 can be calculated numerically, or approximated by a summation over a reasonably fine-mesh partition in the interval [0,1]. The relationships between the derivatives in the line integral in Eq. 5.25 are written in general terms, and are calculated using Eq. 5.65:

$$\begin{cases} \dot{N}(u) = C_{0,N} + C_{1,N}u + C_{2,N}\lambda\cosh(\lambda u) + C_{3,N}\lambda\sinh(\lambda u) \\ \dot{E}(u) = C_{0,E} + C_{1,E}u + C_{2,E}\lambda\cosh(\lambda u) + C_{3,E}\lambda\sinh(\lambda u) \\ \dot{H}(u) = C_{0,H} + C_{1,H}u + C_{2,H}\lambda\cosh(\lambda u) + C_{3,H}\lambda\sinh(\lambda u) \end{cases} \tag{5.75}$$

The measured depth is obtained by adding the arc length to the target depth, as follows:

$$D(u) = D_0 + s(u) \tag{5.76}$$

PROBLEM 5.8

The coordinates and survey data of the initial point at a measured depth of 508 m and vertical depth of 500 m are:

N = 5 m; E = 18 m; inclination = 15°; azimuth = 85°

The target point is at a vertical depth of 1600 m with the coordinates N = 800 m and E = 1000 m. Use the tension parameter 2 to design the well path to reach the target.

Table 5.6 Summary of the Results for the 3D SiT Well Path Computation (Free Inclination and Free Azimuth) for Problem 5.8

u	*H*	*N*	*E*	s	*MD*	α	ϕ
0.00	500.00	5.00	18.00	0.00	508.00	30.00	10.00
0.20	922.15	255.06	117.04	502.38	1010.38	37.01	33.35
0.40	1195.24	439.69	287.39	875.06	1383.06	49.02	51.07
0.60	1375.26	580.48	503.04	1190.15	1698.15	60.88	61.52
0.80	1500.93	695.07	745.36	1485.46	1994.46	68.51	65.74
1.00	1600.00	800.00	1000.00	1778.83	2285.83	71.03	68.32

Solution

Table 5.6 presents the summary of the results of the computation of the 3D SiT well path for the given data. The arc length of the well path is 1778.83 m, and the final measured depth, inclination, and azimuth are, respectively, 2285.83 m, 71.03°, and 68.32°.

Set Inclination and Set Azimuth In this case, the three coordinate functions are set slope SiT functions. There are up to five degrees of freedom (the parameters L_0 and L_1, and three tensions). Using the same tension for all coordinate functions reduces the degrees of freedom to three. These three parameters affect the shape of the 3D well path; in particular, its total length and the varying curvature along the well path. L_0 has a major impact on the initial portion of the well path, while L_1 has a major impact on the final portion of the well path.

Using data for the initial point and the end point of the well path, the additional boundary conditions are calculated as:

$$\begin{cases} \dot{N}_1 = \dot{N}(1) = L_1 \sin\theta_1 \cos\theta_1 \\ \dot{E}_1 = \dot{E}(1) = L_1 \sin\theta_1 \sin\theta_1 \\ \dot{H}_1 = \dot{H}(1) = L_1 \cos\theta_1 \end{cases} \tag{5.77}$$

Proceeding as in the case of cubic well paths, the coefficients for the north, east, and depth coordinate functions are calculated using Eq. 5.71, as:

$$\begin{cases} C_{0,\Gamma} = \dfrac{1}{\Delta f}\left[(\lambda S + 1 - C)\Gamma_0 + (1-c)\Gamma_1 - \left(\dfrac{s}{\lambda} - C\right)\dot{\Gamma}_0 + \left(\dfrac{s}{\lambda} - 1\right)\dot{\Gamma}_1\right] \\ C_{1,\Gamma} = \dfrac{1}{\Delta f}\left[-\lambda S\Gamma_0 + \lambda S\Gamma_1 + (1-C)\dot{\Gamma}_0 + (1-c)\dot{\Gamma}_1\right] \\ C_{2,\Gamma} = \dfrac{1}{\Delta f}\left[S\Gamma_0 + S\Gamma_1 + \left(S + \dfrac{1-C}{\lambda}\right)\dot{\Gamma}_0 - \dfrac{1-C}{\lambda}\dot{\Gamma}_1\right] \\ C_{3,\Gamma} = \dfrac{1}{\Delta f}\left[(1-C)\Gamma_0 - (1-c)\Gamma_1 + \left(\dfrac{s}{\lambda} - C\right)\dot{\Gamma}_0 - \left(\dfrac{s}{\lambda} - 1\right)\dot{\Gamma}_1\right] \end{cases} \tag{5.78}$$

where

$\Gamma = N, E, H$

$\Delta f = \lambda S - 2C + 2$

PROBLEM 5.9

The coordinates and survey data of the initial point at a measured depth of 508 m and vertical depth of 500 m are:

N = 5 m; E = 18 m; inclination = 30°; azimuth = 10°

The target point is at a vertical depth of 1600 m, with the coordinates N = 800 m, E = 1000 m, inclination = 90°, and azimuth = 0°. Use the tension parameter 3 with values $L_0 = 3000$ and $L_1 = 2500$ to design the well path to reach the target.

Solution

Table 5.7 presents a summary of the results of the computation of the 3D SiT well path for the data above. The arc length of the well path is 1888.54 m and the final measured depth is 2395.54 m.

Table 5.7 Summary of the Results for the 3D SiT Well Path Computation (Set Inclination and Set Azimuth) for Problem 5.9

u	*H*	*N*	*E*	*s*	*MD*	α	ϕ
0.00	500.00	5.00	18.00	0.00	508.00	30.00	10.00
0.20	924.81	197.47	155.74	491.19	999.19	34.25	61.27
0.40	1214.44	270.11	400.52	878.42	1385.42	48.52	79.73
0.60	1417.18	325.61	670.78	1221.35	1729.35	58.51	73.39
0.80	1549.10	460.33	898.75	1521.16	2029.16	71.13	40.58
1.00	1600.00	800.00	1000.00	1888.54	2395.54	90.00	0.00

Free Inclination and Set Azimuth In this case, the vertical coordinate function is a free slope SiT function and both the horizontal coordinate functions are set slope SiT functions. Again, there are up to five degrees of freedom (L_0, L_1, and the three tensions). Using the same tension for all the coordinate functions reduces the degrees of freedom to three. Although the parameter L_1 enters only in the expressions for the horizontal coordinates, it will also affect the final inclination. The additional boundary conditions at the initial end are calculated using Eq. 5.72.

Since the inclination α_1 is unknown, and to avoid an iterative process, $L_T = L_1 \sin\theta_1$, which now becomes the second free parameter. The boundary conditions for the horizontal coordinate functions are calculated using the appropriate expressions in Eq. 5.77:

$$\begin{aligned} \dot{N}_1 &= \dot{N}(1) = L_T \cos\phi_1 \\ \dot{E}_1 &= \dot{E}(1) = L_T \sin\phi_1 \end{aligned} \tag{5.79}$$

The coefficients for the north and east coordinate functions are calculated using Eq. 5.73. The coordinate functions are the same as in Eq. 5.75.

PROBLEM 5.10

The coordinates and survey data of the initial point at a measured depth of 508 m and vertical depth of 500 m are:

N = 5 m; E = 18 m; inclination = 30°; azimuth = 10°

The target point is at a vertical depth of 1600 m, with the coordinates N = 800 m, E = 1000 m, and azimuth = 0°. Use tension parameter 3 with values of $L_0 = 3000$ and $L_1 = 2432$.

Solution

Table 5.8 presents the summary of the results of the computation of the 3D SiT well path for the data above. The final inclination is 75.61°. The arc length of the well path is 1848.21 m and the final measured depth is 2355.21 m. Using the final inclination and the value of L_T, L_1 is calculated from:

$$L_1 = \frac{L_T}{\sin\theta_1} = \frac{2432}{\sin 76.61^{\circ}} = 2500$$

Set Inclination and Free Azimuth In this case, the three coordinate functions are set slope SiT functions. Again, there are up to five degrees of freedom (L_0, L_1, and the three tensions). Using the same tension for all coordinate functions reduces the degrees of freedom to three.

Table 5.8 Summary of the Results for the 3D SiT Well Path Computation (Free Inclination and Set Azimuth) for Problem 5.10

u	*H*	*N*	*E*	*s*	*MD*	α	ϕ
0.00	500.00	5.00	18.00	0.00	508.00	30.00	10.00
0.20	909.11	199.32	155.74	478.49	985.49	35.63	60.60
0.40	1167.74	275.59	400.52	843.97	1351.97	52.53	78.97
0.60	1345.37	334.04	670.78	1173.03	1681.03	60.89	73.03
0.80	1481.45	468.27	898.75	1473.80	1981.80	65.81	41.03
1.00	1600.00	800.00	1000.00	1848.21	2355.21	75.61	0.00

The final azimuth is not known and must be determined by setting the curvature of the horizontal projection at the final end to zero. The projection is represented parametrically by the horizontal coordinate functions $N(u)$ and $E(u)$. The condition above is satisfied when[3,7]:

$$\dot{N}(1)\ddot{E}(1) - \ddot{N}(1)\dot{E}(1) = 0 \tag{5.80}$$

A Newton–Raphson Scheme[9] can be used to find the final azimuth, which is the root of a test function defined as ξ, where,

$$\begin{cases} \dot{N}(1) = C_{1,N} + C_{2,N}\lambda\cosh\lambda + C_{3,N}\lambda\sinh\lambda \\ \dot{E}(1) = C_{1,E} + C_{2,E}\lambda\cosh\lambda + C_{3,E}\lambda\sinh\lambda \\ \ddot{N}(1) = C_{2,N}\lambda^2\sinh\lambda + C_{3,N}\lambda^2\cosh\lambda \\ \ddot{E}(1) = C_{2,E}\lambda^2\sinh\lambda + C_{3,E}\lambda^2\cosh\lambda \end{cases} \tag{5.81}$$

Here, $C_{1,N} C_{2,N} C_{3,N}$ and $C_{1,E} C_{2,E} C_{3,E}$ are given by Eq. 5.78.

Since the calculation is rigorous, a recommended initial guess for the Newton–Raphson Scheme is the relative azimuth of the final end horizontal coordinates to the initial end horizontal coordinates:

$$\phi_0 = \arctan\frac{E_1 - E_0}{N_1 - N_0} \tag{5.82}$$

Care must be exercised to find the correct quadrant. The process is stable, and the solution will converge to the correct angle in a few iterations. With the final end azimuth, the well path is obtained using the same steps as in the case from the "Set Inclination and Set Azimuth" section.

Table 5.9 Summary of the Results for the 3D SiT Well Path Computation (Set Inclination and Free Azimuth) for Problem 5.11

u	H	N	E	s	MD	α	ϕ
0.00	500.00	5.00	18.00	0.00	508.00	30.00	10.00
0.20	854.22	249.07	140.39	451.54	959.54	49.52	40.68
0.40	1018.44	429.33	345.40	771.93	1279.93	65.70	54.53
0.60	1127.55	577.29	579.76	1069.93	1577.93	67.41	60.13
0.80	1279.26	702.68	810.04	1374.14	1882.14	49.54	62.42
1.00	1600.00	800.00	1000.00	1763.87	2271.87	20.00	63.14

PROBLEM 5.11

The coordinates and survey data of the initial point at a measured depth of 508 m and vertical depth of 500 m are:

N = 5 m; E = 18 m; inclination = 30°; azimuth = 0°

The target point is at a vertical depth of 1600 m, with the coordinates N = 800 m, E = 1000 m, and inclination = 20°. Use tension parameter 4 with values $L_0 = 3000$ and $L_1 = 2500$ to design the well path to reach the target.

Solution

The initial guess for the end point azimuth is 51.01° and the Newton–Raphson process results in a final end point azimuth of 63.14°. Table 5.9 presents the summary of the results of the computation of the 3D SiT well path for the data above. The arc length of the well path is 1763.87 m and the final measured depth is 2271.87 m.

Sidetracking Wells

The production of oil and gas from a well over a long term leads to casing deformation or damage, cement-seal failure, and increased water production, thereby reducing the oil and gas production and resulting in loss of economic value. In order to enhance the oil recovery and lower the production costs, lateral drilling technology can be implemented. From the original conventional vertical wells, directional wells or horizontal wells can be branched off as sidetracks, or even in the form of subbranches off a branch of the original hole.

Directional Wells

When designing a sidetrack directional well, the sidetrack starting position is usually selected first and then the inclination, azimuth, and coordinates of the starting point are estimated. To make the designed sidetrack well path continuous and smooth, the sidetrack should be tangential to the original well path. Using a design based on the spatial-arc model, the sidetrack directional well path is typically a vertical-build/drop-hold profile,[14] as shown in Figure 5.8.

The tangent of a point s on the well path and the target point t defines a plane Ω in space. Designing a well path is greatly simplified by designing it to be in a single plane.

Assuming the sidetrack starting point to be the origin, a right-hand coordinate system (s-$\xi\eta\zeta$) is set up, where the axis ζ is the direction tangent to the well path, the axis η is the direction normal to the plane Ω, and the axis ξ is perpendicular to axes ζ and η, with the direction the same as the inward-pointing normal of the build (or drop) section. For convenience, the unit vectors along the axes of ζ, η, and ξ are denoted as $\boldsymbol{a}$, $\boldsymbol{b}$, and $\boldsymbol{c}$, respectively.

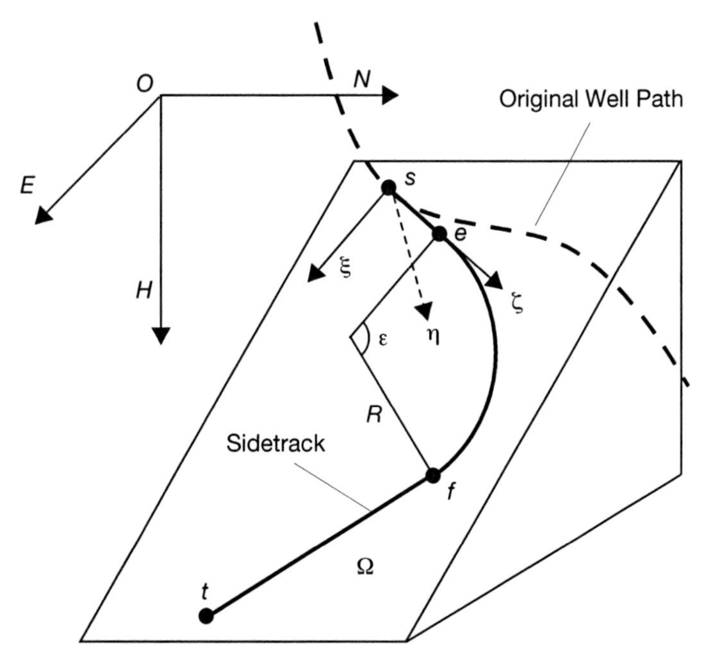

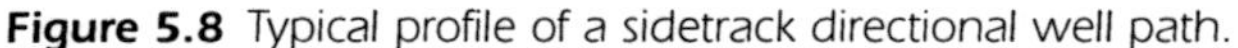

Figure 5.8 Typical profile of a sidetrack directional well path.

According to the theory of differential geometry, the direction cosines of the vector ***c*** in the *O-NEH* coordinate system are:

$$\begin{cases} c_N = \sin\alpha_s \cos\phi_s \\ c_E = \sin\alpha_s \sin\phi_s \\ c_H = \cos\alpha_s \end{cases} \tag{5.83}$$

where

α_s = Lateral position angle, (°)
ϕ_s = Lateral position direction, (°)

If ***d*** is the unit vector of which the direction is from the kick-off point *s* to the target, then:

$$\begin{cases} d_N = \dfrac{N_t - N_s}{d} \\ d_E = \dfrac{E_t - E_s}{d} \\ d_H = \dfrac{H_t - H_s}{d} \end{cases} \tag{5.84}$$

where

$$d = \sqrt{(N_t - N_s)^2 + (E_t - E_s)^2 + (H_t - H_s)^2}$$

N_s, E_s, H_s = North, east, and vertical coordinates of the sidetrack starting point
N_t, E_t, H_t = North, east, and vertical coordinates of the target

Because ***b*** = ***c*** × ***d***, therefore in the *O-NEH* coordinate system the direction cosines of the vector ***b*** are:

$$\begin{cases} b_N = (c_E d_H - d_E c_H)/b \\ b_E = (c_H d_N - d_H c_N)/b \\ b_H = (c_N d_E - d_N c_E)/b \end{cases} \tag{5.85}$$

where

$$b = \sqrt{(c_E d_H - d_E c_H)^2 + (c_H d_N - d_H c_N)^2 + (c_N d_E - d_N c_E)^2}$$

Also because ***a*** = ***b*** × ***c***, ***b*** ⊥ ***c***, the direction cosines for the vector ***a*** are:

$$\begin{cases} a_N = b_E c_H - c_E b_H \\ a_E = b_H c_N - c_H b_N \\ a_H = b_N c_E - c_N b_E \end{cases} \tag{5.86}$$

Therefore, in accordance with the transformation of coordinate systems, it can be written as:

$$\begin{bmatrix} \xi_t \\ \eta_t \\ \zeta_t \end{bmatrix} = \begin{bmatrix} a_N & a_E & a_H \\ b_N & b_E & b_H \\ c_N & c_E & c_H \end{bmatrix} \begin{bmatrix} N_t - N_s \\ E_t - E_s \\ H_t - H_s \end{bmatrix} \tag{5.87}$$

Thus, sidetrack tie-in point s with the coordinate (N_s, E_s, H_s), inclination angle α_s, azimuth ϕ_s, and the coordinate (N_t, E_t, H_t) of the target point t, the coordinate of t (ξ_t,η_t,ζ_t) in s-$\xi\eta\zeta$ coordinate system can be obtained.

Furthermore, over the circular-arc section of the well path, the central angle or the bending angle can be written as[14]:

$$\tan\frac{\varepsilon}{2} = \begin{cases} \dfrac{(\zeta_t - \Delta L_1) - \sqrt{(\zeta_t - \Delta L_1)^2 + \xi_t^{\,2} - 2R\xi_t}}{2R - \xi_t}, & \text{if } 2R \neq \xi_t \\[2ex] \dfrac{\xi_t}{2(\zeta_t - \Delta L_1)}, & \text{if } 2R = \xi_t \end{cases} \tag{5.88}$$

where

R = Radius of curvature, m or ft

ΔL_1 = Initial tangent section length, m or ft

The term under the square root of Eq. 5.88 must be greater than or equal to zero. To satisfy this condition, the bottomhole assembly (BHA) should be selected to ensure an adequate deflection rate. If $(\zeta_t - \Delta L_1)^2 + \xi_t^{\,2} - 2R\xi_t < 0$, the hole curvature is too small and the design cannot be achieved. For well profiles that need small curvature, the curvature can be estimated as:

$$\kappa_{\min} = \frac{5400}{\pi}\frac{2\xi_t}{(\zeta_t - \Delta L_1)^2 + \xi_t^{\,2}} (°)/30\text{ m}$$

or

$$\kappa_{\min} = \frac{18{,}000}{\pi}\frac{2\xi_t}{(\zeta_t - \Delta L_1)^2 + \xi_t^{\,2}} (°)/100\text{ ft} \tag{5.89}$$

Finally, other profile characteristic parameters are[14]:

$$\tan\omega = \left(\frac{a_N}{a_H}\sin\phi_s - \frac{a_E}{a_H}\cos\phi_s\right)\sin\alpha_s \tag{5.90}$$

$$\Delta L_2 = \frac{\pi}{180}R\varepsilon \tag{5.91}$$

$$\Delta L_3 = \sqrt{(\zeta_t - \Delta L_1)^2 + \xi_t^{\,2} - 2R\xi_t} \tag{5.92}$$

where

ω = Initial toolface angle, (°)
ΔL_2 = Circular-arc section length, m or ft
ΔL_3 = Well oblique section length, m or ft

The most simple and commonly used sidetrack profile from a parent wellbore is using arc-straight hold sections.[14]

PROBLEM 5.12

A lateral directional well is to be drilled to a vertical target depth of 1780 m and horizontal displacement of 400 m, with a closure angle of 60°. The lateral drilling kick-off depth from the parent wellbore is at a depth of 1200 m. By interpolation the lateral drilling point parameters are calculated to be: deviation angle α_s = 12°, azimuth = 75°, north coordinate N_s = 5.40 m, east coordinate E_s = 20.16 m, and vertical depth H_s = 1198.54 m. Design the well path selecting a build rate of 3°/30 m.

Solution

The calculation yields the target coordinates: N_t = 200.00 m, E_t = 345.41 m, H_t = 1780.00 m. Since there is no hold section, ΔL_1 = 0. Using the arc-straight hold section profile the vector $\boldsymbol{c}$ = (0.0538, 0.2008, 0.9781), vector $\boldsymbol{d}$ = (0.2802, 0.4697, 0.8372), vector $\boldsymbol{b}$ = (–0.2913, 0.2290, –0.0310), and vector $\boldsymbol{a}$ = (0.6191, 0.7618, –0.1905).

The target location t in the local s-$\xi\eta\zeta$ coordinates is defined by ξ_t = –314.68 m, η_t = 0.00 m, and ζ_t = 64.75 m. The initial tool angle is ω = 335.37°. The arc of the bending angle is ε = 25.982°. The holding section length is ΔL_2 = 269.82 m. Other design results are shown in Table 5.10.

Horizontal Wells

When designing a sidetrack horizontal well, the horizontal well section is usually designed first, followed by the design of the well path from the sidetrack to the first target. Horizontal section design provides the inclination and the azimuth of the target. To design the well path before the target, the sidetrack position should be selected. After it is selected, the inclination, azimuth, and coordinates of the sidetrack can be determined by using the interpolation method. Therefore, when designing the well path of a sidetrack horizontal well, the spatial positions of start and end points and the

Table 5.10 Results for the Design of the Lateral Directional Well for Problem 5.12

Well Depth (m)	α (°)	ϕ (°)	Depth (m)	North (m)	East (m)	Departure (m)	Displacement Azimuth (°)	Location
1200.00	12.00	75.00	1198.54	5.40	20.16	20.87	75.00	Starting point s
1260.00	17.66	67.06	1256.52	10.57	34.58	36.16	73.01	
1320.00	23.48	62.93	1312.68	19.56	53.62	57.08	69.96	
1380.00	29.38	60.37	1366.38	32.29	77.08	83.57	67.27	
1440.00	35.30	58.61	1417.06	48.61	104.70	115.43	65.10	
1469.82	38.26	57.92	1440.94	58.00	119.88	133.18	64.18	End point f
1600.00	38.26	57.92	1543.16	100.81	188.18	213.48	61.82	
1800.00	38.26	57.92	1700.21	166.58	293.10	337.13	60.39	
1901.61	38.26	57.92	1780.00	200.00	346.41	400.00	60.00	Target t

well direction are determined. For horizontal wells and sidetrack horizontal wells, the same situation can be met in real-time, well path correction design during the drilling process, so the design method to be discussed here is also suitable for horizontal well, real-time well path design.[7,8]

In order to satisfy the requirements of the spatial positions of start and end points and the well direction, the well trajectory must include at least two arc sections, and the typical well trajectory is hold-arc (build/drop)-hold-arc-hold as shown in Figure 5.9.

A lateral horizontal well path design should satisfy the following equations[14]:

$$\begin{cases} \sum_{i=1}^{n} \Delta N_i = N_t - N_s \\ \sum_{i=1}^{n} \Delta E_i = E_t - E_s \\ \sum_{i=1}^{n} \Delta H_i = H_t - H_s \end{cases} \tag{5.93}$$

$$\begin{cases} \sum_{i=1}^{n} \Delta \alpha_i = \alpha_t - \alpha_s \\ \sum_{i=1}^{n} \Delta \phi_i = \phi_t - \phi_s \end{cases} \tag{5.94}$$

where

N, E, H = North coordinate (south is negative), east coordinate (west is negative), H is vertical depth, ft or m

α = Inclination angle, (°)

ϕ = Azimuth angle, (°)

t = Target location

n = Number of well profile sections

$\Delta N, \Delta E, \Delta H$ = Increments of the north, east, and depth coordinates over the section, m

In fact, the above constraint equations are not independent of each other. When the constraint equations for the coordinate and the inclination are satisfied, the azimuth constraint equation is also satisfied.[14]

Obviously, the sidetrack horizontal well can be designed if the above constraints equations are solved. However, the relationships between these parameters are very complex, involving many trigonometric functions, and it is difficult to have a reasonable iteration method, as sometimes even divergence in solution may occur. The design and solution method discussed in

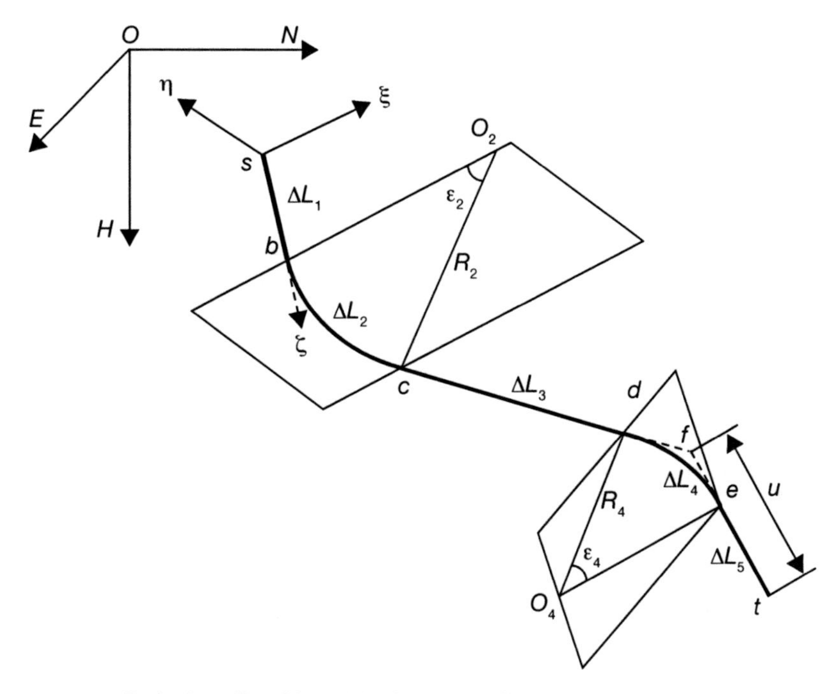

Figure 5.9 Typical profile of lateral horizontal well.

this section not only satisfies Eqs. 5.93 and 5.94, but also avoids solving the constraints equations directly, thereby ensuring convergence and numerical stability.

When designing the well path of a sidetrack horizontal well to reach a target, the known conditions include:

- Well direction and coordinate of the sidetrack point: α_s, ϕ_s, N_s, E_s, H_s
- Well direction and coordinate of the target point: α_t, ϕ_t, N_t, E_t, H_t
- Lengths of the first and final hold sections: ΔL_1, ΔL_5
- Build rates of the build sections: κ_2, κ_4 or R_2, R_4

The steps in the design method for a sidetrack horizontal well are summarized below:

1. Draw a straight line through the target point t, according to the required inclination angle α_t and azimuth ϕ_t. In the direction opposite to that of the well direction, obtain a segment from point t with length u, to find out the position of point f.

2. Assume an initial value of u as u^0, the coordinate of point f is given by:

$$\begin{cases} N_f = N_t - u^0 \sin\alpha_t \cos\phi_t \\ E_f = E_t - u^0 \sin\alpha_t \sin\phi_t \\ H_f = H_t - u^0 \cos\alpha_t \end{cases} \tag{5.95}$$

3. Using Eqs. 5.83 through 5.92, the well path from point s to point f can be designed and the inclination angle $\alpha_c(=\alpha_d)$ and azimuth $\phi_c(=\phi_d)$ at the second hold section can be obtained.

4. Calculate the new value u, which defines the length of the tangent line segment of the second arc section:

$$u = R_4 \tan\frac{\varepsilon_4}{2} \tag{5.96}$$

where the central angle over the second arc section is:

$\cos\varepsilon_4 = \cos\alpha_c \cos\alpha_t + \sin\alpha_c \sin\alpha_t \cos(\phi_t - \phi_c)$

5. The process of iterative calculation yields a value of u closer to the initial value. A convergence criterion of the form $|u - u^0| < \varepsilon$ can be used, where ε is a small number that can be selected for a desired accuracy of convergence.

6. Finally, the parameters of the well path are determined—the straight hold section from point c to point d (ΔL_3) and the second arc section length (ΔL_4) are:

$$\Delta L_3 = \Delta L_3' + \Delta L_5 - u \tag{5.97}$$

$$\Delta L_4 = \frac{\pi}{180} R_4 \varepsilon_4 \tag{5.98}$$

and the toolface angle is given by:

$$\tan\omega_d = \frac{\sin(\phi_t - \phi_c)}{\cos\alpha_c \left[\cos(\phi_t - \phi_c) - \dfrac{\tan\alpha_c}{\tan\alpha_t}\right]} \tag{5.99}$$

Thus, all the characteristic parameters of the well path can be found, and based on the spatial-arc model discussed in Chapter 3, the end points of each section of the well path can be calculated.

PROBLEM 5.13

For a sidetrack horizontal well, the TVD of the starting point t_1 is 1335 m, the horizontal displacement is 316 m, the closure azimuth is 145°, the TVD of final target point t_2 is 1335 m, the horizontal displacement is 670 m, and the closure azimuth is 138°. Drilling from an existing directional well, the sidetrack point is selected at TVD 1025 m; after interpolation, at the sidetrack point, the inclination angle is $\alpha_s = 10°$, the azimuth is $\phi_s = 120°$, the north coordinate is $N_s = -5.44$ m, the east coordinate is $E_s = 9.43$ m, and the TVD $H_s = 1024.37$ m. If a hold-build (drop)-hold-build (drop) well path is used with build rates in the build (drop) sections as $\kappa_2 = 8°/30$ m and $\kappa_4 = 10°/30$ m, design the sidetrack horizontal well with $\Delta L_1 = 30$ m.

Solution

After calculations, the coordinates of the target point t_1 are: $N_{t1} = -258.85$ m, $E_{t1} = 181.25$ m, and $H_{t1} = 1335$ m.

The coordinates of the target point t_2 are: $N_{t2} = -497.90$ m, $E_{t2} = 448.32$ m, $H_{t2} = 1335$ m. According to the selected well path, $\Delta L_5 = 0$. Based on the design method, after iteration, the following data can be obtained: $u = 75.48$ m, $w_b = 39.23°$, $\Delta L_2 = 143.68$ m, $e_2 = 38.31°$, $\Delta L_3 = 151.01$ m, $a_c = a_d = 45.42°$, $f_c = f_d = 152.77°$, $w_d = 330.97°$, $\Delta L_4 = 142.25$ m, and $e_4 = 47.42°$.

Other design results of the sidetrack well path can be found in Table 5.11.

Wells Bypassing Obstructions

To improve the success of drilling and the hydrocarbon recovery rates, sometimes several wells are planned near to each other. In some cases, while drilling such wells, especially from offshore platforms, there is little leeway for positioning or choice of direction. In certain cases obstructions such as salt domes, irretrievable fish, faults, gas gap–bearing formations and existing wells make the design much more complex. Such barriers may require the drilling of either directional or horizontal wells to bypass the obstruction.

Directional Wells

The simplest obstacle model is based on a vertical cylinder or cone method. For example, consider the case of an obstacle created by a previously drilled vertical well. Within a certain range of the previously drilled vertical well, the obstacle can be considered as a vertical cylinder. For an obstacle of a general shape, a circle or an arc is defined in the horizontal projection map, defined as the "detour circle," going around the obstacle to be avoided, as shown in Figure 5.10. Under this condition, the cylindrical spiral model of well path should be used to avoid the obstacle. Since the target position and shape of the obstacle can be found, the known parameters of the obstacle include TVD (H), horizontal displacement (A), and closure azimuth (φ_t) of

Table 5.11 Design Results for the Lateral Horizontal Well for Problem 5.13

Well Depth (m)	Inclination (°)	Azimuth (°)	Depth (m)	North (m)	East (m)	Departure (m)	Displacement Azimuth (°)	Location
1025.00	10.00	120.00	1024.37	–5.44	9.43	10.89	119.98	Start point s
1055.00	10.00	120.00	1053.91	–8.04	13.94	16.10	119.99	b
1080.00	15.73	135.71	1078.28	–11.56	18.19	21.56	122.43	
1198.68	46.42	152.77	1178.87	–62.61	49.91	80.07	141.44	c
1300.00	46.42	152.77	1248.71	–127.87	83.51	152.72	146.85	
1349.69	46.42	152.77	1282.97	–159.88	99.98	188.57	147.98	d
1380.00	55.42	146.83	1302.06	–180.14	111.86	212.05	148.16	
1491.95	90.00	131.83	1335.00	–258.85	181.25	316.00	145.00	target $t_1(e)$
1600.00	90.00	131.83	1335.00	–330.92	261.76	421.93	141.66	
1800.00	90.00	131.83	1335.00	–464.31	410.78	619.94	138.50	
1850.38	90.00	131.83	1335.00	–497.90	448.32	670.00	138.00	target t_2

the target, the horizontal distance between the center of the detour circle and the well location, closure azimuth (φ_M), and radius of the detour circle (r_M). The radius of the detour circle should be larger than the actual size of the obstacle by a certain amount, as a safety margin to ensure that collision is avoided. It is to be noted here that the center of the detour circle may not be the geometrical center of the obstacle, but will be the curvature center in the horizontal projection map. In what follows, for brevity, the section of the well path that detours around an obstacle is referred to as *detour path* or *detour section.*

Horizontal Projection Design As shown in Figure 5.10, if the coordinate system is O-NE rotated around the wellbore with an angle φ_t degree clockwise, such that the axis N passes through the target t and forms a new coordinate system O-xy, then the coordinate of point M is:

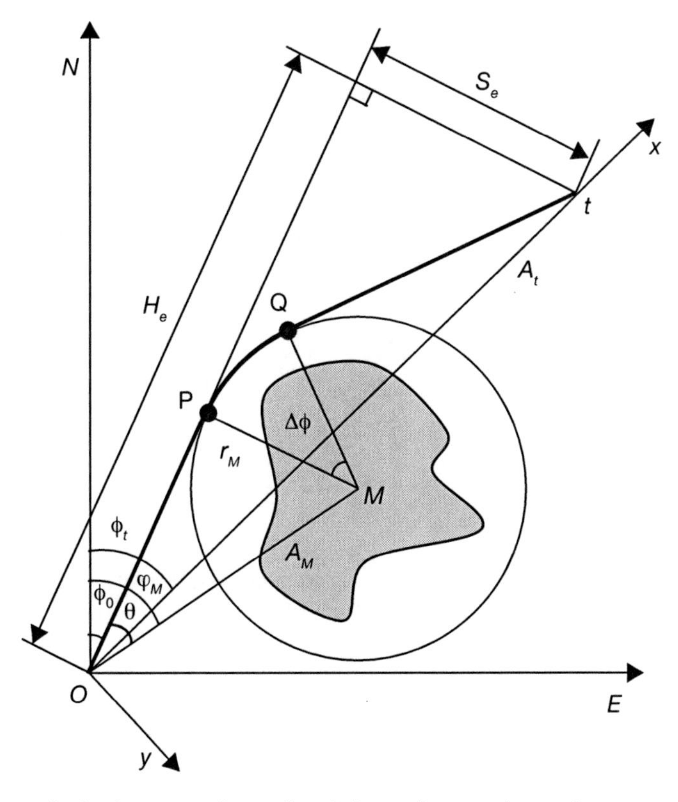

Figure 5.10 Projection map of a well path bypassing an obstruction.

$$\begin{cases} x_M = A_M \cos(\varphi_M - \varphi_t) \\ y_M = A_M \sin(\varphi_M - \varphi_t) \end{cases} \tag{5.100}$$

Here A_M is the horizontal displacement to the center of the circle around the obstruction, and the other related geometrical parameters are depicted in Figure 5.10.

Obviously, only when $0 < x_M < A_t$ and $|y_M| < r_M$, a design is needed to bypass the obstacle; otherwise, there is no need to consider the obstacle and the design becomes a conventional 2D well path.

For the simple case:

$$R_e = \begin{cases} -r_M, & \text{if } y_M < 0 \\ +r_M, & \text{if } y_M \geq 0 \end{cases} \tag{5.101}$$

Thus, when $R_e < 0$ (this case is shown in Figure 5.10), the design should detour the obstacle from the left side (clockwise detour); and when $R_e > 0$, the design should detour the obstacle from the right side (counterclockwise detour).

Therefore, the design for the initial position is:

$$\phi_0 = \varphi_M - \sin^{-1}\left(\frac{R_e}{A_M}\right) \tag{5.102}$$

On the horizontal projection, the well path is a three-step profile, as shown in Figure 5.9. Therefore, the azimuth angle is:

$$\tan\frac{\Delta\phi}{2} = \frac{H_e - \sqrt{H_e^2 + S_e^2 - R_e^2}}{R_e - S_e} \tag{5.103}$$

where

$$\begin{cases} H_e = A_t \cos(\varphi_t - \phi_0) - A_M \cos(\varphi_M - \phi_0) \\ S_e = A_t \sin(\varphi_t - \phi_0) - R_e \end{cases}$$

Then the horizontal length at P, Q, t, respectively, is given by:

$$\begin{cases} S_P = A_M \cos(\varphi_M - \phi_0) \\ S_Q = S_P + \dfrac{\pi}{180} \times R_e \Delta\phi \\ S_t = S_Q + \sqrt{H_e^2 + S_e^2 - R_e^2} \end{cases} \tag{5.104}$$

Vertical Profiled Design While using a vertical profile map to design a directional well path around a barrier, the target point length S_t and the horizontal displacement A_t should be considered. For example, for a single-arc vertical profile design, the maximum deviation angle and the length of the hold section, respectively, are:

$$\begin{cases} \tan\dfrac{\alpha_3}{2} = \dfrac{H_0 - \sqrt{H_0^{\ 2} + S_0^{\ 2} - R_2^{\ 2}}}{R_2 - S_0} \\ \Delta L_3 = \sqrt{H_0^{\ 2} + S_0^{\ 2} - R_2^{\ 2}} \end{cases} \tag{5.105}$$

where

$$\begin{cases} H_0 = H_t - \Delta L_1 \cos\alpha_1 + R_2 \sin\alpha_1 \\ S_0 = S_t - \Delta L_1 \sin\alpha_1 - R_2 \cos\alpha_1 \end{cases} \tag{5.106}$$

In the horizontal projection map, the target horizontal length (S_t) can be obtained from Eq. 5.104. Then, in the vertical well path map, after selecting the well profile type, many parameters need to be determined. In order to find the unknown parameters, the 2D well path design method described in Chapter 4 is used to design the well path with the detour around the obstacle. After calculating the unknowns of the well path, the TVD and horizontal displacement of each of the end points of the different well sections can be calculated.

Well Path Calculations Because various sections of the well profile conform to the cylindrical-helix model, the characteristic parameter varies and it is necessary to consider the positions of the start point P and end point Q located in the vertical projection plot.

As shown in Figure 5.11 for a single-arc profile, if $S_P \le S_a$, then P will be in the straight inclined section of the well path. For this condition, the length to point P is given by:

$$L_P = \frac{S_P}{\sin\alpha_1} \tag{5.107}$$

If $S_a < S_P \le S_b$, the point will be in the build section and the length to the point is given as:

$$L_P = L_a + \frac{\pi R_2}{180}(\alpha_p - \alpha_1) \tag{5.108}$$

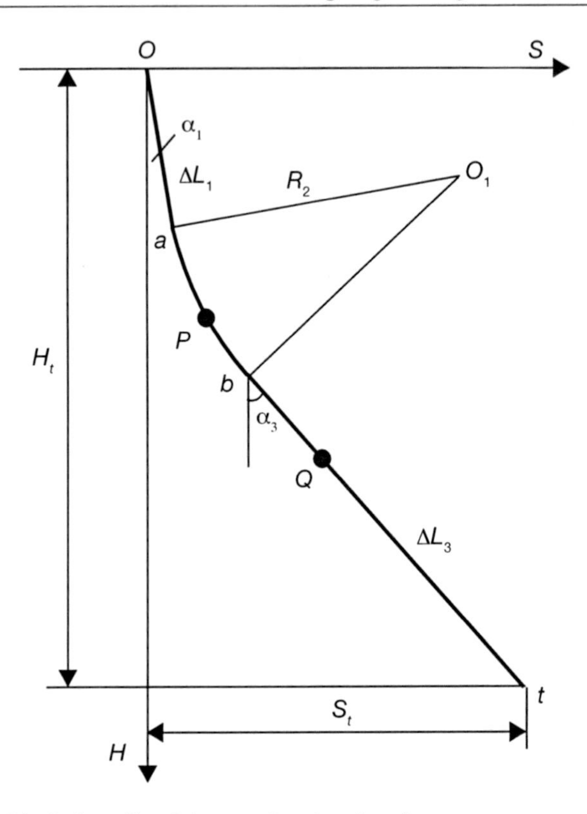

Figure 5.11 Vertical profile of detour directional well.

where

$$\cos\alpha_P = \cos\alpha_1 - \frac{S_P - S_a}{R_2} \tag{5.109}$$

If $S_P > S_b$, point P is in the hold section and the length to point P is:

$$L_P = L_b + \frac{S_P - S_b}{\sin\alpha_3} \tag{5.110}$$

Once the location of point P is determined, it is easier to calculate the location of point Q. For example, in a single-arc profile, if point P is in the inclined section, then point Q has to be in the hold section. Obviously, if there are more sections present in the profile, then the positions of P and Q are very difficult to find out.

According to the section ending points in vertical profile and points P and Q, the well path can be divided into several well sections, each of which has constant curvatures in the vertical profile and the horizontal projection plots. Then, the helical well path model can be used to calculate the well path parameters.

PROBLEM 5.14

A directional well is to be drilled with the target vertical depth H_t = 2000 m, horizontal displacement A_t = 500 m, and displacement angle ϕ_t = 320°. At ϕ_M = 310° and A_M = 120 m, an obstacle is present. A barrier radius of r_M = 80 m is required. If a well profile with straight-build-hold sections is planned to bypass the obstruction and the initial straight section is to be straight vertical, with κ_2 = 8°/100 m and the hold section length ΔL_1 = 1000 m, then design the well path.

Solution

The coordinate of detour center *M* in the *O-xy* coordinate system is:

$$x_M = 120 \times \cos(310° - 320°) = 118.18 \text{ m}$$
$$y_M = 120 \times \sin(310° - 320°) = -20.84 \text{ m}$$

As $0 < x_M < A_t$ and $|y_M| < r_M$, therefore, there is a need to design a detour around the barrier.

Because $y_M < 0$ so $R_e = -r_M = -80$ m, and hence the detour around the obstacle is from the left side (clockwise detour), the initial design direction is:

$$\varphi_0 = 310° - \sin^{-1}\left(\frac{-80}{120}\right) = 351.81°$$

The angle of azimuth change is:

$$H_e = 500 \times \cos(320° - 351.81°) - 120 \times \cos(310° - 351.81°) = 335.46$$

$$S_e = 500 \times \sin(320° - 351.81°) - (-80) = -183.55 \text{ m}$$

$$\Delta\varphi = 2 \times \tan^{-1}\frac{335.46 - \sqrt{335.46^2 + (-183.55)^2 - (-80)^2}}{(-80) - (-183.55)} = -40.76°$$

In the horizontal projection map, the horizontal length at *P*, *Q*, and *t* are:

$$S_P = 120 \times \cos(310° - 351.81°) = 89.44 \text{ m}$$

$$S_Q = 89.44 + \frac{\pi}{180} \times (-80) \times (-40.76°) = 145.35 \text{ m}$$

$$S_t = 146.35 + \sqrt{335.46^2 + (-183.55)^2 - (-80)^2} = 520.28 \text{ m}$$

The well path parameters in the vertical projection are:

$$R_2 = \frac{180 \times 100}{\pi \times 8} = 715.20 \text{ m}$$

$$H_0 = 2000 - 1000 \times \cos 0° + 716.20 \times \sin 0° = 1000 \text{ m}$$

$$S_0 = 520.28 - 1000 \times \sin 0° - 716.20 \times \cos 0° = -195.92 \text{ m}$$

$$\Delta L_3 = \sqrt{1000^2 + (-195.92)^2 - 716.20^2} = 724.88 \text{ m}$$

$$\alpha_3 = 2 \times \tan^{-1} \frac{1000 - \sqrt{1000^2 + (-195.92)^2 - 716.20^2}}{716.20 - (-195.92)} = 33.57°$$

In the vertical projection, the well depth and the corresponding horizontal length at *a*, *b*, and *t* are:

$L_a = 1000$ m

$S_a = 0$ m

$$L_b = 1000 + \frac{\pi}{180} \times 716.20 \times 33.57° = 1419.63 \text{ m}$$

$$S_b = 0 + 716.20 \times (\cos 0° - \cos 33.57°) = 119.45 \text{ m}$$

$L_t = 1419.63 + 724.88 = 2144.51$ m

$S_t = 520.28$ m

Since $S_a < S_P \le S_b$, point *P* is at the build section, and the inclination and depth are:

$$\alpha_P = \cos^{-1}\left(\cos 0° - \frac{89.44 - 0}{716.20}\right) = 28.94°$$

$$L_P = 1000 + \frac{\pi \times 716.20}{180} \times (28.94° - 0°) = 1361.77 \text{ m}$$

And since $S_b < S_Q \le S_t$, point *Q* is at the hold section and its depth is given as:

$$L_Q = 1419.63 + \frac{146.35 - 119.45}{\sin 33.57°} = 1468.28 \text{ m}$$

Thus, for designing this well path, there should be five sections and the associated characteristic parameters are given in Table 5.12. The design results are given in Table 5.13.

Horizontal Wells

Usually, designing horizontal well paths around obstacles is more difficult and complicated. Horizontal wells around obstacles can be designed for three typical situations.

Table 5.12 Characteristic Parameters of the Sidetrack Directional Well Path Bypassing the Obstruction for Problem 5.14

	Section		Vertical Projection		Horizontal Projection	
No.	**Node**	**Well Depth (m)**	**Curvature (°/100 m)**	**Radius of Curvature (m)**	**Curvature (°/100 m)**	**Radius of Curvature (m)**
1	*O-a*	0.00–1000.00	0.00	0.00	0.00	0.00
2	*a-P*	1000.00–1361.77	8.00	716.20	0.00	0.00
3	*P-b*	1361.77–1419.63	8.00	716.20	–71.62	–80.00
4	*b-Q*	1419.63–1468.28	0.00	0.00	–71.62	–80.00
5	*Q-t*	1468.28–2144.51	0.00	0.00	0.00	0.00

Table 5.13 Design Results for the Directional Well Path Bypassing the Obstruction for Problem 5.14

Well Depth (m)	Inclination (°)	Azimuth (°)	Depth (m)	North (m)	East (m)	Horizontal Displacement	Horizontal Length	Displacement Azimuth	Curvature (°/100 m)
0.00	0.00	—	0.00	0.00	0.00	0.00	0.00	—	0.00
1000.00	0.00	(351.81)	1000.00	0.00	0.00	0.00	0.00	—	0.00
1361.77	28.94	351.81	1346.58	88.53	–12.74	89.44	89.44	351.81	8.00
1419.63	33.57	330.31	1396.03	116.76	–22.43	119.46	118.89	349.13	23.31
1468.28	33.57	311.05	1436.57	137.47	–39.39	146.36	143.00	344.01	21.90
2144.51	33.57	311.05	2000.00	383.02	–321.39	520.29	500.00	320.00	0.00

1. *Horizontal Well Around an Obstacle in the Horizontal Section* For designing horizontal wells around an obstacle, inevitably the path needs to be designed in 3D space, as shown in Figure 5.12. Normally, the surface location, the obstacle, and the target are not in the same vertical plane.

2. *Design of a Horizontal Well Path with a Detour Before the Target* The design method for a directional well with a detour section can be used whether the horizontal section is a hold section, or is a complex path such as an arc shape or a ladder shape. After finishing the design of the horizontal section, the sidetrack horizontal well design method can be used to design the subsequent sections. (This part has been covered in the previous chapters and it will not be repeated here.)

PROBLEM 5.15

In a horizontal well, the first target has TVD H_{t1} = 1800 m, horizontal displacement A_{t1} = 280 m, and azimuth φ_{t1} = 180°. The final target has TVD H_{t2} = 1820 m, horizontal displacement A_{t2} = 940 m, and azimuth φ_{t2} = 180°. There is an obstacle at φ_M = 178°, and the detour circle radius is r_M = 80 m. The well is a ladder-shape horizontal well to explore two oil zones at different depths. It requires that the two inclination angles of the well sections in the oil zones be $\alpha_{t1} = \alpha_d$ = 88° and $\alpha_f = \alpha_{t2}$ = 90°. The length of the first oil zone section is $\Delta L_{t1,d}$ = 300 m. Try to design this horizontal well with the detour section.

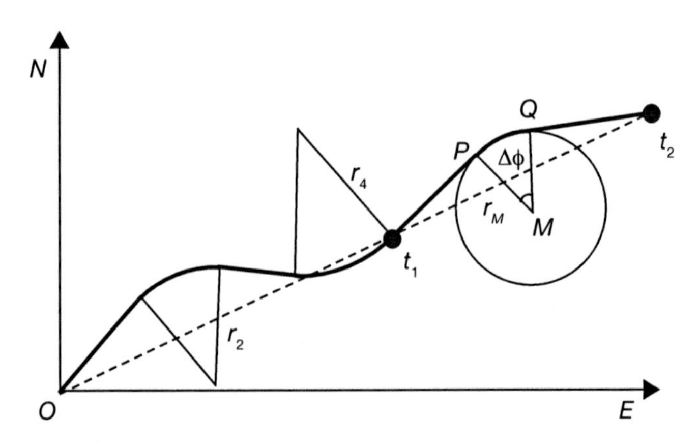

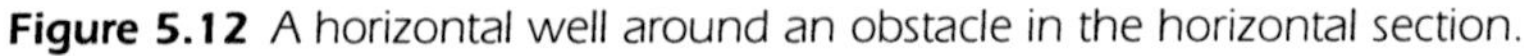
Figure 5.12 A horizontal well around an obstacle in the horizontal section.

Solution

The design and results are as follows:

1. The method for the design of a directional well with a detour section is used to design the horizontal well with a detour section. The horizontal projection mapping can be seen in Figure 5.13. The design results are: $\Delta S_{t1,P} = 412.60$ m, $\Delta S_{t1,Q} = 442.16$ m, $\Delta S_{t1,t2} = 670.20$ m, and $\Delta f_{P,Q} = -21.17°$.
2. The vertical well path of the ladder-shape horizontal well is designed as shown in Figure 5.13. According to the design requirements, on using 1.5°/30 m and 3°/30 m build rates for two build sections, the inclination angle is $\alpha_e = 83.746°$, the section length in the second oil zone is $\Delta L_{f,t2} = 223.137$ m, and two build sections lengths are $\Delta L_{d,e} = 85.08$ m and $\Delta L_{e,f} = 62.54$ m.
3. Based on the design results from Step 2, $\Delta S_{t1,e} = 384.65$ m and $\Delta S_{t1,f} = 447.06$ m are found. Comparing with the results of Step 1, $\Delta S_{t1,e} = 384.65 \text{ m} < \Delta S_{t1,P} = 412.60 \text{ m} < \Delta S_{t1,Q} = 442.16 \text{ m} < \Delta S_{t1,f} = 447.06$ m, so the detour section is the section *EF*.

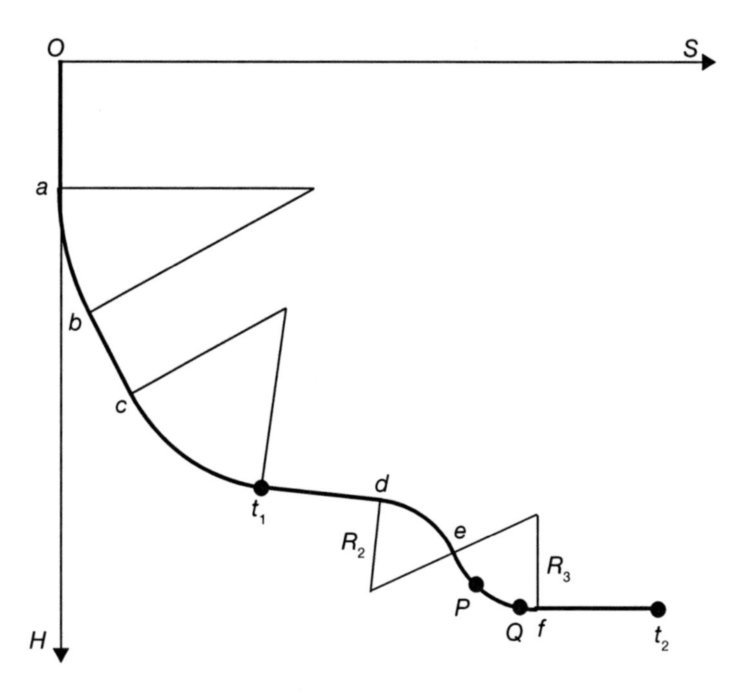

Figure 5.13 Design of a horizontal well with detour.

4. The design method of the sidetrack horizontal well is used to design the well path between the starting point and the first target. On using a two spatial-arc well path, or vertical-build-hold-build well path, the TVD at the kick-off point is 1480 m, the two build rates are 8°/30 m and 10°/30 m, and the inclination and the azimuth at the target are 88° and 187.641°, which ensures that the connection with the horizontal section is smooth. The results are: $\Delta L_{b,c}$ = 155.867 m, $\alpha_b = \alpha_c$ = 40.646°, and $\phi_b = \phi_c$ = 177.13°.
5. According to the well path model, the end points can be calculated and are displayed in Table 5.14.

When the obstacle is located in the last well section, there are two design methods that can be used: From the location of the obstacle, the detour section could be designed first; but from the location of the horizontal well, the horizontal section could be designed first.

If the detour section is designed first (i.e., the well path before the target), the design method for the directional well with a detour section can be used directly. When the well path before the target has been designed, the inclination angle and azimuth at the target are decided. Since the last target usually is not on the tangent line to the well path, a 3D well path design is needed for the horizontal well. Normally, there is no strict condition on the well path direction at the target, so the sidetrack directional well design method discussed earlier in this chapter can be used to design the horizontal well section, as shown in Figure 5.14.

PROBLEM 5.16

In a horizontal well, the TVD of the starting point is H_{t1} = 2000 m, the horizontal displacement is A_{t1} = 285 m, the displacement azimuth is ϕ_{t1} = 60°, the TVD of the final target is H_{t2} = 2010 m, the horizontal displacement is A_{t2} = 620 m, and the displacement azimuth is ϕ_{t2} = 70°. There is an obstacle in the azimuth direction ϕ_M = 68° with a horizontal displacement of A_M = 180 m. If the detour circle radius is r_M = 60 m, design the horizontal well with the detour.

Solution

The design steps and results are as follows:

1. The design method for a directional well with detour is used. The results for the horizontal projection map are: S_P = 169.71 m, S_Q = 202.63 m,

Table 5.14 Design Results for the End Points of the Well Path for Problem 5.15

Well Depth (m)	*Inclination (°)*	*Azimuth (°)*	*Depth (m)*	*North (m)*	*East (m)*	*Horizontal Displacement*	*Horizontal Length*	*Displacement Azimuth*	*Curvature (°/100 m)*
0.00	0.00	—	0.00	0.00	0.00	0.00	0.00	—	0.00
1000.00	0.00	(351.81)	1000.00	0.00	0.00	0.00	0.00	—	0.00
1361.77	28.94	351.81	1346.58	88.53	–12.74	89.44	89.44	351.81	8.00
1419.63	33.57	330.31	1396.03	116.76	–22.43	119.46	118.89	349.13	23.31
1468.28	33.57	311.05	1436.57	137.47	–39.39	146.36	143.00	344.01	21.90
2144.51	33.57	311.05	2000.00	383.02	–321.39	520.29	500.00	320.00	0.00

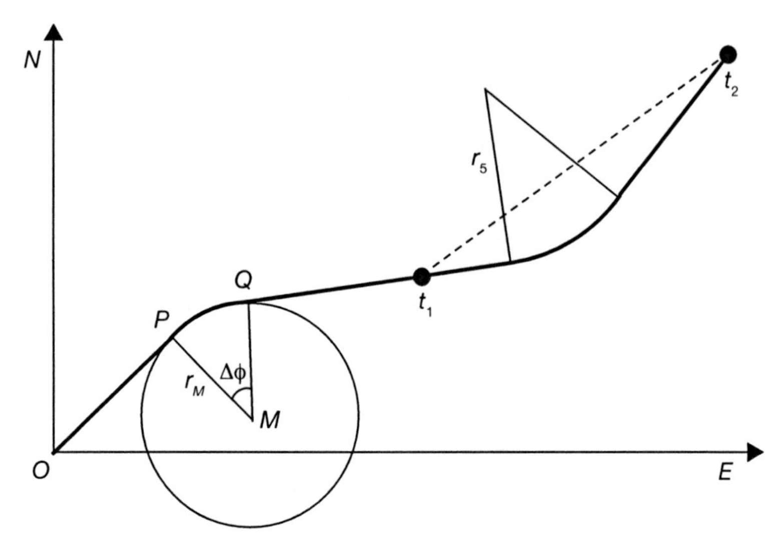

Figure 5.14 Design of a horizontal well path with a detour before the target.

S_{t1} = 294.41 m, and $\Delta\phi_{P,Q}$ = 31.44°. The directional build azimuth is ϕ_a = 48.53° and the target azimuth is ϕ_{t1} = 79.97°.

2. The design of the well profile on the vertical projection map is as shown in Figure 5.15. On using a well path with two build (drop) sections or with hold-build (drop)-hold-build (drop) sections, the kick-off point is 1700 m, and the build rates of the two build (drop) sections are 8°/30 m and 10°/30 m, respectively, and the inclination angle at the target is 88°. The design results are: $\Delta L_{b,c}$ = 148.025 m and $\alpha_b = \alpha_c$ = 49.899°.

3. According to the results from the design in Step 2 the following values are easily calculated: S_b = 75.462 m, S_c = 189.689 m, and S_{t1} = 294.408 m. Comparing with the results of Step 1, and knowing that $S_b < S_P < S_c$, $S_c < S_Q < S_{t1}$, shows that the starting point P of the detour is located in the section bc and the ending point Q is in the section ct_1.

4. The design method for a sidetrack directional well is used to design the horizontal well section. On using a build-hold well path, the build rate is 2°/30 m and the design results are: $\Delta L_{t1,d}$ = 315.52 m, $\alpha_d = \alpha_{t2}$ = 88.343°, and $\phi_d = \phi_{t2}$ = 78.231°.

5. According to the corresponding well path model, the ending points of the well path can be calculated, and the design results for this horizontal well with a detour section are given in Table 5.15.

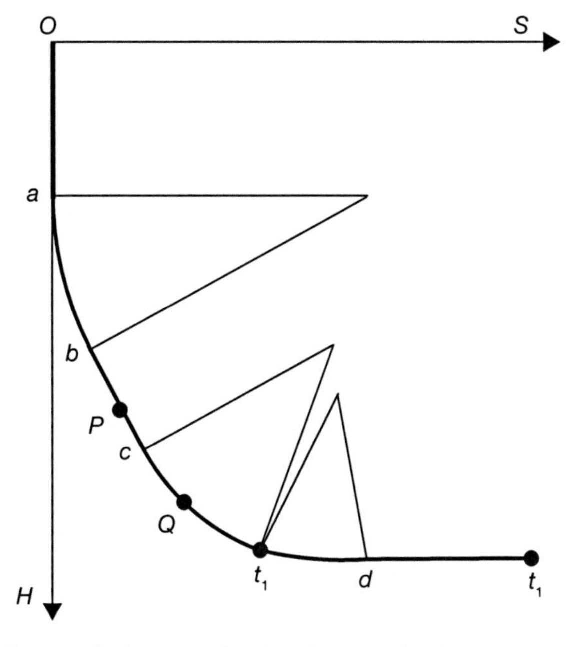

Figure 5.15 Design of a horizontal well to bypass a barrier.

3. *Designing a Horizontal Well Detour Section Prior to Reaching the Target*
When the obstacle is located in the well section just before the target, then if the horizontal well section is designed first and then the well path is designed to reach the target, the inclination angle (α_t) and azimuth (ϕ_t) at the target get decided after finishing the design of the horizontal well section. When designing the well path just before the target, it should not only detour the obstacle, but also make a smooth connection with the target and the horizontal section. Thus, besides the detour section, there should be a build section in the horizontal projection map, as shown in Figure 5.16.

For more details about the detour requirements, refer to the related content in Chapter 4. Based on the left or right direction of detour wells, the sign of r_M changes. So, the initial design azimuth of the detour section is:

$$\phi_0 = \varphi_M - \sin^{-1}\left(\frac{r_M}{A_M}\right) \tag{5.111}$$

In the horizontal projection map, the detour path from the start to the first target is equivalent to a hold-build-hold-build well profile, which can be seen in Figure 5.16. So the azimuth change over the detour section (PQ) is:

Table 5.15 Design Results for the Horizontal Well with Detour for Problem 5.16

Well Depth (m)	*Inclination (°)*	*Azimuth (°)*	*North (m)*	*East (m)*	*Departure (m)*	*Displacement Azimuth*	*Horizontal Length*	*Location*
0.00	0.00	—	0.00	0.00	0.00	—	0.00	Wellhead
1700.00	0.00	(48.53)	0.00	0.00	0.00	—	0.00	a
1887.12	49.90	48.53	50.64	57.29	76.46	48.53	76.46	b
2009.02	49.90	48.53	112.39	127.16	169.71	48.53	169.71	P
2035.15	49.90	67.61	122.90	144.04	189.35	49.53	189.69	c
2051.44	55.33	79.97	126.51	156.44	201.19	51.04	202.63	Q
2149.45	88.00	79.97	142.50	246.82	285.00	60.00	294.41	Target t_1
2176.00	88.34	78.23	147.52	272.87	310.19	61.60	320.94	d
2492.52	88.34	78.23	212.06	582.61	620.00	70.00	637.33	Target t_2

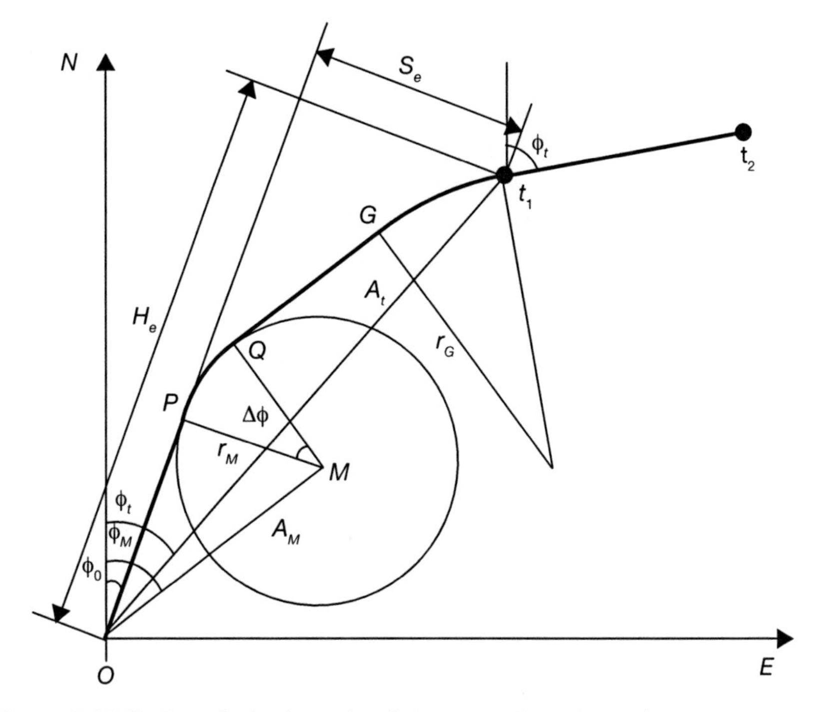

Figure 5.16 Design of a horizontal well detour section prior to the target.

$$\tan\frac{\Delta\phi}{2} = \frac{H_e - \sqrt{H_e^2 + S_e^2 - R_e^2}}{R_e - S_e} \tag{5.112}$$

where

$$\begin{cases} H_e = A_t \cos(\varphi_t - \phi_0) - A_M \cos(\varphi_M - \phi_0) - r_G \sin(\phi_t - \phi_0) \\ S_e = A_t \sin(\varphi_t - \phi_0) - r_M + r_G \cos(\phi_t - \phi_0) \\ R_e = r_M - r_G \end{cases}$$

The length at various points can be calculated as:

$$S_P = A_M \cos(\varphi_M - \phi_0) \tag{5.113}$$

$$S_Q = S_P + \frac{\pi}{180} \times r_M \, \Delta\phi \tag{5.114}$$

$$S_G = S_Q + \sqrt{H_e^2 + S_e^2 - R_e^2} \tag{5.115}$$

$$S_t = S_G + \frac{\pi}{180} \times r_G (\phi_t - \phi_0 - \Delta\phi) \tag{5.116}$$

After obtaining the horizontal length S_t of the first target, the well path can be selected and the vertical section of the well path before the target can be designed. Then, according to the position of the starting point P, ending point Q, and point G in the vertical projection map, the path before the target can be divided into several sections due to the nature of the well path; and well path parameters can be calculated by using the corresponding well path model.

PROBLEM 5.17

In a horizontal well, for the first target, vertical depth is H_{t1} = 1500 m, horizontal displacement is A_{t1} = 280 m, and displacement azimuth is ϕ_{t1} = 50°; and for the final target, vertical depth is H_{t2} = 1500 m, horizontal displacement is A_{t2} = 620 m, and displacement azimuth is ϕ_{t2} = 50°. There is an obstacle in the direction ϕ_M = 68° with the horizontal displacement A_M = 120 m. It is required that the detour-circle radius around the obstacle be r_M = 60 m. Design a horizontal well to detour the obstacle.

Solution

The design steps and results are as follows:

1. The horizontal section can be designed as a hold section, the results of which are: length is $\Delta L_{t1,t2}$ = 340 m, inclination is $\alpha_{t1} = \alpha_{t2}$ = 90°, and azimuth is $\phi_{t1} = \phi_{t2}$ = 50°. The inclination at the target is α_t = 90° and the azimuth is ϕ_t = 50°.

2. Based on the relative positions of the center of the detour circle and the first target, it is easy to decide that the detour should be from the right side (counterclockwise), or the detour-circle radius is r_M = 60 m (positive sign). Thus, the initial azimuth of the detour section is:

$$\varphi_0 = 68° - \sin^{-1}\left(\frac{60}{120}\right) = 38°$$

3. Since a counterclockwise detour is used, the azimuth is 38° and the target inclination is 50°, and the path from point G to first target t_1 should be clockwise. If the curvature on the horizontal projection map is 5°/30 m, then r_G = −343.775 m.

4. Calculating the azimuth change over the detour section PQ:

$$H_e = 280\cos(50° - 38°) - 120\cos(68° - 38°) - (-343.775)\sin(50° - 38°)$$
$$= 241.434 \text{ m}$$

$$S_e = 280\sin(50° - 38°) - 60 + (-343.775)\cos(50° - 38°) = -338.047 \text{ m}$$

$$R_e = 60 - (-343.775) = 403.775 \text{ m}$$

$$\Delta\phi = 2\tan^{-1}\frac{241.434 - \sqrt{241.434^2 + (-338.047)^2 - 403.775^2}}{403.775 - (-338.047)} = 21.941°$$

5. In the horizontal projection map, calculating the horizontal length of each section gives:
   $S_P = 120\cos(68°–38°) = 103.923$ m

$$S_Q = 103.923 + \frac{\pi}{180} \times 60 \times 21.941 = 125.900 \text{ m}$$

$$S_G = 126.900 + \sqrt{241.434^2 + (-338.047)^2 - 403.775^2} = 224.532 \text{ m}$$

$$S_t = 224.532 + \frac{\pi}{180} \times (-343.775) \times (50° - 38° - 21.941°) = 284.180 \text{ m}$$

6. The vertical projection map of the target well path is designed as shown in Figure 5.17. If a hold-build (drop)-hold-build (drop) well profile be chosen, where the target true vertical depth is 1500 m, horizontal displacement is 284.180 m, kick-off point is 1200 m, and the two build rates in the build sections are 8°/30 m and 10°/30 m, respectively, and the inclination angle at the target is 90°, the design results are: $\Delta L_{b,c} = 139.182$ m and $\alpha_b = \alpha_c = 45.575°$. The lengths of the two build sections are $\Delta L_{a,b} = 170.907$ m and $\Delta L_{c,t1} = 133.274$ m.
7. The positions of P, Q, and G in the vertical projection map are found as follows: According to the design results from Step 6, it is easy to calculate $S_b = 64.464$ m, $S_c = 163.863$ m, and $S_{t1} = 284.180$ m. Because $S_b < S_P < S_c$, $S_b < S_Q < S_c$, and $S_c < S_G < S_{t1}$, both the starting point P and the ending point Q of the detour section are located in section bc, and G is located in section c_{t1}. Thus, according to the values of S_P, S_Q, and S_G, the depths of these points can be calculated as follows: $L_P = 1425.159$ m, $L_Q = 1458.332$ m, and $L_G = 1582.449$ m.
8. Calculation of the data for the well path is as follows: According to the depths of the selected points, the well path can be divided into several sections and their starting and ending points can be calculated. Then, based on the well path model, an interpolation method can be used to calculate any position of interest. The design results for these points in the horizontal detour well can be seen in Table 5.16.

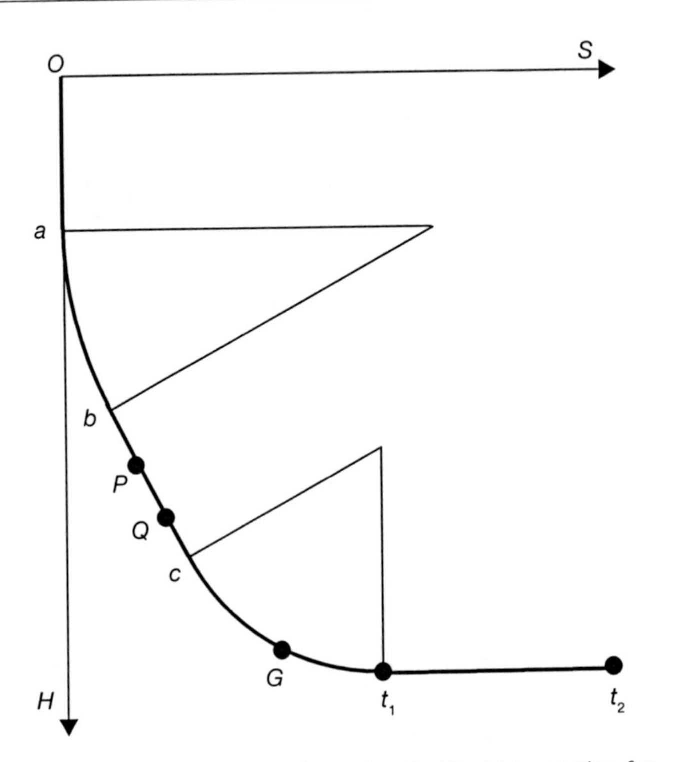

Figure 5.17 The schematic of the horizontal well with detour section for Problem 5.17.

The above-described design of a well with a detour section can only be used for the case of arc-shaped detours around an obstacle on the horizontal projection map. For more general or complicated cases, there is no general design method. The current method used for the general case is to design the well path ignoring the obstacle first, then to calculate the shortest distance between the path and the obstacle to see if it satisfies the design requirements. Then by adjusting the well path and the characteristic parameters, the well path design, obstacle distance calculation, and verification are iterated repeatedly, until the design results satisfy the requirements.

However, it is not necessary that all wells with detours around obstacles should be designed as 3D wells. Wherever conditions permit, they should as far as possible be designed using 2D profiles.

Bit-Walk Wells

Bit walk is the natural tendency of the drill bit to drift sideways while drilling. Generally, a right-hand drift is observed, and usually a left-hand drift is

Table 5.16 Design Results for the Horizontal Well with Detour Section for Problem 5.17

Well Depth (m)	*Inclination (°)*	*Azimuth (°)*	*North (m)*	*East (m)*	*Departure (m)*	*Displacement Azimuth*	*Horizontal Length*	*Location*
0.00	0.00	—	0.00	0.00	0.00	—	0.00	Wellhead
1700.00	0.00	(48.53)	0.00	0.00	0.00	—	0.00	a
1887.12	49.90	48.53	50.64	57.29	76.46	48.53	76.46	b
2009.02	49.90	48.53	112.39	127.16	169.71	48.53	169.71	P
2035.15	49.90	67.61	122.90	144.04	189.35	49.53	189.69	c
2051.44	55.33	79.97	126.51	156.44	201.19	51.04	202.63	Q
2149.45	88.00	79.97	142.50	246.82	285.00	60.00	294.41	Target t_1
2176.00	88.34	78.23	147.52	272.87	310.19	61.60	320.94	d
2492.52	88.34	78.23	212.06	582.61	620.00	70.00	637.33	Target t_2

observed with polycrystalline, diamond-compact bits. The degree of, or tendency to, bit walk depends on the formation dip angle, the bottomhole assembly, bit configuration, and other drilling operational parameters. While designing a well profile, well planners usually add a lead angle while kicking off so that the bit walk is compensated. Effectively designed 3D well paths reduce the frequency of azimuth corrections, reduce the wellbore tortuosity, improve the borehole quality, and reduce the number of trips. Various 3D well paths, including slanted wells, directional wells, and horizontal wells, can be designed to account for the bit-walk rate.

Bit-walk Unit Partition Based on Well Sections

Figure 5.18 shows the typical profile of a bit-walk path for an S-shaped directional well. It is noted that, unlike a spatial arc, a bit-walk section does not lie in a plane in space and the shape of the bit-walk section in the horizontal plot is generally not a simple curve, such as an arc, or a straight line. When there is a complex combination of curves, it is difficult to design the horizontal plot. Eq. 5.117 defines the sum of the coordinate increments over each section, which must be equal to the measured depth through the well path between the wellhead and the target.

The design of a 2D directional S-shaped well is shown in Figure 5.18, and requires the length and inclination of the third section, or hold section (*bc*), as a key parameter. In the case of the design of a 3D directional well, the well designer must determine the initial azimuth or the lead angle as a supplementary parameter.

A typical directional well is shown in Figure 5.18 with the target vertical depth from the wellhead H_t, the horizontal displacement A_t, and the closure angle ϕ_t. The wellbore characteristic parameters are the measured section length ΔL_i, the deviation angle α_i (i is odd), the build rate $\kappa_{\alpha,i}$, the radius of curvature R_i (i is odd), the kick-off point directional angle ϕ_0, and the drift rate $\kappa_{\phi,i}$. These characteristic parameters are mutually interdependent and affect each other, since the design of a 3D profile should consider the well path as a whole.

A 3D well path can be described by the two 2D curves in the vertical well path map and the horizontal projection map. As shown in Figure 5.18, in order to design the vertical well path of the 3D directional well, the horizontal length should be known. However, before designing the horizontal projection map, the horizontal length of the 3D directional well is unknown. In the horizontal map, to guarantee reaching the target depth, it is required not only to have a reasonable initial azimuth, but also suitable

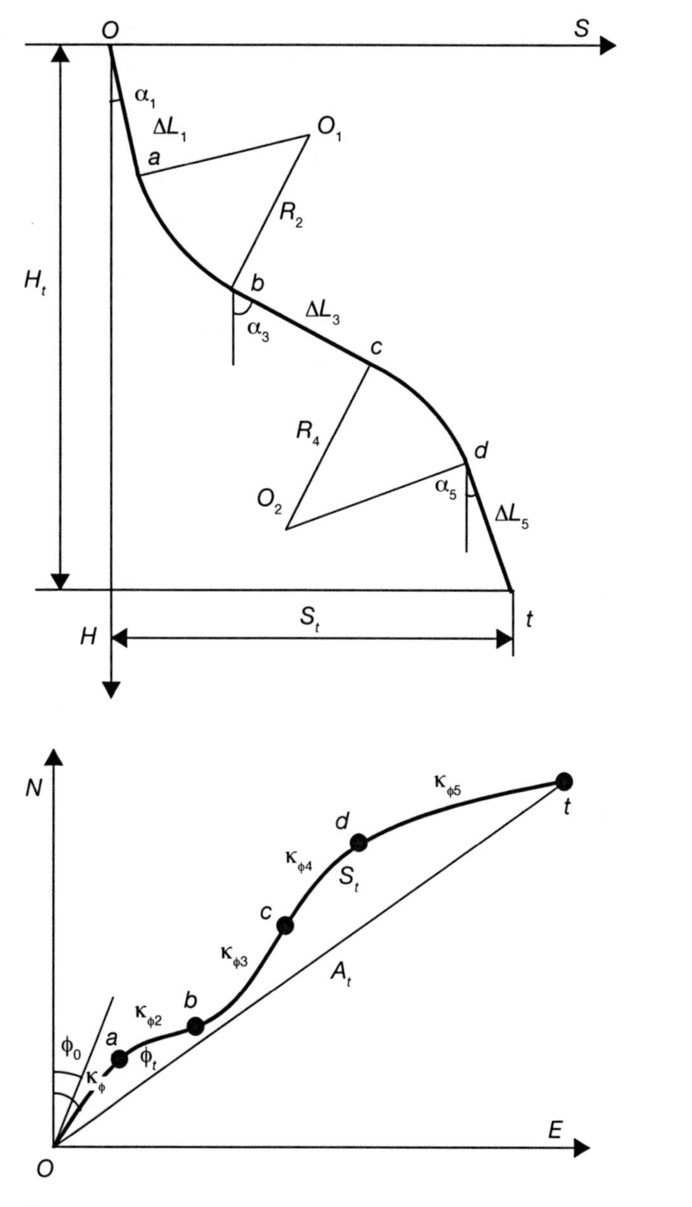

Figure 5.18 Directional well path positions.

horizontal lengths for each section. The horizontal length of each section is related to the section length and the inclination angle of this section, and the horizontal projection map of the well path is usually not made by the straight line and the arc section. Hence, it is difficult to design a 3D directional well using the vertical well path map and the horizontal projection map in a single step.[15-18]

However, the characteristic parameters of the well path sections are not independent, and the well path defined by them must satisfy the requirement of the relative coordinates of the well starting point and the target. In one sentence, the sum of the coordinate increments over the sections from the starting point to the target should be equal to the coordinate difference between the start and the target.

$$\begin{aligned} \sum_{i=1}^{n} \Delta N_i &= A_t \cos \varphi_t \\ \sum_{i=1}^{n} \Delta E_i &= A_t \sin \varphi \\ \sum_{i=1}^{n} \Delta H_i &= H_t \end{aligned} \tag{5.117}$$

where n is the number of well sections.

Using the three constraint equations in Eq. 5.117, three unknowns can be solved for in the design of the 3D drift path of a directional well. The solution has to be found by an iterative method because the system of equations is nonlinear. In principle, a three-level iterative method should be used with three unknowns. However, it is possible to have fewer iterations and improve the convergence by a proper choice of the iterative method.

The natural curve method is the best well path model to describe the azimuth drift. According to the definition, κ_α is positive in the build section and negative in the drop section. κ_ϕ is positive when the azimuth drifts to the right side, κ_ϕ is zero if there is no drift, and κ_ϕ is negative when the azimuth drifts to the left side. Thus, if the build rate (κ_α), azimuth drift rate (κ_ϕ), and section length (ΔL) are given in each well section, the coordinates' increment can be calculated using the natural curve model.

PROBLEM 5.18

In a directional well, the vertical depth is $H_t = 2500$ m, the horizontal displacement of the target is $A_t = 600$ m, and the closure azimuth is $\phi_t = 120°$. An S-shape well profile is used where the kick-off point vertical depth is 1200 m, build rate is 10°/100 m, and drop rate is 6°/100 m. In the hold section of the oil zone, the inclination is 8°, and the increment of the vertical depth is 200 m. If the azimuth drift rates of the build, hold, drop, and oil zone sections are 3°/100 m, 2°/100 m, 0°/100 m, and 2.5°/100 m, respectively, design the azimuth drift path of this directional well.

Solution

Using the design conditions and the constraint Eq. 5.33, it is found that the directional azimuth is $\phi_0 = 104.94°$, the inclination angle of the hold section is $\phi_3 = 42.15°$, and the section length is $\Delta L_3 = 280.05$ m. The end points of the well path are shown in Table 5.17.

Clothoid-Inserted 3D Wells

As discussed in Chapter 4, clothoid can be inserted as transion curves.[19] A clothoid curve is a spiral curve the curvature of which changes linearly from zero to a desired curvature proportional to the arc length. In other words, the radius of curvature at any point of the curve varies as the inverse of the arc length from the starting point of the curve.

Curvature and torsion are given as follows[19]:

$$\kappa(s) = \frac{\pi a}{\sigma^2 + b^2} u \tag{5.118}$$

$$\tau(s) = \frac{\pi b}{\sigma^2 + b^2} u \tag{5.119}$$

As the parameter u varies from $u = 0$ and $u = 2\pi$, the point on the clothoid curve advances in the z direction a distance of $2\pi|b|$, and the x and y components return to their original values.

Supplementary Problems

1. Prove that for the free inclination and free azimuth end condition for a SiT well path segment:

$$\frac{C_{2,H}}{C_{3,H}} = -3$$

2. Using the data in the worked out Problem 5.1 show that using a small tension ($\lambda = 0.01$) results in a well path length of 1805.26 m, final inclination of 75.09°, and final azimuth of 71.32°, similar to the results obtained using a 3D cubic model with the same data.

Table 5.17 Well Path Design Results for Problem 5.18

Well Depth (m)	*Inclination (°)*	*Azimuth (°)*	*North (m)*	*East (m)*	*Horizontal Length*	*Horizontal Departure*	*Displacement Angle*	*Location*
0.00	0.00	—	0.00	0.00	0.00	0.00	—	Wellhead
1200.00	0.00	(104.94)	0.00	0.00	0.00	0.00	—	*a*
1621.47	42.15	117.59	–58.51	135.89	148.15	147.95	113.29	*b*
1901.52	42.15	123.19	–153.53	297.93	336.07	335.16	117.26	*c*
2470.63	8.00	123.19	–283.60	496.80	573.70	572.05	119.72	*d*
2672.60	8.00	128.24	–300.00	519.62	601.81	600.00	120.00	Target *t*

3. A relief well is planned from a nearby well with a starting depth of 2500 ft and a vertical depth of 2400 ft. The coordinates at this depth are $N = 500$ ft, $E = 608$ ft, and azimuth = 10°. The blowing well is to be intersected at a vertical depth of 5000 ft with the coordinates $N = 1800$ ft, $E = 2000$ ft, inclination = 75°, and azimuth = 0°. Design the well path using cubic function as well as SiT function. For SiT functions, use tension parameter 3. $L_0 = 3000$ and $L_1 = 2500$.

4. Prove that for the free inclination and free azimuth end condition for a SiT well path segment:

$$\frac{C_{2,E}}{C_{3,E}} = -1$$

5. Derive the following coefficient for the set inclination and set azimuth end condition for a SiT well path segment:

$$C_{2,H} = \frac{1}{\Delta f}\left[SH_0 + SH_1 + \left(S + \frac{1-C}{\lambda}\right)\dot{H}_0 - \frac{1-C}{\lambda}\dot{H}_1\right]$$

6. Prove that:

$$\begin{cases} x_M = A_M \cos(\varphi_M - \varphi_t) \\ y_M = A_M \sin(\varphi_M - \varphi_t) \end{cases}$$

as defined in Eq. 5.100.

7. Prove that when $R_e < 0$ (as in the case shown in Figure 5.9), the design should detour the obstacle from the left side (clockwise detour); and when $R_e > 0$, the design should detour the obstacle from the right side (counterclockwise detour).

8. Plan an extended-reach well using a straight/vertical build, catenary build, hold-build and hold well path for which the following additional data are provided:
Kick-off point is at 1000 ft with azimuth of 210°
The build section has a build rate angle of 1.5°/100 ft
The catenary section is of 2000 ft with an end angle of 60°
The hold section length is 20,000 ft
The build section continues to 23,600 ft with an end angle of 70°
The build section further continues to 24,000 ft with an end angle of 95° and azimuth of 10°
The final hold section continues to 30,000 ft at an angle of 95°

a. Calculate the dogleg, torsion, toolface angle, and other wellbore parameters at the nodes.
b. How does the results compare with the results obtained on using a cycloid transition curve instead of a catenary curve?

9. Plan an azimuth drift path in an S-shaped directional well where the kick-off depth is at 2000 ft, the horizontal departure to the target is 9000 ft, and the TVD to the target is 15,000 ft. The rate of build and drop are 2°/100 ft. The azimuth drift rate, build-hold-drop rates of the oil sections are 3°/100 ft, 2°/100 ft, 0°/100 ft, and 2.5°/100 ft, respectively.

10. Design a horizontal well to detour an obstacle with the following data:

	First Target	*Final Target*
Vertical depth	5000 ft	5000 ft
Horizontal displacement	1000 ft	2000 ft
Displacement azimuth	230°	230°

Obstacle direction: 280° with a horizontal displacement of 500 ft.
Obstacle radius: 200 ft.

11. Plan a 3D well path (straight vertical-build-hold-build-hold (horizontal) type) with the following given data:
Kick-off point = 1740 ft
Build rate of the first build section = 8°/100 ft
Walk rate of the first build section = 0.6°/100 ft
Build rate of the second build section = 10°/100 ft
Walk rate of the second build section = –5°/100 ft
For the wellhead-to-landing point section:
Total vertical depth = 2000 ft
Total horizontal displacement = 236 ft
Azimuth angle of the horizontal displacement = 33.6°
For the final section:
Vertical depth change = 15 ft
Section horizontal displacement = 500 ft
Azimuth angle of the horizontal displacement = 30°
Bit-walk azimuth change rate = –3°/100 ft

12. Show that principal normals at consecutive points on a well path do not intersect, unless torsion is zero.

13. For well profiles needing small curvature, show that curvature can be estimated using:

$$\kappa_{min} = \frac{5400}{\pi} \frac{2\xi_t}{(\zeta_t - \Delta L_1)^2 + \xi_t^2} (°)/30 \text{ m}$$

or

$$\kappa_{min} = \frac{18{,}000}{\pi} \frac{2\xi_t}{(\zeta_t - \Delta L_1)^2 + \xi_t^2} (°)/100 \text{ ft}$$

References

1. M. Economides, D. Collins, W. Hottman, and J. Longbottom. "Horizontal, Multilateral, and Multibranch Wells in Petroleum Production Engineering." The Petroleum Well Construction Book (November 2005).
2. Azar, J. J., and S. G. Robello. *Drilling Engineering*. Tulsa, OK: Penwell Publishers, 2007.
3. Xiushan, L. *Geometry of Wellbore Well Path*. Beijing, China: Petroleum Industry Press, 2005.
4. Brown, D. E. "Programmed Math Keeps Directional Drilling on Target." *Oil & Gas Journal* 78, no. 12 (1980): 165–167.
5. Xiushan, L., Q. Tongci, S. Zhongguo, et al. "Design of a 3-D Drift Well-path." Beijing. *Acta Petrolei Sinica* 16, no. 4 (1995): 118–124.
6. Xiushan, L., and S. Zaihong. "Technique Yields Exact Solution for Planning Bit-walk Paths." *Oil & Gas Journal* 100, no. 5 (2002): 45–50.
7. Xiushan, L., P. Guosheng, and Z. Xiaoxiang. "An Advanced Method for Planning and Surveying Well Trajectories of Rotary Steering Drilling." *Acta Petrolei Sinica* 24, no. 4 (2003): 81–85.
8. Mitchell, B. *Advanced Oilwell Drilling Engineering Handbook*, 9th ed. Dallas, TX: SPE–AIME, Chapter 3, pp. 355–370.
9. Press, W. H., S. A. Teukolsky, W. T. Vetterling, and B. P. Flannery. *Numerical Recipes in Fortran: The Art of Scientific Computing*, 2nd ed. New York: Cambridge University Press, 1992.
10. Stoker, J. J. *Differential Geometry*. New York: John Wiley & Sons, 1969.
11. Sampaio, Jr., J. H. B. "Planning 3D Well Trajectories," Contribution.
12. Sampaio, Jr., J. H. B. "Planning 3D Well Trajectories Using Cubic Functions." JERT 04–1094 and Sampaio Jr., J. H. B. Contribution.
13. Sampaio, Jr., J. H. B. "Planning 3D Well Trajectories Using Spline-in-Tension Functions." JERT 04–1094 (February 27, 2006).
14. Xiushan, L. *Geometry of Wellbore Trajectory*. Beijing, China: Petroleum Industry Press, 2006.
15. Brown D. E. "Programmed Math Keeps Directional Drilling on Target." *Oil & Gas Journal* 78, no. 12 (1980): 165–167.
16. Xiushan, L., Q. Tongci, S. Zhongguo, et al. "Design of a 3-D Drift Well-path." *Acta Petrolei Sinica Beijing* 16, no. 4 (1995): 118–124.

17. Xiushan, L., and S. Zaihong. "Technique Yields Exact Solution for Planning Bit-walk Paths." *Oil & Gas Journal* 100, no. 5 (2002): 45–50.
18. Xiushan, L., P. Guosheng, and Z. Xiaoxiang. "An Advanced Method for Planning and Surveying Well Trajectories of Rotary Steering Drilling." *Acta Petrolei Sinica* 24, no. 4 (2003): 81–85.
19. Samuel, R. "Ultra Extended-Reach Drilling (*u*-ERD: Tunnel in the Earth)—A New Wellpath Design." SPE-119459,2009 SPE/IADC Drilling Conference, March 17–19, Amsterdam, The Netherlands.

6

Well Trajectory Monitoring

This chapter discusses the principles of trajectory monitoring and surveying, so that the actual shape of a wellbore being drilled is calculated accurately. Well planning and the design of well profiles are based on straight and smooth curved sections fitted between the starting surface location and the final target depth, taking into account the given obstructions and limitations. But during the actual drilling it is difficult, if not impossible, to make the trajectory fully consistent with the designed well path, and therefore, monitoring the trajectory as the well is being drilled is important so that corrective measures are taken in time to meet the engineering and geological requirements. Trajectory monitoring is especially important during the construction of complex wells for precise placement of the wells. This chapter covers some of the methods that can be used to track the well path as it is being drilled and compare it against the designed well path. Survey calculations are part of the necessary basic work for quantitatively monitoring and controlling the wellbore trajectories. The notations used in this chapter are as defined in the previous chapters.

Azimuth Conversion

The azimuth of any direction is the horizontal angle measured in the clockwise direction between the north and that direction. The angle between the north and the horizontal map projection of the tangent at any point of the well profile is called the well profile azimuth.

At each point in the well profile, there are three "north" directions: namely, the geographic north, the magnetic north, and the coordinate north. Because of these three different north directions, there are three different azimuths. For the design, monitoring, and control of well profiles, it is necessary to be able to convert between the different azimuths, and thus, the related knowledge of the declination, the meridian angle of convergence, other azimuth measurements, and azimuth conversions are required.[1]

Geomagnetic Field and Declination

The Earth's magnetic field is a basic physical field extending from the Earth's core to the cosmic space around the Earth. The fundamental data of this field are used in many areas, such as Earth sciences, resource exploration, astronautics and aeronautics, traffic and communication, national defense, earthquake predictions, space weather, and so on. The studies and applications of the Earth's magnetic field and declination are very important for the monitoring and control of well profiles during drilling.

The Composition of the Earth's Magnetic Field

The Earth's magnetic field is similar to the magnetic field created as if a magnetic dipole were positioned at the center of the Earth. The dipole's axis does not coincide with the planet's axis of rotation, and the angle between them is approximately 11.5°. The dipole's axis crosses two points on the Earth's surface, called the Earth's magnetic poles. The one near the geographic North Pole is called the magnetic North Pole, and the other one near the geographic South Pole is called the magnetic South Pole. The magnetic North Pole and the magnetic South Pole of the Earth are named with reference to their geographic locations. In reality, the magnetism of the Earth's magnetic poles would be opposite to that of a magnetic needle. The magnetic field of the Earth's magnetic dipole is shown in Figure 6.1.

The positions of the magnetic North and South Poles keep changing with time, and so in practice the actual locations of the Earth's magnetic poles are based on theoretical estimates. The Earth's magnetic field is a weak magnetic field, and the average magnetic induction on the Earth's surface is about 0.5×10^{-4} T. This magnetic field is created by a variety of magnetic sources, including stable magnetic fields from within the Earth and unstable magnetic fields originating from outside the Earth.

Gauss' theory of the Earth's magnetic field shows that the stable magnetic fields include both fields originating from inside the Earth as well as

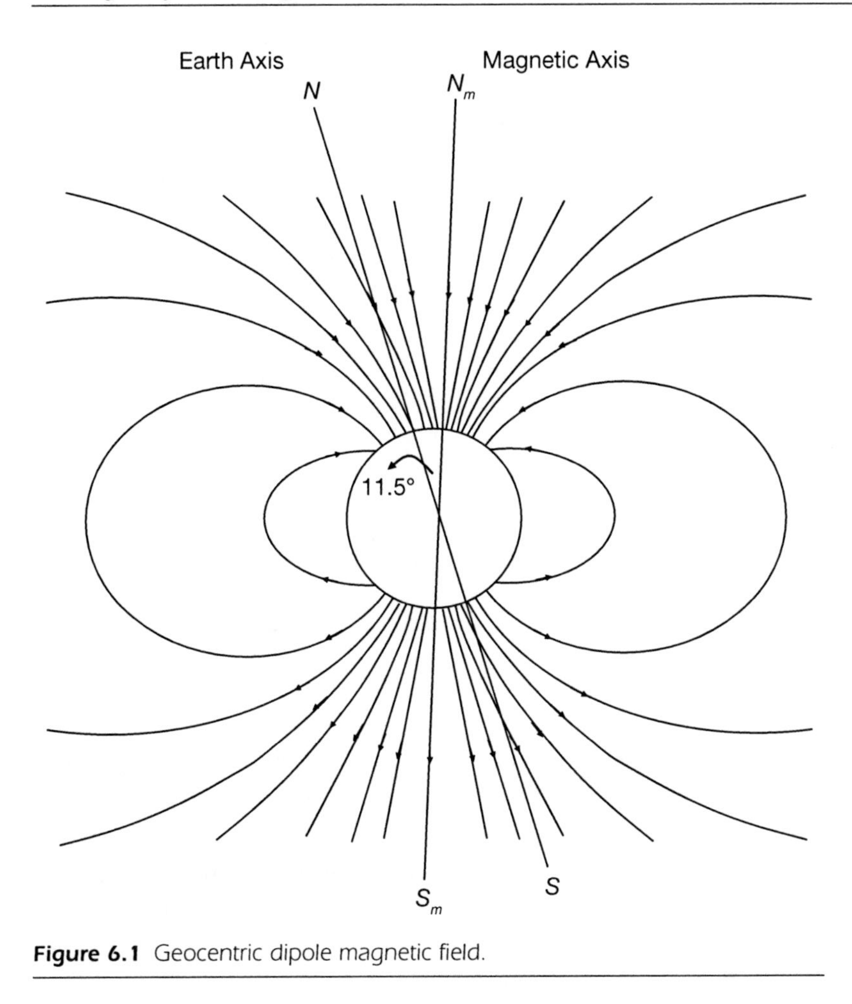

Figure 6.1 Geocentric dipole magnetic field.

fields originating from outside the Earth, and the internal fields contribute to more than 99 percent of the Earth's magnetic induction. Thus, the stable fields, also called the basic magnetic field, are the major part of the Earth's magnetic field, and originate mainly from the inside the Earth.[2,3]

Geomagnetic Elements

The magnetic induction of the Earth's magnetic field ($\boldsymbol{B}_T$) is a vector, having both magnitude and direction, as shown in Figure 6.2. In the geographic Cartesian coordinate system (O-xyz), the direction of the x-axis is north, the y-axis is east, and the z-axis is the vertical pointing downward. H denotes the projection of $\boldsymbol{B}_T$ on the horizontal plane O-xy and is called the horizontal

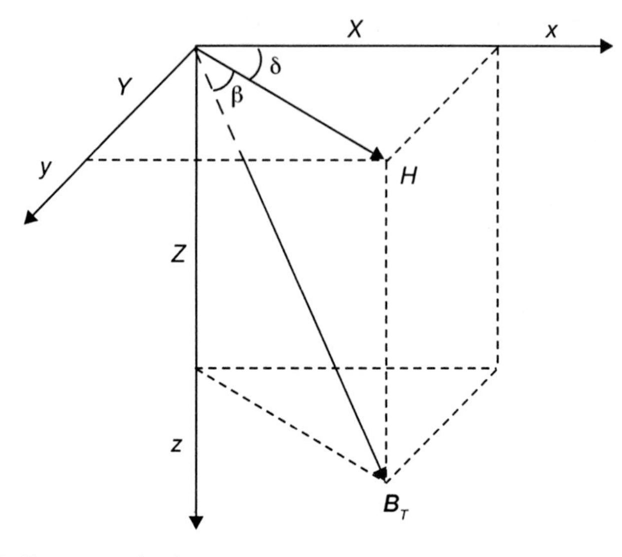

Figure 6.2 Geomagnetic elements.

component of the Earth's magnetic field with direction toward the magnetic north. The horizontal angle between the magnetic north and the geographic north is called the declination (δ), or the magnetic variation, and this is also the angle between the magnetic meridian and the true meridian. Starting from the true meridian, the declination is positive when it is toward the east and negative when it is toward the west. The downward or upward angle that $\boldsymbol{B}_T$ makes with the horizontal plane is called the geomagnetic inclination or the dip angle (β). In the Northern Hemisphere, the north point of the magnetic needle inclines downwards, and in the Southern Hemisphere, the north point of the needle inclines upwards. The geomagnetic inclination is positive when downward and negative when upward.[2,3]

Either using the three components X, Y, Z of $\boldsymbol{B}_T$ in the O-xyz coordinate system, or making use of H, δ, β, the magnitude and the direction of the Earth's magnetic induction can be determined, and thus, these parameters are called the essential elements of the Earth's magnetism or the geomagnetic elements. These parameters are interrelated as follows:

$$\begin{cases} X = H\cos\delta \\ Y = H\sin\delta \\ Z = H\tan\beta \end{cases} \tag{6.1}$$

$$\begin{cases} H^2 = X^2 + Y^2 \\ \tan\delta = \dfrac{Y}{X} \end{cases} \tag{6.2}$$

$$\begin{cases} {B_{\mathrm{T}}}^2 = H^2 + Z^2 \\ B_{\mathrm{T}} = H\sec\beta = Z\csc\beta \\ \tan\beta = \dfrac{Z}{H} \end{cases} \tag{6.3}$$

In order to determine the intensity and the direction of the Earth's magnetic field, at least three independent essential elements should be measured, and in practice, the declination δ is always included. Currently, δ, β, H, Z, and B_T can be measured directly. At a magnetic survey station, the three essential elements recorded by the instruments are usually H, Z, and δ.

Magnetic Field Model

The geodynamic model of the Earth's magnetic field, used in high-altitude physics and in the theory of earthquake prediction, has important applications for the exploration of oil and gas and mineral resources.

Models for the Earth's magnetic field model can be classified as the global model and local models. Since 1968, the International Association of Geomagnetism and Aeronomy (IAGA) has given every five years a set of global International Geomagnetic Reference Fields (IGRF) based on Gauss-Schmidt theory. However, the error in the IGRF values is approximately 10 nT and up to 200 nT in some regions. Thus, many countries have their own local magnetic field models. They are different from the IGRF global model in the sense that the local models do not use the spherical harmonic analysis method, but other methods such as polynomial expansion, torque harmonic analysis, and spherical cap harmonic analysis.

Gauss-Schmidt theory postulates that the Earth's magnetic field satisfies Maxwell's equations, and the basic magnetic field originating from the Earth's interior is the main component, the magnetic potential (U) of which is:

$$\mu_0 U = a\sum_{n=1}^{\infty}\sum_{m=0}^{n}\left(\frac{a}{r}\right)^{n+1}\left(g_n^m \cos m\lambda + h_n^m \sin m\lambda\right)P_n^m\left(\cos\theta\right) \tag{6.4}$$

where

μ_0 = Magnetic permeability
a = Radial distance from the Earth's center
λ = Longitude

θ = Latitude

g_n^m, h_n^m = Gauss spherical harmonic coefficients of the geomagnetic field originating from the Earth's interior

$P_n^m(\cos\theta)$ = Associated legendre function in Schmidt's form.

Schmidt's definition states that:

$$P_n^m(\cos\theta) = \left[\frac{2}{\delta_m} \times \frac{(n-m)!}{(n+m)!}\right]^{\frac{1}{2}} P_n^{(m)}(\cos\theta) \tag{6.5}$$

where

$$P_n^{(m)}(\cos\theta) = \sin^m\theta \frac{d^m}{d(\cos\theta)^m} P_n(\cos\theta)$$

$$\delta_m = \begin{cases} 2, & m = 0 \\ 1, & m \geq 1 \end{cases}$$

The relations between the components of the Earth's basic magnetic field intensity and the magnetic potential are:

$$X = \frac{\mu_0}{r}\frac{\delta U}{\delta\theta} = \sum_{n=1}^{\infty}\sum_{m=0}^{n}\left(\frac{a}{r}\right)^{n+1}\left(g_n^m \cos m\lambda + h_n^m \sin m\lambda\right)\frac{d}{d\theta}P_n^m(\cos\theta) \tag{6.6}$$

$$Y = -\frac{\mu_0}{r\sin\theta}\frac{\delta U}{\delta\lambda} = \sum_{n=1}^{\infty}\sum_{m=0}^{n}\left(\frac{a}{r}\right)^{n+1}\frac{m}{\sin\theta}\left(g_n^m \sin m\lambda - h_n^m \cos m\lambda\right)P_n^m(\cos\theta) \tag{6.7}$$

$$Z = \mu_0\frac{\delta U}{\delta r} = -\sum_{n=1}^{\infty}\sum_{m=0}^{n}(n+1)\left(\frac{a}{r}\right)^{n+1}\left(g_n^m \cos m\lambda + h_n^m \sin m\lambda\right)P_n^m(\cos\theta) \tag{6.8}$$

In particular, for a point on the Earth's surface ($r = a$),

$$X = \sum_{n=1}^{\infty}\sum_{m=0}^{n}\left(g_n^m \cos m\lambda + h_n^m \sin m\lambda\right)\frac{d}{d\theta}P_n^m(\cos\theta) \tag{6.9}$$

$$Y = \sum_{n=1}^{\infty}\sum_{m=0}^{n}\frac{m}{\sin\theta}\left(g_n^m \sin m\lambda - h_n^m \cos m\lambda\right)P_n^m(\cos\theta) \tag{6.10}$$

$$Z = -\sum_{n=1}^{\infty}\sum_{m=0}^{n}(n+1)\left(g_n^m \cos m\lambda + h_n^m \sin m\lambda\right)P_n^m(\cos\theta) \tag{6.11}$$

Thus, using Eqs. 6.1, 6.2, and 6.3, the normal parameters of the geomagnetic field can be calculated.

The polynomial method is simple and has been widely applied, in countries around the world, for the local or regional model of the geomagnetic field. The regional Earth's magnetic field model is widely used due to polynomial method. Four independent magnetic elements (usually H, Z, θ, δ) should be selected to simulate the magnetic field distribution using numerical simulation methods. Otherwise, when different magnetic element combinations are used contradicting results may show up in the Earth's magnetic field map that describes the Earth's magnetic field distribution. In petroleum engineering, the important magnetic field element is dip angle, and so it is feasible to use the single Earth's magnetic element to fit by polynomials.

The geomagnetic elements are expanded in a Taylor series of the form:

$$\begin{aligned} f(U,V) &= a_0 + a_1U + a_2V + a_3U^2 + a_4UV + a_5V^2 \\ &+ a_6U^3 + a_7U^2V + a_8UV^2 + a_9V^3 + \cdots \end{aligned} \tag{6.12}$$

where

$$U = \phi - \phi_0$$
$$V = \lambda - \lambda_0$$

where

ϕ = Latitude at a point on the Earth's surface
λ = Longitude at a point on the Earth's surface
(ϕ_0, λ_0) = Latitude and longitude at the origin of Taylor multinomial expansion, respectively

The above multinomial contains only the spatial variables, and therefore cannot represent the variation of the regional magnetic field pattern with time. To take into account the time variation, an adjustment is made to the annual average of the magnetic declination as follows: If the annual average of the declination changes linearly with time, then:

$$\delta_t = \delta_{t_0} + \dot{\delta}(t - t_0) \tag{6.13}$$

If the annual average of the declination changes nonlinearly with time, a quadratic model can be used:

$$\delta_t = \delta_{t_0} + \dot{\delta}(t - t_0) + \ddot{\delta}(t - t_0)^2 \tag{6.14}$$

where

δ_t = Annual average of the magnetic declination for the year t
δ_{t_0} = Annual average of the magnetic declination for a specific base year t_0

$\dot{\delta}$ = Rate of change with time of the annually averaged magnetic declination for the base year t_0

$\ddot{\delta}$ = Half the rate of acceleration in the change with time of the annually averaged magnetic declination

In order to simplify the calculations, U and V in Eq. 6.12 should be made dimensionless. Moreover, in order to solve the discontinuity problem of the regular polynomial fitting method, the time variable can be added into the polynomial as well.

If magnetic declination data updates are not promptly obtained, there may be errors in the well path calculations. For example, the angular bias error may reach 1°, which for a 1000-m horizontal displacement well would result in a swing error of about 17.45 m. Therefore, timely updates of magnetic declination data are to be ensured for achieving accurate placement and positioning of wells.

Azimuth Conversion Methods

Due to the three different definitions of the north direction there are three different azimuths that can be defined:

- True azimuth ϕ_T: The direction to the geographic North Pole is the true north direction, and the corresponding longitude is also called the true meridian. The azimuth as measured from the true north direction is the true azimuth.
- Magnetic azimuth ϕ_M: The Earth generates a dynamic magnetic field, such that the locations of its magnetic poles change with time. The direction to the magnetic North Pole is called the magnetic north direction. The azimuth as measured from the magnetic north direction is called the magnetic azimuth.
- Coordinate azimuth ϕ_G: In maps based on projections, such as Gauss' projection plane, a coordinate north direction is used. The azimuth measured from the coordinate north direction is the coordinate azimuth.

The relationship between the azimuths is shown in Figure 6.3, and the relationship can be expressed by the following equations (see Chapter 1):

$$\begin{cases} \phi_T = \phi_M + \delta \\ \phi_G = \phi_T - \gamma \\ \phi_G = \phi_M + \delta - \gamma \end{cases} \tag{6.15}$$

where γ = Meridian convergence.

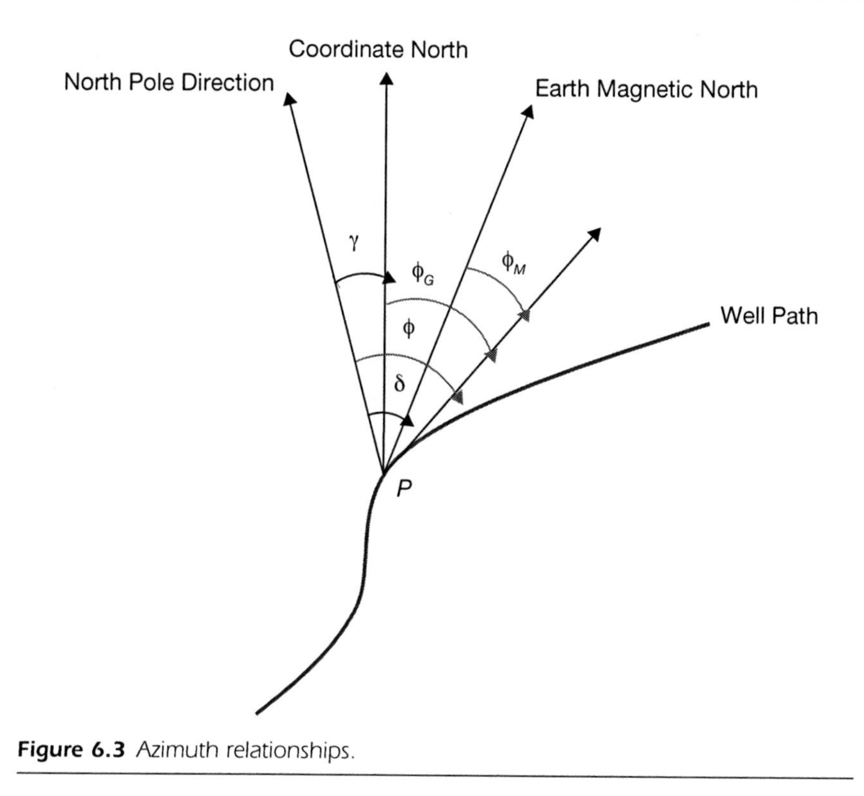

Figure 6.3 Azimuth relationships.

Usually, the drilling industry uses latitudes and longitudes from maps based on Gauss' projection plane for designing the well location and target position. However, the azimuth given by the measurement tools is the magnetic azimuth, and if this azimuth is not converted, the actual well profile would be based on the Earth's magnetic North Pole, which will result in differences between the actual well trajectory and the designed well path. Thus, the designed well trajectory and the actual or real well profile must use the same north reference base.

Because the direction of the magnetic North Pole changes with time, it is not appropriate to use it as the reference north base. If the true north is used, it is necessary to convert the designed well path and the real well profile to the true north coordinate system. The conversion method for the designed trajectory is that, while keeping the same target horizontal displacement, the meridian convergence angle is added to the closure azimuth in the Gauss' projection plane coordinates, which results in the closure azimuth with respect to the true north direction. And the conversion method for the real profile is that the magnetic declination is added to the measured magnetic azimuth so that the coordinates of the real profile can be calculated. If, instead of using the true north direction, the north of the Gauss' projection plane coordinates is used, then the designed well path need not be converted, but it requires conversion of the measured magnetic azimuth to the azimuth in the Gauss' projection plane coordinates for the calculation of the coordinates of the real well profile.[4,5]

The above-mentioned conversion is not only required for the azimuth, but is also needed for the north coordinate and the east coordinate. The azimuth should be converted first, and then the well profile coordinates should be calculated using the converted azimuth. In this way, the azimuth and coordinates of the well profile can be converted for the north direction of the selected map projection coordinate system.

In well trajectory design, monitoring, and control, it is necessary to convert between the different azimuths. However, in earlier times it was considered more important to convert the measured magnetic azimuth to the true azimuth, but not so essential to convert the coordinate azimuth obtained using the geographic north of the Gauss' projection plane map. However, theoretical studies and the field practice show that the meridian convergence angle should not be neglected and that the same north direction should be used for both the designed trajectory and the real well profile. This direction can be either the coordinate north direction of the Gauss' projection plane or the true north direction (with the former one more often being preferred).

Well Survey Calculations

Data Processing

In order to ensure correct and uniform usage of the measured data, the following systematic methodology is followed:

- The trajectory parameters at the survey station points along the well path from top to bottom have the subscript, $i = 1, 2, 3, \ldots, n$.
- Well sections are also subscripted with $i - 1$ for the top end and i for the bottom end.
- When the inclination angle at the first measured point is zero, it is assumed that the well is vertical from the start to the first measured point. When the inclination angle is not zero, it is assumed that the inclination angle to a depth of 25 m above the first measured point is zero and this point is added as a measured point. If the depth of the

first measured point is less than 25 m, it is unnecessary to add this manual point.

- When the inclination angle at the *i*th measured point is zero, then, due to nonexistence of its azimuth, it is assumed for the calculations that the azimuth at this point is equal to the azimuth of the neighboring measured points. In other words, if $\alpha_i = 0$, then when calculating the i measured section, $\phi_i = \phi_{i-1}$ is used; and when calculating the $i + 1$ measured section, $\phi_i = \phi_{i+1}$ is used. If the inclination angles of the two neighboring measured points are both zero, then this section is a vertical section.
- When the inclination angle is very small, the azimuth may change significantly over a section; and even when the inclination is large the azimuth change could become large, due to change in direction around 0°. In such cases, the azimuth change should be properly reduced to be less than 180°. Thus, the azimuth increment and average azimuth in the measured section are:

$$\Delta\phi_i = \begin{cases} \phi_i - \phi_{i-1}, & \text{if} \left|\phi_i - \phi_{i-1}\right| \le 180° \\ \left(\phi_i - \phi_{i-1}\right) - \text{sgn}\left(\phi_i - \phi_{i-1}\right) \times 360°, & \text{if} \left|\phi_i - \phi_{i-1}\right| > 180° \end{cases} \tag{6.16}$$

$$\phi_V = \phi_{i-1} + \frac{\Delta\phi_i}{2} \tag{6.17}$$

where sgn is the sign function (positive or negative). Special care is to be taken if the indicated azimuth change is 180° in a section. The actual azimuth increment and the average azimuth can be decided by comparing with the trend of the azimuth of the neighboring sections.
- The inclinometer reading and the designed well path should always use the same reference north direction.

The calculations of the real well path include calculating the coordinates and torsion of each measured point. The coordinates are with reference to the starting point of the well; that is, the coordinate values of the starting point of the well are set to zero. For the i measured section, the coordinate increment can be calculated and the coordinate of the next measured point also can be obtained. So,

$$\begin{cases} N_i = N_{i-1} + \Delta N_i \\ E_i = E_{i-1} + \Delta E_i \\ H_i = H_{i-1} + \Delta H_i \end{cases} \tag{6.18}$$

Starting from the wellhead point, the calculation proceeds downwards with the incremental calculation of all the coordinates, and the horizontal displacement and closure angle at any survey station are given by[6]:

$$A_i = \sqrt{(N_i - N_0)^2 + (E_i - E_0)^2} \tag{6.19}$$

$$\tan\phi_i = \frac{E_i - E_0}{N_i - N_0} \tag{6.20}$$

where

N_0, E_0 = North and east coordinates of the wellhead, m or ft
N_i, E_i = North and east coordinates of the i measured point, m or ft

In addition, other deviation parameters such as hole curvature, dogleg severity, borehole torsion, and build/drop rate can be calculated at each survey station.

Typical Calculation Methods

Survey measurement tools provide parameters at various survey stations but cannot provide the real trajectory of the well. The various methods proposed to obtain the actual well trajectory are based on some assumptions using typical curves such as cylindrical helix, circular arc, etc. Some methods for describing and calculating the actually drilled wellbore trajectory are explained in detail below and are compared in Table 6.1.

Average Angle Method

The average angle, or averaging angle, method takes the mean of the two measured sets of values of inclination and azimuth at two survey stations and assumes that the wellbore follows a tangential path, as shown in Figure 6.4.[7]

The coordinates at any survey station i are given by:

$$\Delta N_i = \Delta L_i \sin\alpha_V \cos\phi_V \tag{6.21}$$

$$\Delta E_i = \Delta L_i \sin\alpha_V \sin\phi_V \tag{6.22}$$

$$\Delta H_i = \Delta L_i \cos\alpha_V \tag{6.23}$$

where

$$\Delta L_i = L_i - L_{i-1} \tag{6.24}$$

$$\alpha_V = \frac{\alpha_{i-1} + \alpha_i}{2} \tag{6.25}$$

Table 6.1 Typical Inclinometer Calculation Methods

Methods	Contributor	Assumption	Precision Order	Conclusion
Numerical integral	Xiushan Liu, 1991	Inclination and azimuth are cubic multinomials, the coordinates are determined through numerical integral.	1	Perfect
Curve structure	Xiushan Liu, 1993	The coordinates are functions of borehole curvature and torsion at the two survey stations.	2	Excellent
Natural curve	Xiushan Liu, 1993	A 3D curve that the rates of inclination change and azimuth change remain individually constant.	3	Very good
Constant toolface	F. J. Schuh, Guo Boyun, 1992	A 3D curve that borehole curvature and toolface remain individually constant.	4	
Minimum curvature	H. L. Taylor, W. A. Zaremba, 1973	An arc in an inclined plane, the borehole curvature remains constant and borehole torsion remains zero.	4	
Radius of curvature	G. J. Wilson, 1968 Jiying Zheng, 1975	A cylinder-helix curve, the curvatures in a vertical expanded plot and in a horizontal projected plot remain individually constant.	4	
Rectified average angle	Jiying Zheng, 1981	An approximate calculation from the radius of curvature.	5	Good
Chord step	Fuqi Liu, 1986	An arc in an inclined plane, but the measured course length is assumed as its chord.	5	
Average angle	J. E. Edison, 1957	A linear section.	6	Usable
Balanced tangential	J. E. Walstrom, 1971	A polygonal line.	7	

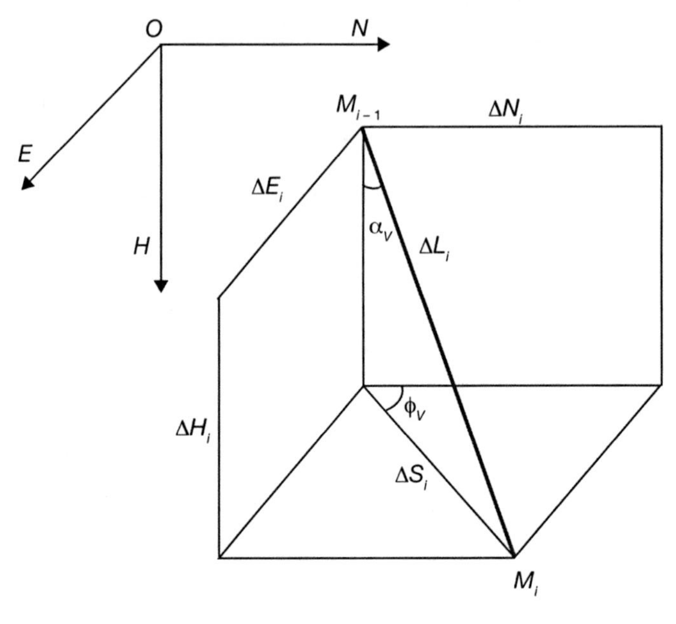

Figure 6.4 Average angle method.

Balanced Tangential Method

In the balanced tangential method, the course length between two survey stations is divided into two halves. It considers the first half of the course length as tangent to the wellbore at the first survey station $i - 1$, and other half as tangent to the wellbore at the second survey station i, as shown in Figure 6.5.[8]

The calculated coordinates are given by:

$$\Delta N_i = \frac{\Delta L_i}{2}\left(\sin\alpha_{i-1}\cos\phi_{i-1} + \sin\alpha_i\cos\phi_i\right) \tag{6.26}$$

$$\Delta E_i = \frac{\Delta L_i}{2}\left(\sin\alpha_{i-1}\sin\phi_{i-1} + \sin\alpha_i\sin\phi_i\right) \tag{6.27}$$

$$\Delta H_i = \frac{\Delta L_i}{2}\left(\cos\alpha_{i-1} + \cos\alpha_i\right) \tag{6.28}$$

Radius of Curvature (Cylindrical-Helix) Method

The radius of curvature method assumes that the wellbore has the shape of a smooth curve described by a circular arc. In the horizontal projection and vertical profile maps, they are described by circular arcs as shown in Figure 6.6. This method is less accurate when a severe dogleg is present in the interval of calculation.[9,10]

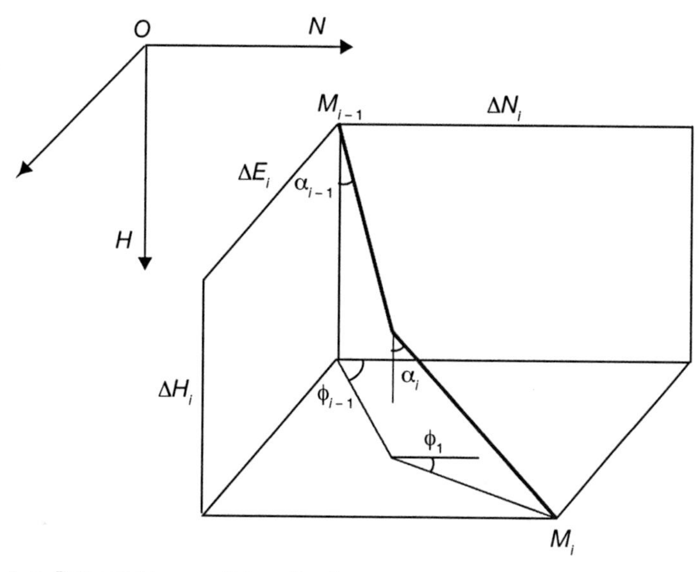

Figure 6.5 Balanced tangential method.

The coordinates using this method are given by:[6,11,12]

$$\Delta N_i = r_i\left(\sin\phi_i - \sin\phi_{i-1}\right) \tag{6.29}$$

$$\Delta E_i = r_i\left(\cos\phi_{i-1} - \cos\phi_i\right) \tag{6.30}$$

$$\Delta H_i = R_i\left(\sin\alpha_i - \sin\alpha_{i-1}\right) \tag{6.31}$$

where

$$R_i = \frac{180}{\pi}\frac{\Delta L_i}{\Delta\alpha_i};\; r_i = \frac{180}{\pi}\frac{R_i}{\Delta\phi_i}\left(\cos\alpha_{i-1} - \cos\alpha_i\right) \tag{6.32}$$

$$\Delta\alpha_i = \alpha_i - \alpha_{i-1} \tag{6.33}$$

where

R = Radius of curvature in the vertical plane, (°)/30 m or (°)/100 ft
r = Radius of curvature on the horizontal projection, (°)/30 m or (°)/100 ft

Obviously, if either $\Delta\alpha_i$ or $\Delta\phi_i$ is zero, then it is not possible to calculate R_i or r_i.[6,8] The following are the general formulas that are applicable for all the cases:

$$\Delta H_i = \begin{cases} \Delta L_i \cos\alpha_i, & \text{if } \Delta\alpha_i = 0 \\ R_i\left(\sin\alpha_i - \sin\alpha_{i-1}\right), & \text{if } \Delta\alpha_i \neq 0 \end{cases} \tag{6.34}$$

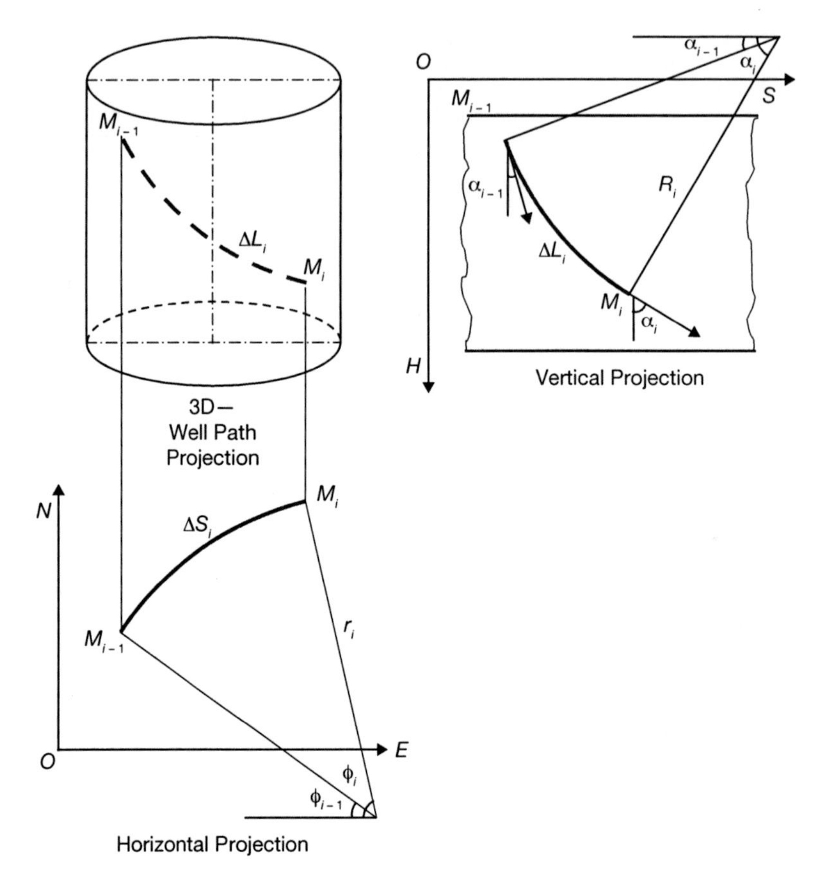

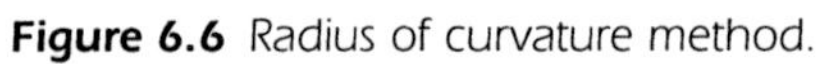
Figure 6.6 Radius of curvature method.

$$\Delta S_i = \begin{cases} \Delta L_i \sin\alpha_i, & \text{if } \Delta\alpha_i = 0 \\ R_i\left(\cos\alpha_{i-1} - \cos\alpha_i\right), & \text{if } \Delta\alpha_i \neq 0 \end{cases} \tag{6.35}$$

$$\Delta N_i = \begin{cases} \Delta S_i \cos\phi_i, & \text{if } \Delta\phi_i = 0 \\ r_i\left(\sin\phi_i - \sin\phi_{i-1}\right), & \text{if } \Delta\phi_i \neq 0 \end{cases} \tag{6.36}$$

$$\Delta E_i = \begin{cases} \Delta S_i \sin\phi_i, & \text{if } \Delta\phi_i = 0 \\ r_i\left(\cos\phi_{i-1} - \cos\phi_i\right), & \text{if } \Delta\phi_i \neq 0 \end{cases} \tag{6.37}$$

where

$$R_i = \frac{180}{\pi}\frac{\Delta L_i}{\Delta\alpha_i}; \quad r_i = \frac{180}{\pi}\frac{\Delta S_i}{\Delta\phi_i} \tag{6.38}$$

Rectified Average Angle Method

The rectified or corrected average angle method is an approximate method to calculate the curvature radius, where the approximation made is that $\Delta\alpha_I$ and $\Delta\phi_i$ in the section are very small. Starting from the equation used for the calculations for the radius of curvature method, applying trigonometric formulas and expanding the sine function in a power series, it is possible to obtain the equations for the rectified average angle method. The results obtained from the rectified average angle method are generally very close to that of the radius of curvature method even though it is an approximate calculation method.

The coordinates calculated using this method are given by:

$$\Delta N_i = f_{b,i}\Delta L_i \sin\alpha_V \cos\phi_V \tag{6.39}$$

$$\Delta E_i = f_{b,i}\Delta L_i \sin\alpha_V \sin\phi_V \tag{6.40}$$

$$\Delta H_i = f_{v,i}\Delta L_i \cos\alpha_V \tag{6.41}$$

where

$$f_{v,i} = 1 - \left(\frac{\pi}{180}\right)^2 \frac{\Delta\alpha_i^{\ 2}}{24} \tag{6.42}$$

$$f_{b,i} = 1 - \left(\frac{\pi}{180}\right)^2 \frac{\Delta\alpha_i^{\ 2} + \Delta\phi_i^{\ 2}}{24} \tag{6.43}$$

In this method it can be seen that the coordinates are obtained by multiplying the coordinates of the average angle method by the coefficients f_{vi} and f_{bi}. Therefore, the method is also called the average angle correction method. Clearly, the correction coefficients f_{vi} and f_{bi} are less than 1, and when the coefficients become equal to 1, this method becomes the same as the average angle method of calculation.

Minimum Curvature Method

The minimum curvature method uses the angles measured at two consecutive survey stations to describe a smooth circular arc representing the wellbore path as shown in Figure 6.7. It uses the dogleg ratio factor given by Eq. 6.47 to smooth the wellbore section.

The coordinates are calculated as:

$$\Delta N_i = \lambda_i\left(\sin\alpha_{i-1}\cos\phi_{i-1} + \sin\alpha_i\cos\phi_i\right) \tag{6.44}$$

$$\Delta E_i = \lambda_i\left(\sin\alpha_{i-1}\sin\phi_{i-1} + \sin\alpha_i\sin\phi_i\right) \tag{6.45}$$

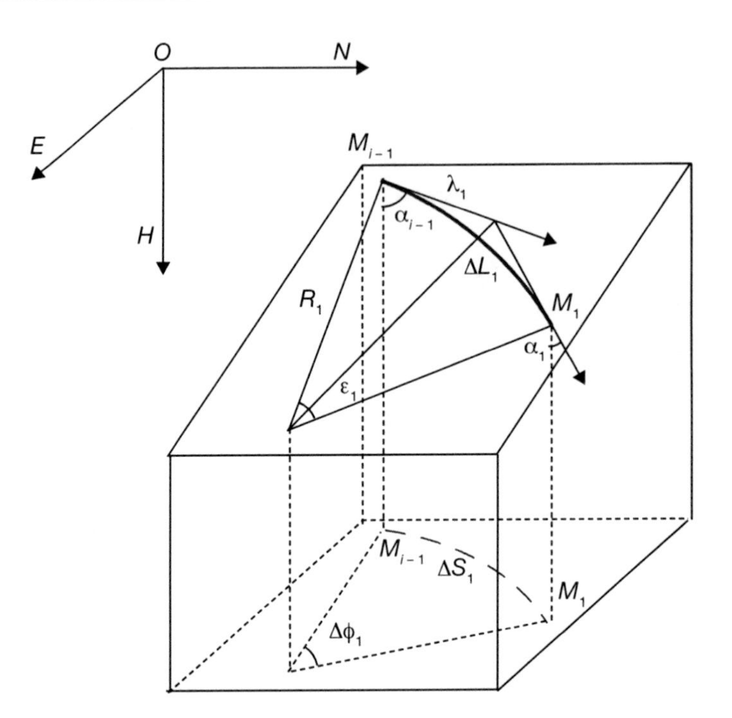

Figure 6.7 Minimum curvature method.

$$\Delta H_i = \lambda_i \left(\cos\alpha_{i-1} + \cos\alpha_i \right) \tag{6.46}$$

where

$$\lambda_i = \frac{180}{\pi} \frac{\Delta L_i}{\varepsilon_i} \tan\frac{\varepsilon_i}{2} \tag{6.47}$$

$$\cos\varepsilon_i = \cos\alpha_{i-1}\cos\alpha_i + \sin\alpha_{i-1}\sin\alpha_i\cos\Delta\phi_i \tag{6.48}$$

Chord Step Method

The chord step method is a hypothetical measurement of the well trajectory for an oblique plane arc in space,[13] as shown in Figure 6.7.

The coordinates are calculated as:

$$\Delta N_i = \lambda_i \left(\sin\alpha_{i-1}\cos\phi_{i-1} + \sin\alpha_i\cos\phi_i \right) \tag{6.49}$$

$$\Delta E_i = \lambda_i \left(\sin\alpha_{i-1}\sin\phi_{i-1} + \sin\alpha_i\sin\phi_i \right) \tag{6.50}$$

$$\Delta H_i = \lambda_i \left(\cos\alpha_{i-1} + \cos\alpha_i \right) \tag{6.51}$$

where

$$\lambda_i = \frac{\Delta L_i}{2} \frac{1}{\cos\frac{\varepsilon_i}{2}} \tag{6.52}$$

Because the spatial well path is located in an inclined plane, the direction lines or the extension lines of the two measurement points on the well path must be across, and the line lengths λ_i of the direction lines between two measurement points and the cross section must be equal. It is not difficult to find that the formulas of the minimum curvature method look similar, but the λ_i are different. Especially, when the bending angle $\varepsilon_i = 0$, both the minimum curvature and descent methods have:

$$\lambda_i = \frac{\Delta L_i}{2} \tag{6.53}$$

For this condition $\alpha_{i-1} = \alpha_i$, $\phi_{i-1} = \phi_i$, which shows this section is a straight line.

Natural Curve Method

The natural curve method uses the inclinations and azimuths at two survey stations and assumes that the rate of change of the inclination and the rate of change of the azimuth remain individually constant over the course length.

The course coordinates are given by:

$$\Delta N_i = \frac{1}{2}\left[F_C\left(A_{P,i}, \kappa_{P,i}, \Delta L_i\right) + F_C\left(A_{Q,i}, \kappa_{Q,i}, \Delta L_i\right)\right] \tag{6.54}$$

$$\Delta E_i = \frac{1}{2}\left[F_S\left(A_{P,i}, \kappa_{P,i}, \Delta L_i\right) - F_S\left(A_{Q,i}, \kappa_{Q,i}, \Delta L_i\right)\right] \tag{6.55}$$

$$\Delta H_i = F_S\left(\alpha_{i-1}, \kappa_{\alpha,i}, \Delta L_i\right) \tag{6.56}$$

where

$$\begin{cases} \kappa_{\alpha,i} = \dfrac{\Delta\alpha_i}{\Delta L_i} \\ \kappa_{\phi,i} = \dfrac{\Delta\phi_i}{\Delta L_i} \end{cases} \tag{6.57}$$

$$\begin{cases} A_{P,i} = \alpha_{i-1} - \phi_{i-1} \\ A_{Q,i} = \alpha_{i-1} + \phi_{i-1} \end{cases} \tag{6.58}$$

$$\begin{cases} \kappa_{P,i} = \kappa_{\alpha,i} - \kappa_{\phi,i} \\ \kappa_{Q,i} = \kappa_{\alpha,i} + \kappa_{\phi,i} \end{cases} \tag{6.59}$$

$$F_C(\theta,\kappa,\lambda)=\begin{cases}\lambda\sin\theta, & \text{if } \kappa=0\\ \dfrac{180}{\pi\kappa}\left[\cos\theta-\cos(\theta+\kappa\lambda)\right], & \text{if } \kappa\neq 0\end{cases} \tag{6.60}$$

$$F_S(\theta,\kappa,\lambda)=\begin{cases}\lambda\cos\theta, & \text{if } \kappa=0\\ \dfrac{180}{\pi\kappa}\left[\sin(\theta+\kappa\lambda)-\sin\theta\right], & \text{if } \kappa\neq 0\end{cases} \tag{6.61}$$

Constant Toolface Angle Method

The constant toolface angle method assumes that in the well section of interest, the well trajectory is a spatial curve with constant curvature and constant reference plane inclination angle.[22,23]

$$\Delta N_i=\begin{cases}R_i\left(\cos\alpha_{i-1}-\cos\alpha_i\right)\cos\phi_i, & \text{when } \alpha_{i-1}=0 \text{ or } \alpha_i=0\\ r_i\sin\alpha_i\left(\sin\phi_i-\sin\phi_{i-1}\right), & \text{when } \Delta\alpha_i=0\\ \displaystyle\int_{L_1}^{L_2}\sin\alpha(L)\cos\phi(L)\,dL, & \text{for the rest of the conditions}\end{cases} \tag{6.62}$$

$$\Delta E_i=\begin{cases}R_i\left(\cos\alpha_{i-1}-\cos\alpha_i\right)\sin\phi_i, & \text{if } \alpha_{i-1}=0 \text{ or } \alpha_i=0\\ r_i\sin\alpha_i\left(\cos\phi_{i-1}-\cos\phi_i\right), & \text{if } \Delta\alpha_i=0\\ \displaystyle\int_{L_1}^{L_2}\sin\alpha(L)\sin\phi(L)\,dL, & \text{for the rest of the conditions}\end{cases} \tag{6.63}$$

$$\Delta H_i=\begin{cases}\Delta L_i\cos\alpha_i, & \text{if } \Delta\alpha_i=0\\ R_i\left(\sin\alpha_i-\sin\alpha_{i-1}\right), & \text{if } \Delta\alpha_i\neq 0\end{cases} \tag{6.64}$$

where

$$R_i=\frac{180}{\pi}\frac{\Delta L_i}{\Delta\alpha_i};\quad r_i=\frac{180}{\pi}\frac{\Delta L_i}{\Delta\phi_i} \tag{6.65}$$

$$\tan\omega_i=\frac{\pi}{180}\frac{\Delta\phi_i}{\ln\dfrac{\tan\frac{\alpha_i}{2}}{\tan\frac{\alpha_{i-1}}{2}}} \tag{6.66}$$

$$\alpha(L)=\alpha_{i-1}+\frac{\Delta\alpha_i}{\Delta L_i}\left(L-L_{i-1}\right) \tag{6.67}$$

$$\phi(L) = \phi_{i-1} + \frac{180}{\pi} \tan\omega_i \times \ln \frac{\tan \frac{\alpha(L)}{2}}{\tan \frac{\alpha_{i-1}}{2}} \quad (6.68)$$

In usual cases, the constant reference plane method should use a numerical integration method to calculate the north coordinate increment ΔN_i and the east coordinate increment ΔE_i. When $\alpha_{i-1} = 0$ or $\alpha_i = 0$, due to nonexistence of the inclination, the value at another measured point has to be used with $\omega_i = 0$. When $\alpha_i = \alpha_{i-1} \neq 0$, there is no inclination change, so $\omega_i = \text{sgn}(\Delta\phi_i) \times 90°$.

With present computer technology, computations are easy even when using a complex method of estimating coordinates. Therefore, reliability and accuracy of inclinometer readings are very important. It can be easily inferred that the tangential method is the least accurate method, whereas the numerical integration method of calculation is the best.

PROBLEM 6.1

Assume that a well path is a cylindrical spiral section. At the top measured point, the inclination is $\alpha_{i-1} = 20°$, the azimuth is $\phi_{i-1} = 100°$, and the curvature in the vertical profile map is 10°/30 m; the curvature in the horizontal projection map is 12°/30 m; and the measured length is $\Delta L_i = 30$ m. Find the well profile parameters.

Solution

After calculations, the following results can be obtained. At the bottom measured point, the inclination and the azimuth are $\alpha_i = 30°$ and $\phi_i = 105.065°$; at the top measured point, the torsion parameters are $\kappa_{\alpha,i-1} = 10°/30$ m, $\kappa_{\phi,i-1} = 4.104°/30$ m, $\kappa_{i-1} = 10.098°/30$ m, and $\tau_{i-1} = 11.421°/30$ m; at the bottom measured point, the torsion parameters are $\kappa_{\alpha,i} = 10.000°/30$ m, $\kappa_{\phi,i} = 6°/30$ m, $\kappa_i = 10.440°/30$ m, and $\tau_i = 14.730°/30$ m. The results of different calculation methods are shown in Table 6.2. For a cylindrical spiral trajectory, the radius of curvature method gives the accurate solution (true values).

PROBLEM 6.2

Assume that a well path is a circular-arc section. At the top measured point, the inclination is $\alpha_{i-1} = 40°$, the azimuth is $\phi_{i-1} = 70°$, the borehole curvature is 12°/30 m, the initial toolface angle is 315°, and the measured length is $\Delta L_i = 30$ m. Find the well profile parameters.

Table 6.2 Cylindrical Spiral Trajectory Calculation Results for Problem 6.1

Calculation Method	*North Coordinate (m)*	*East Coordinate (m)*	*Vertical Depth (m)*	*Horizontal Length (m)*	*Absolute Error (m)*
Average angle	−2.751	12.376	27.189	12.679	0.040
Balanced tangential	−2.840	12.295	27.086	12.630	0.132
Radius of curvature	−2.747	12.357	27.155	12.662	0.000
Rectified average angle	−2.747	12.357	27.155	12.662	0.000
Minimum curvature	−2.848	12.327	27.158	12.656	0.105
Chord step	−2.852	12.344	27.194	12.672	0.113
Natural parameter	−2.781	12.349	27.155	12.662	0.035
Constant toolface	−2.815	12.341	27.155	12.662	0.070
Curve structure	−2.747	12.367	27.199	12.662	0.045
Numerical integral	−2.747	12.357	27.155	12.662	0.000

Note: "Absolute error" is the square roots of the squared sums of the single coordinate errors between the calculated result and the exact solution with north course coordinate, east course coordinate, and course vertical depth.

Solution

After calculations, the following results are obtained. At the bottom measured point, the inclination and the azimuth are $\alpha_i = 49.095°$ and $\phi_i = 58.783°$; at the top measured point, the curvature and the torsion parameters are $\kappa_{\alpha,i-1} = 8.485°/30$ m, $\kappa_{\phi,i-1} = -13.201°/30$ m, $\kappa_{Hi-1} = -20.537°/30$ m, and $\tau_{i-1} = 0°/30$ m; at the bottom measured point, the torsion parameters are $\kappa_{\alpha,i} = 9.588°/30$ m, $\kappa_{\phi,i} = -9548°/30$ m, $\kappa_{Hi} = 12.633°/30$ m, and $\tau_i = 0°/30$ m. The results of different calculation methods are shown in Table 6.3. For a circular-arc trajectory, the minimum curvature method gives the accurate solution (true values).

Table 6.3 Circular-Arc Well Path Calculation Results for Problem 6.2

Calculation Method	*North Coordinate (m)*	*East Coordinate (m)*	*Vertical Depth (m)*	*Horizontal Length (m)*	*Absolute Error (m)*
Average angle	9.096	18.978	21.380	21.045	0.189
Balanced tangential	9.173	18.756	21.313	20.979	0.110
Radius of curvature	9.072	18.928	21.358	21.023	0.173
Rectified average angle	9.072	18.928	21.358	21.023	0.173
Minimum curvature	9.207	18.825	21.391	20.989	0.000
Chord step	9.224	18.859	21.430	21.027	0.055
Natural parameter	9.122	18.904	21.358	21.023	0.121
Constant toolface	9.172	18.880	21.358	21.023	0.073
Curve structure	9.224	18.859	21.430	20.971	0.055
Numerical integral	9.207	18.825	21.391	20.989	0.000

PROBLEM 6.3

Assume that the well path is a circular-arc section. At the top measured point, the inclination is $\alpha_{i-1} = 40°$, the azimuth is $\phi_{i-1} = 353°$, the curvature in the vertical profile map is 10°/30 m, the curvature in the horizontal projection map is 12°/30 m, and the measured length is $\Delta L_i = 30$ m. Find the well profile parameters.

Solution

After calculations, the following results are obtained. At the bottom measured point, inclination and azimuth are $\alpha_i = 50°$ and $\phi_i = 5.0°$; at the top measured point, the curvature and torsion parameters are $\kappa_{\alpha,i-1} = 12.629°/30$ m,

Table 6.4 Natural Curve Well Path Calculation Results for Problem 6.3

Calculation Method	*North Coordinate (m)*	*East Coordinate (m)*	*Vertical Depth (m)*	*Horizontal Length (m)*	*Absolute Error (m)*
Average angle	21.210	–0.370	21.213	21.213	0.096
Balanced tangential	21.017	–0.174	21.132	21.132	0.191
Radius of curvature	21.144	–0.369	21.186	21.186	0.064
Rectified average angle	21.144	–0.369	21.186	21.186	0.064
Minimum curvature	21.109	–0.174	21.225	21.147	0.141
Chord step	21.155	–0.175	21.271	21.193	0.155
Natural parameter	21.146	–0.305	21.186	21.186	0.000
Constant toolface	21.147	–0.240	21.186	21.186	0.065
Curve structure	21.181	–0.318	21.238	21.186	0.064
Numerical integral	21.146	–0.305	21.186	21.186	0.000

$\kappa_{\phi,i-1} = -13.201°/30$ m, $\kappa_{Hi-1} = 18.669°/30$ m, and $\tau_{i-1} = 14.956°/30$ m; at the bottom measured point, the torsion parameters are $\kappa_{\alpha,i} = 13.583°/30$ m, $\kappa_{\phi,i} = -9548°/30$ m, $\kappa_{Hi} = 15.665°/30$ m, and $\tau_i = 11.894°/30$ m. The results of different calculation methods are shown in Table 6.4. For a natural curve well path, the natural parameter method gives the accurate solution (true values).

Summary of Methods

Based on extensive analysis, the following basic conclusions can be drawn[14]:

- The accuracies of the average angle method and the balanced tangential methods are lower.

- The radius of curvature (or cylindrical-helix) method is an approximate calculation method, but the formula is concise and, in general, the calculation error is small.
- The radius of curvature, minimum curvature, constant toolface angle, and the natural curve methods are reasonable choices for typical drilling applications. Generally, the radius of curvature method is suitable for a rotary mode of drilling, whereas the minimum curvature method is suitable for slide drilling.
- The results of the rectified angle method are generally close to those of the radius of curvature method, but it ranks lower only because it is an approximate calculation.
- The curve structure method is inferior to that of the numerical integration method, but superior to that of the other methods. Both methods are described later in this chapter.
- Calculating the course coordinates by numerical integration is the most precise method to simulate and calculate the wellbore trajectory for both rotary and slide modes.

The Real Shape of a Well Trajectory

All conventional methods for calculating the trajectory of a wellbore involve assumptions. Most calculations assume that segments of the trajectory can be approximated as straight lines, polygonal lines, cylindrical helixes, circular arcs, or a combination of these approximate shapes. Conventional methods calculate the course coordinates of a well survey interval according to the predetermined shape for the trajectory. But, what does the real shape of a trajectory in each surveyed interval really look like? What parameters play decisive roles in determining the true shape of a surveyed interval? Is it reasonable to assume that every survey interval has the same kind of approximate shape?

The methods presented in this section can be used to describe the true shape of a wellbore trajectory relative to its survey stations. It also provides a universal equation for a wellbore trajectory in a surveyed interval, and presents an objective approach for describing and calculating a wellbore trajectory in space. The new method does not assume that the shape of wellbore trajectory can be approximated by a combination of typical curves. The study shows that the wellbore curvature and torsion at the survey stations determine the shape of the wellbore trajectory at subsequent intervals, and that the 3D coordinates of the wellbore trajectory in a local coordinate system are, respectively, linear, quadratic, and cubic functions of the curve length. This method yields a continuous wellbore trajectory. In order to develop the shape of the trajectory, two adjacent points in close proximity are considered, as shown in Figure 6.8.

Based on Taylor series with a truncated expansion, the coordinate increments between the two points can be written as:

$$\begin{aligned}\Delta \boldsymbol{r} &= \boldsymbol{r}(L_0 + \Delta L) - \boldsymbol{r}(L_0) \\ &= \dot{\boldsymbol{r}}(L_0)\Delta L + \frac{1}{2!}\ddot{\boldsymbol{r}}(L_0)\Delta L^2 + \frac{1}{3!}\left[\dddot{\boldsymbol{r}}(L_0) + \delta\right]\Delta L^3\end{aligned} \tag{6.69}$$

where $\lim\limits_{\Delta L \to 0} \delta = 0$.

According to differential geometry, the relationship between the unit tangent vector $\boldsymbol{t}$, unit principal normal vector $\boldsymbol{n}$, coordinate vector $\boldsymbol{r}$, and their derivatives can be expressed as:

$$\begin{cases} \dot{\boldsymbol{r}} = \boldsymbol{t} \\ \ddot{\boldsymbol{r}} = \kappa\,\boldsymbol{n} \end{cases} \tag{6.70}$$

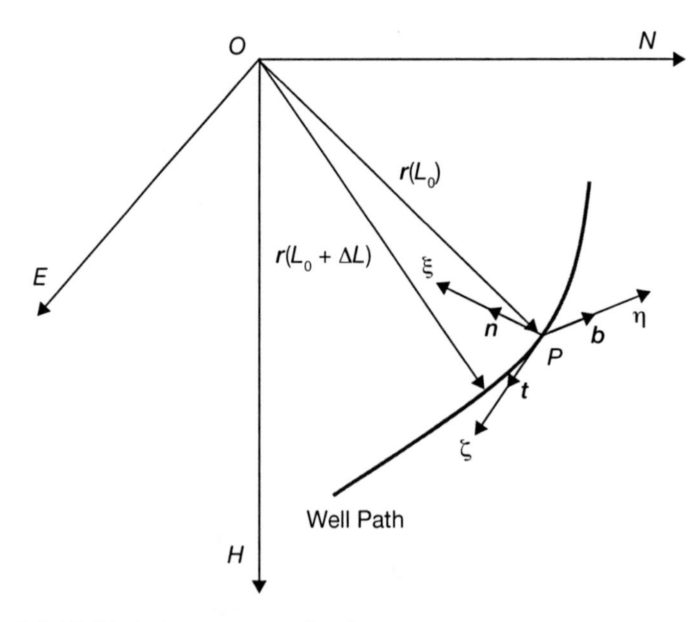

Figure 6.8 Well trajectory on a moving frame.

Further, it can be noted that for the unit binormal vector $\boldsymbol{b}$:

$$\dddot{\boldsymbol{r}} = \dot{\kappa}\,\boldsymbol{n} + \kappa\,\dot{\boldsymbol{n}} = \dot{\kappa}\,\boldsymbol{n} + \kappa\left(-\kappa\,\boldsymbol{t} + \tau\,\boldsymbol{b}\right) = -\kappa^2\,\boldsymbol{t} + \dot{\kappa}\,\boldsymbol{n} + \kappa\,\tau\,\boldsymbol{b} \tag{6.71}$$

As before, dots denote the derivatives.

Substituting the values from Eqs. 6.70 and 6.71 into Eq. 6.69 yields:

$$\boldsymbol{r}\left(L_0 + \Delta L\right) - \boldsymbol{r}\left(L_0\right) = \boldsymbol{t}_0 \Delta L + \frac{1}{2}\kappa_0 \boldsymbol{n}_0 \Delta L^2 + \frac{1}{6}\left(-\kappa_0{}^2 \boldsymbol{t}_0 + \dot{\kappa}_0 \boldsymbol{n}_0 + \kappa_0 \tau_0 \boldsymbol{b}_0 + \delta\right)\Delta L^3 \tag{6.72}$$

where

$$\delta = \delta_1 \boldsymbol{t}_0 + \delta_2 \boldsymbol{n}_0 + \delta_3 \boldsymbol{b}_0$$

In Eq. 6.72, $\boldsymbol{t}_0$, $\boldsymbol{n}_0$, $\boldsymbol{b}_0$, κ_0, and τ_0 represent the values at the point $\boldsymbol{r}(L_0)$.

Algebraic manipulation results in:

$$\begin{aligned}\Delta \boldsymbol{r} = \boldsymbol{r}\left(L_0 + \Delta L\right) - \boldsymbol{r}\left(L_0\right) &= \left[\Delta L + \frac{1}{6}\left(-\kappa_0{}^2 + \delta_1\right)\Delta L^3\right]\boldsymbol{t}_0 \\ &+ \left[\frac{1}{2}\kappa_0 \Delta L^2 + \frac{1}{6}\left(\dot{\kappa}_0 + \delta_2\right)\Delta L^3\right]\boldsymbol{n}_0 + \left[\frac{1}{6}\left(\kappa_0 \tau_0 + \delta_3\right)\Delta L^3\right]\boldsymbol{b}_0\end{aligned} \tag{6.73}$$

On retaining only the first terms in the expressions for the vector components along $\boldsymbol{t}_0$, $\boldsymbol{n}_0$, and $\boldsymbol{b}_0$, Eq. 6.73 simplifies to:

$$\Delta \boldsymbol{r} = \boldsymbol{r}\left(L_0 + \Delta L\right) - \boldsymbol{r}\left(L_0\right) = \Delta L \boldsymbol{t}_0 + \frac{1}{2}\kappa_0 \Delta L^2 \boldsymbol{n}_0 + \frac{1}{6}\kappa_0 \tau_0 \Delta L^3 \boldsymbol{b}_0 \tag{6.74}$$

In the coordinate system P-$\xi\eta\zeta$ (Figure 6.8), where the axes directions are the tangent, the principal normal, and the binormal at point P, the magnitudes of the coordinate increments between two neighboring points can be expressed as[15,16]:

$$\begin{cases}\xi = \dfrac{1}{2}\kappa_0 \Delta L^2 \\ \eta = \dfrac{1}{6}\kappa_0 \tau_0 \Delta L^3 \\ \zeta = \Delta L\end{cases} \tag{6.75}$$

Eqs. 6.74 and 6.75 provide the approximate equations of the well path around point P. Thus, the curvature and the torsion at any point in the well path determine the curve shape around this point completely.

The projections of the well path on the three planes of the moving frame appear respectively as the different parabolas given below[16,17]:

- Well path projection on the normal plane:

$$\begin{cases} \xi = \dfrac{1}{2}\kappa_0 \Delta L^2 \\ \eta = \dfrac{1}{6}\kappa_0 \tau_0 \Delta L^3 \\ \zeta = 0 \end{cases} \tag{6.76}$$

After the elimination of ΔL, Eq. 6.76 can be written as:

$$\eta^2 = \frac{2\tau_0^{\ 2}}{9\kappa_0}\xi^3 \quad (\zeta = 0) \tag{6.77}$$

This is the equation of the semi-cubic parabola shown in Figure 6.9.

- Well path projection on the rectifying plane:

$$\begin{cases} \xi = 0 \\ \eta = \dfrac{1}{6}\kappa_0 \tau_0 \Delta L^3 \\ \zeta = \Delta L \end{cases} \tag{6.78}$$

After the elimination of ΔL,

$$\eta = \frac{1}{6}\kappa_0 \tau_0 \zeta^3 \quad (\xi = 0) \tag{6.79}$$

This is the equation of the cubic parabola shown in Figure 6.10.

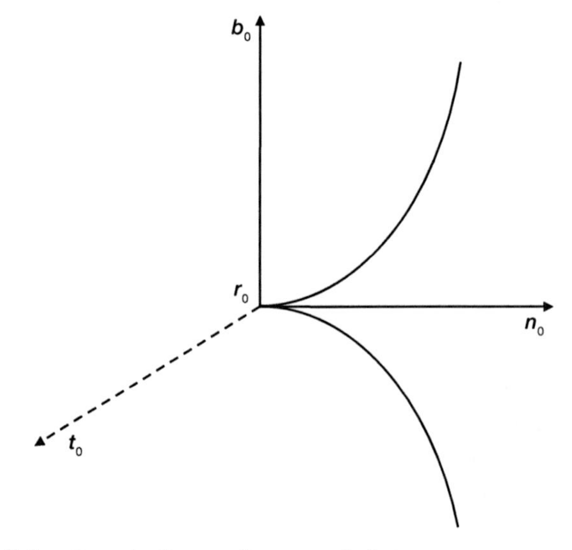

Figure 6.9 Well path projection on the normal plane.

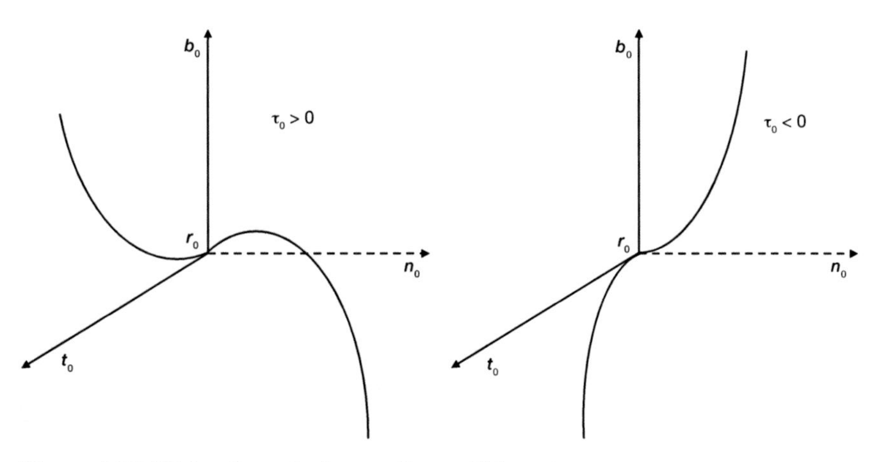

Figure 6.10 Well path projection on the rectifying plane.

- Well path projection on the osculating plane:

$$\begin{cases} \xi = \dfrac{1}{2}\kappa_0 \Delta L^2 \\ \eta = 0 \\ \zeta = \Delta L \end{cases} \tag{6.80}$$

After the elimination of ΔL,

$$\xi = \frac{1}{2}\kappa_0 \zeta^2 \quad (\eta = 0) \tag{6.81}$$

This is the equation of the parabola shown in Figure 6.11.

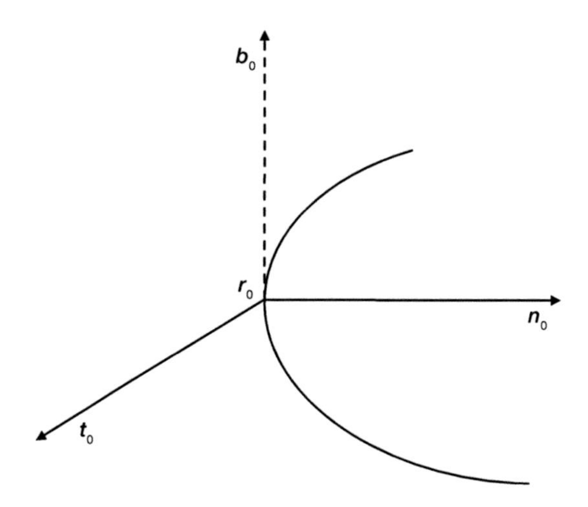

Figure 6.11 Well path projection on the osculating plane.

Based on the above analysis, the following can be concluded[16]:

- The three projection plots of the well path show that the well path penetrates through the normal plane and the osculating plane, but does not penetrate through the rectifying plane.
- The principal normal vector $\boldsymbol{n}_0$ always points in the concave direction of the well path.
- The sign of the wellbore torsion depicts the direction of rotation of the well path. A positive wellbore torsion $\tau_0 > 0$ depicts a right-hand curve, and a negative wellbore torsion $\tau_0 < 0$ depicts a left-hand curve, as shown in Figure 6.12. The impact of the wellbore torsion on the coordinates is shown in Table 6.5.

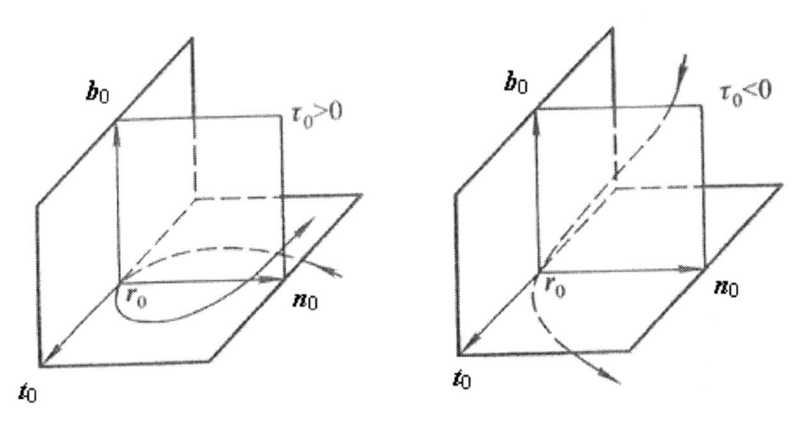

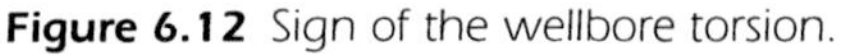
Figure 6.12 Sign of the wellbore torsion.

Table 6.5 Impact of the Wellbore Torsion on the Coordinate Values

τ_0	ΔL	ξ	η	ζ
$\tau_0 > 0$	−	+	−	−
	+	+	+	+
$\tau_0 < 0$	−	+	+	−
	+	+	−	+

Curve Structure Method

A wellbore trajectory generally is a curve in space that is both curved and tortuous. It has also been found that the effect of the torsion is higher than the effect of the curvature on the shape of the trajectory. Evidently, the results for the calculation of the shape of the well trajectory will be more precise when both the hole curvature and the torsion are simultaneously taken into account over the course length. Thus, it is reasonable to simulate a wellbore trajectory using the curve structure method described below. Figure 6.13 illustrates this method in the general *O-NEH* coordinate system.

As mentioned above, the wellbore curvature and the torsion at a given point both determine the shape of wellbore trajectory within a tiny interval. The formulas and techniques for calculating the wellbore curvature and torsion are known. But, the coordinates given by Eqs. 6.74 and 6.75 are in the wellbore coordinate system (P-$\xi\eta\zeta$). Hence, the coordinates require transformation to the general coordinate system (*O-NEH*) as outlined in the steps below.

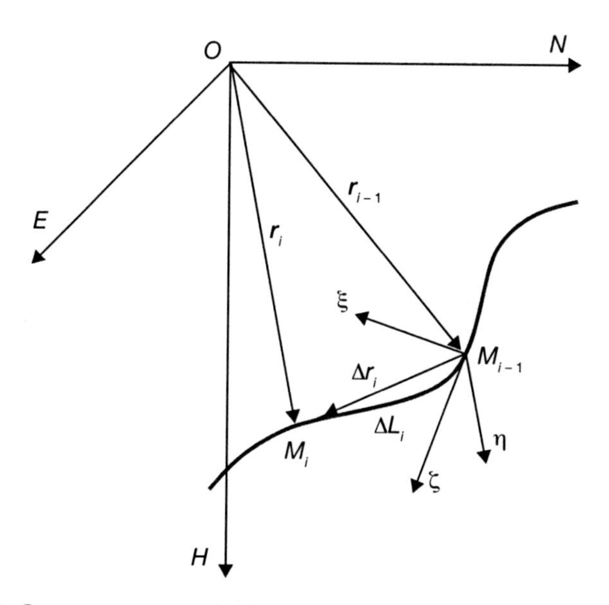

Figure 6.13 Curve structure method.

Step 1. Using the curve structure parameters at the measured points, the increments in the coordinates are calculated over the interval $[L_{i-1}, L_i]$ at the upper survey station as follows:

$$\{\Delta N_i \quad \Delta E_i \quad \Delta H_i\}^{\mathrm{T}} = [T_{i-1}]\{\xi_i \quad \eta_i \quad \zeta_i\}^{\mathrm{T}} \tag{6.82}$$

Where the course coordinates are:

$$\begin{cases} \xi_i = \dfrac{1}{2}\kappa_{i-1}\Delta L_i^{\,2} \\ \eta_i = \dfrac{1}{6}\kappa_{i-1}\tau_{i-1}\Delta L_i^{\,3} \\ \zeta_i = \Delta L_i \end{cases}$$

the transformation matrix is:

$$[T_{i-1}] = \begin{bmatrix} n_{N,i-1} & b_{N,i-1} & t_{N,i-1} \\ n_{E,i-1} & b_{E,i-1} & t_{E,i-1} \\ n_{H,i-1} & b_{H,i-1} & t_{H,i-1} \end{bmatrix} \tag{6.83}$$

and the respective unit vectors are:

$$\begin{cases} n_{N,i-1} = \lambda_{\alpha,i-1}\cos\alpha_{i-1}\cos\phi_{i-1} - \lambda_{\phi,i-1}\sin\alpha_{i-1}\sin\phi_{i-1} \\ n_{E,i-1} = \lambda_{\alpha,i-1}\cos\alpha_{i-1}\sin\phi_{i-1} + \lambda_{\phi,i-1}\sin\alpha_{i-1}\cos\phi_{i-1} \\ n_{H,i-1} = -\lambda_{\alpha,i-1}\sin\alpha_{i-1} \end{cases} \tag{6.84}$$

$$\begin{cases} b_{N,i-1} = -\lambda_{\alpha,i-1}\sin\phi_{i-1} - \lambda_{\phi,i-1}\sin\alpha_{i-1}\cos\alpha_{i-1}\cos\phi_{i-1} \\ b_{E,i-1} = \lambda_{\alpha,i-1}\cos\phi_{i-1} - \lambda_{\phi,i-1}\sin\alpha_{i-1}\cos\alpha_{i-1}\sin\phi_{i-1} \\ b_{H,i-1} = \lambda_{\alpha,i-1}\sin^2\alpha_{i-1} \end{cases} \tag{6.85}$$

$$\begin{cases} t_{N,i-1} = \sin\alpha_{i-1}\cos\phi_{i-1} \\ t_{E,i-1} = \sin\alpha_{i-1}\sin\phi_{i-1} \\ t_{H,i-1} = \cos\alpha_{i-1} \end{cases} \tag{6.86}$$

$$\begin{cases} \lambda_{\alpha,i-1} = \dfrac{\kappa_{\alpha,i-1}}{\kappa_{i-1}} \\ \lambda_{\phi,i-1} = \dfrac{\kappa_{\phi,i-1}}{\kappa_{i-1}} \end{cases} \tag{6.87}$$

Step 2. Using the curve structure parameters of the well path at the lower survey station, the increments in the coordinates are calculated as:

$$\{\Delta N_i \quad \Delta E_i \quad \Delta H_i\}^T = [T_i]\{\xi_{i-1} \quad \eta_{i-1} \quad \zeta_{i-1}\}^T \tag{6.88}$$

where

$$\begin{cases} \xi_{i-1} = -\dfrac{1}{2}\kappa_i \Delta L_i^2 \\ \eta_{i-1} = \dfrac{1}{6}\kappa_i \tau_i \Delta L_i^3 \\ \zeta_{i-1} = \Delta L_i \end{cases} \tag{6.89}$$

The calculation formula for the transformation matrix $[T_i]$ is the same as that of $[T_{i-1}]$ in form and requires replacing the variables for point M_{i-1} in $[T_{i-1}]$ by those for point M_i.

The combination algorithm for adjacent survey stations is used to add and to obtain the increments in the coordinates of the well path over the survey interval $[L_{i-1}, L_i]$, as follows:

$$\begin{aligned} \{\Delta N_i \quad \Delta E_i \quad \Delta H_i\}^T &= [T_{i-1}]\left\{\xi_m^{(i-1)} \quad \eta_m^{(i-1)} \quad \zeta_m^{(i-1)}\right\}^T \\ &+ [T_i]\left\{\xi_m^{(i)} \quad \eta_m^{(i)} \quad \zeta_m^{(i)}\right\}^T \end{aligned} \tag{6.90}$$

where

$$\begin{cases} \xi_m^{(i-1)} = \dfrac{1}{8}\kappa_{i-1} \Delta L_i^2 \\ \eta_m^{(i-1)} = \dfrac{1}{48}\kappa_{i-1} \tau_{i-1} \Delta L_i^3 \\ \zeta_m^{(i-1)} = \dfrac{1}{2}\Delta L_i \end{cases}$$

$$\begin{cases} \xi_m^{(i)} = -\dfrac{1}{8}\kappa_i \Delta L_i^2 \\ \eta_m^{(i)} = \dfrac{1}{48}\kappa_i \tau_i \Delta L_i^3 \\ \zeta_m^{(i)} = \dfrac{1}{2}\Delta L_i \end{cases}$$

In the equation, ($\xi_m^{(i-1)}$, $\eta_m^{(i-1)}$, $\xi_m^{(i-1)}$) are the coordinates of point M_m in the coordinate system $M_{i-1} - \xi\eta\zeta$, and (ξ_m^i, η_m^i, ξ_m^i) are the coordinates of point M_m in the coordinate system $M_i - \xi\eta\zeta$.

Accurate Calculation Method

Effective Method of Well Trajectory Simulation

Engineering and other disciplines have used spline functions for more than 50 years. The spline, made of a slender batten or other elastic materials, was originally a simple device used for drawing a curve. Users lay it on the plate, put some paperweights on it, and form it into the shape of a desired curve. When the bending deflection of the spline is small, the equation of its shape can be expressed by:

$$\kappa(x) = \frac{\ddot{y}(x)}{\left[1+\dot{y}^2(x)\right]^{3/2}} \approx \ddot{y}(x) \tag{6.91}$$

Considering the spline as a small beam, then according to the Euler-Bernoulli bending equation,

$$M(x) = EI \times \kappa(x) \tag{6.92}$$

where

$M(x)$ = Bending moment at x
E = Elastic modulus, psi
I = Moment of inertia, in.^{4}

Using Eq. 6.91:

$$\ddot{y}(x) = \frac{M(x)}{EI} \tag{6.93}$$

In engineering, the spline can be regarded as an elastic, slender beam, and the elementary theory of beams shows that the curve expressed by Eq. 6.93 is the shape of the beam loaded by a series of concentrated forces. The whole curve consists of sectional, cubic curves. The first and second derivatives of the whole curve are continuous, and the points of inflection are at the paperweights, which act as support points. Mathematically, the spline is the curve that a series of sectional cubic curves approach as they try to approximate the elastic curve of a drawing spline. More details about spline functions were discussed in Chapter 5.

Trajectory Spline Function

For many important engineering applications, the mathematical model of the spline is highly accurate and true to nature. Furthermore, a wellbore trajectory is continuous and smooth. Consequently, the drill string in a wellbore is an elastic spline, which in every way simulates a wellbore trajectory.

Depth, inclination, and azimuth are the three essential parameters for determining a wellbore trajectory, and based on them, other parameters needed to simulate a real wellbore trajectory can be calculated. As a rule, the depth is the independent variable among these parameters. For the whole trajectory or for a part of it, it is assumed that there are a series of measured results arranged according to survey depth.

If $\alpha(L)$ and $\phi(L)$ are cubic or lower-order multinomials over the subinterval $[L_i, L_{i+1}]$, and they have continuous first and second derivatives over $[a, b]$, and they satisfy Eq. 6.95, then $\alpha(L)$ and $\phi(L)$ are termed the inclination spline function and the azimuth spline function, respectively, based on the nodes $\{L_i\}$.

The subinterval $[L_i, L_{i+1}]$ $(i = 1, 2, ..., n-1)$ over the interval $[a, b]$ is defined by:

$$a = L_1 < L_2 < ... < L_n = b \tag{6.94}$$

The inclination and azimuth spline functions respectively are given based on the nodes as:

$$\begin{cases} \alpha(L_i) = \alpha_i \\ \phi(L_i) = \phi_i \end{cases} \quad (i = 1, 2, ..., n) \tag{6.95}$$

They are linear functions, and for convenience they can be written as:

$$m_i = \ddot{\alpha}(L_i) \qquad (i = 1, 2, ..., n) \tag{6.96}$$

If $L \in [L_i, L_{i+1}]$, then $[L_i, m_i]$ and $[L_{i+1}, m_{i+1}]$. The two linear functions can be represented as:

$$\ddot{\alpha}(L) = m_i \frac{L_{i+1} - L}{h_i} + m_{i+1} \frac{L - L_i}{h_i} \tag{6.97}$$

where the curved course length is:

$$h_i = L_{i+1} - L_i$$

Integrating twice on both the sides of Eq. 6.97 over the interval of $[L_i, L_{i+1}]$ gives:

$$\dot{\alpha}(L) = -m_i \frac{(L_{i+1} - L)^2}{2h_i} + m_{i+1} \frac{(L - L_i)^2}{2h_i} + b_i \tag{6.98}$$

$$\alpha(L) = m_i \frac{(L_{i+1} - L)^3}{6h_i} + m_{i+1} \frac{(L - L_i)^3}{6h_i} + b_i (L - L_i) + c_i \tag{6.99}$$

where b_i and c_i are the constants of integration.

On interpolation:

$$\begin{cases} \alpha(L_i) = \alpha_i \\ \alpha(L_{i+1}) = \alpha_{i+1} \end{cases} \tag{6.100}$$

Using Eq. 6.100 in Eq. 6.99 yields the constants of integration as:

$$\begin{cases} b_i = \dfrac{\alpha_{i+1} - \alpha_i}{h_i} - \dfrac{h_i}{6}(m_{i+1} - m_i) \\ c_i = \alpha_i - m_i \dfrac{h_i^2}{6} \end{cases} \tag{6.101}$$

From Eqs. 6.99 and 6.101, the spline function $\alpha(L)$ at $[L_i, L_{i+1}]$ is only related to m_i and m_{i+1}. In order to determine the spline functions in the subintervals, m_i needs to be known.

In order to determine m_i, substituting Eq. 6.101 into Eq. 6.98 results in:

$$\dot{\alpha}(L) = -m_i \frac{(L_{i+1} - L)^2}{2h_i} + m_{i+1} \frac{(L - L_i)^2}{2h_i} + \frac{\alpha_{i+1} - \alpha_i}{h_i} - \frac{h_i}{6}(m_{i+1} - m_i) \tag{6.102}$$

where $L \in [L_{i-1}, L_i]$ and $L \in [L_i, L_{i+1}]$, respectively:

$$\dot{\alpha}_-(L_i) = \frac{\alpha_i - \alpha_{i-1}}{h_{i-1}} + \frac{h_{i-1}}{3} m_i + \frac{h_{i-1}}{6} m_{i-1} \tag{6.103}$$

$$\dot{\alpha}_+(L_i) = \frac{\alpha_{i+1} - \alpha_i}{h_i} - \frac{h_i}{3} m_i - \frac{h_i}{6} m_{i+1} \tag{6.104}$$

According to well deviation with spline function definition,

$$\dot{\alpha}_-(L_i) = \dot{\alpha}_+(L_i) \tag{6.105}$$

So, using Eqs. 6.103 and 6.104 in Eq. 6.105 results in:

$$(1 - \mu_i) m_{i-1} + 2m_i + \mu_i m_{i+1} = \lambda_i \qquad (i = 2, 3, ..., n-1) \tag{6.106}$$

Where M (dimensionless) and λ (deg/m^2 or deg/ft^2) are intermediate variables given by:

$$\begin{cases} \mu_i = \dfrac{h_i}{h_{i-1} + h_i} \\ \lambda_i = \dfrac{6}{h_{i-1} + h_i}\left(\dfrac{\alpha_{i+1} - \alpha_i}{h_i} - \dfrac{\alpha_i - \alpha_{i-1}}{h_{i-1}}\right) \end{cases}$$

As there are only $n - 2$ equations available for n unknowns in $(m_1, m_2, ..., m_n)$, there is a need for two more equations for the ends of the simulated well path. There are generally three ways to generate these supplementary equations:

- When the rate of change of the inclination[20,21] can be expressed as a first derivative, and if $\dot{\alpha}_1$ and $\dot{\alpha}_n$ are known, then,

$$\begin{cases} \dot{\alpha}(L_1) = \dot{\alpha}_1 \\ \dot{\alpha}(L_n) = \dot{\alpha}_n \end{cases} \tag{6.107}$$

  Using Eqs. 6.103 and 6.104 provides Eq. 6.108 to determine the boundary conditions:

$$\begin{cases} \dfrac{h_1}{3} m_1 + \dfrac{h_1}{6} m_2 = \dfrac{\alpha_2 - \alpha_1}{h_1} - \dot{\alpha}_1 \\ \dfrac{h_{n-1}}{6} m_{n-1} + \dfrac{h_{n-1}}{3} m_n = -\dfrac{\alpha_n - \alpha_{n-1}}{h_{n-1}} + \dot{\alpha}_n \end{cases} \tag{6.108}$$

- When the curvature of the inclination curve can be expressed as a second derivative, and if $\ddot{\alpha}_1$ and $\ddot{\alpha}_n$ are known, then Eq. 6.109 can determine the boundary equations:

$$\begin{cases} \ddot{\alpha}(L_1) = m_1 = \ddot{\alpha}_1 \\ \ddot{\alpha}(L_n) = m_n = \ddot{\alpha}_n \end{cases} \tag{6.109}$$

  The geometrical meaning is that the curvatures of the well path at the two ends of the well path can be determined.

- Inherent boundary conditions: Using four measured points at the beginning and at the end, and interpolating them with the third-order polynomial $P_3(L)$ and $\overline{P}_3(L)$, and then letting:

$$\begin{cases} \ddot{\alpha}(L_1) = \dddot{P}_3(L_1) \\ \ddot{\alpha}(L_n) = \dddot{\overline{P}}_3(L_n) \end{cases} \tag{6.110}$$

  $P_3(L)$ and $\overline{P}_3(L)$ are both third-order interpolation polynomials, $\alpha(L)$ is also a third-order polynomial in the two measurement sections, and their third differentials are constants, so it can be known from Newton's interpolation polynomial expressions:

$$\begin{cases} \dddot{P}_3(L_1) = 6f[L_1, L_2, L_3, L_4] \\ \dddot{\overline{P}}_3(L_n) = 6f[L_{n-3}, L_{n-2}, L_{n-1}, L_n] \end{cases} \tag{6.111}$$

And with Eq. 6.97:

$$\begin{cases} \dddot{\alpha}(L_1) = \dfrac{m_2 - m_1}{h_1} \\ \dddot{\alpha}(L_n) = \dfrac{m_n - m_{n-1}}{h_{n-1}} \end{cases} \tag{6.112}$$

So, Eq. 6.113 can determine the boundary conditions:

$$\begin{cases} -m_1 + m_2 = 6h_1 \displaystyle\sum_{k=1}^{4} \frac{\alpha_k}{\prod\limits_{\substack{i=1 \\ i \neq k}}^{4} (L_k - L_i)} \\ -m_{n-1} + m_n = 6h_{n-1} \displaystyle\sum_{k=n-3}^{n} \frac{\alpha_k}{\prod\limits_{\substack{i=n-3 \\ i \neq k}}^{n} (L_k - L_i)} \end{cases} \tag{6.113}$$

Undoubtedly, the best way is to give the boundary condition at the first and last survey stations. If cubic interpolation polynomials are used, then Eq. 6.113 can determine the boundary equations.

When one of these boundary conditions is combined with Eq. 6.106, a linear group will emerge. It can be proven that the coefficient matrix of the equation group is not singular. Thus, the equation provides a unique solution for the determination of $\alpha(L)$. Similarly, the azimuth spline $\phi(L)$ function can be determined[18,19,22]:

$$\begin{cases} \alpha(L) = m_i \dfrac{(L_{i+1} - L)^3}{6h_i} + m_{i+1} \dfrac{(L - L_i)^3}{6h_i} + b_i (L - L_i) + c_i \\ \phi(L) = M_i \dfrac{(L_{i+1} - L)^3}{6h_i} + M_{i+1} \dfrac{(L - L_i)^3}{6h_i} + B_i (L - L_i) + C_i \end{cases} \tag{6.114}$$

Where the constants of integration are:

$$\begin{cases} b_i = \dfrac{\alpha_{i+1} - \alpha_i}{h_i} - \dfrac{h_i}{6}(m_{i+1} - m_i), \quad c_i = \alpha_i - m_i \dfrac{h_i^2}{6} \\ B_i = \dfrac{\phi_{i+1} - \phi_i}{h_i} - \dfrac{h_i}{6}(M_{i+1} - M_i), \quad C_i = \phi_i - M_i \dfrac{h_i^2}{6} \end{cases}$$

In the equation, the meanings of and the methods to determine M_i, B_i, and C_i are similar to those of m_i, b_i, and c_i used earlier for the inclination spline functions.

Obviously, when simulating the well path by spline functions, the inclination angle (α) and the azimuth (ϕ) of a section are third-order functions of the well depth L. In the natural curve model, the inclination angle (α) and azimuth (ϕ) vary linearly with well depth L. It can be considered that the natural curve model is a specific case of the spline curve model. This is because for the interval $[L_i, L_{i+1}]$:

$$\dot{\alpha}_i = \dot{\alpha}_{i+1} = \frac{\alpha_{i+1} - \alpha_i}{\Delta L_i} \tag{6.115}$$

Using Eq. 6.108, Eq. 6.116 can be written as:

$$\begin{cases} 2m_i + m_{i+1} = \dfrac{6}{\Delta L_i}\left(\dfrac{\alpha_{i+1} - \alpha_i}{\Delta L_i} - \dot{\alpha}_i\right) \\ m_i + 2m_{i+1} = \dfrac{6}{\Delta L_i}\left(-\dfrac{\alpha_{i+1} - \alpha_i}{\Delta L_i} + \dot{\alpha}_{i+1}\right) \end{cases} \tag{6.116}$$

The solution is:

$$\begin{cases} m_i = \dfrac{6(\alpha_{i+1} - \alpha_i)}{\Delta L_i^2} - \dfrac{2(2\dot{\alpha}_i + \dot{\alpha}_{i+1})}{\Delta L_i} \\ m_{i+1} = \dfrac{2(\dot{\alpha}_i + 2\dot{\alpha}_{i+1})}{\Delta L_i} - \dfrac{6(\alpha_{i+1} - \alpha_i)}{\Delta L_i^2} \end{cases} \tag{6.117}$$

But from Eq. 6.101:

$$\begin{cases} b_i = \dfrac{\alpha_{i+1} - \alpha_i}{\Delta L_i} - \dfrac{\Delta L_i}{6}(m_{i+1} - m_i) \\ c_i = \alpha_i - m_i \dfrac{\Delta L_i^2}{6} \end{cases} \tag{6.118}$$

Substituting Eq. 6.115 into Eqs. 6.117 and 6.118 gives:

$$\begin{cases} m_i = 0 \\ m_{i+1} = 0 \end{cases} \tag{6.119}$$

$$\begin{cases} b_i = \dfrac{\alpha_{i+1} - \alpha_i}{\Delta L_i} \\ c_i = \alpha_i \end{cases} \tag{6.120}$$

At this point Eqs. 6.114 and 6.115 can be simplified as:

$$\alpha(L) = \alpha_i + \frac{\alpha_{i+1} - \alpha_i}{\Delta L_i}(L - L_i) \tag{6.121}$$

$$\kappa_{\alpha,i} = \frac{\alpha_{i+1} - \alpha_i}{\Delta L_i} \tag{6.122}$$

Thus, as long as Eq. 6.115 is satisfied, the build rate of this section will remain constant.

Similarly, if Eq. 6.123 is satisfied, then:

$$\dot{\phi}_i = \dot{\phi}_{i+1} = \frac{\phi_{i+1} - \phi_i}{\Delta L_i} \tag{6.123}$$

Then the azimuth change rate in this section is also constant. Thus, in this section, the rate of change of azimuth remains constant.

If the second-order derivatives are given at the beginning of this section, and

$$\ddot{\alpha}_i = \ddot{\alpha}_{i+1} = 0 \tag{6.124}$$

$$\ddot{\phi}_i = \ddot{\phi}_{i+1} = 0 \tag{6.125}$$

then the well path spline curve will regress to a natural curve.

Numerical Integration Method

The basic idea of the spline interpolation function is to express a spline curve as sectionalized polynomial expressions, to turn the whole function into an useable one and to keep the curve smooth at the nodes.[18,19,22] After determining the inclination spline and azimuth functions using Eq. 6.114, Eqs. 6.126 through 6.128 can be used to calculate the course coordinates based on the differential mode of the wellbore trajectory as follows:

$$\Delta N_i = \int_{L_{i-1}}^{L_i} \sin\alpha(L)\cos\phi(L)\,dL \tag{6.126}$$

$$\Delta E_i = \int_{L_{i-1}}^{L_i} \sin\alpha(L)\sin\phi(L)\,dL \tag{6.127}$$

$$\Delta H_i = \int_{L_{i-1}}^{L_i} \cos\alpha(L)\,dL \tag{6.128}$$

Trajectory Deviation

During drilling it is required not only to keep track of the coordinates of the location and direction of the wellbore bottom, but also to compare the actual well trajectory with the designed well path, so that timely measures are taken to ensure that the target is reached, while maintaining a good-quality wellbore. This demands high-precision control of the well geometry, particularly in the case of directional wells and horizontal wells. Hence, well deviation analysis requires dependable and robust methods to compare and analyze the position and shape of well trajectories.

Projection Plot Method

Projection plots are widely used for well trajectory monitoring. A projection plot has two views: a vertical section view and a horizontal projection view (Figure 6.14). More details about these plots can be found in Chapter 3.

A 2D path is easy to understand and to monitor effectively for the well trajectory. However, in 3D well path designs, the well path will not be located in a vertical plane and selecting the best vertical section plane needs consideration. The traditional practice is to use the vertical plane passing through the wellhead and the target (when there are multiple targets, one is selected) as the vertical projection plane for designing the well path and the drilled trajectories are projected onto this plane for analysis and monitoring. However, this method of projection onto a vertical plane may cause an unrealistic perspective of the relation between the real and the designed path, and this could lead to a false analysis and conclusion. However, this vertical projection map method may distort the relationship between the real trajectory and designed well paths, even lose any comparability, so it may lead to wrong conclusions. Because there is no design plane of the 3D design trajectory, the real well path should project into the curve plane of the design well path to get the vertical projection map, which is convenient to compare the deviation between the real trajectory and designed well paths.

As shown in Figure 6.15, a series of vertical lines through each point in the designed well path can be drawn, and these vertical lines can compose a vertical cylindrical surface as a designed curve plane. If the real well path projects vertically into the designed curve plane, the trajectory curve after projection can be obtained. And then the designed well path and the trajectory curve should be projected into the horizontal plane. In the above projection procedure, a point M on the real well path is projected to point Q in the curve plane, and to point G in the horizontal plane; meanwhile point M can project directly into point F in the horizontal plane without any projection conversion. The projected trajectories in the designed well path and the real path are both the curves in the designed curve plane, so the vertical line through point Q must be across the designed well path at point P, and point P overlaps with point G, which is the projection of Q in the horizontal plane. Therefore, the projected trajectories in the horizontal plane of the

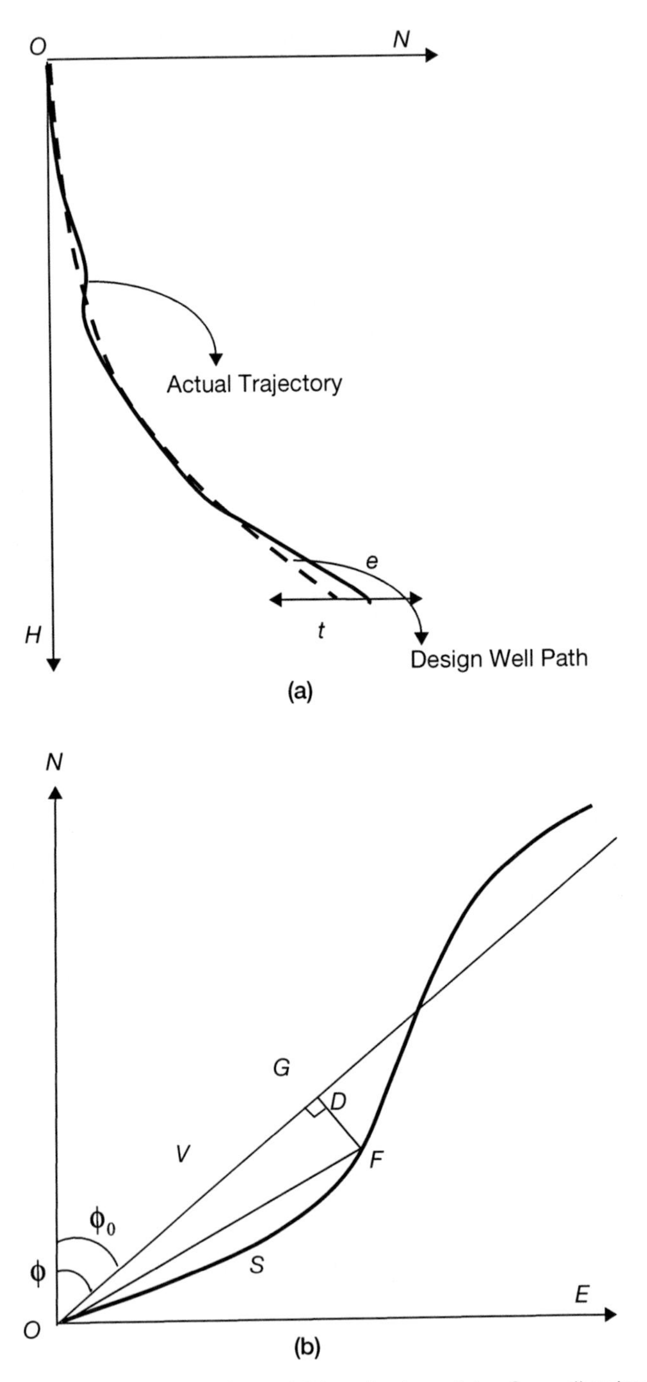

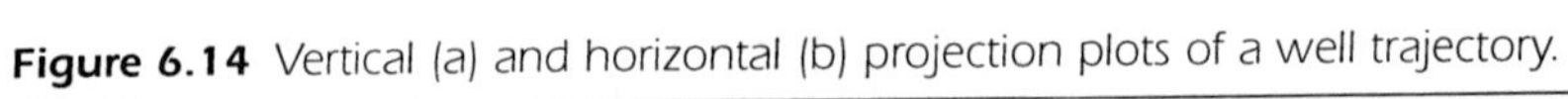
Figure 6.14 Vertical (a) and horizontal (b) projection plots of a well trajectory.

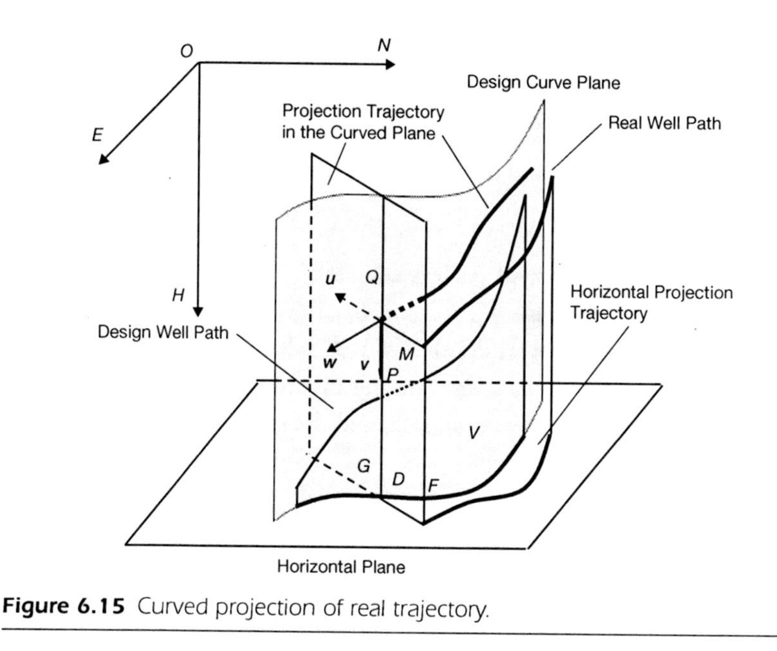

Figure 6.15 Curved projection of real trajectory.

designed and real well path are the same curve, so the displacement of point M in the real path is the length along this curve from point G to the projected point of the wellhead. In this way, if expanding the designed curve plane into a plane, the vertical projection map of the well trajectory can be obtained. Obviously, this vertical projection map is the vertical profile of the designed well path, and to the real well path, it is the vertical profile of the projection on the curve plane. The essence of this curve plane projection method is, for a point M in the real well path, to find the reasonable reference point P on the designed well path, and to analyze the deviation of the real well path by calculating the displacement and the horizontal deviation.

This surface projection method, in essence, provides the degree of deviation of the drilling point M from the design point P in the designed well path. Points M, Q, and P form a vertical plane W. In the W plane, the unit vectors of the projections of the curved plane can be denoted in the MQ direction and in the vertical direction as u and v, respectively, so that the unit vectors can be written as:

$$\boldsymbol{u} = -\sin \phi_P \times \boldsymbol{i} + \cos \phi_P \times \boldsymbol{j} \tag{6.129}$$

$$\boldsymbol{v} = \boldsymbol{k} \tag{6.130}$$

where $\boldsymbol{i}, \boldsymbol{j}$, and $\boldsymbol{k}$ are unit coordinate vectors in the O-NEH coordinate system.

Because $\boldsymbol{u}$ and $\boldsymbol{v}$ are perpendicular to each other, the unit normal vector of plane Ω is:

$$\boldsymbol{w} = \cos\phi_P \times \boldsymbol{i} + \sin\phi_P \times \boldsymbol{j} \tag{6.131}$$

With the M and P points, the Ω plane equation is:

$$(N_M - N_P)\cos\phi_P + (E_M - E_P)\sin\phi_P = 0 \tag{6.132}$$

A point M in the real well path is selected, so that the coordinates of point M are known. Because N_P, E_P, and ϕ_P of point P in the designed well path are functions of the well depth, point P is located by Eq. 6.132. Further, the horizontal length to point P along the designed path is the vertical displacement of point M in the real path, and the horizontal displacement of point M is[1]:

$$D_M = \sqrt{(N_M - N_P)^2 + (E_M - E_P)^2} \tag{6.133}$$

In the curved plane projection, the projection of point M of the real path is point Q in the designed curved plane, and then a vertical line can be drawn from point Q to cross the designed well path at point P. Obviously, if the well path is two dimensional, then the surface projection method reduces to a planar view.

Normal Plane Scanning

As noted above, it is necessary to analyze, design, and monitor the drilled well trajectory and compare it with the actually designed well trajectory, using common projection plots and also the vertical surface projection method. This projection method maps the real and designed well paths onto vertical and horizontal projections, and analyzes the deviation of the real and designed well paths by comparing the vertical and horizontal displacements. Besides this, it is also possible to analyze and express the deviation of the real and designed well path in 3D space, by using a method called the normal plane scanning method,[19,20] or traveling cylindrical scanning method.[21]

As shown in Figure 6.16, a normal plane is made at a point P in the designed well path, which cuts across the real path at a point M. The coordinates of points P and M can be represented by the vectors $\boldsymbol{r}_P$ and $\boldsymbol{r}_M$. Because the vector $\boldsymbol{r}_{PM}$ from point P to point M is located in the normal plane Ω, it is perpendicular to the unit tangent vector $\boldsymbol{t}_P$ at point P of the designed well path. So,[15,20]

$$(\boldsymbol{r}_M - \boldsymbol{r}_P) \times \boldsymbol{t}_P = 0 \tag{6.134}$$

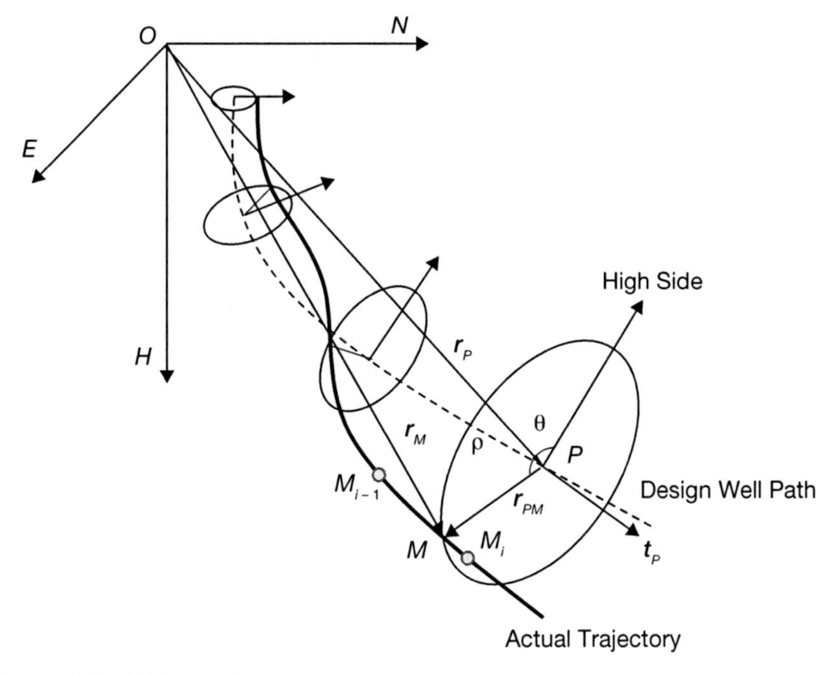

Figure 6.16 Well scanning.

because,

$$\boldsymbol{t}_P = \sin\alpha_P \cos\phi_P \boldsymbol{i} + \sin\alpha_P \sin\phi_P \boldsymbol{j} + \cos\alpha_P \boldsymbol{k} \tag{6.135}$$

Therefore,

$$a(N_M - N_P) + b(E_M - E_P) + c(H_M - H_P) = 0 \tag{6.136}$$

where

$$a = \sin\alpha_P \cos\phi_P$$
$$b = \sin\alpha_P \sin\phi_P$$
$$c = \cos\alpha_P$$

If P is a designed node or point, the well path parameters can be directly obtained, otherwise interpolation may be required. For the position of a certain point P of the designed well path, if the point of intersection M of the normal plane with the real well path needs to be calculated, it is necessary to decide to which interval M belongs. So,

$$f(L) = a(N - N_P) + b(E - E_P) + c(H - H_P) \tag{6.137}$$

Putting the coordinates of two neighboring points into Eq. 6.137, if $f(L_{i-1}) \times f(L_i) \leq 0$, the actually drilled target e is located in the interval $[L_{i-1}, L_i]$. In the particular cases that $f(L_{i-1}) = 0$ or $f(L_i) = 0$, then either point M_{i-1} or point M_i is point M. Generally, point M would be located within the interval of M_{i-1} and M_i, but not exactly at a measured point, so an interpolation should be used to find out the coordinates of point M.[18]

For any measured point M in the real well path, in order to analyze its deviation from the design, a normal plane to the designed well path passing through point M, is to be made, and then the corresponding point P of the designed well path can be obtained. In short, whether the reference point is in the designed well path or in the real well path, the designed well path should be taken as the base, and the normal plane should be perpendicular to the designed well path.[1]

If a point M of the real well path is located on the normal plane through point P of the designed well path, then two parameters, scanning distance and scanning angle, can be used to express the relative position of these two points. The scanning distance ρ is the distance between points P and M, and the scanning angle θ is the angle that starts from the wellbore high side at point P of the designed well path and rotates clockwise to point M. Obviously, the scanning angle is located at the normal plane of point P and it seems similar to a reference plane angle. So, according the conversion relationship between the wellbore coordinate system P-xyz and the coordinate system O-NEH, there is[15,19]:

$$\rho = \sqrt{x_M^{\ 2} + y_M^{\ 2}} \tag{6.138}$$

$$\tan\theta = \frac{y_M}{x_M} \tag{6.139}$$

where

$$\begin{bmatrix} x_M \\ y_M \\ z_M \end{bmatrix} = \begin{bmatrix} \cos\alpha_P \cos\phi_P & \cos\alpha_P \sin\phi_P & -\sin\alpha_P \\ -\sin\phi_P & \cos\phi_P & 0 \\ \sin\alpha_P \cos\phi_P & \sin\alpha_P \sin\phi_P & \cos\alpha_P \end{bmatrix} \begin{bmatrix} N_M - N_P \\ E_M - E_P \\ H_M - H_P \end{bmatrix}$$

Thus, for each specific location P, the distance and the scanning angle can be calculated from the corresponding scanning. If point P is taken as the origin of a polar coordinate system, the scanning distance as the radial coordinate, and the scanning angle as the polar angle, the position of point M can be located in this polar coordinate system.

During scanning with the normal plane, point P moves along the designed well path, and point M moves correspondingly along the real well

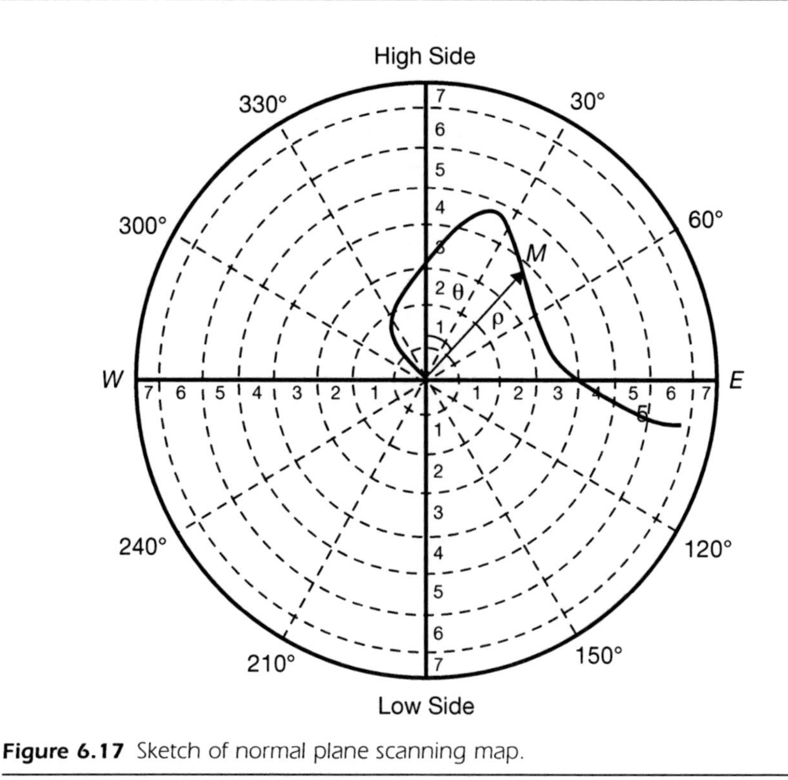

Figure 6.17 Sketch of normal plane scanning map.

path. This is the relationship between points M and P in the normal plane, which is shown in the normal plane scanning map of the real and designed well paths given in Figure 6.17.

Off-Target Distance

Vertical, directional, horizontal, and extended-reach wells have stipulated target areas based on which drilling trajectory is planned, so that the designed target is reached within a certain tolerance. The target area shape, the well trajectory, and the quality of the wellbore are some of the key indicators to be considered for specifying the tolerance. If the reservoir shape is complex, the target area and shape also become complex. The target plane is defined as a plane, not necessarily normal to the designed well path, which contains the designed target point t, and which the actually drilled well path intersects to give the real target point e. The distance between the two target points is the off-target distance. Therefore, a common and versatile target plane equation is needed for the calculation of the real target point and the off-target distance that can be used for all types of wells.

Target Plane Equation

It is extremely important to reach positively the target point and the target area, and for this the well trajectory should be controlled in such a way that the target is reached within a certain range of error. Vertical wells and directional wells generally have a specified circular target area in the horizontal plane. Horizontal wells use a parallelepiped region, as shown in Figure 6.18. Another type of complex horizontal well design is the ladder, or staircase, which will have a curved trajectory. Depending on the complexity of the

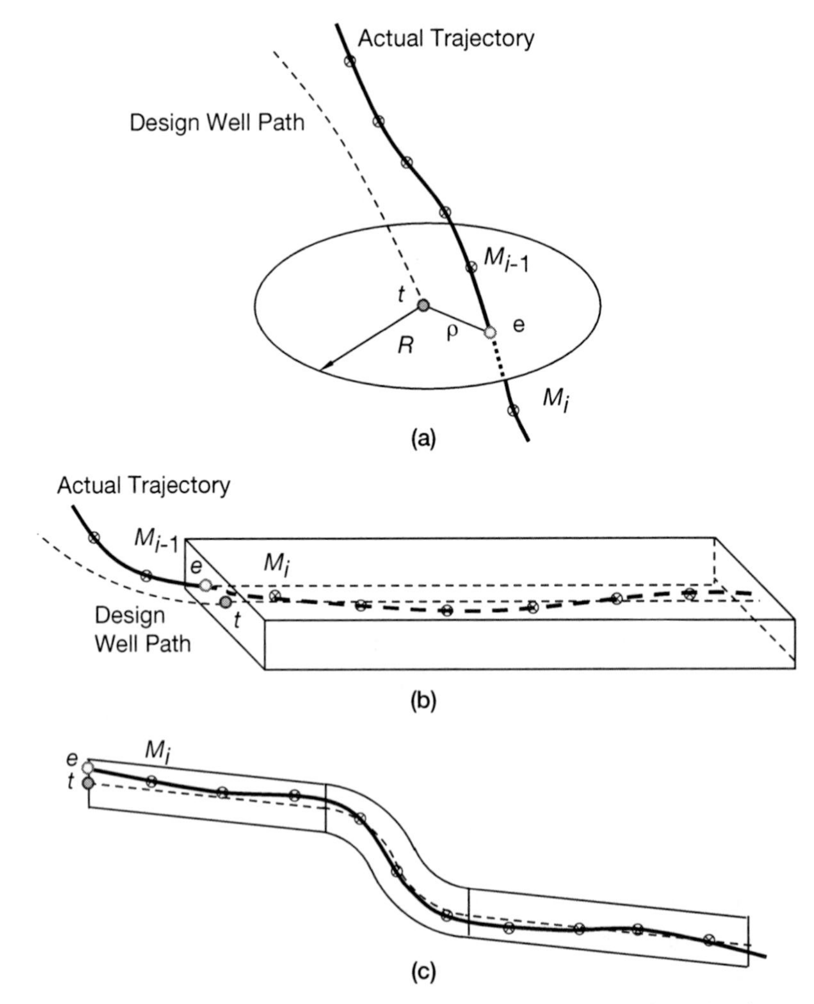

Figure 6.18 Target area shapes: (a) directional trajectory, (b) horizontal trajectory, and (c) step target with horizontal trajectory.

reservoir shapes, the shape of target also differs, but the majority of the target areas are simple geometrical shapes.

In order to control effectively the well trajectory for positively reaching the target area, it is essential to have critical control points along the wellbore. In normal circumstances, a vertical well or a directional well will have one control point, while horizontal wells will have two control points. Designed well paths will pass through these control points, whereas the actual trajectory will have certain tolerances. The target point t in the designed well path, and the actual target point e achieved in the drilled well path, are separated by a distance ρ in the target plane, as shown in Figure 6.19. So, the off-target distance is not just the shortest spatial distance between the real well path and the designed target point t, but it is the distance in the target plane between the designed target point and the point of intersection e of the real well path with the target plane, and it has to be calculated at each target.

Usually, the target plane of a directional well is a horizontal plane, and the target plane of a horizontal well is a vertical plane. However, for the designed well path at the target, the inclination of a directional well is often not 0°, and the inclination of a horizontal well is not always 90°. Thus, the designed well path at the target is usually not perpendicular to the target plane, or the tangent to the designed well path at the target does not coincide with the normal to the target plane, as shown in Figure 6.19. Even though the target

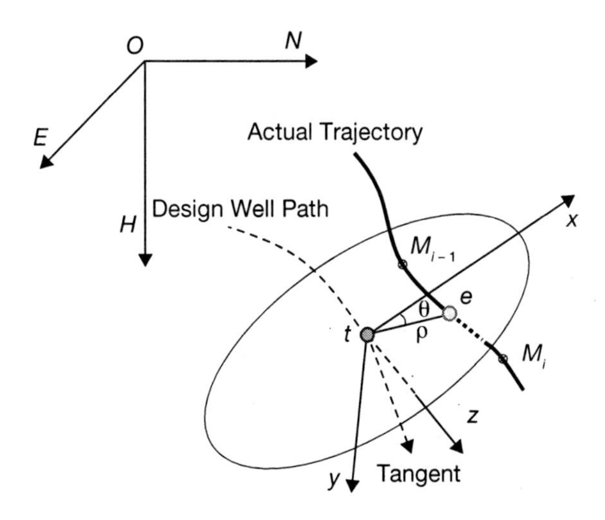

Figure 6.19 Target plane position.

plane is often a horizontal plane or a vertical plane, the study of an arbitrarily oriented target plane is required, as it will be applicable for all the cases.

The coordinates of the designed target point determine the location of the target plane, and the orientation of the target plane can be used to determine the direction of its normal. Therefore, referring to the definitions of the inclination angle and azimuth, the angle between the normal to the target plane and the vertical direction, and the angle between the projection of the normal to the target plane onto the horizontal plane and the true north can be defined as the normal inclination angle and the normal azimuth of the target plane. Obviously, the normal inclination angle and the normal azimuth of the target plane are similar to the inclination angle and azimuth of the well path, but their meaning and values are different.

In some cases, the normal inclination angle and the normal azimuth of the target plane may be equal to the inclination angle and azimuth of the designed well path at the target. Such instances happen when the target plane is normal to the well path at the target, as for example when a directional well enters the oil zone vertically or a horizontal well extends into a horizontal oil zone etc.

As shown in Figure 6.19, the target point t is taken as the origin in the target plane with the z-axis taken normal to the target plane, the x-axis taken toward the high-side direction, and the right-hand law used to define the y-axis, thereby establishing a t-xyz coordinate system. Because the target plane has the selected target t, its normal inclination angle and the normal azimuth angle, respectively, are α_z and ϕ_z. Therefore, the target plane's equation is:

$$(N-N_t)\sin\alpha_z\cos\phi_z+(E-E_t)\sin\alpha_z\sin\phi_z+(H-H_t)\cos\alpha_z=0 \quad (6.140)$$

Target Distance Calculation

The key to calculating the off-target distance is to find out the coordinates of the real target. The position of the real target is the point of intersection of the real well path and the target plane. The target plane equation is given by Eq. 6.140 and if the real well path equation is also available, then the real target coordinates can be obtained by solving these two equations. However, the real well path is characterized by a series of points, and so it is necessary to first establish which measured well section contains the target, and then the real well path equation can be determined. Thus,

$$f(L)=(N-N_t)\sin\alpha_z\cos\phi_z+(E-E_t)\sin\alpha_z\sin\phi_z+(H-H_t)\cos\alpha_z \quad (6.141)$$

Putting the coordinates of two neighboring points into Eq. 6.141, when $f(L_{i-1}) \times f(L_i) \leq 0$, the target e is located in the interval $[L_{i-1}, L_i]$. In usual cases, the target e is located within the section defined by the points M_{i-1} and M_i, but not exactly at a measured point, so interpolation should be used to find out the target coordinates.

Whichever inclination calculation method is used, the real well path in an interval can be always expressed as a function of the well depth L. In this way, putting these functions into Eq. 6.141, the depth L_e and the coordinates (N_e, E_e, H_e) of the target point e can be found. Further, the off-target distance can be calculated using:

$$\rho = \sqrt{(N_e - N_t)^2 + (E_e - E_t)^2 + (H_e - H_t)^2} \tag{6.142}$$

In order to find the relationship between the real target point e and the designed target point t, besides the off-target distance, another parameter in the target plane must be known (i.e., the angle between the vector te and the x-axis, called the deflection angle, which is denoted by θ). According to the coordinate transformation relationship[23]:

$$\begin{bmatrix} x_e \\ y_e \\ z_e \end{bmatrix} = \begin{bmatrix} \cos\alpha_z \cos\phi_z & \cos\alpha_z \sin\phi_z & -\sin\alpha_z \\ -\sin\phi_z & \cos\phi_z & 0 \\ \sin\alpha_z \cos\phi_z & \sin\alpha_z \sin\phi_z & \cos\alpha_z \end{bmatrix} \begin{bmatrix} N_e - N_t \\ E_e - E_t \\ H_e - H_t \end{bmatrix} \tag{6.143}$$

Because the real target point e is located in the target plane, therefore $z_e = 0$. Using Eqs. 6.143 and 6.140, the distance ρ between the designed and the real target points and the deflection angle are, respectively, given by:

$$\rho = \sqrt{x_e^2 + y_e^2} \tag{6.144}$$

$$\tan\theta = \frac{y_e}{x_e} \tag{6.145}$$

Thus, the location of the target e is determined by (ρ, θ) or (x_e, y_e).

Application of the Calculation Method

The coordinates of the target t determine the position of the target plane, and α_z and ϕ_z decide how the plane is oriented in space. Usually, the orientation of the target plane is related to the type of well, but not necessarily related to the direction of the real well path at the target. In most cases, the target plane of a directional well is horizontal, and that of a horizontal well is vertical.

For a vertical target plane, $\alpha_z = 90°$ and, based on Eqs. 6.149 and 6.143, the target plane equation and the equation for the real target point e are:

$$\left(N - N_t\right)\cos\phi_z + \left(E - E_t\right)\sin\phi_z = 0 \tag{6.146}$$

$$\begin{bmatrix} x_e \\ y_e \\ z_e \end{bmatrix} = \begin{bmatrix} 0 & 0 & -1 \\ -\sin\phi_z & \cos\phi_z & 0 \\ \cos\phi_z & \sin\phi_z & 0 \end{bmatrix} \begin{bmatrix} N_e - N_t \\ E_e - E_t \\ H_e - H_t \end{bmatrix} \tag{6.147}$$

From Eqs. 6.146 and 6.147, for this condition, the x-axis is normal upwards, opposite to the H axis direction; while the y-axis and the z-axis are located in the horizontal plane, and $z_e = 0$. Thus, when calculating the target distance in a vertical target plane, the Eqs. 6.140 through 6.145 are used by setting $\alpha_z = 90°$. Thus, x_e and y_e represent the horizontal and vertical coordinates of the real target e with respect to the designed target t.

For a horizontal target plane $\alpha_z = 0°$. For this condition, the target plane equation and the equation for the real target point e can be simplified as:

$$H_e - H_t = 0 \tag{6.148}$$

$$\begin{bmatrix} x_e \\ y_e \\ z_e \end{bmatrix} = \begin{bmatrix} \cos\phi_z & \sin\phi_z & 0 \\ -\sin\phi_z & \cos\phi_z & 0 \\ 0 & 0 & 1 \end{bmatrix} \begin{bmatrix} N_e - N_t \\ E_e - E_t \\ H_e - H_t \end{bmatrix} \tag{6.149}$$

When $\alpha_z = 0°$, ϕ_z does not exist. However, if ϕ_z takes a certain value, some special cases can be determined. For example, if $\phi_z = 0$, then,[23]

$$\begin{bmatrix} x_e \\ y_e \\ z_e \end{bmatrix} = \begin{bmatrix} 1 & 0 & 0 \\ 0 & 1 & 0 \\ 0 & 0 & 1 \end{bmatrix} \begin{bmatrix} N_e - N_t \\ E_e - E_t \\ H_e - H_t \end{bmatrix} \tag{6.150}$$

The above equation indicates that translation of the coordinate system *O-NEH* to the target point gives the coordinate system *t-xyz* with the axis directions unchanged. Thus, x_e and y_e represent the target distance between e and t in north and east directions, and the deflection angle θ represents the angle with the north direction, as shown in Figure 6.20(a).

If $\phi_z = \phi_t$:

$$\begin{bmatrix} x_e \\ y_e \\ z_e \end{bmatrix} = \begin{bmatrix} \cos\phi_t & \sin\phi_t & 0 \\ -\sin\phi_t & \cos\phi_t & 0 \\ 0 & 0 & 1 \end{bmatrix} \begin{bmatrix} N_e - N_t \\ E_e - E_t \\ H_e - H_t \end{bmatrix} \tag{6.151}$$

Eq. 6.151 shows that translation of the coordinate system O-*NEH* to the target point t, followed by rotation through an angle ϕ_t, gives the coordinate system *t-xyz*. In this case, x_e is the distance of deviation between the real tar-

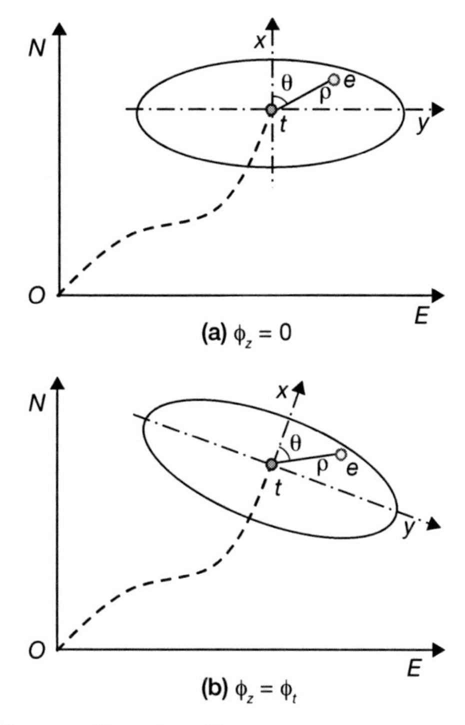

Figure 6.20 Significance of target position.

get point *e* and the designed target point *t* along the direction with azimuth ϕ_t, while y_e represents the distance of deviation between the target points in the direction perpendicular to the direction with azimuth ϕ_t, and the deflection angle θ is also measured relative to the direction with azimuth ϕ_t, as shown in Figure 6.20(b). In the case of a 2D design, because the target plane normal azimuth ϕ_t is equal to designed azimuth, x_e is the distance of deviation between *e* and *t* along the direction of the designed azimuth, and y_e is the distance of deviation between *e* and *t* in the direction perpendicular to the designed azimuth.

Sometimes the target plane may be perpendicular to the designed well path at the target point, which can be called a normal target plane. In this case, the normal to the target plane points along the direction of the tangent to the designed path, or $\alpha_z = \alpha_t$, and $\phi_z = \phi_t$. So, the target plane equation and the target coordinates are:

$$(N - N_t)\sin\alpha_t \cos\varphi_t + (E - E_t)\sin\alpha_t \sin\varphi_t + (H - H_t)\cos\alpha_t = 0 \tag{6.152}$$

$$\begin{bmatrix} x_e \\ y_e \\ z_e \end{bmatrix} = \begin{bmatrix} \cos\alpha_t \cos\phi_t & \cos\alpha_t \sin\phi_t & -\sin\alpha_t \\ -\sin\phi_t & \cos\phi_t & 0 \\ \sin\alpha_t \cos\phi_t & \sin\alpha_t \sin\phi_t & \cos\alpha_t \end{bmatrix} \begin{bmatrix} N_e - N_t \\ E_e - E_t \\ H_e - H_t \end{bmatrix} \tag{6.153}$$

Evidently, Eq. 6.153 is more general in regards to the target plane definition and can be used for a horizontal target plane or a vertical target plane, as well as a normal target plane. Therefore, the method of calculating the target distance is general and has a universal applicability.

Target Distinction

Calculation of target parameters, such as the distance of deviation, is also important for reasons other than discovering whether the trajectory target point *e* is within the desired target range. Usually, the geometrical shape of the target area is simple and easy to define. For example, for horizontal wells, a vertical rectangular parallelepiped geometry is generally used with height and width in the vertical and horizontal direction. However, from a more general point of view, the designed target point *t* may not be located at the center of the target area, and also the geometry of the target area may require a rotation of the normal to the target plane by an angle.

As shown in Figure 6.21, assuming that the coordinate of the geometric center of the target area is (x_c, y_c) under the coordinate system *t-xy*, and the standard geometry of the target area rotates δ° around the target normal plane, the coordinate of the target point *e* under coordinate system *c*-ξη is:

$$\begin{cases} \xi_e = (x_e - x_c)\cos\delta + (y_e - y_c)\sin\delta \\ \eta_e = -(x_e - x_c)\sin\delta + (y_e - y_c)\cos\delta \end{cases} \tag{6.154}$$

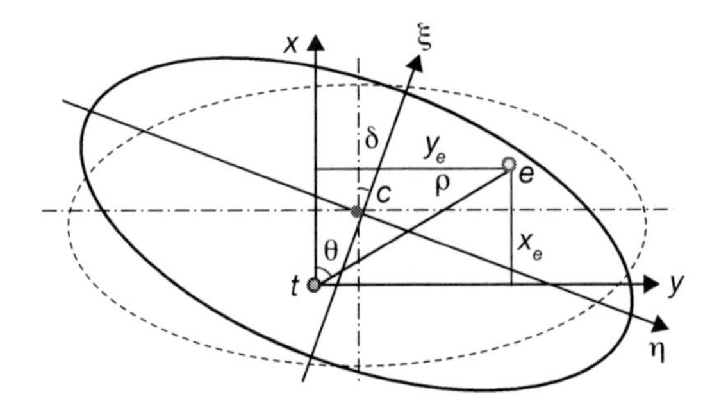

Figure 6.21 Target description.

Similarly, using the c-$\xi\eta$ coordinate system, it is easy to find out whether or not the real target point e is in the target area. As an example, for an elliptical target area the target condition to be satisfied is given as[23]:

$$\frac{\xi_e^2}{b^2} + \frac{\eta_e^2}{a^2} \le 1 \tag{6.155}$$

where a and b are the major and minor axes of an ellipse.

The above calculations form the basis for multiple target and anti-collision estimates, which are discussed in detail in Chapter 7.

Supplementary Problems

1. Based on the inclination spline function and azimuth spline function, prove that:

$$\begin{cases} \kappa_\alpha = \dfrac{\alpha_{i+1} - \alpha_i}{\Delta L_i} - \dfrac{\Delta L_i}{6}\left(m_{i+1} - 2m_i\right) \\ \dot{\kappa}_\alpha = m_i \end{cases}$$

$$\begin{cases} \kappa_\phi = \dfrac{\phi_{i+1} - \phi_i}{\Delta L_i} - \dfrac{\Delta L_i}{6}\left(M_{i+1} - 2M_i\right) \\ \dot{\kappa}_\phi = M_i \end{cases}$$

2. Assume a course to be a spline curve with the length of 90 m. The parameters at the upper survey station are: $\alpha_i = 50°$, $\phi_i = 200°$, $\kappa_{\alpha,i} = 8°/30$ m, and $\kappa_{\phi,i} = -5°/30$ m. The parameters at the lower survey station are: $\alpha_{i+1} = 50°$, $\phi_{i+1} = 200°$, $\kappa_{\alpha,i+1} = 8°/30$ m, and $\kappa_{\phi,i+1} = -5°/30$ m. Using the curve structure method, calculate the wellbore coordinates, curvature, and torsion.

3. Assume a course to be a cylindrical-helix curve. The inclination and azimuth at the upper survey station are: $\alpha_i = 20°$ and $\phi_i = 100°$. The inclination and azimuth at the lower survey station are: $\alpha_{i+1} = 30°$ and $\phi_{i+1} = 110°$. The length of the course is $\Delta L_i = 30$ m. Calculate the wellbore coordinates, curvature, and torsion.

4. Assume a course to be an arc in an inclined plane. The inclination and azimuth at the upper survey station are: $\alpha_i = 40°$ and $\phi_i = 70°$. The inclination and azimuth at the lower survey station are: $\alpha_{i+1} = 49°$ and $\phi_{i+1} = 58°$. The length of the course is $\Delta L_i = 30$ m. Calculate the wellbore coordinates, curvature, and torsion.

5. Assume a course to be a natural curve. The inclination and azimuth at the upper survey station are: $\alpha_i = 40°$ and $\phi_i = 353°$. The inclination and azimuth at the lower survey station are: $\alpha_{i+1} = 50°$ and $\phi_{i+1} = 5°$. The length of the course is $\Delta L_i = 30$ m. Calculate the wellbore coordinates, curvature, and torsion.

References

1. Xiushan, L. *Geometry of Wellbore Trajectory*. Beijing, China: Petroleum Industry Press, 2006.
2. Backus, G., R. Parker, and C. Constable. *Foundations of Geomagnetism*. Cambridge, UK: Cambridge University Press, 1996.
3. Wenyao, X. *Geomagnetism*. Beijing, China: Seismological Press, 2003.
4. Xiushan, L. "Naturalization of Azimuth Angles and Coordinates in Directional Drilling." Renqiu, China. *Oil Drilling & Production Technology* 29, no. 4 (2007): 1–5.
5. Xiushan, L., and W. Jiping. "A Method for Monitoring Wellbore Trajectory Based on the Theory of Geodesy." Dezhou, China. *Petroleum Drilling Techniques* 35, no. 4 (2007): 1–5.
6. Rushan, L., Z. Yueyun, L. Xiushan, et al. "Well Path Planning and Survey Calculation for Directional Drilling (SY/T 5435—2003)." Released by China National Economic and Trade Committee, 2003.
7. Edison, J. E. "Engineering Planning and Supervision of Directional Drilling Operations." *Journal of Petroleum Technology* 9, no. 11 (1957): 16–19.
8. Walstrom, J. E., R. P. Harvey, and H. D. Eddy. "Directional Survey Methods: The Balanced Tangential Method, A Comparison of Various Methods." SPE Paper No. 3379, SPE 46th Annual Fall Meeting, New Orleans, Louisiana, October 3–6, 1971.
9. Wilson, G. J. "An Improved Method for Computing Directional Surveys." *Journal of Petroleum Technology* (August 1968): 871–876.
10. Callas, N. P. "Computing Directional Surveys with a Helical Method." *SPE Journal* (December 1976): 327–336.
11. Azar, J. J., and S. G. Robello. *Drilling Engineering*. Tulsa, OK: Penwell Publishers, 2007.
12. Zhiyong, H. *Design and Calculation of Directional Drilling*, 2nd ed. Dongying, China: China University of Petroleum Press, 2007.
13. Fuqi, L. "Chord Step Method for Calculating the Real Trace of Well Bore." Chengdu, China. *Natural Gas Industry* 6, no. 4 (1986): 40–46.
14. Xiushan, L., and S. Zaihong. "Numerical Approximation Improves Well Survey Calculation." *Oil & Gas Journal* 99, no.15 (2001): 50–54.
15. Xiushan, L., W. Shan, and J. Zhongxuang. *Designing Theory and Describing Method for Wellbore Trajectory*. Harbin, China: Heilongjiang Science and Technology Press, 1993.

16. Xiushan, L., and S. G. Robello. "Actual 3D Shape of Wellbore Trajectory: An Objective Description for Complex Steered Wells." SPE 115714, Annual Technical Conference, Denver, Colorado, September 21–24, 2008.
17. Xiushan, L. "Objective Description and Calculation of Drilled Wellbore Trajectories." Beijing, China. *Acta Petrolei Sinica* 28, no. 5 (2007): 128–132, 138.
18. Xiushan, L., Z. Daqian, G. Lingdi, et al. "How to Simulate Actual Well Trajectories with Spline Function." Anda, China. *Journal of Daqing Petroleum Institute* 15, no.1 (1991): 46–51.
19. Zhiyong, H., and N. Xiuxu. "A New Plotting for Directional Drilling—Normal-plane-scanning Chart." Dongying, China. *Journal of the University of Petroleum* 14, no. 3 (1990): 24–30.
20. Xiushan, L., and C. Zhangzhi. "Description and Calculation of Relative Positions of Wellbore Trajectories." Guanghan, China. *Drilling & Production Technology* 22, no. 3 (1999): 7–12.
21. Thorogood, J. L., and S. J. Sawaryn, "The Traveling-Cylinder Diagram: A Practical Tool for Collision Avoidance." SPE Drilling Engineering 6, no.1 (1991): 31–36.
22. Xiushan, L., A. Chi, and W. Xinqing. "Value Intercalation Method of Wellbore Track." Renqiu, China. *Oil Drilling & Production Technology* 19, no. 2 (1997): 11–14.
23. Xiushan, L. "The Universal Method for Calculating Off-Target Distances." Renqiu, China. *Oil Drilling & Production Technology* 30, no.1 (2008).

7

Anti-Collision and Error Analysis

In the previous chapter, the survey calculations and the different methods used for monitoring a well trajectory were explained, including the projection plot options. Proper monitoring of the location and placement of new wells is important for avoiding collision with any existing well. This becomes more complex when drilling a well in a multiple-wells environment, in which the chances of the new well colliding with an existing well are higher. Collision with an existing well could result in loss of production, and other unwanted problems; and so, a rigorous procedure and consistent methodology are required to avoid such incidents. The principles introduced in the previous chapter, and the basic steps described earlier for comparing an actually drilled well trajectory with the designed well path, can be also used for avoiding collision with already drilled wells. Survey methods and anti-collision techniques are important aspects of well planning, not only to meet multitarget geological objectives, but also to have the ability to drill wells in close proximity to other wells within a constrained space.

This chapter treats two fundamental aspects: the first being the theoretical methods needed for the analysis and the planning to avoid well collision, and the second aspect concentrates mainly on error analysis and probability risk analysis used in well collision studies. Inaccuracies from the survey data of the previous wells, and the deviation of the existing well from its planned well path, increase the risk of well collision. This chapter discusses the mathematical treatment of these uncertainties and their use in well collision risk analysis.

Anti-Collision

As just noted, in an environment with a cluster of wells, designing and placing a well without colliding with an existing well is extremely important. While designing such a well it is necessary to consider its distance to all the adjacent wells. For a relief well, it is important to intersect the existing well at the desired depth in order to carry out relief operations. Whether it is a conventional well or a relief well, it is necessary to map the relative position of the new well with respect to all the existing adjacent wells.[1-3] In the treatment below, the new well is called the reference well and the existing well is called the offset well.

Let P denote a point on the new or reference well trajectory with the position vector $\boldsymbol{r}_P$. As shown in Figure 7.1, the designed well path for the reference well and the real well path of the existing or offset well can be treated as continuous and smooth spatial curves. Hence, there always exists a spherical surface with center at P, which can be drawn tangent to the tra-

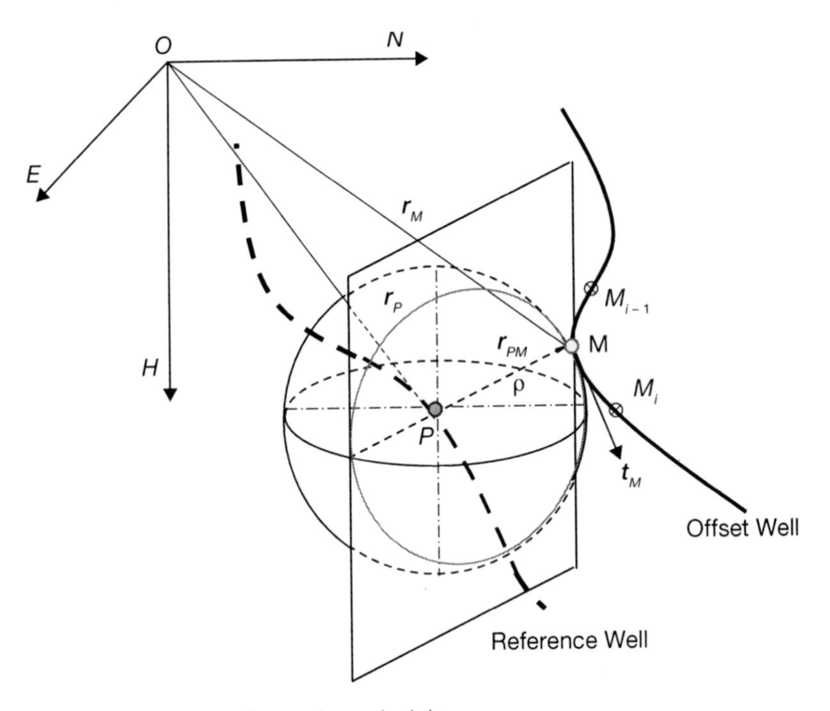

Figure 7.1 Adjacent well scanning principle.

jectory of the offset well. Let the spherical surface have the radius of curvature ρ, and let the surface touch the trajectory of the offset well at point M with the position vector $\boldsymbol{r}_M$. Clearly, $\rho = |\boldsymbol{r}_M - \boldsymbol{r}_P|$. The unit tangent vector along the offset well path at point M is denoted by $\boldsymbol{t}_M$, and so:

$$(\boldsymbol{r}_M - \boldsymbol{r}_P) \times \boldsymbol{t}_M = 0 \tag{7.1}$$

Eq. 7.1 can be written in scalar form as:

$$a(N_M - N_P) + b(E_M - E_P) + c(H_M - H_P) = 0 \tag{7.2}$$

where

$$a = \sin\alpha_M \cos\phi_M$$
$$b = \sin\alpha_M \sin\phi_M$$
$$c = \cos\alpha_M$$

Eq. 7.2 should be satisfied by point M of the offset well path that is nearest to point P of the reference well path. From another perspective, the distance from point P of the reference well, with coordinates N_P, E_P, and H_P, to any point of the offset well, with coordinates N, E, and H, is given by:

$$\rho = \sqrt{(N - N_P)^2 + (E - E_P)^2 + (H - H_P)^2} \tag{7.3}$$

When ρ is the shortest distance, it must satisfy:

$$\frac{d\rho}{dL} = 0 \tag{7.4}$$

The coordinates N, E, and H of any point in the offset well are functions of the well depth L, and so by substituting Eq. 7.3 into Eq. 7.4, point M nearest to point P can be obtained as:

$$a(N_M - N_P) + b(E_M - E_P) + c(H_M - H_P) = 0 \tag{7.5}$$

It can be seen that Eqs. 7.2 and 7.5 are exactly the same.

It is to be noted that in some special cases, the starting point and ending point of the well may not satisfy Eqs. 7.2 and 7.5, and so, besides the calculations using the above method, these two points should be separately compared.

Whether using the designed well path or the measured points of the already drilled trajectory, the offset well can be described by a series of calculated or measured points. When calculating the shortest distance between the reference well and the offset well described by these discrete data points, it needs to be checked in which section of the offset well the nearest point M actually lies. Currently, there are three methods of calculation with different criteria that are used to determine this.[1,4]

Method 1. Insert the coordinates of the calculated points, or the measured points, of the offset well sequentially into Eq. 7.3 and find the nearest point Q to point P of the reference well. Then perform interpolations for points within the two neighboring well sections of point Q, and thus find the nearest point M of the offset well to point P of the reference well.

Method 2. Insert the coordinates of the two neighboring calculated points or measured points of each section of the offset well sequentially into Eq. 7.6. If $f(L_{i-1}) \times f(L_i) \leq 0$, then point M is located in the interval $[L_{i-1}, L_i]$. Then, use interpolation to calculate the exact location of point M in the section $[L_{i-1}, L_i]$.

$$f(L) = a(N - N_P) + b(E - E_P) + c(H - H_P) \tag{7.6}$$

Method 3. For every section of the offset well the two neighboring calculated points or measured points are interpolated to find the minimum distance to point P of the reference well. Comparing all the sections, the smallest value of the nearest distance is selected.

The first method is commonly used, as it satisfies the engineering requirements, and is theoretically rigorous enough, as shown in Figure 7.2. There are five calculated or measured points A, B, C, D, and E of the offset well that are used to study the relationship with point P of the reference well. It is not difficult to find out that even though point C is the nearest offset point from the reference point P, the real nearest point M in the offset well from point P is located in the section AB.

The second method is an improved version of the first method that takes into account all the coordinate values of the offset well to determine the sec-

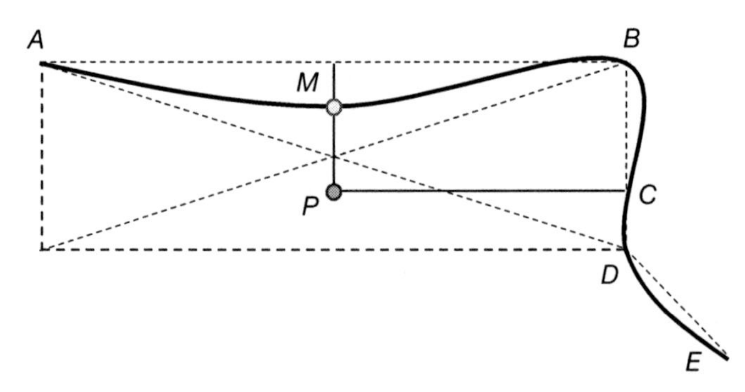

Figure 7.2 Relationship between the nearest offset point and the shortest distance (a special case).

tion containing point M. This method compares the well path trajectory change of the reference well; and, therefore, has a very high reliability, and also the computation time is shorter than the first method.

The third method is, theoretically, the most rigorous and computationally intensive; however, with today's computers, such computations can be done efficiently.

Obviously, as point P is moved along the reference well path, a series of nearest points M in the offset well can be obtained. By calculating the distance between each point P and the corresponding point M, the shortest-distance scanning map can be drawn to display clearly the variation of the shortest distance and its trend as a function of the distance along the well path.

The traditional shortest-distance scanning map is similar to the normal-plane scanning map with the scanning radius as the shortest distance. The scanning angle is the angle between the horizontal projection of the scanning radius and the north direction (for convenience, it will be called the horizontal scanning angle). No doubt this method has some uncertainty, for as shown in Figure 7.1, for a given point P, the position of point M cannot be determined uniquely by the shortest-distance scanning map. A vertical plane is drawn passing through the line $\overline{PM}$ so that its cross-section with the sphere is a circle. The vertical line through point P divides the circle into two half-circles. On one semi-circle, the azimuth of any point connected with point P is equal to the horizontal scanning angle; and on the other semi-circle, the azimuth of any point connected with point P has a 180° difference from the horizontal scanning angle. Because the distance between any point on the circle and point P is equal to the shortest distance ρ, this type of shortest-distance scanning map cannot describe the relationship of relative positions of point M and point P.

It is known that to determine the relative position of two points in space, three parameters are required. The third parameter that is used, in addition to the shortest distance and the horizontal scanning angle, is the horizontal inclination, which is the angle between the scanning radius and the horizontal plane. So, the shortest distance in space between the offset well and the reference well in the spherical coordinate system can be expressed with these three parameters. The horizontal scanning angle θ_h, and the horizontal inclination β_h, are given by[1,4]:

$$\tan\theta_h = \frac{E_M - E_P}{N_M - N_P} \tag{7.7}$$

$$\tan\beta_b = \frac{H_M - H_P}{\sqrt{(N_M - N_P)^2 + (H_M - H_P)^2}} \tag{7.8}$$

Since in the spherical coordinate system, the shortest-distance scanning map is a 3D curve, for the convenience of engineering applications, it is decomposed into two plane curves in two polar coordinate systems—the curve of the shortest distance versus the horizontal scanning angle, and the curve of the shortest distance versus the horizontal inclination, as shown in Figure 7.3.

There is another way to express the shortest-distance scanning map[1,5] using the normal plane to the reference well trajectory through point P as follows. The normal scanning angle is defined as the angle between the projection of the scanning radius onto the normal plane and the direction of the reference well path at point P. The normal inclination is defined as the angle between the scanning radius and the normal plane. Thus, the expressions for the normal scanning angle θ_n and the normal inclination β_n are:

$$\tan\theta_n = \frac{y_M}{x_M} \tag{7.9}$$

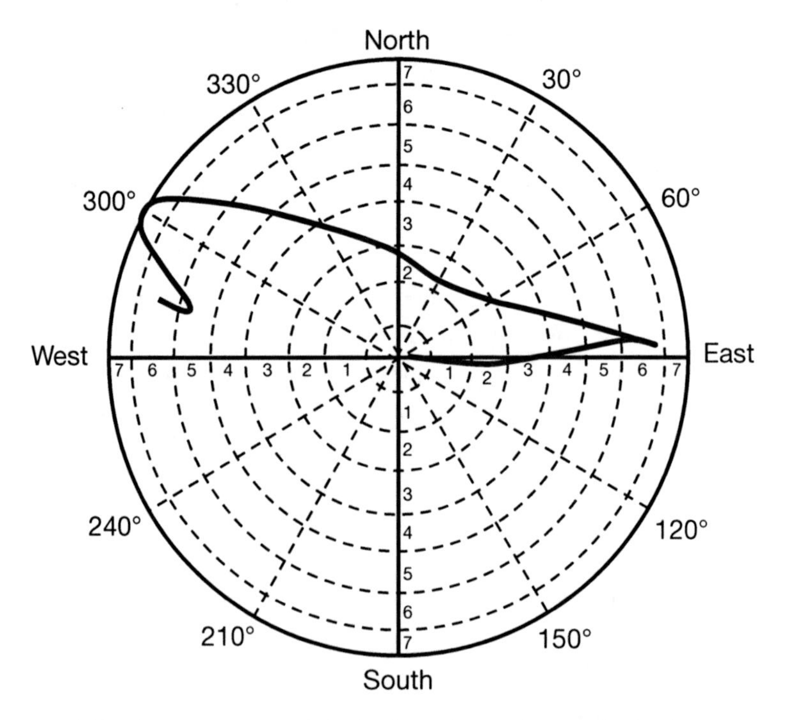

Figure 7.3 Shortest distance and the horizontal scanning angle chart.

$$\sin\beta_n = \frac{z_M}{\rho} \tag{7.10}$$

where

$$\begin{Bmatrix} x_M \\ y_M \\ z_M \end{Bmatrix} = \begin{bmatrix} \cos\alpha_P\cos\phi_P & \cos\alpha_P\sin\phi_P & -\sin\alpha_P \\ -\sin\phi_P & \cos\phi_P & 0 \\ \sin\alpha_P\cos\phi_P & \sin\alpha_P\sin\phi_P & \cos\alpha_P \end{bmatrix} \begin{Bmatrix} N_M - N_P \\ E_M - E_P \\ H_M - H_P \end{Bmatrix}$$

Clearly, in these two situations, the reference planes are different and have a clear, relative translation. The ranges of the parameters are as follows: the scanning angle range is [0, 360°), and the inclination range is (–90°, 90°], so the curve of the shortest distance versus the horizontal inclination, or the normal inclination, ranges only over a semi-circle. The maps are shown in Figures 7.3 and 7.4.

In summary, by using the concepts of the scanning angle and the inclination angle, the shortest-distance scanning map in the spherical coordinates system can be decomposed into two plane curves in two polar coordinate systems, which enhances the scientific reliability, integrity, and availability of the visual map. Even though the visual map has inherent advantages, the table of calculated data should still be prepared and used to obtain numerical values of the parameters quickly and accurately. The use of the calculated data table is also required as the actual distances to the nearby wells and the trends of its change with depth are critically important.

The accuracy of the calculations for the shortest distance between two wells depends on the step size. If the step size is too large, important intermediate

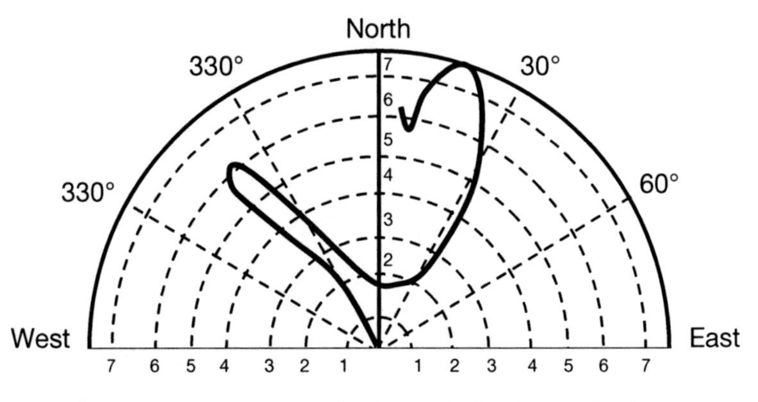

Figure 7.4 Shortest distance and the horizontal inclination angle chart.

data may get omitted, which may even lead to wrong results. Therefore, more calculation points must be used in the intervals of the "dangerous sections," where the chances of collision are higher.

Collision Avoidance[5]

Rule-Based Collision Avoidance

Most oil companies have adopted directional drilling collision avoidance rules based on stringent controls to prevent surface collisions and the consequent human and environmental damage. Deep intersections have several characteristics, which require a separate treatment as compared to shallow intersections. Separation factor is the key parameter used for analyzing situations for rule-based collision avoidance.

- Minimum separation refers to the minimum allowed distance between the reference well and the offset well. In other words, it is the minimum distance of approach to the offset well by the reference well. During planning, a conservative (major-risk rule) value of the allowed distance of approach by a reference well to an offset well is used. If this value is not economically feasible, then a special dispensation is applied (minor-risk rule) to the offset well, which enables reduction of the minimum separation distance, making the drilling feasible.
- Separation factor (SF) collision avoidance (ratio based) uses the ratio of the separation distance *S* divided by the minimum separation to facilitate the comparison. The different separation factors for a planned well (reference) against an existing well (offset) should be correlated against different levels of tolerance. Common levels of tolerance are given in Table 7.1.

Table 7.1 Collision Tolerance

SF	Collision Tolerance	Action
<1	Should not be drilled	Must be sidetracked at shallow depth
<1.25	Action should be taken	Nearby wells should be shut-in
<1.5	Can be tolerated for trajectories and not for plan	Dispension may be allowed
>1.5	Allowed to drill	Closely monitored

Separation Factor Collision Avoidance

There are several methods used for surface collision avoidance, and the most common method of determining a minimum separation is to add the radii of the projections of the uncertainty ellipsoids of the offset well and the reference well together. The minimum separation between two wells, as shown in Figure 7.5, is given by:

$$\text{Minimum separation} = e_r + e_o \tag{7.11}$$

$$\text{Separation factor } SF = \frac{S}{(e_r + e_o)} \tag{7.12}$$

where

S = Separation distance between well centers

e_r = Error given by the radius of the uncertainty ellipsoid of the reference well, as projected in the direction of closest approach

e_o = Error given by the radius of the uncertainty ellipsoid of the offset well as projected in the direction of closest approach

The radii of the uncertainty ellipses are taken at a specified confidence (sigma) level. Two sigma are most commonly used, and are roughly equivalent to 95% confidence in one dimension. Depending on the situation, higher levels of confidence can be used.

Other modifications to the separation factor anti-collision method are:

- Using the major axis of the ellipsoid as the ellipse dimension.
- Addition or subtraction of casings diameters.
- Combining the covariance matrices of the ellipsoids to produce a single relative error ellipsoid.

Ellipsoid Major Axis

In this method, the largest dimension of the ellipsoid is taken as the ellipse size to be used in estimating the separation factor, as shown in Figure 7.6. In shallow intersections, the errors may be nearly symmetrical, so the difference

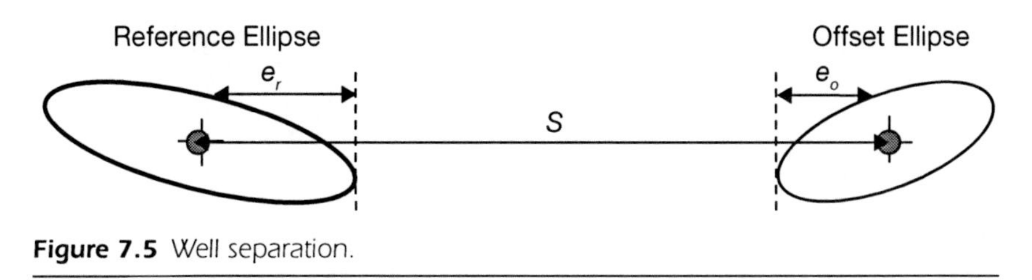

Figure 7.5 Well separation.

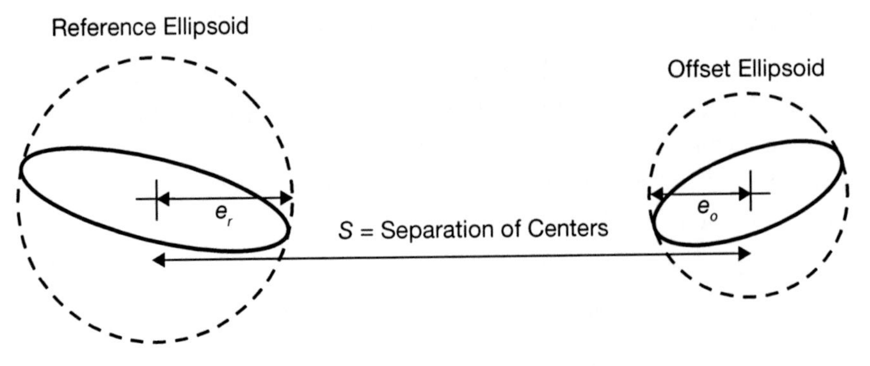

Figure 7.6 Well separation using largest ellipsoid dimension.

is negligible. In deep intersections, the errors are dominated by the azimuth errors, resulting in overly conservative estimates.

Casing Diameters

In this method of estimation, the casing radius is taken into consideration for calculating the minimum separation distance (Figure 7.7). The separation factor used in this method is modified as follows:

$$SF = \frac{S}{\left[e_r + e_o + \left(\frac{d_r + d_o}{2}\right)\right]} \tag{7.13}$$

where

d_r = Hole diameter of the reference wellbore at the depth of interest
d_o = Casing diameter of the offset wellbore

Alternatively, the casing radius of the offset well and the hole diameter of the reference wellbore could be treated as physical distances and subtracted

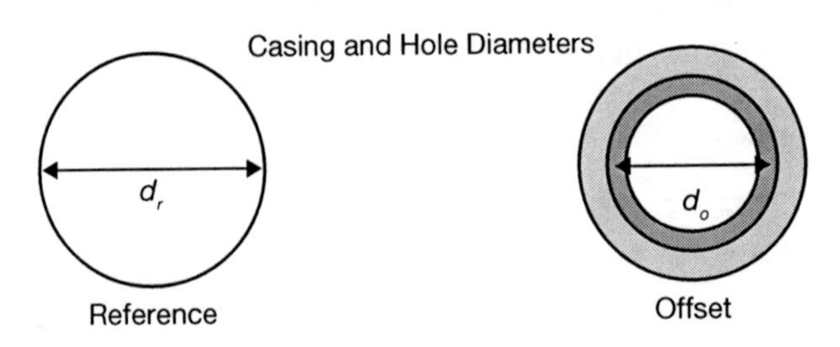

Figure 7.7 Well separation using casing/open-hole diameters.

from the separation distance (S). Using this option, the separation factor is given by:

$$SF = \frac{S - \left(\frac{d_r + d_o}{2}\right)}{\left[e_r + e_o\right]} \tag{7.14}$$

Combined Ellipsoids

In order to obtain an exact assessment of the position error of the reference well relative to the offset well, the uncertainty ellipsoids are combined (Figure 7.8) by a root sum square summation. This is done by adding the elements of the two covariance matrices together. The significant difference between this method and the earlier methods is the statistical combination of errors before the calculation.

Using this method, the separation factor is given by:

$$SF = \frac{S - \left(\frac{d_r + d_o}{2}\right)}{f \times e_t} \tag{7.15}$$

where

e_t = Combined error ellipsoid

f = Scaling factor to increase the confidence level to an acceptable level

A common value would be 2.8, which would scale the one sigma ellipse size to the equivalent of 2 × 2 sigma ellipses used in more traditional methods.

Separation Factor and Probability

The estimated separation factor can be converted into a probability by integrating the tail of the normal probability distribution, as shown in Figure 7.9.

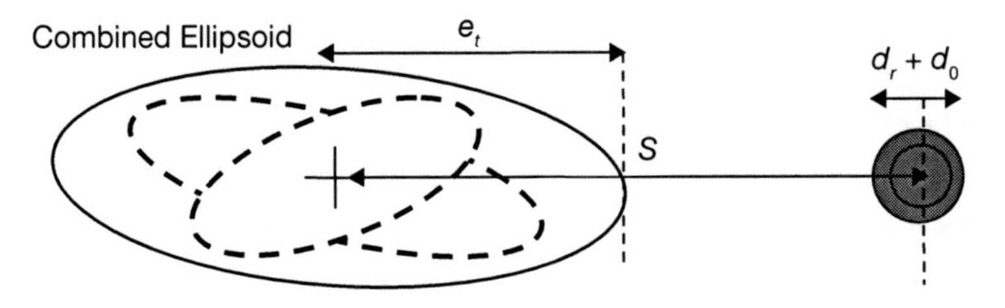

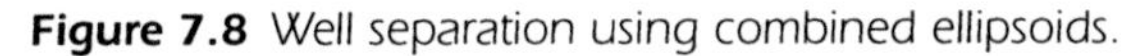

Figure 7.8 Well separation using combined ellipsoids.

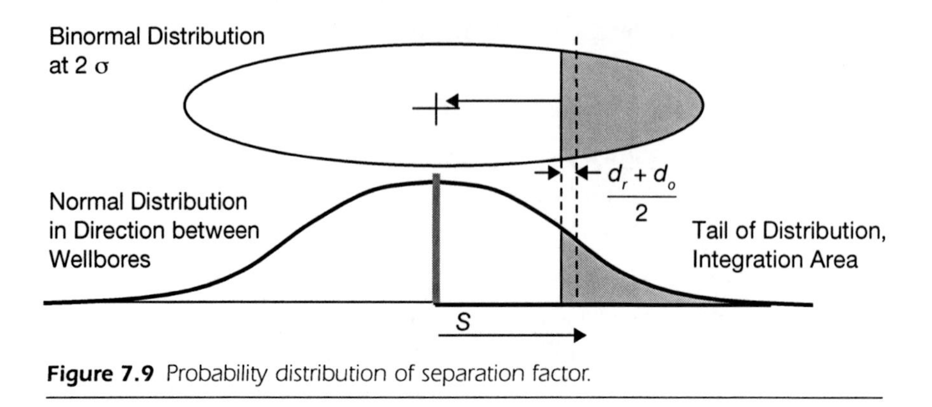

Figure 7.9 Probability distribution of separation factor.

Risk formulas use a function, which is an integration of the normal probability distribution function, and is given by:

$$\Phi(x) = \frac{1}{\sqrt{2\pi}} \int_{-\infty}^{x} e^{-\frac{x^2}{2}} dx \tag{7.16}$$

where

x = SF (separation factor)

In the case of two ellipsoids of equal size with two sigma dimensions, an SF of 1 will yield a collision probability of approximately 1 in 400. An SF of 1.5 will yield a collision probability of approximately 1 in 80,000.

Risk-Based Collision Avoidance

Evaluation of the minimum separation based on risk is an important anti-collision technique used while constructing a well. The following steps are used to get a minimum separation distance based on risk, which in turn gives the separation factor.

1. Decide on a transition depth where risk-based techniques can be used.
2. Decide a level of acceptable risk of collision that can be tolerated in drilling below that depth.
3. Determine the closest approach (CA) vector between the two wellbore centers.
4. Determine the errors at the depth of interest of the offset well and reference well, and combine them statistically.

5. Determine the single error value that can be sectioned from the ellipsoid at the depth of interest, in the vector direction of the minimum distance between the two wells.
6. Determine the effective casing plus hole diameters. Increase these diameters if the angle of incidence of the two wells is very small to account for "multiple hits."
7. Determine the minimum separation distance between the two wells. This is the minimum allowable distance that a reference well can approach an offset well.
8. The wells should not be allowed to approach closer than an SF of 1.0. There may be different safe drilling rules applied to various levels of SF greater than 1.0. For instance, offset wells should be shut down if they are closer than an SF of 1.5.

The stages in this process are discussed further below.

Transition Depth

There is an acceptable vertical depth in the zone of drilling, where a collision below that depth can be assessed not to cause surface damage or escape of hydrocarbons. This vertical depth may be at an intermediate casing point, or a point below which a subsurface safety valve (SSSV) is set for existing production and injection wells. For the reference well, it will be a depth where a reservoir (or offset well) pressure kick can be safely detected and controlled. Above that depth, the traditional ratio or no-go-type rules will apply, as collisions above this depth could potentially result in well control problems.

Level of Acceptable Risk

The economic risk of a collision at a deep position comes from the loss of the offset well plus the cost to control and sidetrack the current well and the potential loss of production. It is necessary to follow a standard shut-in/bleed-off procedure for a production or injection well. For rapidly drilled wells and for ones that have a short lifespan, the level of economic risk may be low. At this level of risk and dealing with the position uncertainty below the transition depth it is possible to drill through an offset wellbore.

Closest Approach Scan Methods

The commonly used scanning methods to determine the closest point on an offset wellbore for a range of depths on the reference wellbore are:

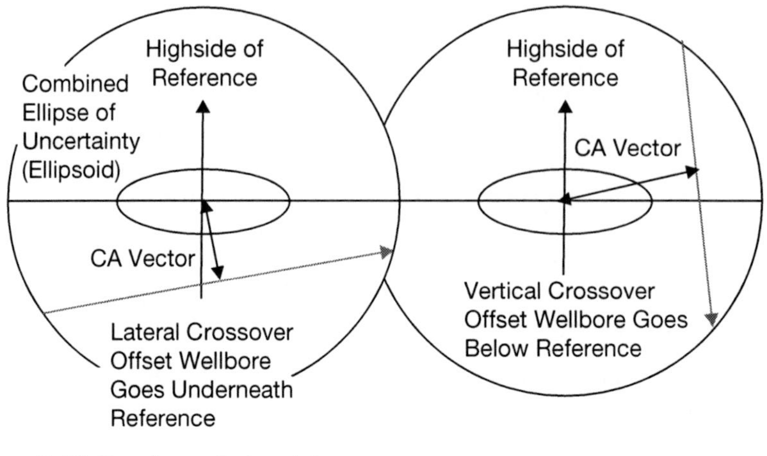

Figure 7.10 Traveling cylinder plot.

- Scan from regular depths on the reference well to find the closest point on the offset well. This produces a vector between the wells, which is perpendicular to the offset well.
- Scan from regular depths on the offset well back to the closest point on the reference well. This produces a closest-distance approach vector for the wells that are perpendicular to the reference wellbore and are designed to represent offset wellbores correctly in a traveling cylinder diagram.

Figure 7.10 shows the plot of the two traveling cylinders with the two types of approach vectors, one perpendicular to the offset wellbore and the second vector being perpendicular to the reference wellbore.

Survey Errors

The Industry Steering Committee on Wellbore Surveying Accuracy (ISCWSA) provides a model that provides comprehensive and standard directional survey errors for both magnetic and gyroscopic instruments, and a method for accumulating and combining the errors to determine the position uncertainty. The position uncertainty of a point on a wellbore is described by an ellipsoid of uncertainty. The uncertainty ellipsoids of the reference and offset wellbores are used for the determination of the overall error at the well depth of the closest approach vector. A covariance matrix provides the mathematical description of the uncertainty ellipsoid as a function of the depth.

Combined Errors of Reference and Offset Wells

To get a correct assessment of the survey error of the reference well relative to the offset well, the uncertainty ellipsoids must be combined by the root sum square summation method. This is done by adding the components of the two covariance matrices together. This assumes that the survey errors of the reference well are independent of the survey errors of the offset well. This may not necessarily be the case, and the ISCWSA method does provide for the management of global-type errors. An example would be when the same magnetic declination correction is applied to the two surveys of the converging wellbores. The crustal component of the declination error would cause both wells to be deviated. If they converged from opposite directions, the error will be compounded. The combination of the covariance matrices is given by:

$$C = C_r + C_o \tag{7.17}$$

where

C_r, C_o = Covariance matrices
C = Matrix formed by addition of the elements of the covariance matrices

The covariance matrix C is a 3 × 3 symmetric matrix representing the combined 3D uncertainty ellipsoid of the relative position of the reference and offset wells, and is given by:

$$C = \begin{bmatrix} xx & xy & xz \\ yx & yy & yz \\ zx & zy & zz \end{bmatrix} \tag{7.18}$$

where xx is the square of the error, or variance, in the x direction, and xy would be a covariance, which gives the correlation between the cross-axis error values, and which represents orientation information.

Relative well errors are handled in the same way, because the measurement of their surface location is independent to downhole instrument measurements.

Sectioning the Uncertainty Ellipsoid

To obtain the error of the position of a point of a well in a particular direction, it is necessary to extract a one-dimensional (1D) error value from the 3D surface of the uncertainty ellipsoid along the vector in the direction of interest. This procedure is referred to as "sectioning the uncertainty ellipsoid." The combined error of the position of the reference well relative to the offset well is important for collision avoidance, as it enables calculation

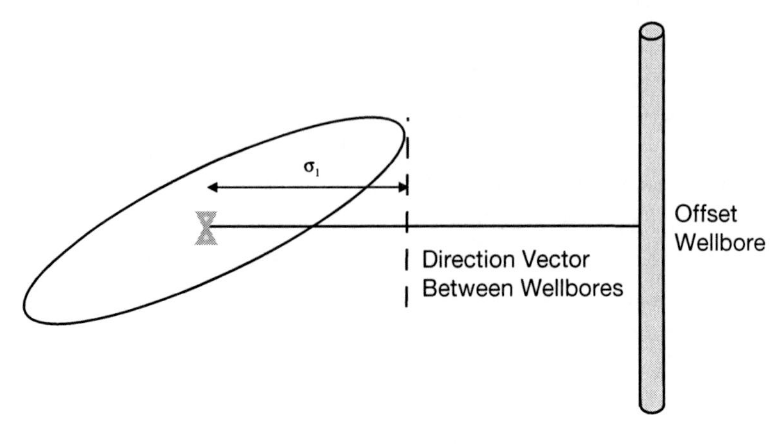

Figure 7.11 Combined ellipsoid of both wellbores at one sigma.

of the error in the direction of closest approach between the reference and offset wells. It would be reasonable to use the direction of the closest approach vector between the two wellbores to section the ellipsoid. There could be two types of CA vectors depending on the scan method.

- For CA scans from the reference well, it is a vector from the reference well to the offset well that is always perpendicular to the direction of the offset well.
- For CA scans from the offset well, it is a vector from the offset well to the reference well that is always perpendicular to the direction of the reference well.

Unfortunately, both of these vectors can cause highly misleading results for scan depths that are close to, but not quite, at the depth of the closest approach, as these methods do not accurately give the direction of the wellbores relative to each other.

The closest approach vector that is perpendicular to both the reference and offset well directions is used for the risk-based analysis. It is equally applicable to both types of scanning methods because the direction of the approach vector will be perpendicular to both the wellbores at the depth of the minimum closest approach.

The cross product of the reference and offset wellbore directions is used to determine the vector to section the ellipsoid, and is given by:

$$\boldsymbol{u} = \boldsymbol{R} \times \boldsymbol{O} \tag{7.19}$$

where

$\boldsymbol{R}$ = Vector direction of the reference wellbore
$\boldsymbol{O}$ = Vector direction of the offset wellbore

This method assumes that the directions of the wellbores relative to each other are accurately known.

The error in the direction between the two wells is expressed as:

$$\sigma_1 = \sqrt{\boldsymbol{u}^T \boldsymbol{C}_1 \boldsymbol{R}} \tag{7.20}$$

where

$\boldsymbol{u}$ = Vector between the wells
$\boldsymbol{C}_1$ = Covariance matrix

Effective Casing and Hole Diameters

To derive the collision risk at a depth, the combined diameters of the reference and offset wellbores must be properly estimated. The hole diameter is used for the reference well and the casing size is used for the offset wellbore, as discussed earlier. The size is extracted from the casing design details for the depths of interest. The formula for minimum separation gives the sum of the reference hole diameter and offset casing diameter as $d_r + d_o$.

Minimum Separation Distance

The minimum separation distance along the direction vector between the two wells is the minimum allowable distance that a reference well can approach an offset well. Several analyses have been attempted to use more refined methods for expressing the probability of collision, but when the results were compared to the minimum separation distance method, the differences are found to be negligible. The minimum separation distance method, has two advantages:

- It can be used to determine minimum separation distance.
- It is conservative for the case when the casing diameters are large and the combined survey errors leave little margin. This method is also applicable for surface collision analysis.

Figure 7.12 shows the derivation of the minimum separation distance.

The shaded rectangle area can be approximated as the probability:

$$P \approx (d_r + d_o) f\left(\frac{d_r + d_o}{2}\right) = \frac{d_r + d_o}{\sigma\sqrt{2\pi}} \exp\left\{-\frac{\left[S - \left(\frac{d_r + d_o}{2}\right)\right]^2}{2\sigma^2}\right\} \tag{7.21}$$

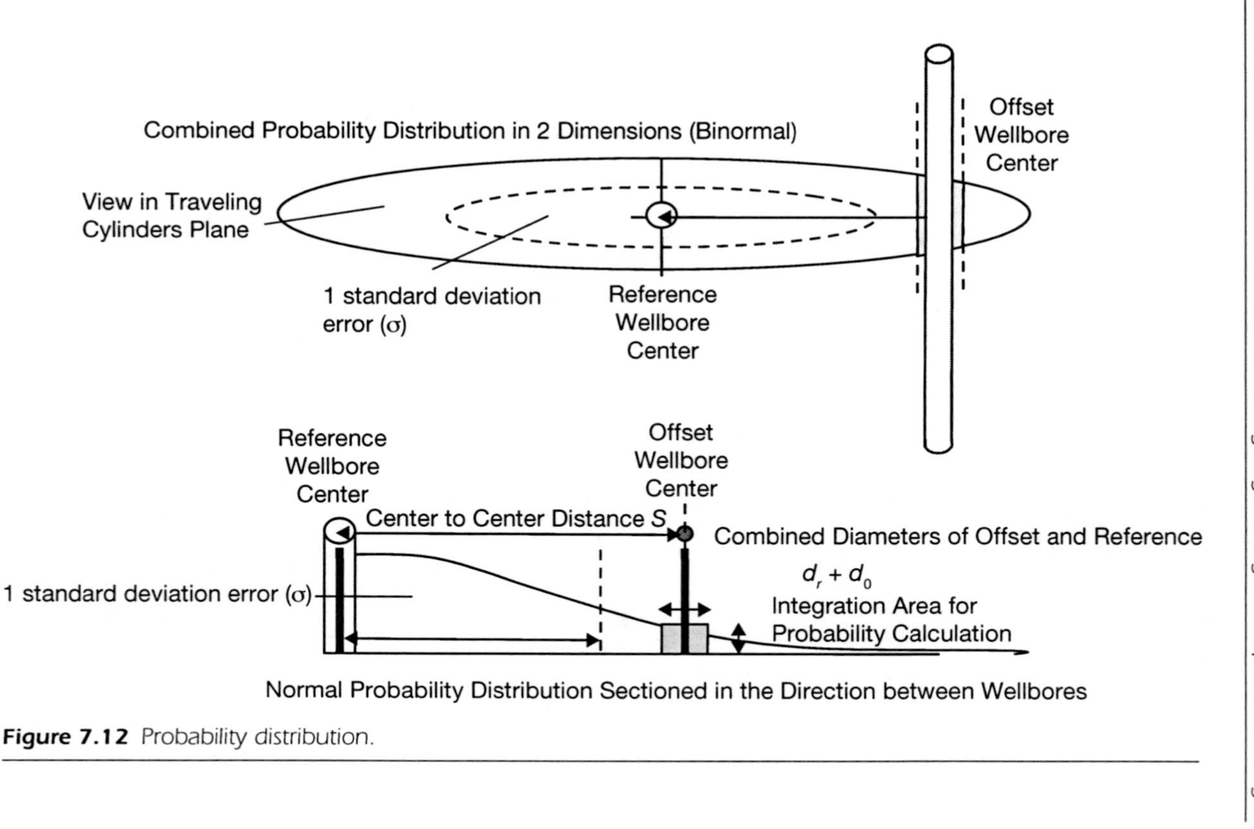

Figure 7.12 Probability distribution.

Mathematical manipulation of Eq. 7.21 yields the separation S as:

$$S = \sigma\sqrt{2\ln\left(\frac{d_r + d_o}{P\sigma\sqrt{2\pi}}\right)} + \frac{d_r + d_o}{2} \tag{7.22}$$

Vectors for Sectioning the Ellipsoid

All the collision avoidance methods require a 1D error value to be extracted from a 3D error surface (covariance matrix) along a vector in the direction of interest. Obtaining this vector is a bit more complex than it seems, because an anti-collision scan goes through a range of depths on the reference well, while scanning for the closest point on the offset well. The real closest approach vector is always perpendicular to both the wells, when the wellbores are continuous. Other points are not at right angles to both the wells. These odd angles on the offset well can lead to unreliable results when analyzing the errors, which further leads to unrealistic risk factors. There are three choices for this vector:

1. A vector directly from the point on the reference well to the point on the offset well. This is the preferred option for a 3D closest approach scan and the vector is given by:

   $$\boldsymbol{u} = \boldsymbol{R} - \boldsymbol{O} \tag{7.23}$$

   where

   $\boldsymbol{R}$ = Position on the reference wellbore

   $\boldsymbol{O}$ = Position on the offset wellbore

2. A vector in a plane perpendicular to the reference well and in the toolface direction of the offset well. This is the preferred option for a traveling cylinders scan and the vector is given by:

   $$\boldsymbol{u} = \begin{bmatrix} \cos\alpha\cos\phi\cos\omega - \sin\phi\sin\omega \\ \cos\alpha\sin\phi\cos\omega + \cos\phi\sin\omega \\ -\sin\alpha\cos\omega \end{bmatrix} \tag{7.24}$$

   where

   α = Inclination of the reference wellbore

   ϕ = Azimuth

   ω = Toolface angle from the reference to the offset wellbore

3. The cross product, called the projected vector, is used to determine the vector to section the ellipsoid, and is given by:

   $$\boldsymbol{u} = \boldsymbol{R} \times \boldsymbol{O} \tag{7.25}$$

At the closest approach, all three methods yield the same vector. The third method predicts the direction of the vector at the closest approach, where the two wells become parallel from their actual directions at the depth of interest. There are problems in resolving the cross product where the wellbores are nearly parallel; and, in such cases, the perpendicular vector method may be used instead.

The following are two examples clarifying the above considerations:

1. A high-angle reference well that has significant azimuth turn crosses a low-angle well. Due to the latency of the azimuth, the error aligns along the depth axis of the well, which has no significance for collision avoidance. But the direct vector method 1 does include it for depths before and after the closest point, so the irregular errors are propagated.

2. Two high-angle wells cross over each other. Actually, both methods 1 and 2 vectors lead to unrealistically high errors where they should not. The nonclosest approach vectors lead to the inclusion of the large azimuth error; whereas, if one were to deliberately drill on the high side or the low side of another well, then the inclination-type error should only be considered. This is known as the "pedal curve" problem.

Pedal Curve

The pedal curve problem described in the previous example is due to the major axes of the ellipses crossing over (Figure 7.13). It gives a false sense of security before and after the closest point of approach. The projected vector will provide a more constant evaluation of the errors.

The pedal curve method seems to assume that there is no knowledge of the direction of the reference well relative to the offset well (it can approach from any direction). Yet the position error has been quantified to the confidence level as given by the uncertainty ellipse shape. In reality, there is a great deal more confidence in the direction of the wells for a deep intersection. The errors in inclination and azimuth measurements are generally not cumulative with depth, unlike error in position (with the exception of continuous gyros or inertial instruments). See Figure 7.14.

Normal Probability Distribution for Risk Level

Calculation of risk level is important and can be directly calculated with the right error value (σ), if the probability distribution in two dimensions can be integrated. This is a slightly "more correct" method to get the risk, than the minimum separation method (Figure 7.15).

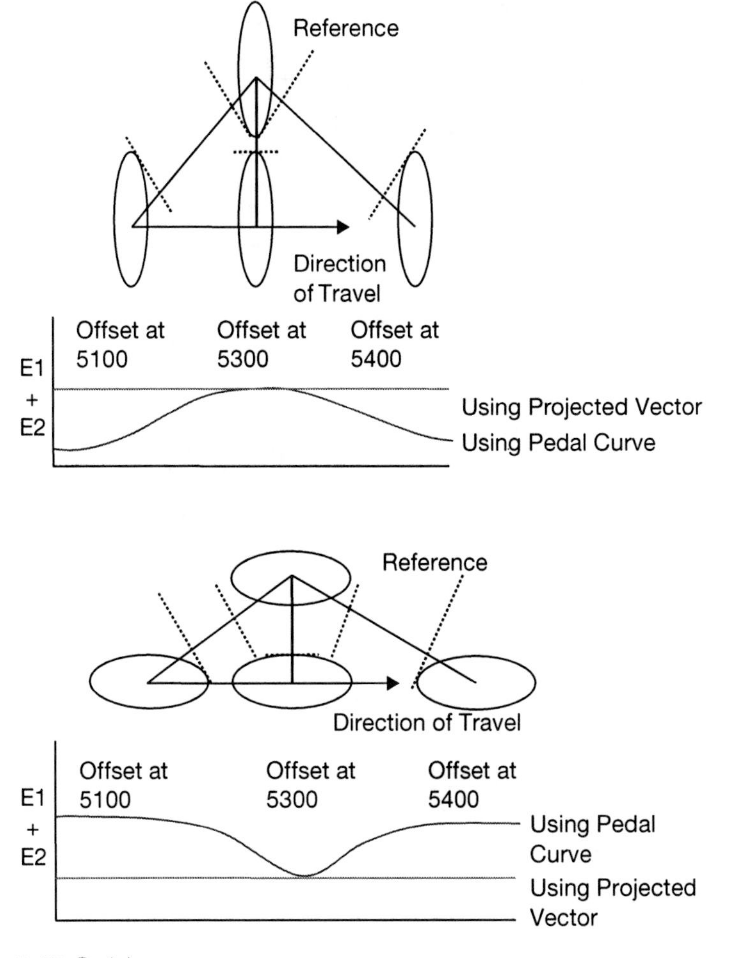

Figure 7.13 Pedal curve.

Integration formulas use a function, which is an integration of the normal probability distribution and is given by:

$$\Phi(x) = \frac{1}{\sqrt{2\pi}} \int_{-\infty}^{x} e^{-\frac{x^2}{2}} dx \tag{7.26}$$

where

$$x = \frac{S}{\sigma}$$

σ = One standard deviation error in the direction between the two wells

S = Separation between the two wells

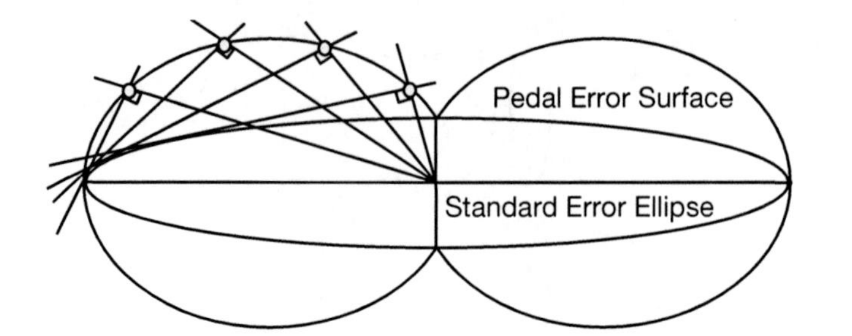

Figure 7.14 Pedal curve error surface.

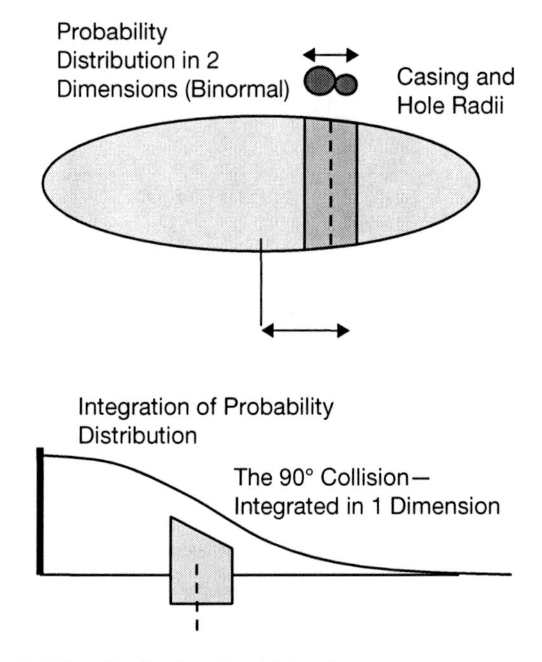

Figure 7.15 Probability distribution for risk level.

Using this function, the result of the following integration for a perpendicular intersection can be obtained:

$$P = \Phi\left(\frac{S-C}{\sigma}\right) - \Phi\left(\frac{S+C}{\sigma}\right) \tag{7.27}$$

where

$$C = \frac{d_r + d_o}{2}$$

is the sum of the radii of the reference wellbore hole size and the offset wellbore casing size at the depth of interest.

P, the risk number, gives the number of reference wells that can be drilled under similar conditions before the probability of one well colliding with the other well reaches unity. If there are two similar closest points to an offset well from a reference well, then effectively the overall chances of hitting the offset well increases. Risk method can also be used for drilling a relief well for blowout control.

Anti-Collision Plots

Some of the anti-collision plots that are commonly used are:

- Projection view plots
- Ladder view plots
- Traveling cylinder plots
- Spider view plots
- Separation factor view plots

Projection plots are extensively described in Chapter 3, and consist of vertical section and horizontal projection plots. A ladder view anti-collision plot gives the center-to-center separation between the reference well to one or more offset wells along the wellbore depth, as shown in Figure 7.16. It can be used to assess the risk of collision of the drilled well with the offset wells. The reference well is considered as the x-axis and, wherever the center-to-center separation from an offset well becomes zero, it indicates that the particular offset well at the corresponding depth is at risk of colliding with the reference well. It allows one to compare the collision risk of multiple wells as shown in Figure 7.17.

A traveling cylinder anti-collision plot is commonly used to provide the polar position of the offset wells, with respect to the reference well trajectory center. As shown in Figure 7.17, it gives the distance to the offset well usually measured at an angle from the high-side of the reference well path or the north direction and the largest radius provides the scan limit. A spider view anti-collision plot provides the plan view of the wells to be considered with north(+)/south(–) on the y-axis and east(+)/west(–) on the x-axis as shown in Figure 7.18. The coordinates can be either in local or grid coordinates, depending on the necessity. Separation factor anti-collision plots are also useful for collision risk monitoring. They plot the separation factor between the reference well and one or more offset wells.

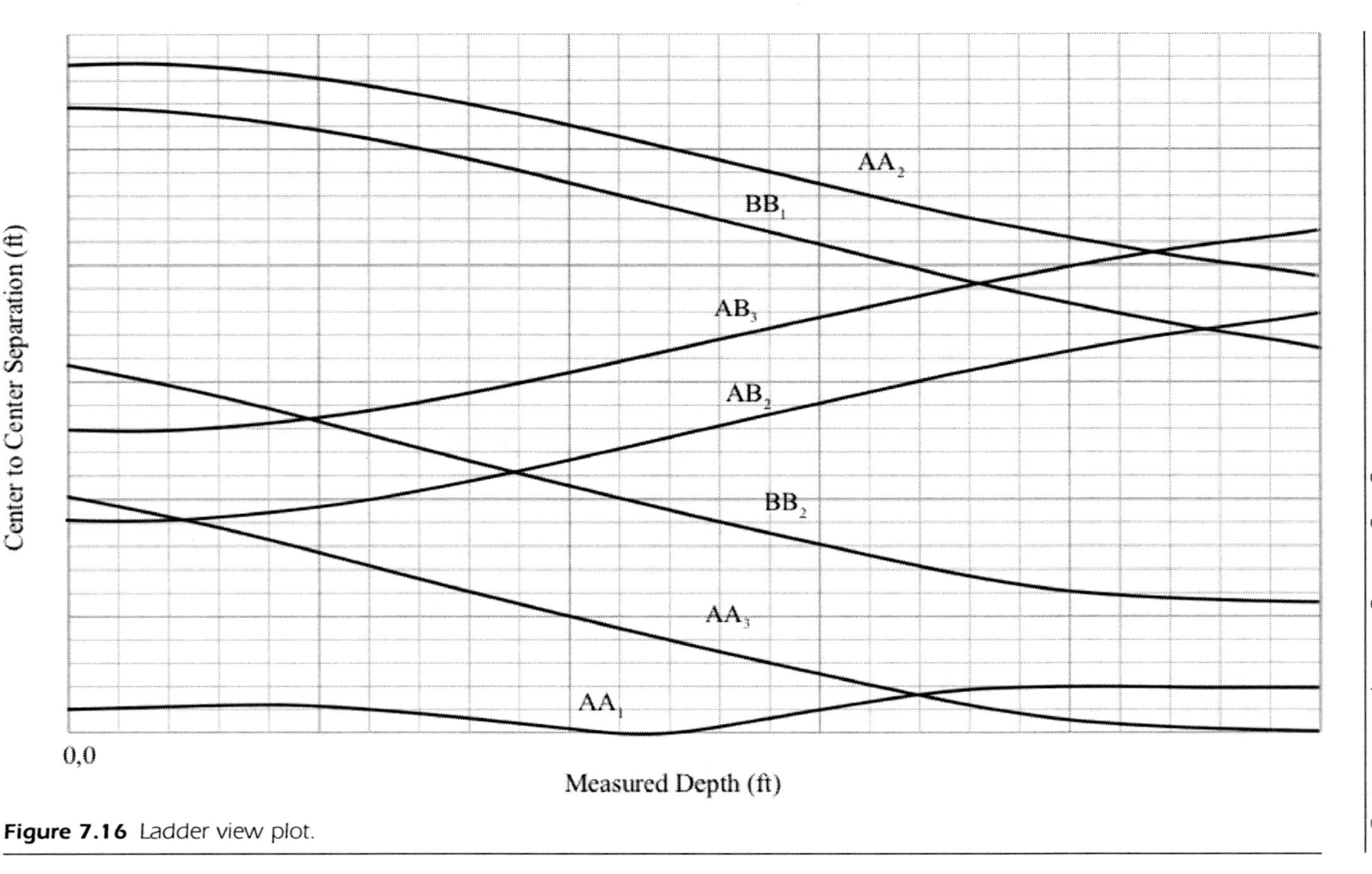

Figure 7.16 Ladder view plot.

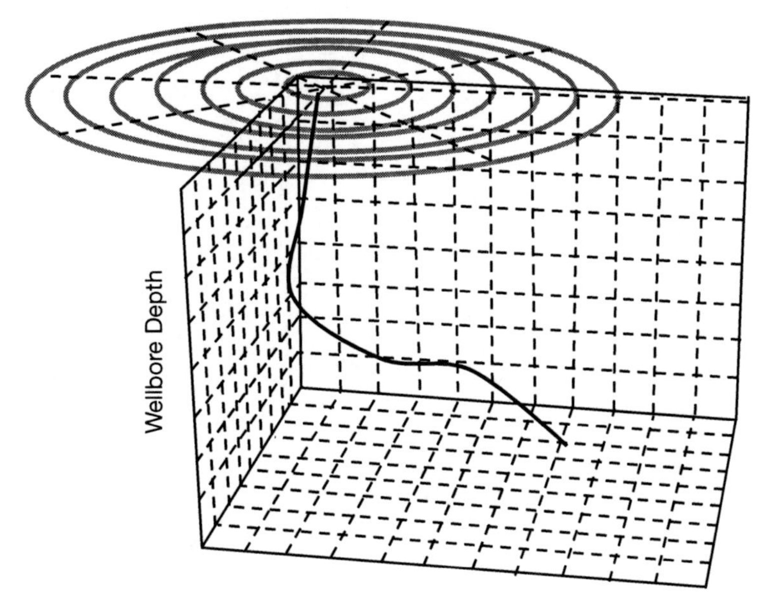

Figure 7.17 Traveling cylinder plot.

Tortuosity

Tortuosity is one of the critical factors to be considered for designing complex directional well trajectories, for having accurate build rates, for precise steering in thin reservoirs, and for planning extended-reach wells.

Tortuosity plays an important role in anti-collision design, as well as collision risk analysis of the well placement. Increased evidence of hole spiraling resulting in tortuosity has been reported.[6-8] The hole spiraling, or threading, has a serious impact on the well construction and completion process.

Tortuosity is usually expressed in °/100 ft, similar to the units used for dogleg severity. The calculation of "running tortuosity" is the station-to-station summation of the total curvature normalized to a standard wellbore course length between survey stations.

Methods for Applying Continuous Tortuosity to a Well Path

Tortuosity is expressed as the difference between the actual and planned curvatures, divided by the respective distance between the survey stations, and can be written as:

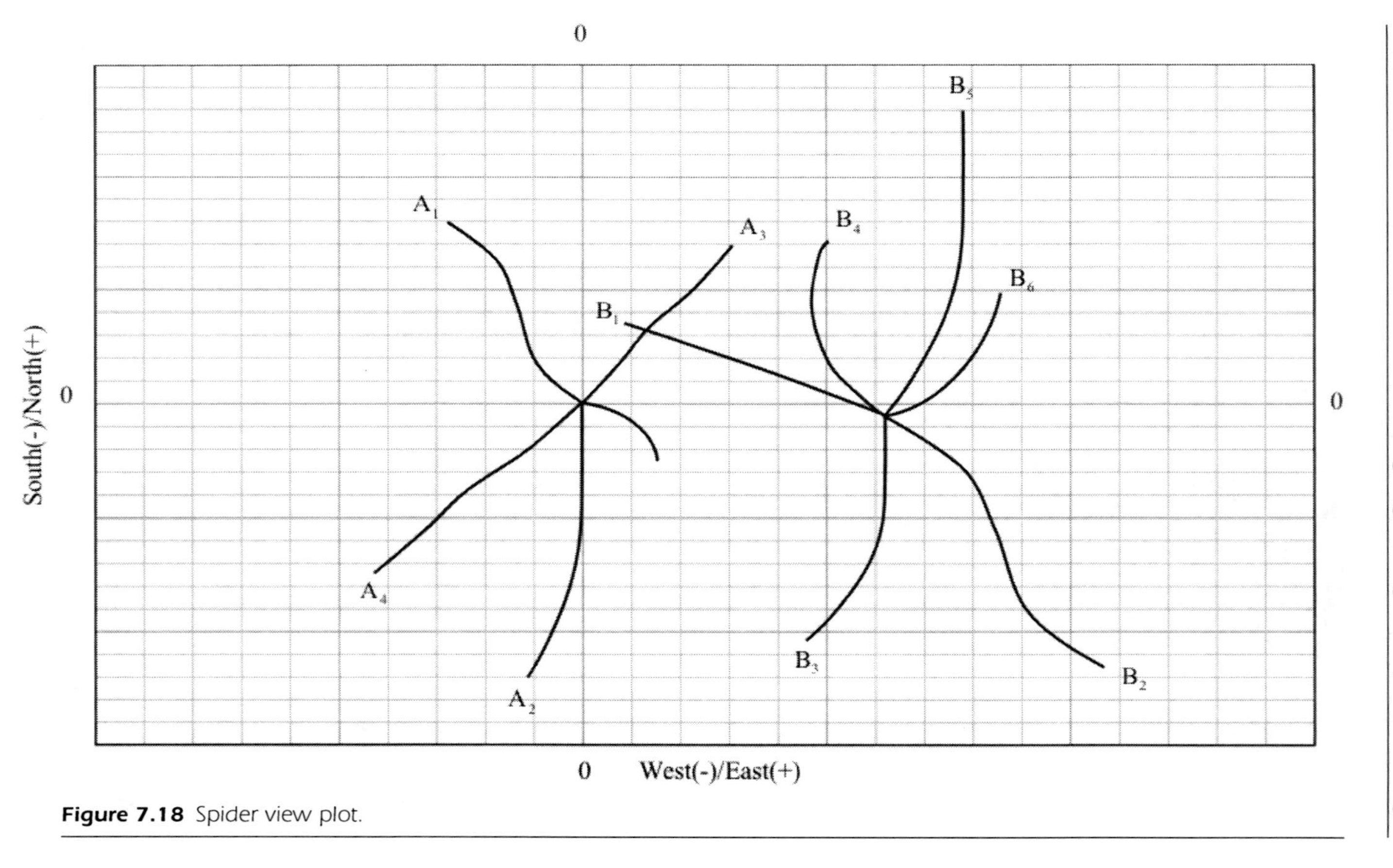

Figure 7.18 Spider view plot.

$$\Gamma = \frac{\sum_{j=1}^{n} DLS_r - \sum_{i=1}^{m} DLS_o}{MD_j - MD_i} \text{ (°)/100 ft} \tag{7.28}$$

where

DLS = Dogleg severity, (°)/100 ft
r = Reference well
o = Offset well
MD = Measured depth

As there are only discrete points of survey measurements along a well path, it is necessary to artificially apply a rippling, or undulation on the well path, to obtain a continuous tortuosity. This is done by one of the following four methods:

- Sine wave method
- Helical method
- Random inclination–dependent azimuth method
- Random inclination and azimuth method

Sine Wave Method

This method modifies the inclination and azimuth of the survey point based on the concept of a sine wave–shaped ripple running along the wellbore using a specified magnitude (amplitude) and period (wave length). The change of the inclination angle is modified using the following relationship:

$$\Delta\alpha = \sin\left(\frac{MD}{P} \times 2\pi\right) \times M \tag{7.29}$$

where

MD = Measured depth
P = Period
M = Magnitude

The magnitude M is the maximum variation that will be applied to the inclination and azimuth of the original well path. Further, the inclination angle is modified so that it does not become less than zero, since negative inclination angles are not allowed. The new inclination angle and azimuth are given as follows:

$$\alpha_i = \alpha_{i-1} + \Delta\alpha$$

$$\phi_i = \phi_{i-1} + \Delta\alpha + \psi_{ac} \tag{7.30}$$

where

α = Inclination angle
ϕ = Azimuth
ψ_{cvc} = Cross-vertical correction

Also, while applying artificial tortuosity, it is to be ensured that the measured depth of the survey point is not an exact integer multiple of the period, as

$$\Delta\alpha = \sin\left(\frac{\Delta MD}{P}2\pi\right) = 0$$

PROBLEM 7.1

Calculate the new angle and azimuth when the sine wave tortuosity method is applied to the following survey data:

Well measured survey depth = 3725 ft
Inclination = 3.25°
Azimuth = 165°

Use pitch = 1000 ft.

Solution

Calculating the change in the inclination angle:

$$\Delta\alpha = \sin\left(\frac{3725}{1000}2\pi\right)\times 1 = -0.99$$

α_i = 3.25 – 0.99 = 2.26°
ϕ_i = 165 – 0.99 + 0 = 164.01
If $\alpha_i < 0$, then the cross-vertical correction = 180°
$\alpha_i = |\alpha_i|$ and if $\alpha_n \geq 0$, then the cross-vertical correction = 0°

Helical Method

This method modifies the inclination and azimuth of the survey points by superimposing a helix along the well path, using a specified magnitude (radius of the cylinder in the parametric equation) and period (pitch). This method uses the circular-helix defined by:

$$f(\alpha) = a\cos(\alpha) + a\sin(\alpha) + b\alpha$$

The generalized parametric set of equations for the helix superimposed on the well path is given by:

$$x(\alpha) = M\cos(\alpha)$$
$$y(\alpha) = M\sin(\alpha)$$

$$z(\alpha) = \frac{P}{2\pi}\alpha \tag{7.31}$$

Random Inclination–Dependent Azimuth Method

This method applies a random variation to the survey inclination and azimuth within the magnitudes specified. The random numbers used may be between –1.0 to +1.0. In this method, the azimuth variation is inversely proportional to the inclination, resulting in a higher inclination with lower azimuth variation, and lower inclination with higher azimuth variation. The change in inclination angle, new inclination angle, and new azimuth are given by:

$$\Delta\alpha = \zeta \times \delta \tag{7.32}$$

where

$$\delta = \frac{\Delta MD}{P} M$$

ζ = Random number

$$\alpha_i = \alpha_{i-1} + \Delta\alpha$$

$$\phi_i = \phi_{i-1} + \frac{\Delta\alpha}{2\sin\alpha_i} + \psi_{\alpha c}$$

PROBLEM 7.2

Calculate the new angle and azimuth when the random inclination–dependent azimuth tortuosity method is applied to the following survey data:

Well measured survey depth = 3725 ft
Inclination = 3.25°
Azimuth = 165°

Well measured survey depth = 3900 ft
Inclination = 5.15°
Azimuth = 166°

Use pitch = 100 ft. Use the random number = 0.375.

Solution

$$\Delta\alpha = \frac{0.375 \times (3900 - 3725)}{100} \times 1 = 0.66$$

$$\alpha_i = 5.15 + 0.66 = 5.81°$$

$$\phi_i = 166 + \frac{0.66}{2\sin 5.81} + 0 = 169.26$$

Random Inclination and Azimuth Method

This method is similar to the random inclination–dependent azimuth method, except that the azimuth variation is taken to be independent of the inclination. The new angle and azimuth are given as follows:

$$\alpha_i = \alpha_{i-1} + \Delta\alpha$$

$$\phi_i = \phi_{i-1} + \Delta\alpha + \psi_{cvc} \tag{7.33}$$

$$\Delta\alpha = \frac{\Delta MD}{P} M$$

where

ψ_{cvc} = Cross-vertical correction

PROBLEM 7.3

Calculate the new angle and azimuth when the random inclination and azimuth tortuosity is applied to the following survey data:

Well measured survey depth = 3725 ft
Inclination = 3.25°
Azimuth = 165°

Well measured survey depth = 3900 ft
Inclination = 5.15°
Azimuth = 166°

Use pitch = 100 ft. Use the random number = 0.375.

Solution

$$\Delta\alpha = \frac{0.375 \times (3900 - 3725)}{100} \times 1 = 0.66$$

$\alpha_i = 5.15 + 0.66 = 5.81°$
Since $\alpha_i > 0$, the cross-vertical correction = 0°
$\phi_I = 166 + 0.66 + 0 = 166.66°$

Calibration of Tortuosity Factors

Quantitative estimation of the tortuosity factors are important not only for the analysis of the torque and drag but also for the anti-collision calculations. The following steps describe how to calculate tortuosity factors that better represent the actual curved wellbore:

1. Collect the relevant drilling data. These include the original well plan (casing program, directional plan, drill-string designs, etc.) and actual drilling data (well configuration, definitive survey data, mud logging data, and any downhole torque and weight-on-bit measurements, etc.).
2. For each hole section, calibrate the friction factor through the casing, based on the actual drilling data obtained while drilling out of the previous casing shoe, and the definitive survey data without any tortuosity correction.
3. Repeat step 2 for the open hole in each of the hole sections. Note that the friction factor through the casing should be that obtained from step 2.
4. Once the appropriate friction factors have been established for both the cased and the open holes, these can be used in conjunction with the original planned well profile to define the most appropriate tortuosity factor for the well by matching the torque and drag predictions with the actual drilling data.

Absolute and Relative Tortuosity

The calculation of running tortuosity is the station-to-station summation of the total curvature, normalized to a standard course length (the units used are the same as for the DLS), and absolute tortuosity is the tortuosity of the initial well path before the artificial tortuosity is applied.

Absolute tortuosity at the survey point n, of the initial well path before artificial tortuosity was applied is given by the following equation:

$$\Gamma_{(abs)_n} = \left(\frac{\sum_{i=1}^{i=n} \alpha_{adj}}{D_n + \Delta D_n} \right) \times 100°/100 \text{ ft} \tag{7.34}$$

where

$\alpha_{dj} = \alpha_i + \Delta D_i \times \delta_i$ is the dogleg adjusted summed total inclination angle
i = survey station index
δ = dogleg severity, (°)/100 ft
ΔD = distance between the survey stations, feet

In order to quantify how the well trajectory has changed after applying the artificial tortuosity, a relative tortuosity is defined that is the tortuosity of the wellbore relative to the absolute tortuosity, and is as follows:

$$\Gamma_{(rel)_n} = \left(\Gamma_{(abs)_n}\right)_{tor} - \left(\Gamma_{(abs)_n}\right)_{notor} \quad (°)/100\text{ ft}$$

where

$\left(\Gamma_{(abs)_n}\right)_{tor}$ = Absolute tortuosity with artificial tortuosity applied, (°)/100 ft

$\left(\Gamma_{(abs)_n}\right)_{notor}$ = Absolute tortuosity with no artificial tortuosity applied, (°)/100 ft

So it can be inferred that the relative tortuosity will be zero, unless artificial tortuosity has been applied.

Discussion of Multiple Hits

It is often perceived that, the longer the distance over which the reference wellbore is in proximity to an offset wellbore, the higher the risk of collision. In fact, if the wells are perfectly parallel, then whatever the distance over which they are in proximity, it makes no difference to the risk of collision. However, in normal drilling practice the wells do wander about due to natural drift, and are corrected back to the planned well path, and this results in tortuosity. Under such conditions, there is an increased chance of multiple intersections with each intersection having nearly the same level of risk.

The approach to computing the increased risk is to find the increased area that is integrated in the "binormal" distribution (i.e., the amount of extra coverage of the well path within the elliptical shape), given that the total integration area under the curve is unity. Figure 7.19 shows two tracks

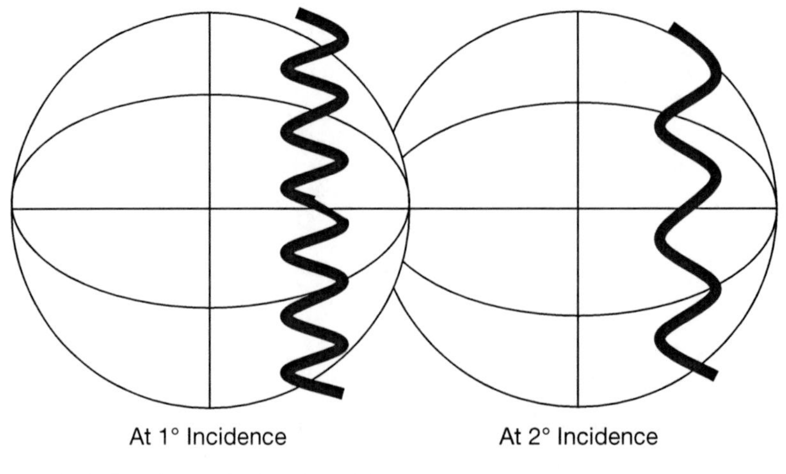

Figure 7.19 Traveling cylinder plot with tortuosity.

with the same proximity and tortuosity, but different angles of incidence, one at 1° and the other at 2°. It can be seen that the 1° track covers more integration area within the elliptical shape.

Using results from a tortuosity study of offset wells, it appears that the average tortuosity for intermediate hole sizes approximates to a sinusoidal wave with 0.5°/100 ft amplitude and a period of 500 ft. In the case of anti-collision calculations, if the reference well is taken as stationary, then the relative effect on the offset well is an amplitude of 1°/100 ft and a period of 500 ft. To be conservative, a dogleg of 2°/100 ft is used in this analysis. Figure 7.19 shows the effect at two different angles of incidence. The total effect of this is not that significant. An angle of incidence of 0.7° doubles the integration area of the wells, and therefore, nearly doubles the probability of interception.

To convert this tortuosity into a ratio that can be used in the minimum separation formula, the increased integration area can be viewed as a "thickening" of the hole and casing diameters.

Low angle of incidences are likely to be found in three types of scenarios:

1. Surface wells will inevitably start parallel. These need not be considered, because they will be above the depth required for considering collision risk analysis. Standard "high-risk" collision procedures will be based on methods that account for the increased risk of multiple hits.
2. Twinned wells may be defined as where one well deliberately drilled parallel to another well for production reasons. The increased risk of multiple hits should be included for these wells.
3. Relief wells are deliberately drilled to pass and then converge to a parallel well, and in this case, the increased chance of multiple hits is an advantage.

Some of the equations used for the analysis of multiple hits are given below:

$$\alpha = \arc\tan\frac{P}{4r} \tag{7.35}$$

where

P = Period
r = Radius

The amplitude is given by:

$$\phi = 2r(1-\cos\alpha) \tag{7.36}$$

The angle of incidence is given in terms of the dot product of the reference well and the offset well vectors:

$$I = \arc\cos(R \bullet O) \tag{7.37}$$

The period in the traveling cylinders plot is: $P \sin \alpha$.
The average angle in the offset well as shown in Figure 7.19 is:

$$AA = \arc\tan\left(\frac{2M}{P \sin \alpha}\right) \tag{7.38}$$

The relative thickening in terms of the average angle is given as:

$$RT = \frac{1}{\cos AA} \tag{7.39}$$

Bias-Type Errors

The ISCWSA survey error model allows for bias-type errors as an option. Opinion is divided on their usefulness and validity. In collision avoidance analysis, they may only serve to confuse the process. The argument against use of such errors is that, if there is a known bias, then it should be corrected in the processed survey data. The ISCWSA error model allows for two types of bias: (1) wire line or drill-string stretch on the depth, and (2) magnetic interference on the azimuth. If this bias treatment is valid, then BHA sag should be treated as a bias error and not as a survey correction.

However, if the policy of the organization requires that bias errors be tracked, then it is suggested that the bias should be subtracted from the center-to-center distance of the two wells only in the case where the bias causes the wells to become closer, rather than farther apart. The most common source of survey bias is magnetic interference. This should cause the magnetic survey of a location to place it north of a gyro survey of the same location. There are rare circumstances when magnetic interference will cause the opposite effect. So, the method used errs on the side of caution.

The magnitude of the bias error for the two wells is determined along the vector used for bisecting the ellipsoids:

$$Sb = Max(0, u.(B_o - B_r)) \tag{7.40}$$

where

u = Vector direction used for bisecting the ellipsoids
B_r, B_o = Positional bias vectors of the reference and offset wells, respectively

Modeling Uncertainty[9]

Uncertainty modeling provides a method of determining how much error was made when an offset well was drilled. If the well section is straight and has an inclination I and an azimuth A, then both are potentially in error.

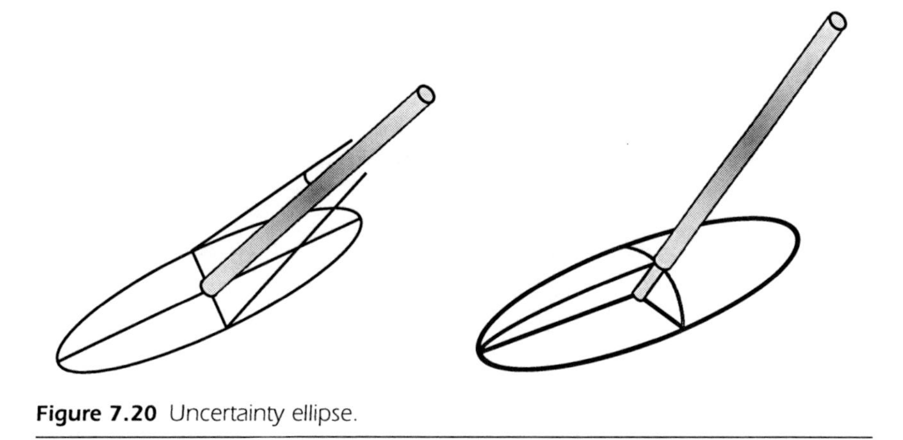

Figure 7.20 Uncertainty ellipse.

An uncertainty ellipse (Figure 7.20) is formed around the wellbore where the lateral dimension is proportional to the azimuth error, and the high-side dimension is proportional to the inclination error. If the azimuth were to be more accurate than the inclination, the uncertainty ellipse would be thinner across the wellbore. On the other hand, more typically, the azimuth is less accurate than the inclination, creating an ellipse with a larger lateral dimension. The final shape is like an almond, elliptical in all three orthogonal planes.

Determining the Size and Shape

A very simple rule of thumb estimates the spatial error from an angular error as follows: 1° in angle creates about 2% in distance. For example, if a line of 1000 ft was measured on a bearing of 90° ±1°, the final point would be in error by approximately ±20 ft (Figure 7.21).

The correct answer would be closer to 1000 sin (1) = 17.45 m, but as a conservative estimate the thumb rule is acceptable and easy to calculate. It is not true that errors are proportional to the measured depth. In surveying, systematic (i.e., unchanging) errors propagate in proportion to distance traversed from the origin.

PROBLEM 7.4

A target is to be drilled with measurement-while-drilling (MWD) tools with the following typical accuracies. The measured depth is accurate to 2 ft/1000, the inclination to ±0.3°, and the azimuth to ±1°. Develop the approximate ellipse of uncertainty when the target is reached.

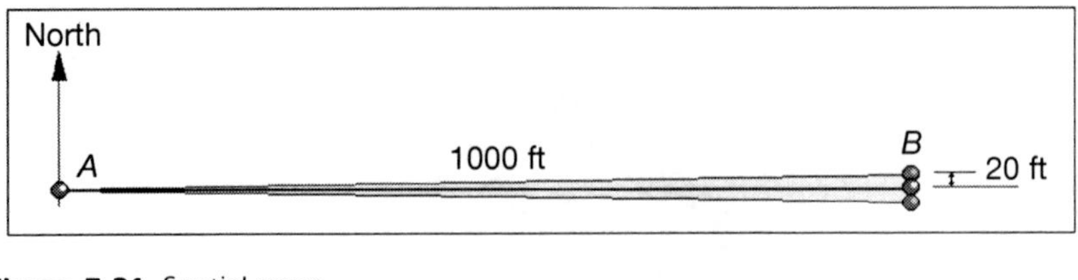

Figure 7.21 Spatial error.

Solution

In this example, the effect of the measured depth error of 2 ft/1000 can be determined without knowing the planned trajectory. The accumulated error at the target will be the same, no matter what path is taken to get there. Using the fact that the survey error propagates proportional to distance from the origin, the axis of the uncertainty ellipse created by the depth error is a line of 10 ft (2 ft/1000 × 5000 ft distance) toward and away from the origin, as shown in Figure 7.22.

To clarify by an example, if the well was drilled as a horizontal well with 4000 ft vertical drilling and 3000 ft horizontal drilling after a sharp build to horizontal, the total measured depth would be 7000 ft. Assume that the depth error causes overshooting by 2 ft/1000.

In this case, the depth error would be

$$\frac{2}{1000} \times 7000 = 14 \text{ ft}$$

but the effect on the target position would be to overshoot the vertical by 8 ft and overshoot the horizontal by 6 ft, creating a positional error of 10 ft on the axis, as shown in Figure 7.23.

The effect of the inclination error can be similarly calculated using the simple rule. Inclination only affects the vertical plane, and as the distance from the origin is 5000 ft in this plane, so the positional error will be at right angles to the space vector and will have a magnitude of 0.6% of 5000 ft or 30 ft (Figure 7.24).

The third axis will be in the horizontal plane and is due to the azimuth error of ±1°. This will produce an ellipse axis across the wellbore of approximately 2% of 3000 ft or 60 ft. It is quite common for the azimuth error to be the dominant error in a 3D ellipse of uncertainty.

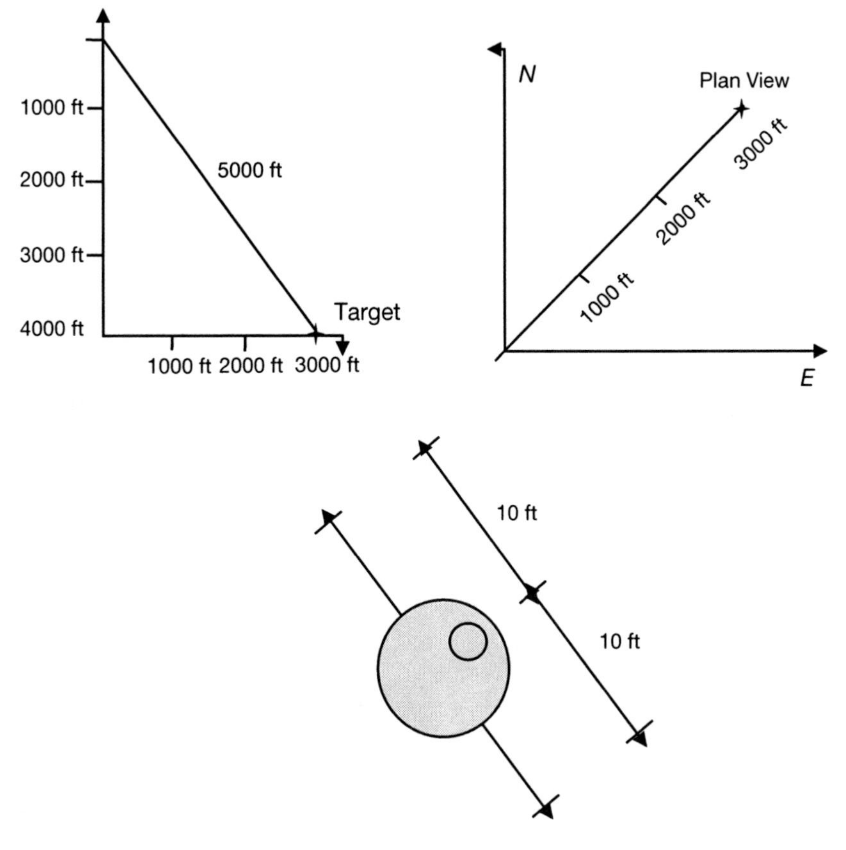

Figure 7.22 Uncertainty axis.

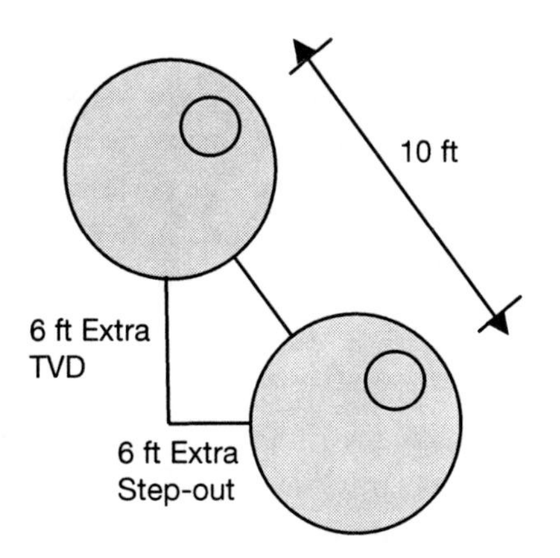

Figure 7.23 Positional error.

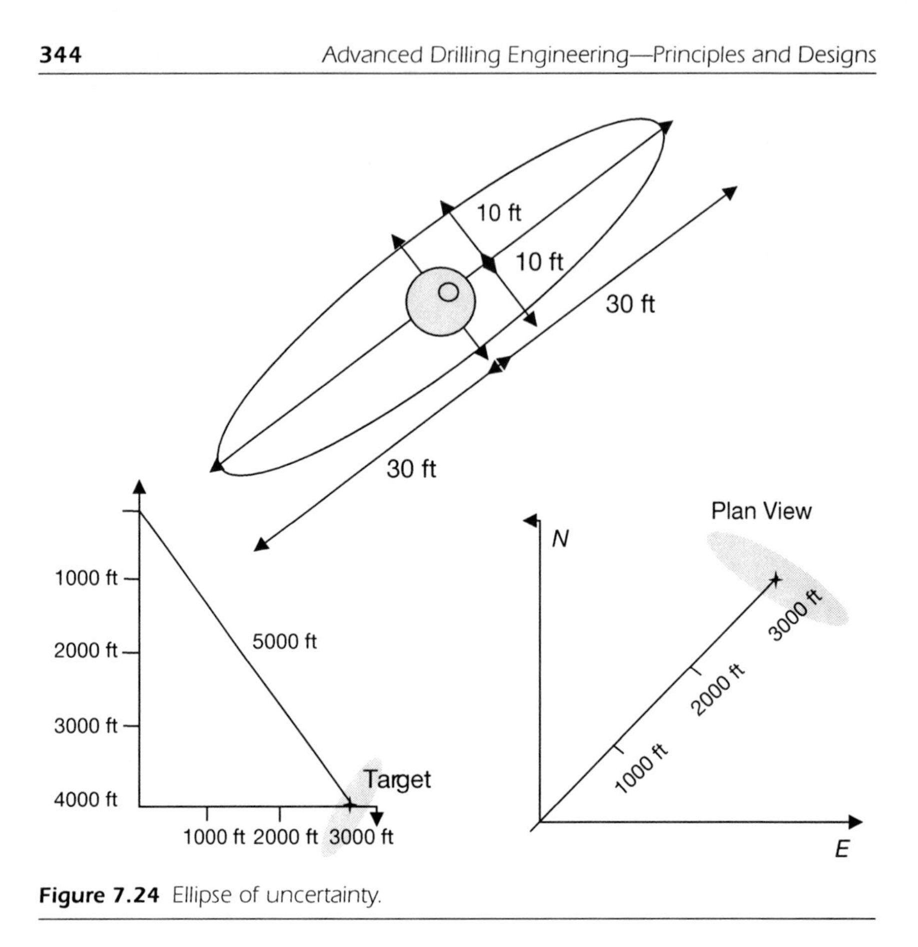

Figure 7.24 Ellipse of uncertainty.

Covariance Matrix

As the name suggests, the covariance matrix of a set of parameters is a measure of how the parameters vary together. The variance of an observation is simply the square of its deviation, but the covariance is the product of the deviations of two different parameters. The covariance helps to measure the correlation between the parameters, which shows up as a "skew" in the uncertainty ellipse. The covariance matrix is also called the variance–covariance matrix, as the diagonal terms give the variance of the parameters and the off-diagonal terms give the covariance (Figure 7.25), where

μx, μy = Average values of the parameters x and y, respectively
δx, δy = Respective errors in x and y
$\delta x \delta y$ = Covariance of the parameters x and y

Covariance is positive when δx and δy have the same sign.

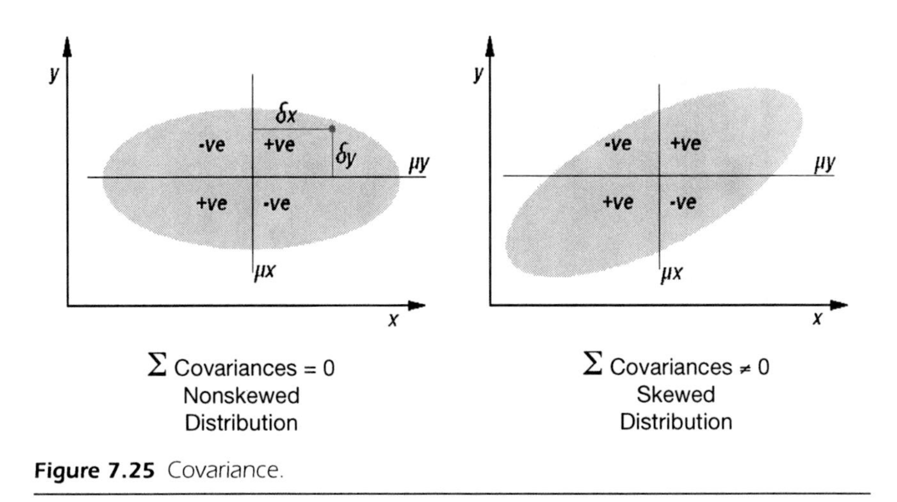

Figure 7.25 Covariance.

The same concept is also applicable in 3D, and the covariance matrix of the north, east, and TVD (vertical) position parameters permits the derivation of the dimensions and orientation in space of the uncertainty ellipsoid.

In order to build up the covariance matrix, it is necessary to know how each source of error affects the observations, and how the observation error affects the position of the well path in the north, east, and TVD directions.

These error sources are multiple and varied. In the ISCWSA error model, there are dozens of error sources. Each error affects a given survey station in a different way. For example, the accuracy of knowledge of the magnetic north direction would be a simple azimuth effect. This will not have any effect on the observations of either inclination or measured depth. Weighting functions are used to determine the effect that any error source would have on the measured depth, inclination, and azimuth, respectively. As an example, the weighting functions for a compass reference error are (0, 0, 1), which signifies that this error has no effect on the measured depth, no effect on the inclination, and all its effect on the azimuth.

However, a tool misalignment due to a bent housing in the assembly might affect inclination or azimuth depending on the toolface. In this case, the weighting function would be (0, cos ω, sin ω). A drill pipe stretch error would have a weighting function of (1, 0, 0) as might be expected, and a sag correction error would have a weighting function of (0, 1, 0).

Determination of the effect of the observation error on the well path position is important, and a measured depth error could affect the north, the east, and the TVD position parameters, as shown in Figure 7.26, whereas an azimuth error would affect only the north and east position parameters.

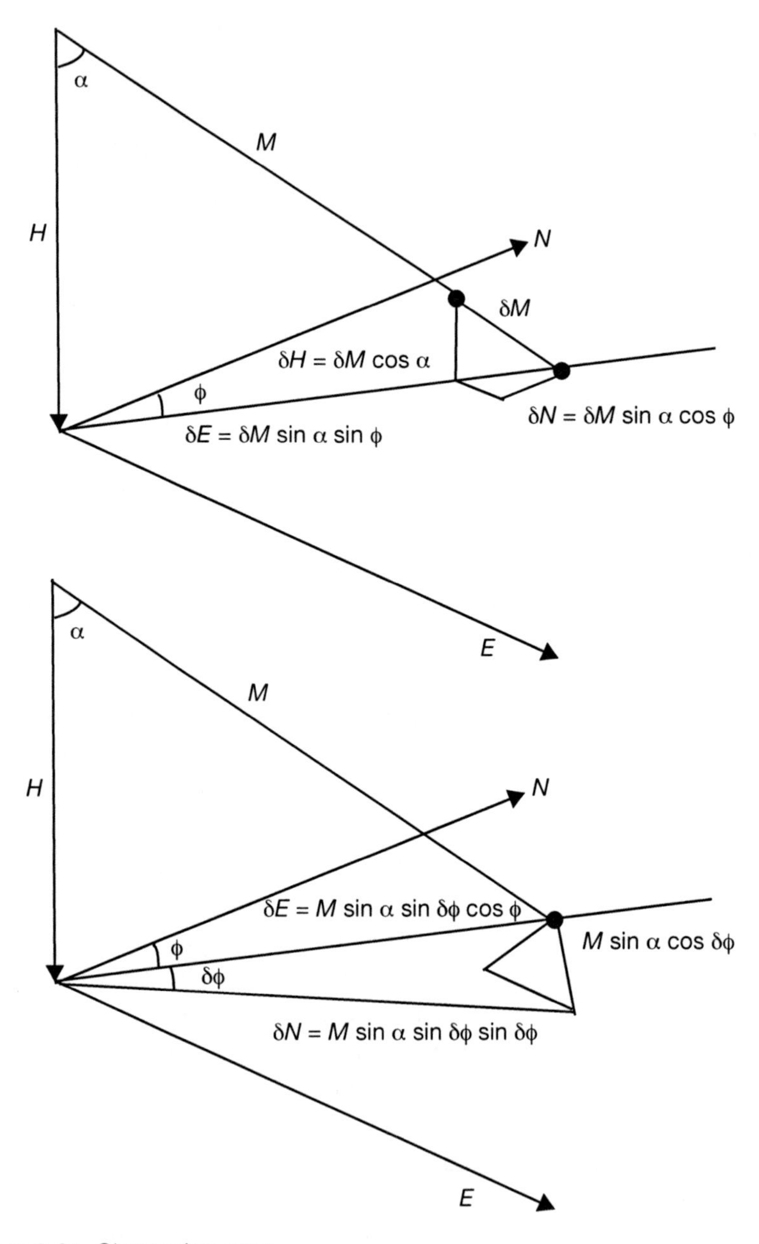

Figure 7.26 Observation error.

The variance–covariance matrix is given by:

$$\begin{bmatrix} \sum \delta N^2 & \sum \delta N \delta E & \sum \delta N \delta H \\ \sum \delta E \delta N & \sum \delta E^2 & \sum \delta E \delta H \\ \sum \delta H \delta N & \sum \delta H \delta E & \sum \delta H^2 \end{bmatrix} \quad (7.41)$$

When constructing the matrix, care has to be taken to ensure that errors that correlate from one survey station to the next are directly added to the delta values before summing, but errors that can vary from one survey to the next are summed in their product form. This ensures that random and systematic effects are propagated correctly. Some errors also continue from one survey leg to another, and some even from one well to another, such as the error in magnetic north. Once the matrix is complete, a technique is required to derive the size and orientation of the three main axes. The eigenvectors of the covariance matrix describe the "attitude vectors" of the three axes and the eigenvalues give their length. The mathematics of obtaining the eigenvectors and eigenvalues of the 3×3 covariance matrix is straightforward, and provides the dimensions and orientation of the uncertainty ellipsoid.

Collision Risk

It does not necessarily follow that, if the two uncertainty ellipses of two wells touch each other, then there is always a high risk of collision. Normally, three standard deviations are used for the size of the uncertainty ellipses, which means that the probability of being outside the ellipse is very small. When determining the collision risk, the separation factor is often used and calculated in a plane at right angles to the offset well. A separation factor of one would indicate that the ellipses were just touching. As an example, if the hole size occupied one-third of the ellipse, the probability of collision of two wells, with just touching uncertainty ellipses, would be less than 1 in 300,000. But the risk rises rapidly as the ellipses overlap. The separation factor is an excellent way of drawing attention to high-risk areas quickly, and then more detailed analysis may be needed to determine the safety of continuing to drill.

Figure 7.27 shows the general steps followed for anti-collision calculations of wells.

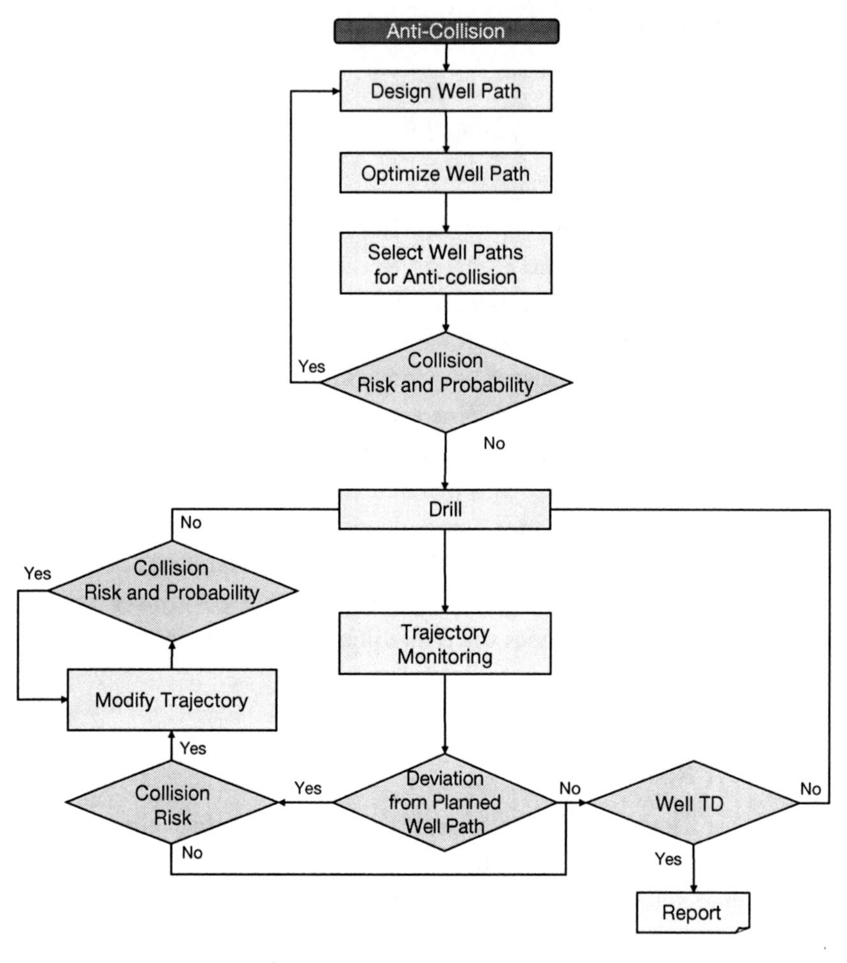

Figure 7.27 Anti-collision flow chart.

Supplementary Problems

1. For the given data find out whether the planned well will collide with the existing drilled well.

   Planned well (straight-build-hold) N/S = –12.6 ft; E/W = 25.9 ft

MD	*Inclination (°)*	*Azimuth (°)*
3200	0	0
4000	0	0
8300	30	140
8900	30	140
10,500	90	140

   Drilled well N/S = 0 ft; E/W = 0 ft

Plot the following views:
a. Traveling cylinder plot
b. Ladder view plot
c. Separation factor plot
d. Spider plot

2. Calculate the relative tortuosity at the survey point of 7990 ft, with the following given data and with the starting survey station at 7900 ft.

Survey without tortuosity:

MD (ft) – D	*Inclination (°)* α	*Azimuth (°)* ϕ
7800	58.22	0
7900	58.22	0
7990	58.22	0
8100	58.22	0

Survey after tortuosity is applied:

MD (ft) – D	*Inclination (°)* α	*Azimuth (°)* ϕ
7800	58.54	0.32
7900	53.78	359.57
7990	58.71	0.49
8100	53.72	359.51

3. Consider two well path designs. Prove that, in order that the principal normals of a well path be binormals of another, the relationship

$$\left(\frac{\kappa^2}{\kappa^2+\tau^2}\right) = \text{Constant}$$

where

τ = torsion

κ = curvature

References

1. Xiushan, L., and C. Zhangzhi. "Description and Calculation of Relative Positions of Wellbore Trajectories." Guanghan, China. *Drilling & Production Technology* 22, no. 3 (1999): 7–12.
2. Thorogood, J. L., and S. J. Sawaryn. "The Traveling-Cylinder Diagram: A Practical Tool for Collision Avoidance." *SPE Drilling Engineering* 6, no. 1 (1991): 31–36.
3. Xiushan, L. "The Universal Method for Calculating Off-Target Distances." Renqiu, China. *Oil Drilling & Production Technology* 30, no. 1 (2008).
4. Xiushan, L., and S. Yinao. "Description and Application of Minimum Distance between Adjacent Wells." Beijing, China. *China Offshore Oil and Gas (Engineering)* 12, no. 4 (2000): 31–34.

5. Contributed by Dale Jaimeson.
6. Luo, Y., K. Bharucha, R. Samuel, and F. Bajwa. "Simple Practical Approach Provides a Technique for Calibrating Tortuosity Factors." *Oil & Gas Journal* 15 (2003).
7. Samuel, R., K. Bharucha, and Y. Luo. "Tortuosity Factors for Highly Tortuous Wells: A Practical Approach." SPE 92565 SPE/IADC Drilling Conference, February 23–25, Amsterdam, The Netherlands.
8. Gaynor, T., D. Hamer, D. Chen, D. Stuart, and B. Comeaux. "Tortuosity versus Micro-Tortuosity—Why Little Things Mean a Lot." SPE 67818 SPE/IADC Drilling Conference, Amsterdam, The Netherlands, February 27–March 1, 2001.
9. Contributed by Jerry Codling.

8

Well Path Optimization

The purpose of this chapter is to develop optimization methodology and calculation techniques to select well path geometrical parameters, such as length, deviation angle, well path profile, etc. Selection of target location and reasonable drilling plan, such as horizontal, multilateral, and multi-branch wells, are paramount. Well path construction will essentially use the theory presented in the previous chapters, coupled with reservoir characteristics. More complex reservoir models than presented in this chapter can be used, and the choices depend on the complexity. In this chapter, some new type of well profiles are also discussed, and the quantification is based on minimum energy principle. In addition to planning the survey, calculations evidently require an effective mathematical model in a 3D plane, so that well position to the desired target is analyzed and achieved with minimal cost. Several choices of objective function exist for designing well paths that include minimum cost, maximum production, minimum torque, minimum drag, etc. In some wells, there may be several criteria to be satisfied simultaneously with the design constraints. A number of other optimization methods can be used for solving different well path problems.

Derivative Analysis

Intrinsic geometrical attribute, derivative of the curvature, curvature squared, borehole torsion, torsion squared, and sum of the square root of curvature squared and torsion squared provide some of the calculation methods to quantify how the borehole twists and turns. The conventional method of plotting the curvature versus depth may not give more details about the borehole path. The method presented is simple differentiation between two survey stations $i-1$ and i, using three sequential neighboring survey station data. The change in curvature, curvature squared, etc., give more one-to-one correspondence on the steepest change in the slope, due to the dependence on the curvature or torsion between the well paths, in addition to the fact they are obscured in the curvature or torsion plots. The calculation may be improved by using the weighted mean between the survey station data at the point of interest. Weighting produces a better estimate of the smoothness well path and interpretation of the quality of the wellbore, and can be done using mean weighting, midpoint rule, or least-squared regression methods. Least-square regression[1-3] uses all the preceding survey station data and the following survey data point, whereas the weighted mean uses the preceding and successive survey data points. Derivatives for curvature squared, square of curvature derivative, torsion squared, or combinations of such terms with mean weighting are as follows:

$$\left(\frac{d\kappa}{dD}\right)_i = \frac{\left[\left(\frac{\Delta\kappa_{i-1}}{\Delta D_{i-1}}\right)\Delta D_{i+1} + \left(\frac{\Delta\kappa_{i+1}}{\Delta D_{i+1}}\right)\Delta D_{i-1}\right]}{\left(\Delta D_{i-1} + \Delta D_{i+1}\right)} \tag{8.1}$$

$$\left(\frac{d\tau}{dD}\right)_i = \frac{\left[\left(\frac{\Delta\tau_{i-1}}{\Delta D_{i-1}}\right)\Delta D_{i+1} + \left(\frac{\Delta\tau_{i+1}}{\Delta D_{i+1}}\right)\Delta D_{i-1}\right]}{\left(\Delta D_{i-1} + \Delta D_{i+1}\right)} \tag{8.2}$$

$$\left(\frac{d\kappa^2}{dD}\right)_i = \frac{\left[\left(\frac{\Delta\kappa^2_{i-1}}{\Delta D^2_{i-1}}\right)\Delta D_{i+1} + \left(\frac{\Delta\kappa^2_{i+1}}{\Delta D^2_{i+1}}\right)\Delta D_{i-1}\right]}{\left(\Delta D_{i-1} + \Delta D_{i+1}\right)} \tag{8.3}$$

$$\left(\frac{d\tau^2}{dD}\right)_i = \frac{\left[\left(\frac{\Delta\tau^2_{i-1}}{\Delta D^2_{i-1}}\right)\Delta D_{i+1} + \left(\frac{\Delta\tau^2_{i+1}}{\Delta D^2_{i+1}}\right)\Delta D_{i-1}\right]}{\left(\Delta D_{i-1} + \Delta D_{i+1}\right)} \tag{8.4}$$

$$\left(\frac{d\left(\kappa^2+\tau^2\right)}{dD}\right)_i = \frac{\left[\left(\frac{\Delta\left(\kappa^2_{i-1}+\tau^2_{i-1}\right)}{\Delta D^2_{i-1}}\right)\Delta D_{i+1} + \left(\frac{\Delta\left(\kappa^2_{i+1}+\tau^2_{i+1}\right)}{\Delta D^2_{i+1}}\right)\Delta D_{i-1}\right]}{\left(\Delta D_{i-1} + \Delta D_{i+1}\right)} \tag{8.5}$$

where

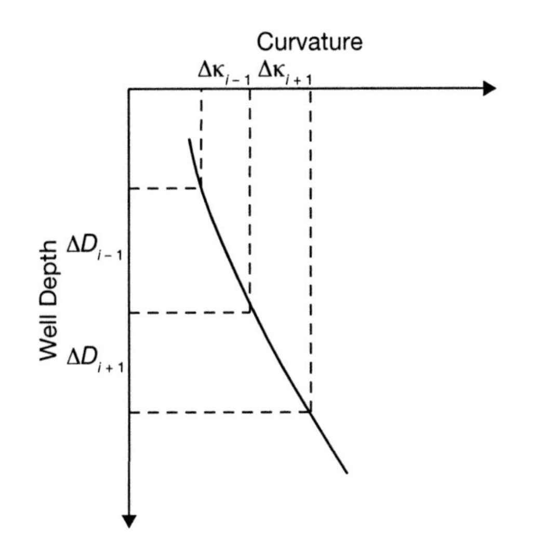

Figure 8.1 Right and left derivative slopes.

D = Measured depth at the survey stations
i = Sequence of survey stations
κ = Curvature
τ = Borehole torsion

$$\frac{\Delta\kappa_{i-1}}{\Delta D_{i-1}} \text{ and } \frac{\Delta\kappa_{i+1}}{\Delta D_{i+1}}$$

are the right and left derivative slopes of the curvature, respectively, as shown in Figure 8.1.

The derivative analysis may be useful to determine the depth of the occurrence of the unevenness in the well path that may cause tool failure, and this method will provide more details using continuous survey tools. It also provides accurate depicting of the well trajectory with continuous survey data.

Optimization of Total Strain Energy of the Well Path

This criterion for measuring the borehole quality is based on the physical reasoning rather than the geometric meaning. The nonlinear curve modeling a thin elastic beam is known as the minimum energy curve, and is characterized by bending the least while passing through a given set of points. It is considered to be excellent criterion considering the simplicity for producing smooth curves. Hence, this method describes the minimum energy of the wellbore path. The added advantage is that it emphasizes the undulation of the well path curvature of the sharp well paths. The strain energy (E) of the wellbore path is given as the arc length integral of the curvature squared.

$$E = \int_0^{\ell} \kappa(x)^2 dx \tag{8.6}$$

The inclusion of the torsion parameter as the arc length integral of the torsion squared makes it more comprehensive, and can be given as[4]:

$$E = \int_0^{\ell} \left(\kappa(x)^2 + \tau(x)^2\right) dx \tag{8.7}$$

Minimization of the total energy of the curve will result in less torque and drag during various operations.

Length Optimization for Horizontal Wells

Horizontal wells are drilled to improve the hydrocarbon production. Considerable attention is being focused on the feasibility and economics of horizontal wells and lateral well designs. In order to improve cost effectiveness and productivity from the formation, different designs and methodologies have been introduced in the past. In order to compare the productivity indices of vertical wells to that of horizontal wells, various parameters, such as length, diameter of the wellbore, number of wells, arrangement of wells, well spacing, and the ratio of horizontal to vertical permeabilities, are important. Development and completion of the field and recovering the maximum amount of hydrocarbon requires unified and mutual coordination between different departments, such as geology, drilling, reservoir, and completion.

Horizontal Well Length

Due to fluid frictional losses, there is a pressure drop in horizontal pipe when the fluid flows inside. If the horizontal wellbore pressure drop is less than the reservoir pressure, it will lead to the cessation of production from the heal portion of the well. Thus, the horizontal wellbore frictional pressure losses not only lower the production of oil, but also increase the drilling and completion costs. Other problems, such as gas and water coning, may also result.

Fluid flow from the reservoir to the wellbore and the coupling relation is shown in Figure 8.2. Considering the horizontal direction along the well-

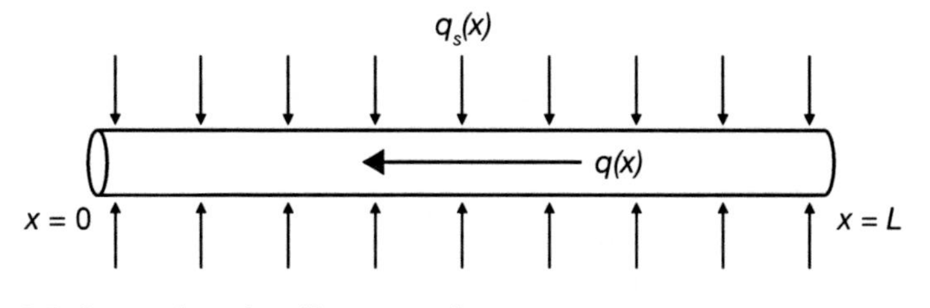

Figure 8.2 Reservoir and wellbore coupling.

bore as x-axis with the coordinates origin and the length L, the production flowrate is given as:

$$q(x) = J_s(x)\left[P_e - P_{wf}(x)\right] \tag{8.8}$$

where

P_e = Constant external boundary pressure, psi
$P_{wf}(x)$ = Bottomhole flowing pressure at the length x, psi
$J_s(x)$ = Productivity index of the wellbore, bbl/d-psi

A horizontal drainage well increases production, because its pay zone exposure area is larger than that of a vertical well drilled in the same reservoir. The steady-state inflow equation for a horizontal well in oilfield units can be given as[2]:

$$q_h = \frac{(0.00708)k_h h \Delta p}{\mu B_o \ln\left[\dfrac{a + \sqrt{a^2 - (L/2)^2}}{L/2}\right] + \dfrac{h}{L}\ln\left(\dfrac{h}{2r_w}\right)} \text{STB/day} \tag{8.9}$$

Valid for $L > h$ and $\dfrac{L}{2} < 0.9 r_{eh}$

where

k_h = Horizontal permeability, md
h = Reservoir height, ft
Δp = Pressure drop from the drainage radius to the wellbore, psi
μ_o = Crude oil viscosity, cP
B_o = Formation volume factor, RB/STB
L = Length of horizontal well, ft
a = Half major axis of drainage ellipse, as given by

$$a = 0.5L\left[0.5 + \sqrt{0.25 + \frac{1}{\left(0.5L / r_{eb}\right)^4}}\right]^{0.5}$$

where

r_{eb} = Equivalent outer radius of the drainage area, ft
r_w = Wellbore radius, ft

For the case where

$$\frac{L}{2} < 1$$

the value of a is approximately equal to r_{eb}.

The modified flow equation, which takes into consideration the ratio of horizontal to vertical permeability

$$\left(\beta = \sqrt{\frac{k_b}{k_v}}\right)$$

is given as[5]:

$$q_b = \frac{(0.00708)k_o b\Delta p}{\mu B_o \ln\left[\frac{a + \sqrt{a^2 - (0.5L)^2}}{0.5L}\right] + \frac{\beta b}{L}\ln\left(\frac{\beta b}{2r_w}\right)} \text{ for } L > \beta b \tag{8.10}$$

The steady-state flowrate for a vertical well can be given as:

$$q_v = \frac{(0.00708)k_o b\Delta p}{\mu B_o \ln\left(\frac{r_e}{r_w}\right)} \tag{8.11}$$

where

r_w = Drainage radius, ft

The productivity index ($\mathcal{J}$) defined as

$$\frac{q}{\Delta p} \text{ (m}^3\text{/day/kPa or bbl/day/psi)}$$

can be used for calculating the optimum horizontal length for optimum production.

So the productivity index for a horizontal well is given by:

$$\mathcal{J}_b = \frac{(0.00708)k_o b}{\mu B_o \ln\left[\frac{a + \sqrt{a^2 - (0.5L)^2}}{0.5L}\right] + \frac{\beta b}{L}\ln\left(\frac{\beta b}{2r_w}\right)} \tag{8.12}$$

The ratio of the productivity index of a horizontal well to vertical well is obtained by dividing Eq. 8.10 by Eq. 8.11:

$$\frac{J_h}{J_v} = \frac{\ln\left(\frac{r_e}{r_w}\right)}{\mu B_o \ln\left[\frac{a+\sqrt{a^2-(0.5L)^2}}{0.5L}\right] + \frac{\beta h}{L}\ln\left(\frac{\beta h}{2r_w}\right)} \tag{8.13}$$

The ratio of productivity index of horizontal wells of different wellbore radius is given by:

$$\frac{J_{h1}}{JI_{h2}} = \frac{\ln\left[\frac{a+\sqrt{a^2-(0.5L)^2}}{L/2}\right] + \frac{\beta h}{L}\ln\left(\frac{\beta h}{2r_w}\right)_2}{\ln\left[\frac{a+\sqrt{a^2-(0.5L)^2}}{0.5L}\right] + \frac{\beta h}{L}\ln\left(\frac{\beta h}{2r_w}\right)_1} \tag{8.14}$$

More complex analysis can be done by including a skin factor to account for the effect of formation damage.

PROBLEM 8.1

Evaluate the length of a horizontal well with the following reservoir parameters:

Reservoir thickness = 100 ft; wellbore radius = 5.5 in

Horizontal and vertical permeabilities = 50 md

Crude oil viscosity = 0.6 cP; formation volume factor = 12.5 RB/STB

Ratio of horizontal to vertical productivity index = 5

The vertical well spacing is 50 acres.

Solution

Assuming a circular drainage area, the drainage radius is

$$\sqrt{\frac{50\times 43{,}560}{\pi}} = 832.62 \text{ ft}$$

The half major radius is:

$$a = 0.5L\left[0.5+\sqrt{0.25+\frac{1}{(0.5L/832.62)^4}}\right]^{0.5}$$

The ratio of the productivity indices is

$$5 = \frac{\ln\left(\dfrac{832.62}{0.45833}\right)}{0.6 \times 1.25 \ln\left[\dfrac{a + \sqrt{a^2 - (0.5L)^2}}{0.5L}\right] + \dfrac{1 \times 100}{L} \ln\left(\dfrac{1 \times 100}{2 \times 0.45833}\right)}$$

Solving the above equations for L results in the horizontal length of L = 1132 ft.

Placement and Length of Horizontal Section

Placement of the horizontal section in the reservoir is as important as the length of the horizontal section. If the horizontal section of the well is off-centered, the production capability affects the required horizontal length of the well, besides the wellbore diameter. The productivity index of the off-centered well is given as[5]:

$$J_{ho} = \frac{(0.00708) k_o h}{\mu B_o \ln\left[\dfrac{a + \sqrt{a^2 - (0.5L)^2}}{0.5L}\right] + \dfrac{\beta h}{L} \ln\left[\dfrac{\left(\dfrac{\beta h}{2}\right)^2 + \beta^2 e^2}{\left(\dfrac{\beta h r_w}{2}\right)}\right]} \tag{8.15}$$

where

e = Horizontal eccentricity, m or ft

The portion of the well and the length in thick reservoirs, with or without a gas cap or aquifer, should be planned meticulously, so that energy to drive the oil production is increased and water production is delayed. The optimum half-spacing of a well drilled at the same level from the parent wellbore can be given as[6]:

$$x_e^{opt} = h\sqrt{\frac{k_h}{k_v}} \tag{8.16}$$

PROBLEM 8.2

The well is located 10 ft from the bottom of the payzone. Evaluate the length of a horizontal well, using the reservoir parameters given in Problem 8.1.

Solution

Eccentricity e = 50 – 25 ft = 25 ft

$$a = 0.5L\left[0.5 + \sqrt{0.25 + \frac{1}{(0.5L/832.62)^4}}\right]^{0.5}$$

The ratio of the productivity indices is

$$5 = \frac{\ln\left(\dfrac{832.62}{0.45833}\right)}{0.6 \times 1.25 \ln\left[\dfrac{a + \sqrt{a^2 - (0.5L)^2}}{0.5L}\right] + \dfrac{1 \times 100}{L} n \left[\dfrac{\left(\dfrac{1 \times 100}{2}\right)^2 + 1^2 \times 25^2}{\left(\dfrac{1 \times 100 \times 0.4583}{2}\right)}\right]}$$

Solving the above equations for L results in the horizontal length L = 1167 ft. It can be seen that the length of the horizontal section has increased due to the offset placement.

Completion Design

Even though the horizontal wells provide larger reservoir contact area, the cost of drilling and completing such wells is higher than the vertical wells. Optimum placement and length of the horizontal section also depends on the completion design and method. If a slotted completion is planned, the optimum length of the horizontal length will be different from a conventional completion design, as the inflow and wellbore hydrodynamics are different. The modeling should involve the slot design and number of slots and configuration. Some of the factors, such as wellbore configuration, number of wells in case of a multiple-wells environment, reservoir parameters, and type of completion, should be taken into account for the optimum well profile design. This will ensure not only maximum production, but also minimum well cost.

The ratio of the flow with and without friction can be generalized as below, so that optimum length can be calculated:

$$\frac{Q_f}{Q_{nf}} = \frac{Q_f}{J_s L_{\text{opt}} \Delta P} = x \tag{8.17}$$

where

Q_f = Flow with friction
Q_{nf} = Flow without friction
x = Factor

Number of Wells

To offset the increased cost of drilling horizontal wells, several wells can be drilled as shown in Figure 8.3. For dual lateral wells[7] or multilateral wells from the same parent wellbore, the total oil production can be given as[5]:

$$Q_h = \frac{0.007078 k_h h \Delta p}{\mu_o B_o \left[\ln\left(\frac{F r_e}{L}\right) + \left(\frac{h}{nL}\right) \ln\left(\frac{h}{2\pi r_w}\right) \right]} \tag{8.18}$$

where

n = Number of wells
F = 4, 2, 1.86, and 1.78 for n = 1, 2, 3, and 4, respectively

When the wells are placed at a different elevation (Figure 8.3) the production can be given as[5]:

$$Q_h = \frac{0.007078 k_h h \Delta p}{\mu_o B_o \left[\ln\left(\frac{F r_e}{L}\right) + \left(\frac{h}{mnL}\right) \ln\left(\frac{h}{2\pi m r_w}\right) \right]} \tag{8.19}$$

where

m = Number of elevations
H = Thickness drained by each drain hole ($h = mH$)

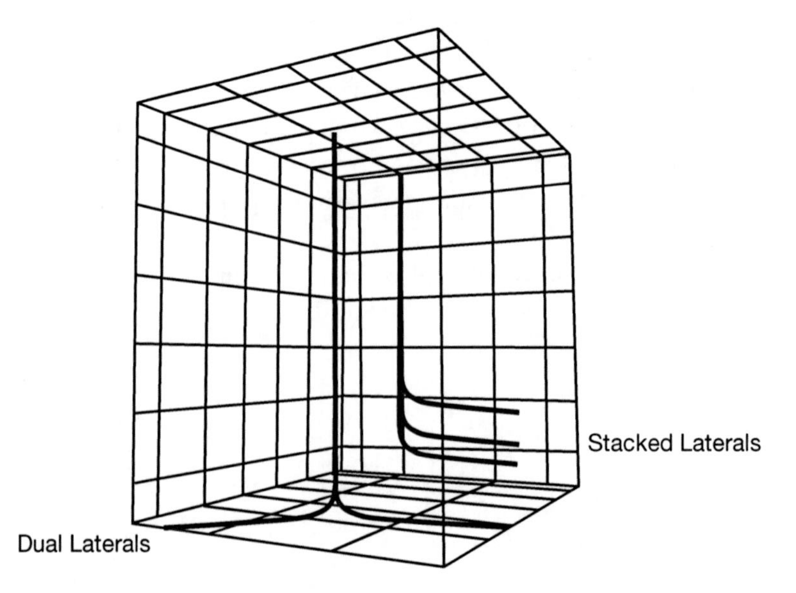

Figure 8.3 Dual and multilateral wells.

Length and Direction of Horizontal Section

Wellbore profile, target depth, horizontal length, or the length of the wellbore into the hydrocarbon-bearing formation should consider the formation dip and fracture so that it meets other design requirements for drilling, completion, and logging. Theoretically, it can be seen that the longer the horizontal length and bigger the hole diameter, the greater the productivity index will be, and thereby the higher the productivity. However, selection of the hole diameter depends on the casing program, completion technique (including tubing size), drilling rig capacity, and other factors, such as borehole stability, hole-cleaning conditions, and drilling costs. Therefore, the length of the horizontal well is not much restricted, and the selection of the length is very important as it greatly influences the well path design.

Apart from the length, the direction of the well inside the formation, as well as the formation reservoir properties are important to ensure higher production results. Inclination and slope of the formation, referred to as the formation dip, is also necessary to place the well accurately. The formation dip can be determined based on the data from offset wells, seismic data, core data, logging data, etc. In order to exploit the reservoir more efficiently, the horizontal well section should be parallel with the reservoir layer, and it is often located in the middle of the reservoir layer.

The direction of the horizontal section includes inclination and azimuth. When the horizontal section is parallel with the reservoir layer, the inclination and the azimuth of the horizontal section should have a certain relationship, or one can be decided by another. Because the designed azimuth is usually the same as, or perpendicular to, the azimuth of the reservoir layer, the azimuth of the horizontal section is decided first, followed by the inclination calculation.

As shown in Figure 8.4, if $\boldsymbol{u}$ is the unit vector to describe the reservoir below direction, $\boldsymbol{v}$ is the unit vector of the well direction in the horizontal section, $\boldsymbol{w}$ is the unit vector perpendicular to $\boldsymbol{u}$ in the reservoir layer, and $\boldsymbol{n}$ is the unit vector of the direction of the normal of the reservoir layer plane, then[8,9]:

$$\begin{cases} \boldsymbol{u} = \cos\beta\cos\psi\ \boldsymbol{i} + \cos\beta\sin\psi\ \boldsymbol{j} + \sin\beta\ \boldsymbol{k} \\ \boldsymbol{v} = \sin\alpha\cos\phi\ \boldsymbol{i} + \sin\alpha\sin\phi\ \boldsymbol{j} + \cos\alpha\ \boldsymbol{k} \\ \boldsymbol{w} = -\sin\psi\ \boldsymbol{i} + \cos\psi\ \boldsymbol{j} \end{cases} \tag{8.20}$$

where

β = Reservoir formation dip (°)
ψ = Direction of the declining reservoir (°)

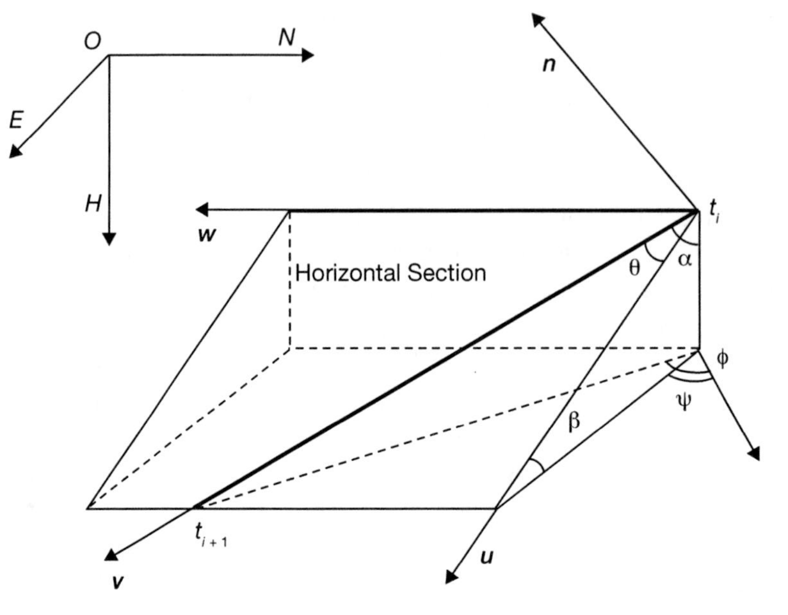

Figure 8.4 Schematic of the horizontal section.

α = Inclination angle (°)

ϕ = Azimuth angle (°)

$\boldsymbol{i}, \boldsymbol{j}, \boldsymbol{k}$ = Coordinate unit vectors in N, E, H coordinates, respectively

Because $\boldsymbol{n}$ is vertical to $\boldsymbol{u}$ and $\boldsymbol{w}$, and so $\boldsymbol{n} = \boldsymbol{w} \times \boldsymbol{u}$. Therefore,

$$\boldsymbol{n} = \sin\beta\cos\psi\,\boldsymbol{i} + \sin\beta\sin\psi\,\boldsymbol{j} - \cos\beta\,\boldsymbol{k} \tag{8.21}$$

Since the horizontal well plane is maintained parallel to the reservoir section, that is $\boldsymbol{v} \perp \boldsymbol{n}$. Therefore, $\boldsymbol{v} \times \boldsymbol{n} = 0$.

$$\sin\alpha\cos\phi \times \sin\beta\cos\psi + \sin\alpha\sin\phi \times \sin\beta\sin\psi + \cos\alpha \times (-\cos\beta) = 0 \tag{8.22}$$

After rearranging,

$$\sin\beta\cos(\phi - \psi) \times \sin\alpha = \cos\beta \times \cos\alpha \tag{8.23}$$

Using Eq. 8.23, three types of solutions can be obtained for the inclination angle[9,10]:

$$\alpha = \tan^{-1}\left(\frac{1}{c}\right) \tag{8.24}$$

$$\alpha = 90° - \tan^{-1} c \tag{8.25}$$

$$\alpha = 2\tan^{-1}\left(\sqrt{1+c^2} - c\right) \tag{8.26}$$

where

$c = \tan\beta \cos(\phi - \psi)$

It can be seen that Eqs. 8.24 and 8.26 are mathematically equivalent, especially when $\beta = 0$ or $|\phi - \psi| = 90°$, $c = 0$, and so $\alpha = 90$.

Moreover, in the tilted reservoir layer, according to the angle θ between the directional of the horizontal well section and the tilted direction of the reservoir layer, the inclination angle and azimuth of the horizontal well section can be calculated.

Because,

$$\cos\theta = \frac{\boldsymbol{u} \times \boldsymbol{v}}{|\boldsymbol{u}||\boldsymbol{v}|} \tag{8.27}$$

Using Eq. 8.20 to substitute the respective vectors in Eq. 8.27 and rearranging, yields:

$$\cos\theta = \cos\alpha \sin\beta + \sin\alpha \cos\beta \cos(\phi - \psi) \tag{8.28}$$

Employing the vertical relationships from Eqs. 8.23 and 8.28, it can be written as:

$$\begin{cases} \cos\alpha = \sin\beta\cos\theta \\ \cos(\phi - \psi) = \dfrac{1}{\tan\alpha\tan\beta} \end{cases} \tag{8.29}$$

The angle θ can be defined as starting from the tilted direction $\boldsymbol{u}$ of the reservoir layer, going along the normal direction $\boldsymbol{n}$ and rotating clockwise until the direction $\boldsymbol{v}$ of the horizontal well section, the angle is positive; otherwise, it is negative. Based on this definition and $|\theta| \le 180°$, Eq. 8.29 can be written as[9]:

$$\begin{cases} \alpha = \cos^{-1}(\sin\beta\cos\theta) \\ \phi = \psi + \operatorname{sgn}(\theta)\cos^{-1}\left(\dfrac{1}{\tan\alpha\tan\beta}\right) \end{cases} \tag{8.30}$$

where

$$\operatorname{sgn}(\theta) = \begin{cases} 1, & \text{if}\,\theta > 0 \\ 0, & \text{if}\,\theta = 0 \\ -1, & \text{if}\,\theta < 0 \end{cases}$$

PROBLEM 8.3

A reservoir formation dip angle is estimated to be 5°, with a tilt position of 100° and the horizontal design position direction is 80°. Estimate the horizontal well angle, if the horizontal direction and the reservoir hole under the direction of the tilting angle is 40°. Calculate the deviation angle and azimuth.

Solution

Given data:

$\beta = 5°;\ \psi = 100°;\ \phi = 80°;\ \theta = 40°$

$c = \tan 5° \cos(80° - 100°) = 0.082212$

Using the three equations to calculate α:

$$\alpha = \tan^{-1}\left(\frac{1}{0.082212}\right) = 85.3°$$

$$\alpha = 90° - \tan^{-1}(0.082212) = 85.3°$$

$$\alpha = 2 \times \tan^{-1}\left(\sqrt{1+0.082212^2} - 0.082212\right) = 85.3°$$

Substituting $\alpha = 85.3°$ and $\phi = 80°$ into Eq. 8.40 yields:

$$\theta = \cos^{-1}\left[\cos 85.3° \sin 5° + \sin 85.3° \cos 5° \cos(80° - 100)\right] = 19.93°$$

According to the sign convention θ takes the negative value, and so $\theta = -19.93°$. Using this included angle, the deviation angle and azimuth are calculated using Eq. 8.42; that is:

$$\alpha = \cos^{-1}\left[\sin 5° \cos(-19.93°)\right] = 85.3°$$

$$\varphi = 100° + \operatorname{sgn}(-19.93°)\cos^{-1}\left(\frac{1}{\tan 85.3° \tan 5°}\right) = 80°$$

when

$\theta = 40°$

$$\alpha = \cos^{-1}\left[\sin 5° \cos(40°)\right] = 86.172°$$

$$\varphi = 100° + \operatorname{sgn}(40°)\cos^{-1}\left(\frac{1}{\tan 86.172° \tan 5°}\right) = 140.108°$$

Using the azimuth in Eqs. 8.36 through 8.38, the well deviation angle and the well included angle can be computed with Eq. 8.40:

$$c = \tan 5° \cos(140.108° - 100°) = 0.0669$$

$$\alpha = \tan^{-1}\left(\frac{1}{0.0669}\right) = 86.172°$$

$$\alpha = 90° - \tan^{-1}(0.0669) = 86.172°$$

$$\alpha = 2 \times \tan^{-1}\left(\sqrt{1+0.0669^2} - 0.0669\right) = 86.172°$$

Upon substitution, θ can be found as before:

$$\theta = \cos^{-1}\left[\cos 86.172° \sin 5° + \sin 86.172° \cos 5° \cos(140.108° - 100)\right] = 40°$$

From the above, results can be seen that the well inclination calculated from Eqs. 8.36 and 8.38 are identical; that is, no matter when one uses the given azimuth ϕ to calculate inclination α and included angle θ, or uses the included angle θ to calculate the inclination α and azimuth ϕ, the results will be consistence.

General Utility Planning Methods

The quest for exploring hydrocarbons in deeper frontiers necessitates complex well path designs. Since the goal of drilling is to tap the oil from the complex reservoirs, and subsequently, the well path design requirements are different, the type of profiles, number of well path designs, and, thereby, combinations increase. These methods often use specific designs for realization of the well profile, conforming to the constraints and conditions of a particular design.

Faced with these complex and complicated profiles for the complex well construction, it is always advantageous to have a general design method that can be used for various conditions, thereby optimizing well path designs.[10-13] The general arc design method presented in this section, with the characteristic well path parameters and constraint equations, will help in designing complex well paths for complex well construction.

Universal Arc Profile

Currently, there are more than ten well path designs for various directional wells, horizontal wells, and special wells. For example, the directional well can use a J-shape or S-shape well path, and the horizontal well can use a build-hold-build or build-hold-build-hold-build well path, or even arc-shape and step-shape horizontal sections, etc. Thus, the variety of the well paths leads to many design methods.

Even for S-shape and build-hold-build trajectories, there is a similar form of design formula,[12] but it can't cover arc-shape (such as, build-hold-build-hold-build or step-shape horizontal sections), so it can't be used everywhere. In fact, a 2D arc-shape well path is composed of vertical, hold, horizontal, build, and drop sections, and vertical and horizontal sections can be considered as special cases of the hold section. Besides above, by defining the radius of curvature, a build section as a positive number and that of a drop section as a negative number, it gave an equivalent method for build and drop sections. So, the normal 2D arc-shape well section of a well path is hold-build-hold-hold-build-hold, as shown in Figure 8.5.

For an arbitrary well path, it is only required to know the number of sections n and other characteristic parameters. For example, in a J-shape well path, $n = 3$; in a S-shape or build-hold-build well path, $n = 5$; in an arch-shape horizontal section well path, $n = 3$; in a step-shape horizontal section, $n = 5$ and $\Delta L_3 = 0$; and in a hold-build-hold-build-hold well path, $n = 7$ and $\Delta L_5 = 0$. In fact, the common three-build well path should be a hold-build-hold-build-hold-build-hold well path.

The common arc-shape well path starts and ends with a hold section. The order number of any hold section is odd and the arc section is even, so the number of total sections should be an odd number. Build and drop sec-

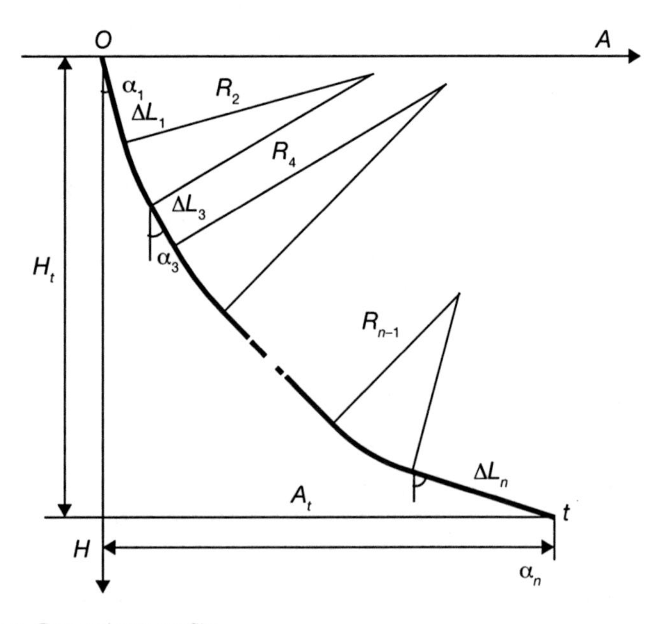

Figure 8.5 General arc profile.

tions can be called arc-shape sections, of which the build or drop trend can be expressed by the curvature and the sign of the curvature radius. The typical well path can be described by the amount of the arc sections. A single-arc well path includes a J-shape well path and an arch-shape horizontal well path; a double-arc well path includes an S-shape well path, double-build well path, and step-shape horizontal well path; and a triple-arc well path includes a triple-build well path and three continuous arc well paths; etc.

Characteristic Parameters and Constraint Equations

The characteristic parameters of a hold section are the section length and the inclination angle, and the parameters of an arc section are the curvature, the curvature radius, and the start and end inclination angles. Because of the alternative of the hold and arc sections, and continuity and smoothness of the whole well path, the start and end inclination angles must be equal to the inclination angles of the neighboring sections.

If a standard well path is composed by n sections, then the numbers of hold sections and arc sections are

$$\frac{n+1}{2} \text{ and } \frac{n-1}{2}$$

and the total number of the well path parameters of the well path is

$$\frac{3n+1}{2}$$

For example, the section number of a double-arc well path is $n = 5$, the number of the hold sections is 3, the number of arc sections is 2, and the parameters' number totals 8.

Obviously, these parameters determine the shape of the well path. After deciding all the parameters, the well path can be determined correspondingly, and vice versa. The design of the well path is to determine the well path parameters based on the design conditions and requirements.

Irrespective of well path and number of sections, the summations of the TVD and horizontal displacement of each section are equal to the total TVD and total horizontal displacement[8,13]:

$$\begin{cases} \sum_{i=1}^{n} \Delta H_i = H_T \\ \sum_{i=1}^{n} \Delta A_i = A_T \end{cases} \tag{8.31}$$

where

i = Section number
n = Total number of sections
T = Total

The summation goes for all sets of $\{H_i\}$ and $\{A_i\}$. The right side of the equation gives the total vertical depth and total horizontal displacement at any depth or section of the well path.

For well paths with frequently occurring profiles composed of circular-arc profiles with n sections, Eq. 8.31 can be modified as:

$$\begin{cases} \sum_{i=1}^{\frac{n+1}{2}} \Delta L_{2i-1} \cos\alpha_{2i-1} + \sum_{j=1}^{\frac{n-1}{2}} R_{2j} \left(\sin\alpha_{2j+1} - \sin\alpha_{2j-1}\right) = H_t \\ \sum_{i=1}^{\frac{n+1}{2}} \Delta L_{2i-1} \sin\alpha_{2i-1} + \sum_{j=1}^{\frac{n-1}{2}} R_{2j} \left(\cos\alpha_{2j-1} - \cos\alpha_{2j+1}\right) = A_t \end{cases} \tag{8.32}$$

where

i = Section number for straight sections
j = Section number for circular-arc sections
n = Total number of sections
t = Total

Universal Method for Planning Various Well Paths

Two-dimensional well path constraints are composed by TVD and horizontal displacement in two equations, so two unknowns can be found out. The previous well path design often considered the section length and inclination angle of a section as the unknowns, but these two unknowns could be selected arbitrarily. From this point of view, the old design method only gives a special case of the well path constraints.

The interactive well path design method can select arbitrarily two parameters as unknowns. The concept is that the parameters can be verified by each other and optimized. That is, two parameters are selected to perform the design and can also be optimized. For example, for the directional well single-arc well path, the kick-off depth and build-rate are the unknowns, so the reasonable kick-off point and required build rate can be determined. In another way, the kick-off depth and hold section length, or build rate and hold section length can be selected as unknowns. This interactive design method can avoid blind trial and manual disturbance, so it can optimize the well path design.

Numerical Design Method

The key of the well path design is to determine the parameters satisfying the constraint equations. In mathematics, for a continuous function $f(x)$, the solution of equation $f(x) = 0$ can be found by two steps:

1. First, assume a possible value as the initial solution. Assume that there is only one solution ξ of the equation $f(x) = 0$ in the interval $[a,b]$. If starting from the left side $x = a$, and selecting a positive number h as the step length, for every step, a value of $f(x)$ can be calculated until x is equal to x_0, which satisfies $f(x_0)\,f(x_0 + h) \leq 0$. If the solution ξ is between x_0 and $x_0 + h$, x_0 or $x_0 + h$ can be the initial guess for the equation $f(x) = 0$. Especially, if $f(x_0)\,f(x_0+h) = 0$, x_0 or $x_0 + h$ is the solution of the equation $f(x) = 0$.
2. Refine the initial solution, until finally finding the approximate satisfying precision. The refining methods include bisection, simple iteration, Newton, and decent methods. The iteration method is a common method to solve an equation or equations, which uses a formulation again and again to make the approximation more and more precise.

Suppose the equation $f(x) = 0$ in the interval $[a,b]$ has only one root ξ, also such that $f(a)\,f(b) \leq 0$. The equation can be changed to an equivalent form $x = g(x)$ of $f(x) = 0$.

If $|g'(x)| \leq q < 1$ is satisfied during the whole interval of $[a,b]$, any value x_0 in $[a,b]$ can be approximated as the initial root, ξ, of the equation, the iterative step is $x_k = g(x_{k-1})$.

If the iteration converges, then

$$\lim_{n\to\infty} x_n = \xi \tag{8.33}$$

If x_n is such that $f(x) = 0$ and root ξ is the approximate root, then the convergence error estimate can be given as:

$$|\xi - x_n| \leq \frac{q}{1-q}|x_n - x_{n-1}| \tag{8.34}$$

The error estimate indicates that as long as the estimation $|x_n - x_{n-1}|$ is very small, $|\xi - x_n|$ guarantees the error ε so that the $|x_n - x_{n-1}| < \varepsilon$.

However, there are two constraint equations in Eq. 8.46, and two unknowns can be solved. For example, when the unknowns are ΔL_p and α_q, according to the constraints equations, the equivalent forms of the TVD equation and horizontal equation $\Delta L_p = g(\Delta L_p)$ and $\alpha_q = h(\alpha_q)$ need to be derived, then use the following steps for the iteration:

1. Assume an estimated initial value of ΔL_p, or $\Delta L_{p,i} = \Delta L_{p,0}$, when $i = 0$.
2. Assume an estimated initial value of α_q, or $\alpha_{q,j} = \alpha_{q,0}$, when $j = 0$.
3. According to the equation $\alpha_{q,j} = h(\alpha_{q,j-1})$, calculate $\alpha_{q,j}$.
4. For the given precision ε_α of the inclination angle, if $|\alpha_{q,j} - \alpha_{q,j-1}| < \varepsilon_\alpha$ is satisfied, go to Step 5; otherwise, $j = j + 1$, and go back to Step 3 to recalculate.
5. According to the equation $\Delta L_{p,i} = g(\Delta L_{p,i-1})$, calculate $\Delta L_{p,i}$.
6. For the given precision $\varepsilon\Delta_L$ of the section length, if $|\Delta L_{p,i} - \Delta L_{p,i-1}| < \varepsilon\Delta_L$ is satisfied, the iteration ends; otherwise, $i = i + 1$, and go back to Step 2 until the accuracy is satisfied.

By using the numerical method to design the well path, it is necessary to derive an iterative formula from constraints according to the unknowns, and to guarantee the convergence of the iterative formula. For a same unknown, sometimes there may be several iterative formulas, but their convergent speeds may be different and the fastest formula should be selected. The greatest difficulty of the numerical method is the initial guess. There are trigonometric functions in the constraints, so the solutions are not unique, and the solution interval has to be estimated correctly.

Analytical Design Method

A well profile of the 2D constraint equations contains only two equations, which is the lowest order of the equation for which a direct method is very effective. The so-called direct method is to find a solution directly, solving the unknown parameters according to the constraint equations for the calculation of the formula. A general profile of the 2D profile comprised of straight and circular arcs includes a characteristic parameter length of the hold or tangent section ΔL, inclination angle α, and radius of curvature of the arc R. These three characteristic parameters constitute six kinds of solution combinations, namely:

- ΔL-ΔL solution combination
- R-R solution combination
- α-α solution combination
- ΔL-R solution combination
- R-α solution combination
- α-ΔL solution combination

For a dual-arc profile (S-type and double-build profiles), the traditional design method solves the third section for the tangent angle, and this belongs to the α-ΔL solution combination group.

Undoubtedly, constraint equations can be derived for all the categories of the solution groups. For example, the solution for the tangent length with the subscript p with the inclined angle and radius of curvature with the subscript q in the solution group ΔL-R can be written as:

$$\Delta L_p = \frac{H_0\left(\cos\alpha_{q-1} - \cos\alpha_{q+1}\right) - A_0\left(\sin\alpha_{q+1} - \sin\alpha_{q-1}\right)}{\cos\left(\alpha_{q-1} - \alpha_p\right) - \cos\left(\alpha_{q+1} - \alpha_p\right)} \tag{8.35}$$

$$R_q = \frac{A_0\cos\alpha_p - H_0\sin\alpha_p}{\cos\left(\alpha_{q-1} - \alpha_p\right) - \cos\left(\alpha_{q+1} - \alpha_p\right)} \tag{8.36}$$

where

$$H_o = H_t - \sum_{\substack{i=1 \\ i\neq\frac{p+1}{2}}}^{\frac{n+1}{2}} \Delta L_{2i-1}\cos\alpha_{2i-1} - \sum_{\substack{j=1 \\ j\neq\frac{q}{2}}}^{\frac{n-1}{2}} R_{2j}\left(\sin\alpha_{2j+1} - \sin\alpha_{2j-1}\right)$$

$$A_o = S_t - \sum_{\substack{i=1 \\ i\neq\frac{p+1}{2}}}^{\frac{n+1}{2}} \Delta L_{2i-1}\sin\alpha_{2i-1} - \sum_{\substack{j=1 \\ j\neq\frac{q}{2}}}^{\frac{n-1}{2}} R_{2j}\left(\cos\alpha_{2j-1} - \cos\alpha_{2j+1}\right)$$

In a standard well profile with n sections, if the number of characteristic parameters is denoted as m and the number of solution combinations as k, then

$$m = \frac{3n+1}{2} \tag{8.37}$$

$$k = \frac{m(m-1)}{2} = \frac{1}{8}\left(9n^2 - 1\right) \tag{8.38}$$

Because $3n$ is an odd number, m is an integer; there must be an even number of (m) and (m – 1), so k is also an integer. For example, a one-arc well path (straight vertical/inclined-build-hold) has three sections, or $n = 3$, so there are five characteristic parameters (m), and ten possible solution combinations (k); a two-arc well path (straight inclined-build-hold-build/drop-hold) has five sections, or $n = 5$, so $m = 8$ and $k = 28$; in a three-arc well path (straight inclined-build-hold-build/drop-build/drop-hold), $n = 7$, $m = 11$, and $k = 55$.

Several simplified special trajectories can be obtained by decreasing the number of hold sections. If one hold section is removed, the unknown length of this hold section doesn't exist and so the number of the characteristic parameters becomes one less. If (s) hold sections are removed, the number of the characteristic parameters becomes (s) less. Therefore, after simplification, the number of the well path characteristic parameters (m') and the number of possible solutions (k') can be related with those in the original well path as:

$$m' = m - s = \frac{3n+1}{2} - s \tag{8.39}$$

$$k' = k - \frac{s}{2}(3n - s) \tag{8.40}$$

Since ($3n$) is an odd number, there must be one even number between (s) and ($3n - s$) and so (k') is an integer.

For example, the ladder-shape horizontal well path is a two-build (drop) well path, or hold-build (drop)-hold-build (drop)-hold, a special form that is one hold section less, or $s = 1$. In a two-build (drop) well path, $n = 5$, $m = 8$, and $k = 28$; so there are $m' = 7$ characteristic parameters in the ladder-shape horizontal well path and $k' = 21$ possible number of the solutions; build (drop)-hold-build (drop) is a special form that is two hold sections less than a normal two-build (drop) well path, so $m' = 6$, and $k' = 15$. The three-build (drop) well path, hold-build (drop)-build (drop)-build (drop) is a special form of normal three-build (drop) well path that is two hold sections less, so $m' = 9$ and $k' = 36$. And so on.

The well profile and its main characteristics parameters for the number of solution combinations are shown in Table 8.1.

In brief, by using build or drop sections effectively, the various directional wells and horizontal wells can be equivalently considered a unified form of a general build/drop well path. This type of well path is composed of hold sections and build/drop sections. Because the characteristic parameters of hold and build/drop sections are explicit, according to the continuity and first derivative continuity of the well path, the total number of the well characteristic parameters can be obtained. Based on the above-proposed theory and method of well path construction two characteristic parameters may be arbitrarily selected as the design parameters and such parameters may be optimized and used for a variety of well paths composed by hold and build/drop sections. This new well path design method can make the drilling engineering design more standard with wide universal applicability.

Table 8.1 Major Well Profile and Its Solution Portfolio

Profile Type	Profile Shape	Section Number	Simplified	Parameters	Solution Combination
Hold Profile	Line	1	/	2	1
Single Arc Profile	Arc	1	2	3	3
	Line Arc	2	1	4	6
	Arc Line	2	1	4	6
	Line Arc Line	3	0	5	10
Consecutive Double Arc Profile	Arc Arc	2	3	5	10
	Line Arc Arc	3	2	6	15
	Arc Arc Line	3	2	6	15
	Line Arc Arc Line	4	1	7	21
Double Arc Profile	Arc Line Arc	3	2	6	15
	Line Arc Line Arc	4	1	7	21
	Arc Line Arc Line	4	1	7	21
	Line Arc Line Arc Line	5	0	8	28
Consecutive Three Section Arc Profile	Arc Arc Arc	3	4	7	21
	Line Arc Arc Arc	4	3	8	28
	Arc Arc Arc Line	4	3	8	28
	Line Arc Arc Arc Line	5	2	9	36
Three Section Arc Profile	Arc Line Arc Line Arc	5	2	9	36
	Line Arc Line Arc Line Arc	6	1	10	45
	Arc Line Arc Line Arc Line	6	1	10	45
	Line Arc Line Arc Line Arc Line	7	0	11	55

PROBLEM 8.4

At the target of a direction well, with TVD = 2500 m and horizontal displacement = 900 m, when using a vertical-build-hold-drop-hold well trajectory profile, there are 28 solutions in combination of possibility, shown in Table 8.2.

Solution

This example shows that two arbitrary parameters can be chosen as unknown parameters to design a well path, and it also lists the results of all possible solutions. For example, the tenth solution shows that when ΔL_1= 500 m, κ_2 = 6°/100 m, ΔL_3 = 800 m, κ_4 = –4°/100 m, ΔL_5 = 300 m, and α_5 = 12°, two unknowns can be solved, α_1 = 0.18° and α_3 = 38.34°. Obviously, α_1= 0.18° is not a reasonable solution, so the known parameters should be adjusted to calculate again. Of course, sometimes using another solution can be more convenient. However, this does not depict that there is any inherent problem of choosing α_1 and α_3 as unknowns, but the problem is that this combination is not practical under this condition. In fact, even for the traditional unknown combination ΔL_3 and α_3, if the known parameters are not reasonable, the design can also be unreasonable, or no solution at all.

Overall, every combination can lead to a solution, but only some of them are practically reasonable during the field well path design. Moreover, the sensitivity of parameters in each solution is different. Thus, after optimizing parameters, the accuracy of the design results should be checked.

Smoothening the Path

Usually the trajectories, designed with constant curvature, are well-defined arcs connecting the transition between the tangent sections. Even though the transition between the tangent section and build section or tangent section and drop section appears to be smooth, there will be discontinuity and a lot of stresses in the tubulars will be caused. Discontinuity is obvious when two circular arcs, or one tangent section and circular arc, or circular-arc and tangent sections are joined together. An option is to use different build curves based on continuous build; even with these designs, there exists a discontinuity in the transition zones. To avoid the discontinuity and curvature, bridge curves or transition curves can be used. Transition curves that are used are defined as the curve segments connecting the tangent section of the well paths to the build or drop sections. Several transition curves can be used.

Table 8.2 Interactive Design Method of Analysis for Problem 8.4

Index Number	1st Section	2nd Section	3rd Section / 4th Section	5th Section	1st Section	2nd Section	3rd Section	4th Section
	ΔL_1 m	α_1 (°)	κ_2 (°)/	ΔL_3 m	α_3 (°)	κ_4 (°)/	ΔL_5 m	α_5 (°)
1	**504.59**	**0.03**	6.00	780.00	34.70	-4.00	300.00	12.00
2	**528.48**	0.00	**6.38**	780.00	35.00	-4.00	300.00	12.00
3	**510.69**	0.00	6.00	**762.21**	35.00	-4.00	300.00	12.00
4	**503.82**	0.00	6.00	780.00	**34.71**	-4.00	300.00	12.00
5	**519.59**	0.00	6.00	780.00	35.00	**-4.19**	300.00	12.00
6	**544.13**	0.00	6.00	780.00	35.00	-4.00	**250.92**	12.00
7	**539.53**	0.00	6.00	780.00	35.00	-4.00	250.00	**11.78**
8	500.00	**0.04**	**5.90**	775.00	34.70	-4.00	300.00	12.00
9	500.00	**0.04**	6.00	**792.19**	34.50	-4.00	300.00	12.00
10	500.00	**0.18**	6.00	800.00	**34.34**	-4.00	300.00	12.00
11	500.00	**0.24**	6.00	800.00	34.30	**-3.98**	300.00	12.00
12	500.00	**0.27**	6.00	800.00	34.30	-4.00	**302.78**	12.00
13	500.00	**0.34**	6.00	800.00	34.30	-4.00	300.00	**11.85**
14	500.00	0.00	**6.03**	**794.04**	34.50	-4.00	300.00	12.00
15	500.00	0.00	**6.06**	800.00	**34.43**	-4.00	300.00	12.00
16	500.00	0.00	**5.91**	800.00	34.50	**-4.13**	300.00	12.00
17	500.00	0.00	**6.30**	800.00	34.50	-4.00	**319.32**	12.00
18	500.00	0.00	**6.15**	800.00	34.50	-4.00	300.00	**11.77**
19	500.00	0.00	6.00	**789.75**	**34.55**	-4.00	300.00	12.00
20	500.00	0.00	6.00	**795.28**	34.50	**-4.03**	300.00	12.00
21	500.00	0.00	6.00	**793.48**	34.50	-4.00	**298.20**	12.00
22	500.00	0.00	6.00	**792.83**	34.50	-4.00	300.00	**12.05**
23	500.00	0.00	6.00	800.00	**34.46**	**-4.05**	300.00	12.00
24	500.00	0.00	6.00	800.00	**34.41**	-4.00	**295.05**	12.00
25	500.00	0.00	6.00	800.00	**34.39**	-4.00	300.00	**12.16**
26	500.00	0.00	6.00	800.00	34.50	**-4.09**	**304.72**	12.00
27	500.00	0.00	6.00	800.00	34.50	**-4.09**	300.00	**11.91**
28	500.00	0.00	6.00	800.00	34.50	-4.00	**279.70**	**11.50**

Note: Shading indicates solved unknown parameters.

In the past, several mathematicians and physicists have studied the properties of several curves. Out of which, the Cornu's spiral or Euler spiral (also known as linarc) are of great interest to the authors, due to the very nature of the special properties of the curve. In fact, Euler described several properties of the curve, including the curve's quadrature, which is also widely called the Fresnel spiral. Insertion of clothoid sections between the tangent and build or drop section of the well path will result in curvature continuity. This curvature bridge will alleviate the drag problems that will enable the design engineers to extend the reach with the given mechanical limitations. Accuracy of the clothoid spiral matches the cubic spiral to more than eight digits.

Clothoid Curve

Clothoid curves are spiral curves, the curvatures of which change linearly from zero to a desired curvature with respect to the arc length. In other words, the radius of curvature at any point of the curve varies as the inverse of the arc length from the starting point of the curve.

$$R \propto \frac{1}{L} \text{ or } L_1 \times R_1 = L_2 \times R_2 = \ldots = L_n \times R_n = \sigma \tag{8.41}$$

$$\kappa(s) = \kappa(0) + \sigma s \tag{8.42}$$

where

R = Radius of curvature
L = Length of the curve
κ = Curvature
σ = Sharpness of the curve
s = Arc length of the curve

The clothoid curves can be parametrically given as:

$$\begin{aligned} f(\ell) &= (C_f(\ell), S_f(\ell)) \\ C_f(\ell) &= \xi \int_0^{\ell} \cos\left(\frac{\pi u^2}{2}\right) du \\ S_f(\ell) &= \xi \int_0^{\ell} \sin\left(\frac{\pi u^2}{2}\right) du \end{aligned} \tag{8.43}$$

where ξ is the characteristic parameter.

The following are called the Fresnel sine and cosine integrals:

$$\begin{cases} \text{Fresnel } C_f(\ell) = \int_0^{\ell} \cos\left(\frac{\pi u^2}{2}\right) du \\ \text{Fresnel } S_f(\ell) == \int_0^{\ell} \sin\left(\frac{\pi u^2}{2}\right) du \end{cases} \tag{8.44}$$

Since it is not possible to obtain a closed-form solution to the above equations, several approximate numerical computations are presented in the literature using Taylor, power series, and Maclaurin expansions. A simple expression using Maclaurin expansions and the coordinates can be expressed in terms of the length of the spiral arc as[14]:

$$\begin{aligned} y &= \frac{L^3}{6\xi^2} - \frac{L^7}{336\xi^6} + \frac{L^{11}}{42240\xi^{10}} + \ldots \\ x &= L - \frac{L^5}{40\xi^4} + \frac{L^9}{3456\xi^{10}} + \ldots \end{aligned} \tag{8.45}$$

Omitting higher-order terms,[14] it can be written as (which is of cubic parabola in nature):

$$y = \left(6\xi^2 x\right)^{\frac{1}{3}} \tag{8.46}$$

Based on the properties of the clothoid curve, the relationship between the curvature and the scale parameter can be given as:

$$L_1 \times R_1 = L_2 \times R_2 = \ldots = L_n \times R_n = \xi^2 \tag{8.47}$$

The clothoid spiral can be expressed using computationally efficient rational approximations,[15] within an error of 1.7×10^{-3}, which would be sufficient for well path designs:

$$\begin{cases} C_f(\ell) \approx \frac{1}{2} - R(\ell)\sin\left[\frac{\pi}{2}\left(A(\ell) - \ell^2\right)\right] \\ C_f(\ell) \approx \frac{1}{2} - R(\ell)\cos\left[\frac{\pi}{2}\left(A(\ell) - \ell^2\right)\right] \end{cases} \tag{8.48}$$

where

$$R(\ell) = \frac{0.506\ell + 1}{1.79\ell^2 + 2.054\ell + \sqrt{2}}$$

$$A(\ell) = \frac{1}{0.803\ell^3 + 1.886\ell^2 + 2.5424\ell + 2}$$

Also, it can be noted that the tangents at the connection point between the clothoid spiral and the straight segment are the same. It has been found that the Cornu's spiral path reduces the lateral stresses on the tubulars that pass through the section. Necessary conditions for the curvature bridges are that, at the start of the build section, the spiral should have the same curvature and end with the same curvature that fits to the tangent section, bridging the circular-arc section. In the same way, the spiral section should end with the curvature for that of the tangent section.

Figure 8.6 illustrates well paths with and without a clothoid spiral in a curved section. It can be seen that the well paths consist of the following sections with curvature bridging:

- Fresnel spiral arc from the kick-off depth
- Circular arc with maximum curvature κ_{max}
- Fresnel spiral arc including partial tangent section
- Partial tangent section
- Fresnel spiral arc including partial hold section
- Hold horizontal section

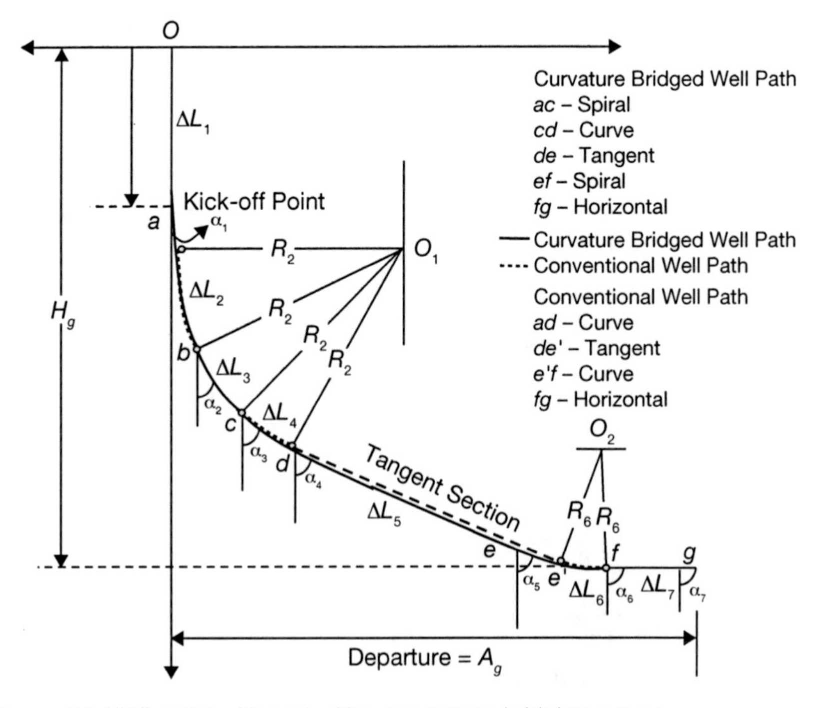

Figure 8.6 Well paths with and without curvature bridging curves.

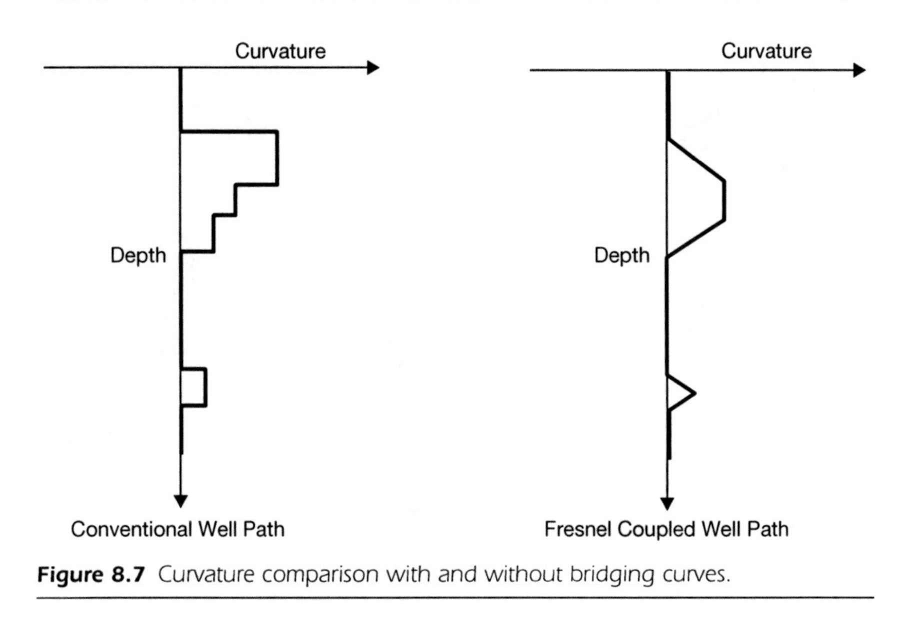

Figure 8.7 Curvature comparison with and without bridging curves.

Figure 8.7 is a simple diagram that illustrates the curvature of the wellbore with and without a clothoid spiral in the curved sections, as well as straight sections along the wellbore. It can be seen that the curvature change between the sections is smooth.

Curvature Bridging[4]

Curvature bridging is important for the torque-and-drag and fluid mechanics analysis, such as cuttings transport and swab and surge wellbore pressure calculations. Figure 8.8 illustrates the curvature bridging of a standard S-type well. The figure is embedded with the commonly used method, as well as the clothoid spiral arcs. The well path consists of the following sections with the clothoid arc lengths:

- Fresnel spiral arc from the kick-off depth—build section
- Fresnel spiral arc including drop and hold sections

The well path consists of the following sections with the circular-arc lengths:

- Straight vertical/inclined section up to kick-off depth
- Circular arc from the kick-off depth—build section
- Second circular arc—build section

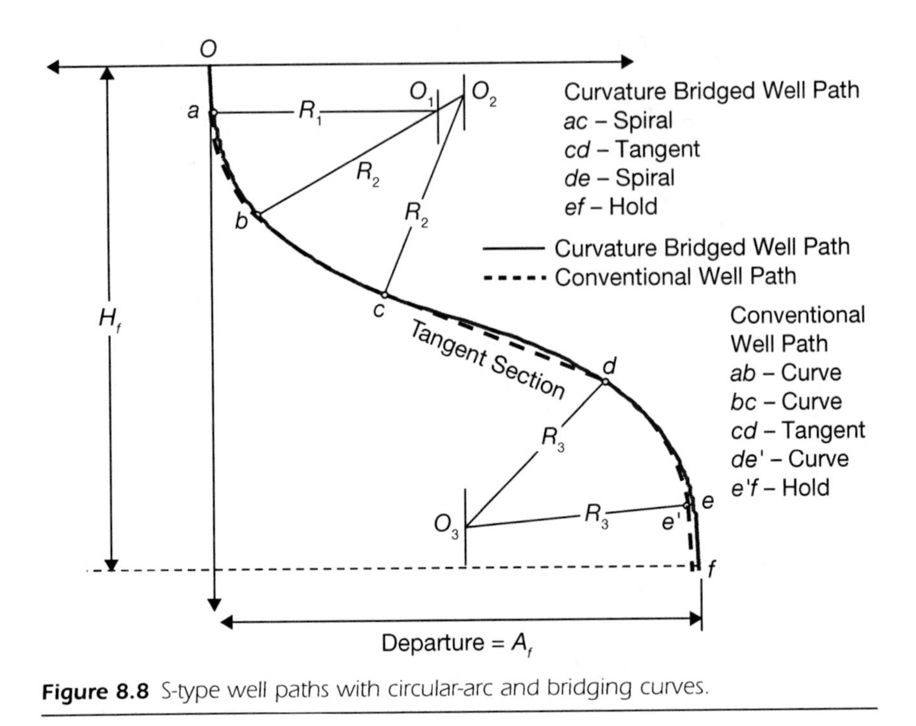

Figure 8.8 S-type well paths with circular-arc and bridging curves.

- Tangent section
- Circular arc—drop section
- Hold section

Figure 8.9 illustrates the curvatures for both the options. It can be seen that the curvature bridging is smooth with Fresnel spiral-arc wellbore paths. Figure 8.10 shows the flow chart for optimizing the well paths, using the clothoid curves and minimum-energy criterion.

Well Path Optimization with Cubic Function

Details of the well path using the cubic function are described in detail in Chapter 5.[16] The free parameters of the well path allow several optimization options. For example, changes in L_0 or L_1 affect the arc length and curvatures along the well path. It is evident that the arc length has a minimum value (the Euclidean distance between the initial and the final ends of the well path). However, reducing the arc length of the well path generally increases significantly the curvature close to the ends.

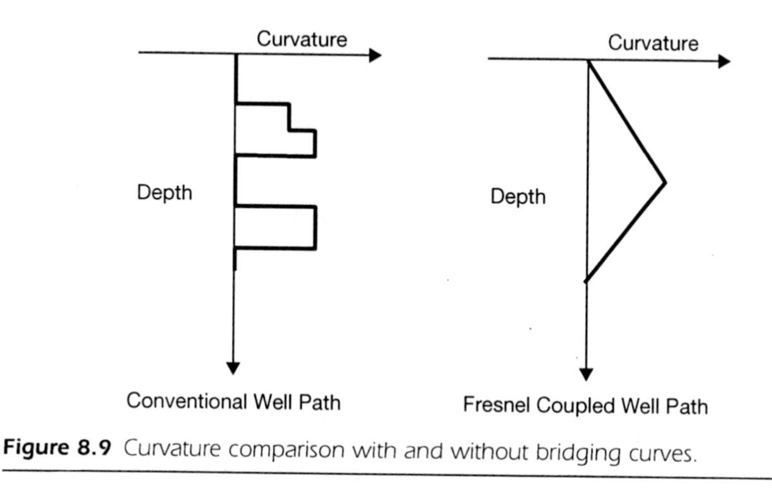

Figure 8.9 Curvature comparison with and without bridging curves.

It is desirable to avoid points with large curvatures, in order to reduce torque, drag, and wear of drilling and production equipment. One possible method of optimization is to minimize the standard deviation of the curvature along the well path. For trajectories with two or more degrees of freedom, a gradient descent scheme can be used. For trajectories with only one degree of freedom, a golden section or a parabolic interpolation scheme is appropriate.

PROBLEM 8.5

As an example, a curvature optimization can be performed on the data of the set inclination and azimuth case (Problem 5.3).

For simplicity, we consider $L_0 = L_1$, which, therefore, reduces the degree of freedom to one. By minimizing the standard deviation of the curvature, the value $L_0 = L_1 = 2412.19$ was found.

Solution

Table 8.3 presents the computation results. In the optimized well path, the maximum dogleg severity is 8.37°/30 m at the measured depth of 2029 m. This should be compared with the maximum dogleg severity of 5.01°/30 m at 2036 m for the non–optimized well path presented before.

If L_0 and L_1 are treated as two independent parameters, the optimized well path is obtained for $L_0 = 2045.90$ $L_1 = 2808.68$. In this case, the maximum dogleg severity is 3.98°/30 m at 1968 m.

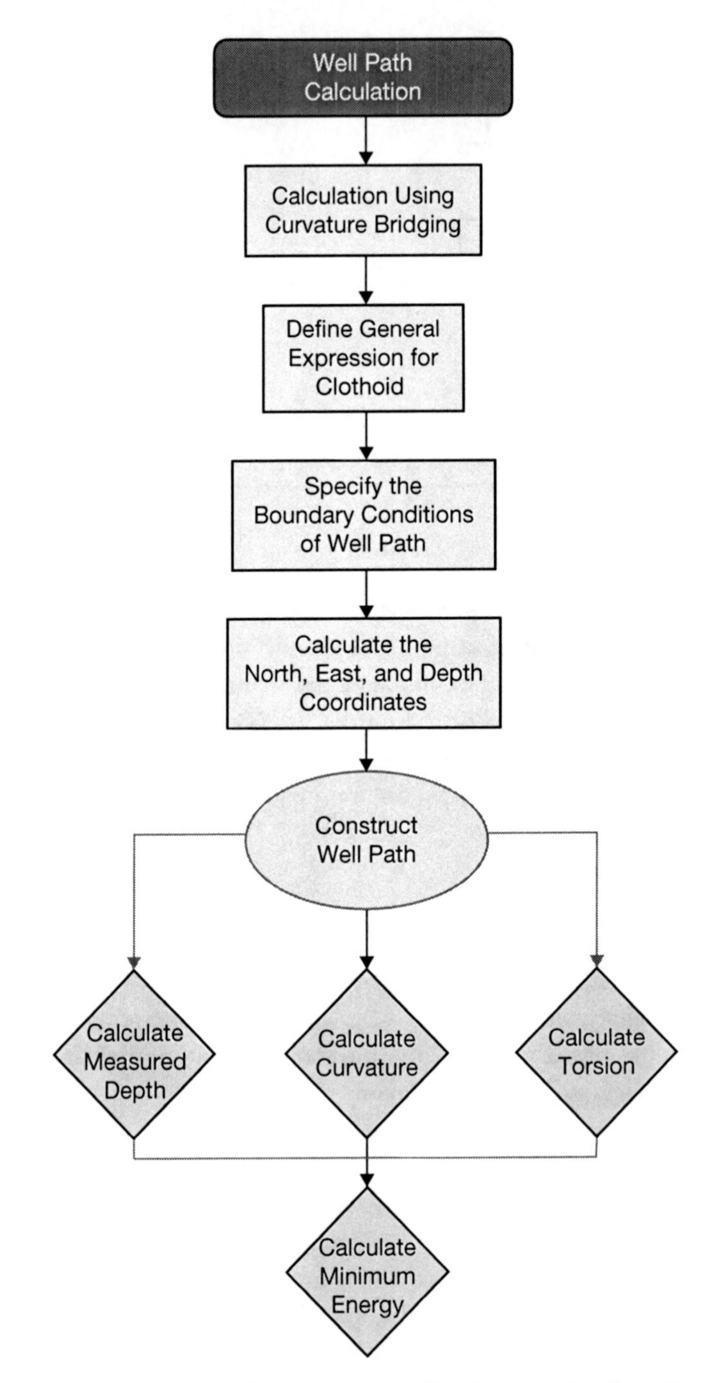

Figure 8.10 Clothoid and minimum-energy well path calculation flow chart.

Table 8.3 Results for Problem 8.5

u	V	N	E	s	MD	INCL	AZIM	DLS	TF	dir
0	500	5	18	0	508	30	10	1.86	112.94	0.8
0.08	661.45	85.67	50.03	183.63	691.63	27.71	34.57	2.01	88.41	0.88
0.16	811.31	141.77	109.02	354.44	862.44	30.34	57.02	1.86	64.46	0.78
0.24	949.25	179.45	189.57	518.75	1026.75	35.66	71.17	1.53	45.72	0.58
0.32	1074.94	204.91	286.3	679.48	1187.48	41.38	78.27	1.18	27.92	0.35
0.4	1188.02	224.31	393.82	836.78	1344.78	46.56	80.58	0.94	3.36	0.04
0.48	1288.16	243.83	506.74	989	1497	51.08	79.24	0.98	29.16	-0.31
0.56	1375.01	269.64	619.66	1133.84	1641.84	55.23	74.41	1.47	52.98	-0.46
0.64	1448.23	307.91	727.2	1269.58	1777.58	59.62	65.68	2.41	62.89	-0.45
0.72	1507.5	364.83	823.97	1396.78	1904.78	65.15	52.68	3.64	64.79	-0.38
0.8	1552.45	446.57	904.57	1520.51	2028.51	72.48	36.32	4.37	63.72	-0.27
0.88	1582.75	559.3	963.63	1651.87	2159.87	80.61	19.63	3.71	63.33	-0.15
0.96	1598.07	709.19	995.73	1806.39	2314.39	87.42	5.64	2.35	64.85	-0.04
1	1600	800	1000	1897.37	2405.37	90	0	1.77	66.13	0

Well Path Optimization with SiT Method

Details of the SiT method can be seen in Chapter 5.[16] The free parameters of the well path allow several optimizations. For example, changes in L_0 or L_1 affect the arc length and curvatures along the well path. It is evident that the arc length has a minimum value (the Euclidean distance between the initial and the final points of the well path). However, reducing the arc length of the well path generally significantly increases the curvature close to the ends.

PROBLEM 8.6

As an example, a curvature optimization can be performed using the data from Problem 5.9. The minimization is performed on the standard deviation of the curvatures along the well path calculated at the partition points. The independent variables are L_0, L_1. The tension was kept constant for the three coordinate functions ($\lambda = 2$).

The optimized well path for $\lambda = 2$ is obtained for $L_0 = 2111.27$ and $L_1 = 2956.23$. In this case, the maximum dogleg severity is 8.01°/30 m at 1990 m of measured depth.

Other goals can be used for optimization. For example, the minimization of the standard deviation of curvatures can be weighted with the inverse of the measured depth, to obtain lower curvatures in the upper portion of the well path at the expense of higher curvatures in the lower portion. The tension parameter can also be used to obtain the minimum length for a maximum DLS.

Bit-Walk Unit Partition Based on Rock Strata

The formation properties are the objective factors influencing the azimuth drift, so the selection of the drilling tools and techniques is usually based on the formation properties. If the formation structure (formation declination or direction), the formation lithology and the formation anisotropy are the main factors to influence the azimuth drift, and the azimuth drift unit should be divided based on the formations. For the thick formation with a large change in formation lithology and anisotropy, the formation can be divided into sublayers. According to the azimuth drift, the coupling relationship of the inclination angle with the azimuth angle can be seen in Figure 8.11. One formation may cover only part of one section (build section, hold section, or drop section), and so this section can be divided into two or more than two units; or it may cover the whole formation layer, so it is equal

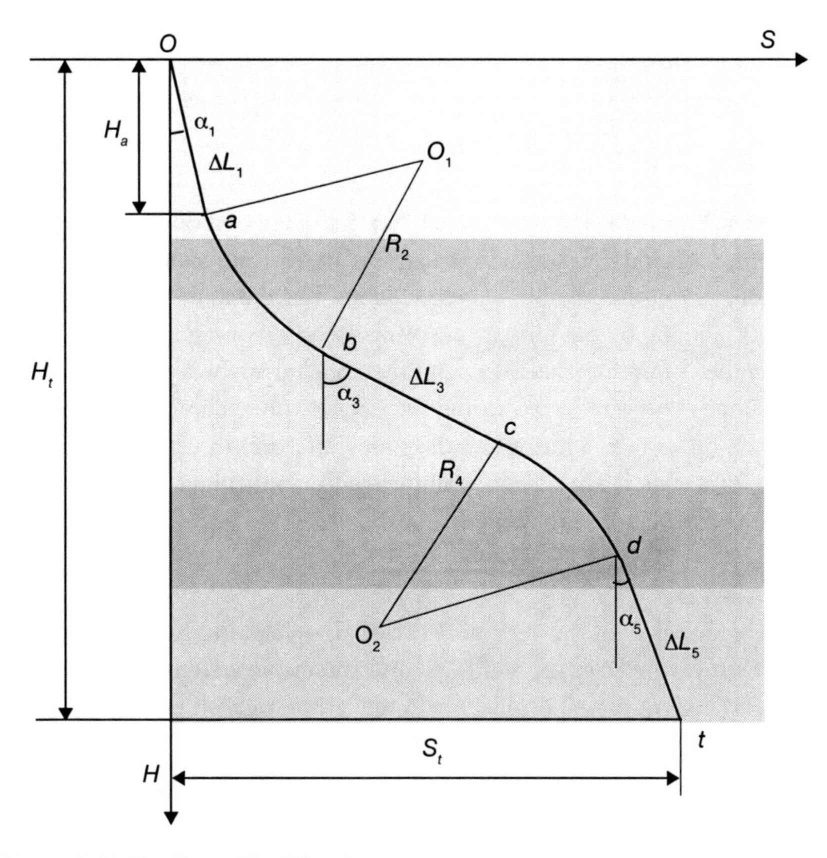

Figure 8.11 Partition with drift units.

to the well section unit; or it may cover more than one section and separates some section(s).

So when dividing the azimuth drift units by formations, the coupled relationship of inclination and azimuth should be considered (i.e., when dividing the well path into more detailed sections according to inclination and azimuth, the unit sections can be obtained). If keeping the build rate and the azimuth change rate constant in each section unit, the azimuth drift path can be described and calculated by the natural curve model of the well path.[12,17-19]

For the well path from the surface to the target, after coupling build units and azimuth units, if there are totally n calculation units, the sum of the coordinate increments in each unit should satisfy:

$$\left(\sum_{i=1}^{n} \Delta N_i\right)^2 + \left(\sum_{i=1}^{n} \Delta E_i\right)^2 = A_t^2 \tag{8.49}$$

$$\sum_{i=1}^{n} \Delta E_i = \tan\phi_t \sum_{i=1}^{n} \Delta N_i \tag{8.50}$$

$$\sum_{i=1}^{n} \Delta H_i = H_t \tag{8.51}$$

In fact, Eqs. 8.48 through 8.50 and Eq. 5.117 are equivalent, though they look different, and the constraint equations are basically same. However, the value of n is not the number of well sections, but the number of calculation units. If dividing by the formation layers, the value of n is often larger than the number of the well sections, and the coupled analysis of the build unit and azimuth unit is required during the iterative procedure, then calculation units can be known. Thus, when designing an azimuth drift path, dividing by the formation layers is more complicated than dividing by well sections, and more calculations are required.

PROBLEM 8.7

In a directional well, the vertical depth is H_t = 2500 m, the horizontal displacement of the target A_t = 600 m, and the displacement azimuth is ϕ_t = 120°. If the two-arc well profile is selected, at the kick-off point, the vertical depth is 1200 m, the build rate is 10°/100 m, the drop rate is 6°/100 m, the inclination of hold section at the oil zone is 8°, and the increment of the vertical depth is 200 m. If dividing by formation layers, when the vertical depth is 1200~1300 m, the azimuth drift rate is 2°/100 m; when the vertical depth is 1400~1500 m, the azimuth drift rate is 1.5°/100 m; when the vertical depth is 1500~1600 m, the azimuth drift rate is 2.5°/100 m; when the vertical depth is 1600~1800 m, the azimuth drift rate is 2°/100 m; when the vertical depth is 1800~2200 m, the azimuth drift rate is 3°/100 m; when the vertical depth is 2200~2500 m, the azimuth drift rate is 2°/100 m; otherwise, there is no azimuth drift. Design the directional well azimuth drift path.

Except the azimuth drift, this example is the same as Example 5.18. According to the constraints and design conditions, obtain that the azimuth is ϕ_0 = 111.50°, the inclination angle of hold section is α_3 = 42.73°, and the section length is ΔL_3 = 267.00 m. The ending point of each section of the well path can be seen in Table 8.4.

When planning a horizontal well, normally the horizontal well section is designed first and then the path from the surface to the first target is designed. Whether the design is the horizontal hold section, the ladder-shaped horizontal section, or the more complicated horizontal well section, the above method can be used to design 3D azimuth drift horizontal wells. However,

Table 8.4 Design Results of Azimuth Drift Path for Problem 8.7

Depth (m)	Inclination (°)	Azimuth (°)	N (m)	E (m)	TVD (m)	Displacement (m)	Node
0.00	0.00	—	0.00	0.00	0.00	0.00	Wellhead
1200.00	0.00	(111.50)	0.00	0.00	1200.00	0.00	*a*
1300.51	10.05	109.49	–3.03	8.26	1300.00	8.79	
1404.30	20.43	109.49	–12.12	33.94	1400.00	36.04	
1515.74	31.57	111.16	–29.11	79.66	1500.00	84.81	
1627.32	42.73	113.95	–54.98	141.77	1588.79	152.05	*b*
1642.58	42.73	114.33	–59.21	151.21	1600.00	162.39	
1894.32	42.73	119.37	–136.34	303.57	1784.91	332.78	*c*
1914.66	41.51	119.77	–143.08	315.44	1800.00	346.37	
2371.25	14.12	133.47	–265.60	486.68	2200.00	554.43	
2473.19	8.00	135.51	–279.27	500.64	2300.00	573.27	*d*
2675.15	8.00	139.55	–300.00	519.62	2500.00	600.00	Target *t*

when designing the well path before the horizontal section, because the well direction has been fixed at the time of designing the horizontal section, it is to be noted that another constraint of the azimuth requirement has to be satisfied. Thus, three constraints are required to design the azimuth drift path for a directional well, whereas four constraints are required for a horizontal well.

Supplementary Problems

1. Evaluate the length of a horizontal well with the following reservoir parameters:
   Reservoir thickness = 100 ft
   Wellbore radius = 5.5"
   Horizontal permeability = 100 md
   Vertical permeability = 50 md
   Crude oil viscosity = 0.6 cP
   Formation volume factor = 12.5 RB/STB
   Ratio of horizontal to vertical productivity index = 5
   The vertical well spacing is 50 acres.

References

1. Curtis, L. J., I. Martinson, and R. Buchta. "Lifetimes of Excited Levels in PI-PV." *Fhysica Scripta* 3(1971): 197–202.
2. Juliano, T., V. Domnich, T. Buchheit, and Y. Gogotsi. "Numerical Derivative Analysis of Load-Displacement Curves in Depth-Sensing Indentation." Mat. Res. Soc. Symp. Proc. Vol. 791 © 2004 Materials Research Society.
3. Bourdet, D., J. A. Ayoub, and Y. M. Pirard. "Use of Pressure Derivative in Well-Test Interpretation." SPE Formation Evaluation, June 1989: 293–302.
4. Samuel, R. "Ultra Extended Reach Drilling (*u*-ERD): Tunnel in the Earth—A New Wellpath Design." SPE-119459, SPE/IADC Drilling Conference, March 17–19, 2009, Amsterdam, The Netherlands.
5. Joshi, S. D. *Horizontal Well Technology*. Tulsa, OK: Pennwell Books.
6. Ehlig-Economides, C. A., G. R., Mowat, and C. Corbett. "Techniques for Multibranch Well Trajectory Design in the Context of a Three-Dimensional Reservoir Model." SPE 35505, European 3-D Reservoir Modeling Conference, April 16-17, 1996.
7. Ellis, C. A., and R. Samuel. "Drilling Short Radius Dual Lateral Wells in Oklahoma: An Operator's Experience." SPE 37493, Production Operation Symposium, Oklahoma City.
8. Xiushan, L. *Geometry of Wellbore Trajectory*. Beijing, China: Petroleum Industry Press, 2006.
9. Xiushan, L., and S. Zaihong. "Practical Method for Design of Horizontal Well Trajectory." *Oil Drilling & Production Technology* 16, no. 1 (1994): 5–8, 23.
10. Xiushan, L. "Study on Arched Horizontal Well Design." *Natural Gas Industry* 26, no. 6 (2006): 68.65.
11. Wiggins, M. L., and H. C. Juvkam-Wold. "Simplified Equations for Planning Directional and Horizontal Wells." SPE Paper No. 21261, SPE Eastern Regional Meeting, Columbus, Ohio, October 31–November 2, 1990.
12. Xiushan, L., L. Rushan, and S. Mingxin. "New Techniques Improve Well Planning and Survey Calculation for Rotary-Steerable Drilling." IADC/SPE Paper No. 87976, IADC/SPE Asia Pacific Drilling Technology Conference and Exhibition, Kuala Lumpur, Malaysia, September 13–15, 2008.
13. Chi, A., L. Xiushan, and W. Jun. "Universal Equations and Their Application to Wellbore Trajectory Design." Daqing, China. *Journal of Daqing Petroleum Institute* 22, no. 4 (1998): 23–26.
14. Brandse, J., M. Mulder, and M. M. van Paassen. "Clothoid-Augmented Trajectories for Perspective Flight-Path Displays." *International Journal of Aviation Psychology* Volume 17. Available at http://www.informaworld.com/smpp/35862053-84996811/title~content=t775653651~db=all~tab=issueslist~branches=17-v17, Issue 1, January 2007, 1–29.
15. Heald, M. A. "Rational Approximations for the Fresnel Integrals." *Mathematics of Computation* 44, 170 (1985), 459–461.
16. Contribution by Jorge Sampiao.

17. Xiushan, L., Q. Tongci, S. Zhongguo, et al. "Design of a 3-D Drift Well Path." Beijing, China. *Acta Petrolei Sinica* 16, no. 4 (1995): 118–124.
18. Xiushan, L., and S. Zaihong. "Technique Yields Exact Solution for Planning Bit-Walk Paths." *Oil & Gas Journal* 100, no. 5 (2002): 45–50.
19. Xiushan, L., P. Guosheng, and Z. Xiaoxiang. "An Advanced Method for Planning and Surveying Well Trajectories of Rotary Steering Drilling." *Acta Petrolei Sinica* 24, no. 4 (2003): 81–85.

9

Measurement Tools

In the drilling process, in order to effectively monitor and control well trajectory, it is necessary to measure the inclination and direction of the wellbore in a timely manner. The determination of the real, or approximately real, trajectory of a wellbore requires surveying the wellbore with special probes to determine the inclination, azimuth, and possibly the toolface orientation at several depths along the trajectory (survey stations), where the measured depth is known. With the nature of the complex design and well construction methods, the process is supported with several downhole tools with advanced technologies. In the last decade, rapid advancements have been made in this area. The tools are capable of providing their position at any time with the real time data, so that corrective measures may be taken then and there. This chapter covers a few advanced downhole tools used in the well construction process. They can provide accurate data with high data transfer rates. Also, this chapter provides a brief description of expandable technology, and how its application in the oil industry enables the drilling of wells, increases the production of wells being drilled, and lowers the cost of well construction by better addressing drilling hazards and reducing nonproductive time (NPT).

Survey Tools

Surveying[1] a wellbore is accomplished by running tools along the hole, and registering or transmitting to the surface the relative position of the tool, with respect to a frame of reference, which may include the Earth's gravitational field, the Earth's magnetic field or other inertial reference, or even measuring the distortion or bending of the housing of the probe caused by the curvature of the borehole. Each method has its own advantages, disadvantages, applicability, and limitations.

The required survey station density (the average distance between the survey stations) depends on several factors, like the cost of the service, the daily rate of the rig (some equipments require stopping the drilling operation for a period), and the expected accuracy of the calculated trajectory. The accuracy of the tool used to survey evidently affects the accuracy of the calculated trajectory; however, the use of inaccurate probes cannot be compensated with higher survey density. Coordinates, inclination, and azimuth of points of the trajectory between the survey stations must be interpolated. Therefore, the greater the survey station density, the more accurate the interpolated values will be. Recently, technology of surveying the wellbore has been continuously imitated to map the wellbore trajectory with greater accuracy.

Classification of Survey Tools[2]

Permanent records of hole locations can have significant impact on future drilling and anti-collision calculations. The available measuring instruments to determine the wellbore course (inclination and azimuth) may be categorized into two groups:

- Magnetic survey instruments
- Gyroscopic survey instruments

Both groups have single-shot and multishot survey capability. They can be dropped, lowered on a wireline, or be part of measuring-while-drilling (MWD) packages. The basic components of both systems consist of a time device and compass, or an electronic sensor device. The single-shot instrument records only one point at a given well depth during the measuring process, while the multishot instruments record several points along the well depth.

Magnetic Survey Tools

Magnetic survey tools, whether electronic or mechanical, utilize the magnetic field to determine the hole path. The Earth's magnetic field is described by the following components, which vary considerably with the geographic location and must be known for the area:

- Magnetic north
- Vertical and horizontal components of the local magnetic field

- Total field strength of local magnetic field
- Variation between true north and magnetic north
- Dip angle of local magnetic field in reference to the measured horizontal field

The mechanical survey instrument is based on the compass principle and therefore uses only the Earth's local horizontal magnetic component to reference magnetic north.

Electronic magnetic survey instruments use magnetometers to measure the Earth's magnetic field and accelerometers to measure the Earth's gravitational field. These electronic survey tools can be used to measure hole inclination and azimuth and toolface. They are used in MWD tool packages. The output data are transmitted to the surface through what are called pressure mud pulse telemetry systems, and then decoded to actual data.

Magnetic Single-Shot Instruments

The purpose of the magnetic single-shot instrument is to perform one measurement at a given depth. Single-shot instruments are often used by the directional engineer to track the bit's progress while drilling is under way. Normally, a measurement is made at every 9 or 12 joints corresponding to approximately 270 ft to 360 ft of depth advance. The reason for this average density is the time needed to perform the measurement, which increases with depth. Due to the magnetic principle to determine the azimuth, the magnetic single-shot instrument must be positioned inside nonmagnetic drill collars.

The tool consists of a compass and clockwork mechanism, mounted on gimbals with the bottom side weighted to maintain the gravitational vertical reference, as it moves freely in its housing. It is lowered in the borehole to the measurement depth, and at a preset time, the clock locks all the moving parts, and the device is retrieved from the borehole to read the inclination and magnetic azimuth value at that depth. The inclination is measured to within one degree, and the azimuth is read to the nearest half degree. It can be run in an open hole on a wireline, or inside the drill string. Because of the magnetic compass, the instrument must be positioned in front of a nonmagnetic drill collar to avoid interference with the Earth's magnetic field.

A more precise equipment uses photograph shots to record the inclination and azimuth. The instrument consists of three basic units: a timing device or a motion sensor unit, a camera section, and an angle-indicating unit (inclination and azimuth). The timing device is used to operate a camera at a predetermined time, but since it is quite difficult to accurately predict the time element involved in lowering the instrument, a "motion sensor" was developed to eliminate needless downtime. It is an electronic device that operates an electrical light system in the camera, just seconds after motion has ceased. A third type senses when the probe has reached the nonmagnetic section of the drill string and starts the measurement process. The camera itself is prefocused and loaded with a film disc made of a special heat-resistant material. The angle-indicating unit combines a magnetic compass and a plumb bob, which gives the angle of inclination from vertical.

In actual operation, the timing device is first set to a predetermined time and the instrument is assembled in a protective barrel. The probe is run on a wire line or dropped into the well on a dull bit. When it reaches the bottom, the timing device operates the camera and a photographic record is made of the relative positions of the compass and plumb bob. The probe is then returned to the surface, where the film is developed and placed in a magnifying reader that shows the exact direction and inclination of the borehole.

In high-inclined or horizontal wells, the probe is pumped toward the bit at a slow flow rate. An increase in the surface pressure indicates that the probe has reached the bottom (the probe lands on a baffle plate located at the nonmagnetic drill collar). There are several different types of angle-indicating units. They differ between manufacturers, and also in the range of use. Some magnetic single-shot instruments can also register the toolface of a deflection tool. The angle unit is basically the same as in the regular magnetic single-shot instruments, and, in addition to the inclination and azimuth, a hand indicates the toolface side and the angle of the tool.

Magnetic Multishot Instruments

The purpose of the magnetic multishot instrument is to perform a series of measurements along a length of the open hole. The multishot instrument is normally run at the end of the drilling of a phase of the well. Multishot instruments are used by the directional engineer to obtain a more accurate trajectory by increasing the density of the survey stations. It is usual to perform the calculations solely using the readings from the multishot instrument, and compare the trajectory thus obtained with that obtained with the single-shot instruments. Since the trajectory calculated with the multishot instrument is considered more accurate, it is taken as the final borehole trajectory. After a multishot instrument is run and the trajectory calculated, the very last station is used as the starting point (a new tie-in) for the calculation using single-shot instruments for the next phase.

The multishot instrument uses the same principles of the single-shot instrument to sense the inclination and azimuth (the multishot does not register the toolface). The difference is that the multishot instrument records multiple survey stations in a film strip as the drill string is pulled out of the hole. The instrument may be equipped with a timing device, synchronized with a surface watch. Normally, the timer is set to take a shot every 20 or 25 seconds (the time needed for the angle sensors to stabilize when the drill string is motionless). In a trip operation, normally it takes 60 to 90 seconds to pull out one stand (for a three-joints stand, which corresponds to approximately 90 ft), and 2 to 3 minutes to disconnect, rack the stand, and pick up the next stand. During this cycle, 3 to 5 shots will be blurry (because of the movement of the drill string), and another 3 to 6 will be steady. The directional engineer or technician records the time of each stand cycle (including the depth) to compare with the series of shots. It is usual during the operation to extend time in slips for 1 or 2 minutes (before picking up the next stand) every ten stands, to serve as a time stamp for a perfect synchronization of time with depth. The film used to register the readings is sensitive to high temperatures. Therefore, in very deep wells, or in high thermal gradient regions, a heat shield is used. A heat shield can provide heat protection up to 500° F for four hours. Both single-shot and multishot instruments are conventional equipment (mechanical inclination units, magnetic compasses, and photographic recording system). High-end instruments use inertial accelerometers combined with magnetometers to measure the inclination and magnetic azimuth. The readings are stored in solid-state memories and downloaded into a computer at the surface.

Gyro Survey Tools

Gyro survey tools are generally used when the accuracy of the magnetic survey system may be affected by the presence of magnetic objects, such as casings, or the inability of placing the instrument in the proper location in the bottomhole assembly (BHA), or because of geographic location. There are three basic types of gyro systems:

- Free gyro
- Rate gyro
- Inertial navigation system

A free gyro system is the first survey instrument used and dates back to the late 1920s. It consists of a motor-driven spinning mass, referred to as the rotor, that is mounted in a set of gimbals.

A rate gyro system, also referred to as an inertial-grade gyro, has a very accurate drift rate of 0.01° per hour. Therefore, it can be used to detect the Earth's rotation and allows the calculation of the Earth's spin axis, and hence, the geographic north.

The inertial navigation survey system differs from other traditional survey instruments, where it measures movements of hole inclination and azimuth through the 3D Cartesian coordinates, x, y, and z to yield a plotted well path trajectory. This system uses a group of gyros to orient the system to the north and accelerometers to detect movements in the x, y, and z planes.

Although the inertial navigation system is thought to be the most accurate, the accuracy of all surveys are affected by the following:

- Calculation methods
- Well depth
- Tool axial misalignment
- Magnetic interferences in the case of magnetic survey instruments

Gyroscope Effect

A gyroscope is a highly balanced wheel. If the wheel is put to rotate about its principal axis at high angular speed, an angular moment L parallel to the axis of rotation is created (arbitrarily oriented with the right-hand rule). If ω is the angular speed and I is the polar moment of inertia of the wheel, the angular moment of inertia is given by:

$$L = I\omega \tag{9.1}$$

The relationship between L and the torque T is given as:

$$T = \frac{dL}{dt} \tag{9.2}$$

If the torque is parallel to L, it changes only the magnitude of T. If the torque is perpendicular to L, a change in direction of L will occur (and consequently a change in the direction of ω). If no torque is applied to the wheel, the vector L remains the same with respect to any inertial system of reference.

Conventional Gyroscope ("Gyros")

If a wheel is mounted on a frictionless two-axes gimbals and rotated at high speed, and if the frame is moved around, the wheel maintains its direction with respect to a system of reference (Figure 9.1). If the axis is put horizon-

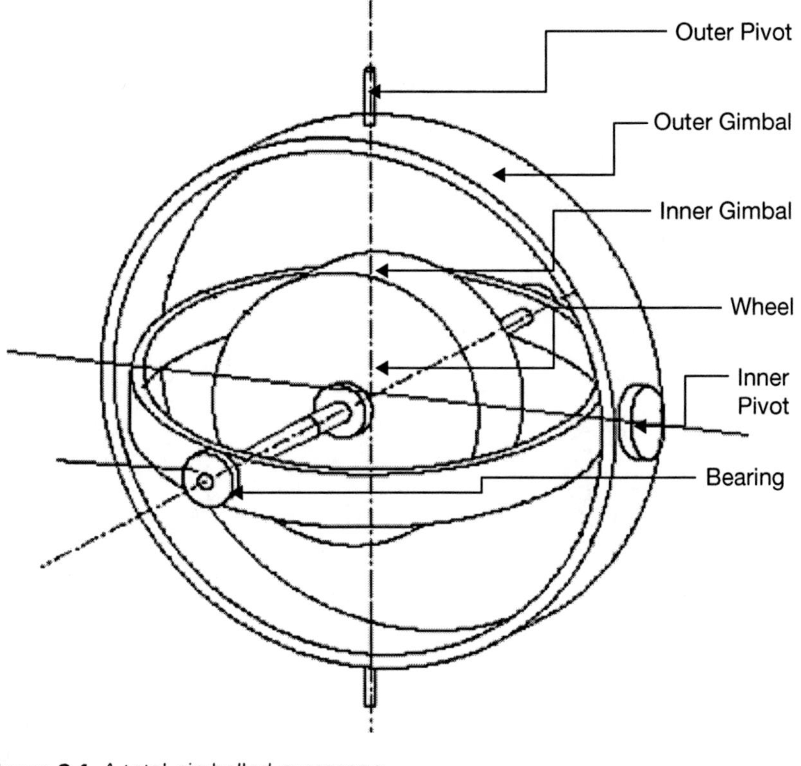

Figure 9.1 *A total gimballed gyroscope.*

tal aligned with the E-W direction, the axis gradually turns as the Earth rotates. Six hours later, it will be pointing perpendicular to the Earth's axis of rotation. If the axis is oriented parallel to the Earth's rotation axis, as shown in Figure 9.2, it will remain parallel to it as the Earth rotates. This principle is used to drive a compass card similarly to the magnetic compass in a magnetic single-shot or multishot instrument.

Typical angular velocity in a gyroscope compass wheel is of about 40,000 rpm, which creates a relatively large angular moment of inertia. The gimbals' pivots are high-precision jeweled bearings. Even so, since there is no perfect frictionless bearing, torque is created when the gimbals' axes rotate, as the probe moves along the borehole. These friction torques cause small changes in the magnitude and direction of the angular moment of inertia. Torque in the wheel bearings causes a reduction in the magnitude of L; that is, a reduction in the rotation speed of the rotor. However, torque in the inner pivot causes a rotation of the outer gimbal, and torque in the outer pivot causes a rotation of

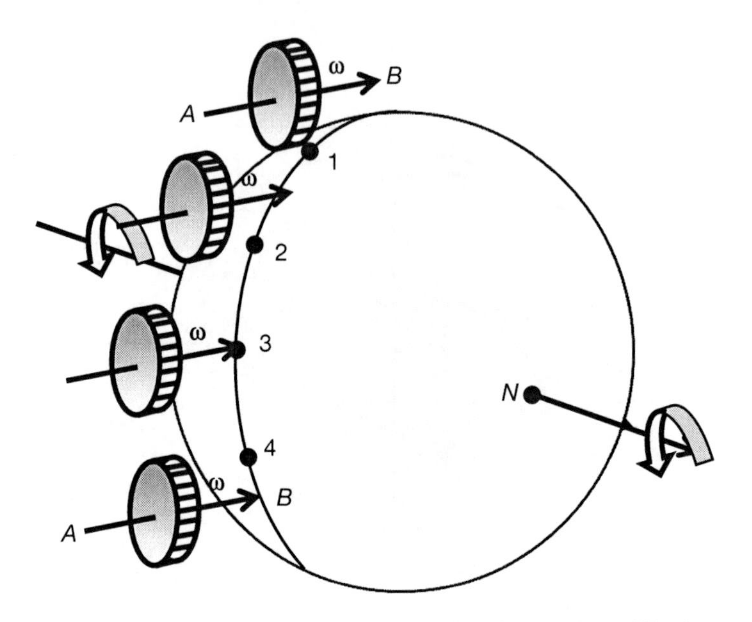

Figure 9.2 A total gimballed gyroscope subjected to the rotation of Earth.

the inner gimbal. Also, unavoidable small off-center and unbalanced wheels cause additional torques. These torques create unpredictable small drifts of the wheel, and consequently, a drift of the gyro card. To minimize these torques, the wheel axis should be oriented perpendicularly to the prevailing direction of the borehole trajectory. In addition to these unpredictable drifts, a relatively predictable drift occurs due to the rotation of the Earth. Conventional gyros are limited to a maximum drift (normally 6° per hour) in bench tests, corrected for the drift due to Earth rotation. Also, tests run during the operations help to account more accurately for the unpredictable drift.

Single-shot gyros are primarily used for deflection tool orientation, and also used to check results from single-shot instruments when it is suspected that magnetic interference exists. Multishot gyros are commonly used to survey cased boreholes or open boreholes in the presence of magnetic interference. Gyroscopic instruments are very sensitive, and should be handled with extreme care. They must be frequently tested, since the jeweled bearing wears out quickly and must be adjusted to fine tolerance, or changed.

North-Seeking Gyroscope (Gyrocompass)

The north-seeking gyroscope is a system that uses the rotation of the Earth to orient its wheel axis toward the geographic north. It is a self-orienting device

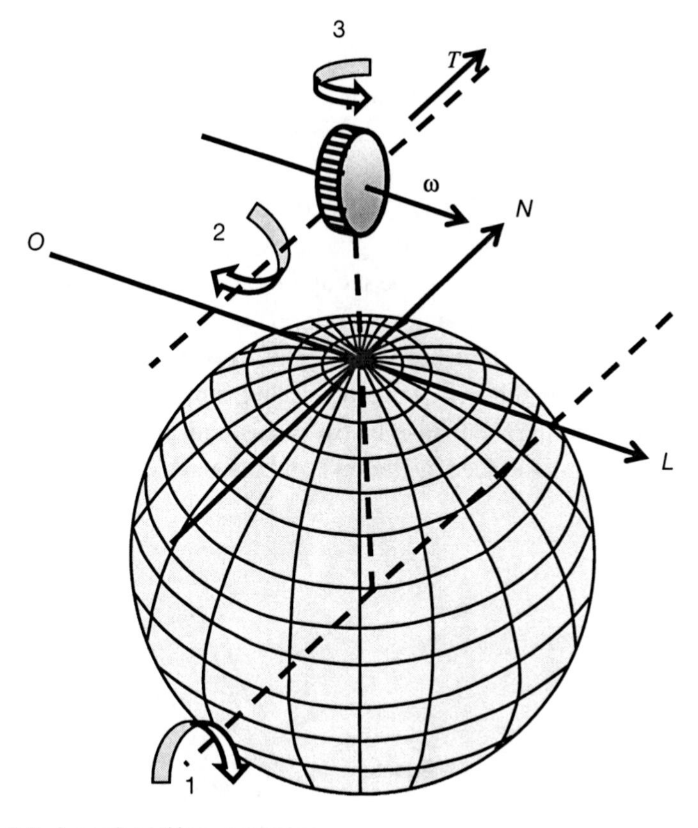

Figure 9.3 A north-seeking gyroscope.

and, therefore, does not require orienting to a given direction. It is a device largely used in ships and airplanes to provide precise geographic orientation. The system is based in the following principle: The gyroscope has its axis kept in a horizontal plane (by means of a pendulum, or floating in a liquid, such as mercury). If the axis is oriented in the E-W direction, for example, the Earth's rotation (arrow 1) will apply a torque T in the direction shown by arrow 2. This torque will make the vertical axis of the gyroscope turn in the direction shown by arrow 3, toward geographic north.

Several other inertial devices exist. High technology, like the Ferranti inertial navigation system (stabilized inertial platform), uses three accelerometers and two north-seeking gyroscopes mounted in an inertial platform, and three motors provide the correct torque in each axis of the gimbals. Because of the size (about 10 inches) north-seeking gyros are limited to use inside 13⅜-inch or larger casings.

Surveying Principle

As shown in Figure 9.4 three mutually perpendicular acceleration sensors are used to measure the gravitational field component, and three mutually perpendicular magnetic flux gate sensors are used to measure the magnetic field component.

In order to describe the relationship between the data measured by acceleration and flux gate sensors, and the geometrical parameters of the well path, two coordinate systems need to be set up first. If true north is considered as the x-axis, east as the y-axis, and the z-axis pointing to the center of the Earth can be decided by the right-hand rule, the Cartesian coordinate system O-XYZ can be set up in this way. If the well drilling direction is defined as the z-axis, the x-axis is perpendicular to the z-axis and points to the bending direction of the tool (decided by the key), and the y-axis can be determined by

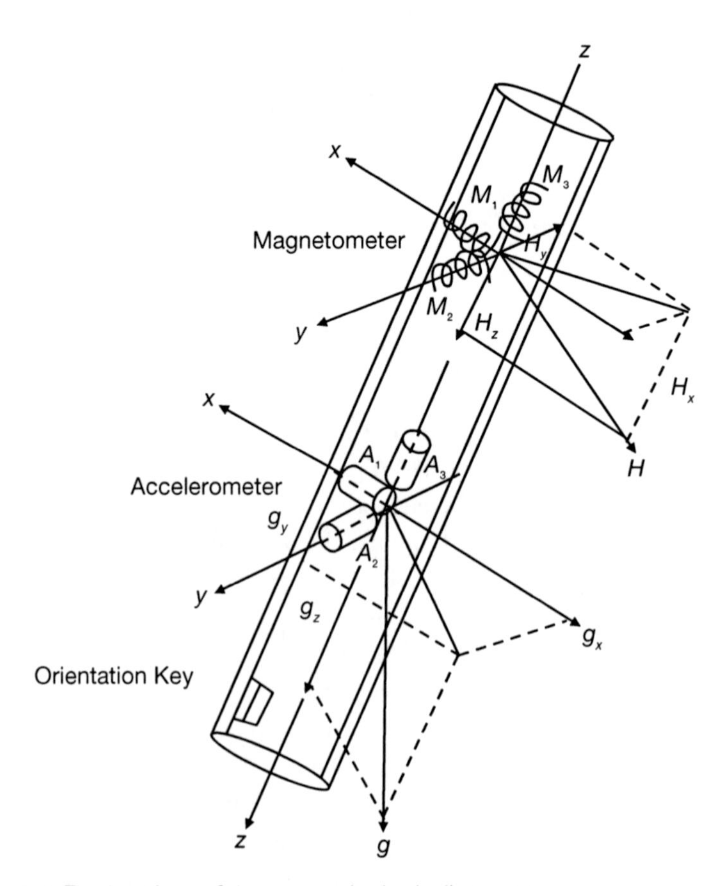

Figure 9.4 The locations of the sensors in the inclinometer.

the right-hand rule, the Cartesian coordinate system p-xyz is set up. The sensors of the acceleration and flux gate are along with these three axes.

The wellbore parameters, such as inclination angle α, azimuth ϕ, and reference plane inclination ω, are as shown in Figure 9.5. Among them, reference plane inclination ω is the angle rotating clockwise around the z-axis from the high side of the wellbore to the x-axis, which locates in the normal plane of the z-axis.

From Figure 9.5, the components of gravity acceleration g in the coordinate systems O-XYZ and p-xyz have the relationship as[2,3]:

$$\begin{Bmatrix} g_x \\ g_y \\ g_z \end{Bmatrix} = [T] \begin{Bmatrix} g_X \\ g_Y \\ g_Z \end{Bmatrix} \tag{9.3}$$

and

$$\begin{cases} T_{11} = \cos\alpha\cos\phi\cos\omega - \sin\phi\sin\omega \\ T_{12} = \cos\alpha\sin\phi\cos\omega + \cos\phi\sin\omega \\ T_{13} = -\sin\alpha\cos\omega \end{cases}$$

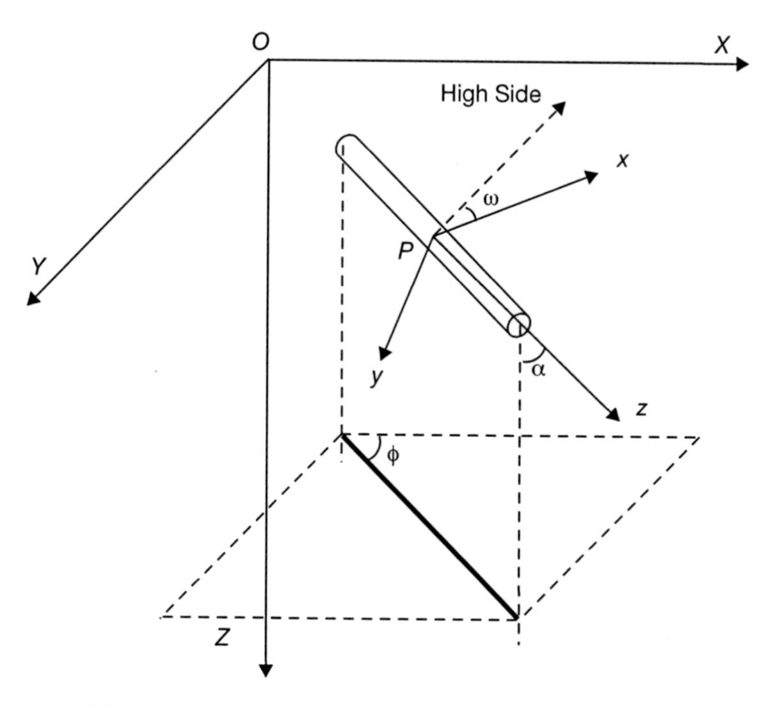

Figure 9.5 Wellbore geometric parameters.

$$\begin{cases} T_{21} = -\cos\alpha\cos\phi\sin\omega - \sin\phi\cos\omega \\ T_{22} = -\cos\alpha\sin\phi\sin\omega + \cos\phi\cos\omega \\ T_{23} = \sin\alpha\sin\omega \end{cases}$$

$$\begin{cases} T_{31} = \sin\alpha\cos\phi \\ T_{32} = \sin\alpha\sin\phi \\ T_{33} = \cos\alpha \end{cases}$$

where

g_x, g_y, g_z are the components of gravity acceleration g in the coordinate system p-xyz

g_X, g_Y, g_Z are the components of gravity acceleration g in the coordinate system O-XYZ

Because of $g_X = g_Y = 0$ and $g_Z = g$, Eq. 9.3 reformats as:

$$\begin{cases} g_x = -g\sin\alpha\cos\omega \\ g_y = g\sin\alpha\sin\omega \\ g_z = g\cos\alpha \end{cases} \tag{9.4}$$

According to Eq. 9.4[2,3]:

$$\begin{cases} \tan\alpha = \dfrac{\sqrt{g_x{}^2 + g_y{}^2}}{g_z} \\ \tan\omega = -\dfrac{g_y}{g_x} \end{cases} \tag{9.5}$$

At this time, the reference plane inclination is called the gravity reference plane inclination.

Due to the same reasons, for the components of the Earth's magnetic field, there are:

$$\begin{Bmatrix} H_x \\ H_y \\ H_z \end{Bmatrix} = [T]\begin{Bmatrix} H_X \\ H_Y \\ H_Z \end{Bmatrix} = [T]\begin{Bmatrix} H\cos\delta \\ H\sin\delta \\ 0 \end{Bmatrix} \tag{9.6}$$

where

H = Earth's magnetic field intensity

δ = Dipole

After rearranging,

$$\begin{cases} H_x = H\cos\alpha\cos\omega\cos(\phi-\delta) - H\sin\omega\sin(\phi-\delta) \\ H_y = -H\cos\alpha\sin\omega\cos(\phi-\delta) - H\cos\omega\sin(\phi-\delta) \\ H_z = H\sin\alpha\cos(\phi-\delta) \end{cases} \tag{9.7}$$

According to the former two equations in Eq. 9.7, it is obtained:

$$\begin{cases} H_x\sin\omega + H_y\cos\omega = -H\sin(\phi-\delta) \\ H_x\cos\omega - H_y\sin\omega = H\cos\alpha\cos(\phi-\delta) \end{cases} \tag{9.8}$$

Then,

$$(H_x\cos\omega - H_y\sin\omega)\cos\alpha + H_z\sin\alpha = H\cos(\phi-\delta) \tag{9.9}$$

Using Eqs. 9.8 and 9.9,

$$\tan(\phi-\delta) = -\frac{H_x\sin\omega + H_y\cos\omega}{(H_x\cos\omega - H_y\sin\omega)\cos\alpha + H_z\sin\alpha} \tag{9.10}$$

The x-axis points to true north, so the azimuth ϕ is the true azimuth ϕ_T, and ϕ-δ is the magnetic azimuth ϕ_M. Thus,[2,3]

$$\tan\phi_M = -\frac{H_x\sin\omega + H_y\cos\omega}{(H_x\cos\omega - H_y\sin\omega)\cos\alpha + H_z\sin\alpha} \tag{9.11}$$

When the inclination is very small, $\sin\alpha \approx 0$, $\cos\alpha \approx 1$, and the azimuth almost does not exist, so Eq. 9.7 is:

$$\begin{cases} H_x = H\cos(\omega-\delta) \\ H_y = -H\sin(\omega-\delta) \\ H_z = 0 \end{cases} \tag{9.12}$$

Thus, the magnetic reference plane inclination ω_M is:

$$\tan\omega_M = \tan(\omega-\delta) = -\frac{H_y}{H_x} \tag{9.13}$$

Overall, the inclination angle α can be calculated by Eq. 5.31; when the inclination is big, using the gravity reference inclination ω, it is calculated by Eq. 5.31; when the inclination is small, the magnetic reference inclination ω_M should be in use by Eq. 5.39. The steps for calculating the magnetic azimuth ϕ_M include: first, calculate the inclination angle; then, according to how big the inclination is, decide to use the gravity reference plane or the magnetic reference plane, and calculate the reference plane inclination; finally, calculate the magnetic azimuth by Eq. 5.37.

PROBLEM 9.1

An MWD accelerometer's scale factor is 3 mA/g. In a deep well, the accelerometer's readings are $g_x = -2$ mA, $g_y = 1$ mA, and $g_z = 2$ mA, and the magnetometer's details are $H_x = -0.1077$ G*, $H_y = 0.2$ G, and $H_z = 0.45$ G. Calculate the deviation angle, azimuth angle, and toolface angle.

Solution

Since

$$\sqrt{g_x^{\ 2} + g_y^{\ 2} + g_z^{\ 2}} = 3 \text{ mA}$$

the accelerometer works fine.

The inclination angle is:

$$\alpha = \tan^{-1}\left(\frac{\sqrt{g_x^{\ 2} + g_y^{\ 2}}}{g_z}\right) = 48.19°$$

Because the inclination is big, the gravity reference plane should be used, or G is the unit of magnetic intensity, called Gauss, 1 $G = 10 - 4T$ (Tesla):

$$\omega = \tan^{-1}\left(-\frac{g_y}{g_x}\right) = 26.57°$$

The magnetic bearing:

$$\varphi_M = \tan^{-1}\left[-\frac{H_x \sin\omega + H_y \cos\omega}{\left(H_x \cos\omega - H_y \sin\omega\right)\cos\alpha + H_z \sin\alpha}\right] = 328.29°$$

MWD Tools

Measurement-while-drilling (MWD) tools are used to measure various downhole parameters while drilling. MWD tools can be used to measure well trajectory, drilling parameters, other physical parameters (such as pressure, temperature), and to orient downhole motors and rotary steerable tools. MWD tools use accelerometers, which measure local acceleration, and magnetometers, which measure the strength of Earth's magnetic field. Some MWD tools measure formation parameters, such as resistivity, porosity, sonic velocity, and gamma ray. These tools are commonly referred to as logging-while-drilling (LWD) tools. Figure 9.6 shows a typical MWD system.

The tool consists of three main components:

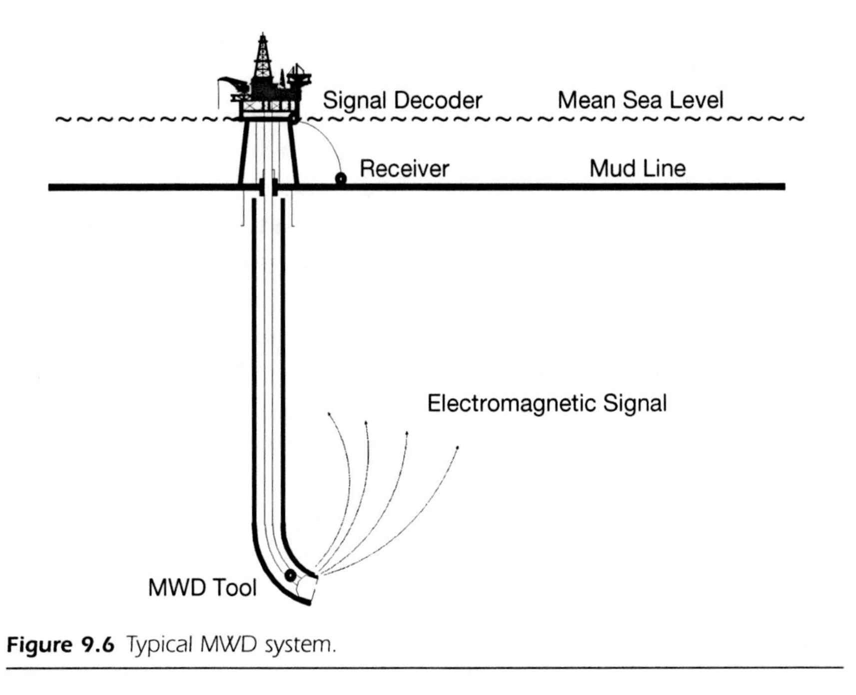

Figure 9.6 Typical MWD system.

- Downhole sensor
- A method to send the signal
- Surface equipment to decode the signal

Large varieties of sensors have been developed for evaluation of the data. A pressure transducer installed in the stand pipe receives the signal, which is further decoded. Filters are used to extract only the relevant data. The weight of a drilling mud plays an important role in mud pulse telemetry. Eq. 9.14 is used to calculate the pressure differential of a mud pulser:

$$\Delta p = \frac{\rho Q^2}{1085 A^2} \tag{9.14}$$

where

ΔP = Pressure pulse differential, psi
ρ = Weight of mud, lb/gal
Q = Flow rate, gal/min
A = Flow area, in^2

MWD tools are usually powered by the battery of a generator powered by the circulating fluid. The advantage of the generator is that it can power the instrument (sensors, electronics, and pulser) for a very long period. Battery-powered MWD tools have a limited operation life and must be replaced frequently. On the other hand, the power unit causes a large pressure drop in the circulation fluid. The generator, powered by a fluid turbine, is also very sensitive to the solids in the drilling fluid. Some battery-powered MWD tools can be retrieved to the surface through the interior of the drill string, have the batteries replaced, and be repositioned again in the bottom of the drill string.

The procedure to measure and send the downhole measurements varies between manufacturers, but is essentially the same. When drilling is interrupted to add a new joint or stand, the MWD tool records the inclination and azimuth. When circulation is reestablished, the data are sent to the surface via the waves generated by the pulser. When used with a bottomhole motor, the toolface is continuously measured and the results sent to the surface. A high degree of control is obtained with this equipment.

The daily rate of MWD tools has dropped considerably since the 1990s. The use of MWD tools associated with bottomhole motors gave origin to the complex tool called "steerable systems." This was an essential ingredient in the advance to more complex trajectories (multitargets, designer wells, horizontal wells, etc.). The improvement of the telemetry system (in particular, the pressure modulator) and in the sensor technology, increased the capabilities of the MWD to beyond the directional drilling purposes. Several other sensors can now be included in the MWD probe; in particular, sensors to log the formations while drilling and measure other parameters (gamma ray, resistivity, annular pressure, temperature, weight-on-bit (WOB), torque at the bit, noise, etc.). It is common to call these instruments LWD—logging-while-drilling. More recently, telemetry became a two-way method of communication.

Telemetry System

Telemetry technology is a communication process of data streaming between transmitting and receiving in inaccessible locations. The data transfer is between a configured transmitter and a receiver, using different transmission mediums and carriers. It can be a wired or wireless communication system. The following four types of common borehole telemetry methods used for the collection of drilling parameters, production data, and petrophysical data are discussed.

- Mud pulse telemetry
- Acoustic telemetry

- Electric telemetry
- Electromagnetic telemetry

Borehole Mud Pulse Telemetry

Mud pulse is a telemetry method that transmits a number of pulses, or pressure surges, which can be detected and decoded at the surface. The information received from the pulser is encoded in binary format. The occurrence of a pulse is a binary 1 and the absence of a pulse is a binary 0. The pressure pulses are propagated through the mud at the speed of sound, approximately 3000–4000 feet per second. Figure 9.7 shows the raw data of the recorder.

Some of the notable limitations of the mud pulse telemetry are the compressible nature of the drilling fluids and the mud pulse shape. Mud pulse timing and the peak pulse amplitude depend on the drilling fluid properties, and are degraded when the fluid is compressible.

There are three kinds of pulses: positive pulse, negative pulse, and continuous wave.

- *Positive Pulse:* These pulsers are of two types, based on the method of moving the pulser valve into the desired position. Direct drive pulsers use motors and gearboxes or solenoids that change the valve position. Indirect drive pulsers use the drilling fluid hydraulic forces and downhole hydraulic pressure circuits to control the position of a moveable piston. Both these methods cause a positive pressure pulse.
- *Negative Pulse:* These pulsers use a valve, where a mechanical obstruction is placed between the drill-string bore and an annular flow port.

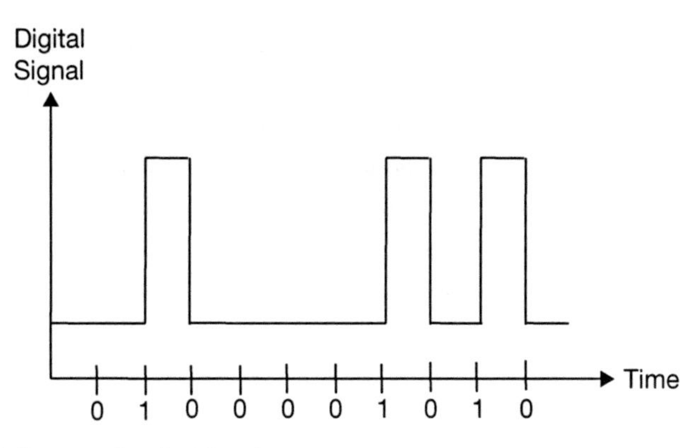

Figure 9.7 Raw mud pulse signal.

When the valve is actuated, the annular port allows a small volume of fluid to escape from the drill string to the annulus, resulting in a pressure drop, and thereby a negative pulse is generated.

- *Continuous Wave:* These pulsers have a downhole valve that consists of fixed and rotating flow restrictors. The rotor is continuously rotated relative to the stator. By varying the speed of the rotation of the rotor, varying frequency in the positive pressure pulses is generated.

Borehole Acoustic Telemetry

Borehole acoustic telemetry, a wireless communication method, uses the drill string, coiled tubing, or production tubing as the propagation medium, and acoustic or stress wave as the data carrier. Also, it can be used for data transfer during drilling or production operations.

This method of transmission uses high-frequency signal generators coupled to the drill string to transmit downhole information to the surface. A downhole transmitter placed at the bottom of the string digitally encodes sensor data, and sends it to the surface in the form of acoustic waves via the drill string, which are further decoded at the surface.

It also provides duplex or bidirectional communication with higher bandwidth for data transmission, with which commands can also be sent from the surface to the downhole tools, in addition to receiving data from the downhole instruments. A field-tested LWD acoustic telemetry system[4] consists of an acoustic wave generator sub above the LWD tool. The stress waves are transmitted through the drill pipe and received at the surface through the receiver. Unlike the borehole telemetry, the downhole drilling tools do not need any modification for this telemetry, and have been proven to be successful for underbalanced drilling (UBD) operations. The estimated data transfer rate is found to be between 50 and 100 bits per second (bps).[5] These frequencies would allow acoustic waves to transmit much more data more frequently through the drill string. This method of transmission is very useful for UBD applications, as the mud pulse telemetry is ineffective in the compressible fluids, due to drastic attenuation of the mud pulse signal. However, due to the large number of pipe joints and different types of downhole tools, acoustic waves undergo partial transmission and reflection at each joint, resulting in suppression of the signals that are transmitted.

Borehole Electric Telemetry

The telemetry drill-pipe method of data transmission was first proposed as early as 1939, which included direct electrical contact between drill pipes.

Borehole electric telemetry, a wired communication method, uses the electric wires that are embedded in the drill string (Figure 9.8). It also provides duplex or bi-directional communication with a higher bandwidth for data transmission, and is much higher than the acoustic and mud pulse telemetry. The system tested so far could achieve 57,000 bps.[6] Typical drill-string telemetry is shown in Figure 9.8, which consists of downhole measurement, or steering tools, an interfacing sub that transmits and receives data, pipe embedded with electric circuitry for data transmission, an amplifier sub, and a data-extraction interface.

The telemetry pipe is an important component in the electric telemetry system, which has embedded high-speed cable connected to the coils through a small conduit inside the pipe connection (Figure 9.9). This modification neither alters the pipe properties nor the pipe geometry. The connection, data cable, and the inductive coil are shown in the figure. Even though the entire system is completely wired, it still has power losses that require a booster system to amplify the signal.[8] Amplifier subs need to be added to the drill string every 1500–2000 ft to amplify or boost the signal. In addition to the wired drill pipe, there are other downhole drilling tools, such as crossover subs, reamer, stabilizer, and jars. Accelerators require

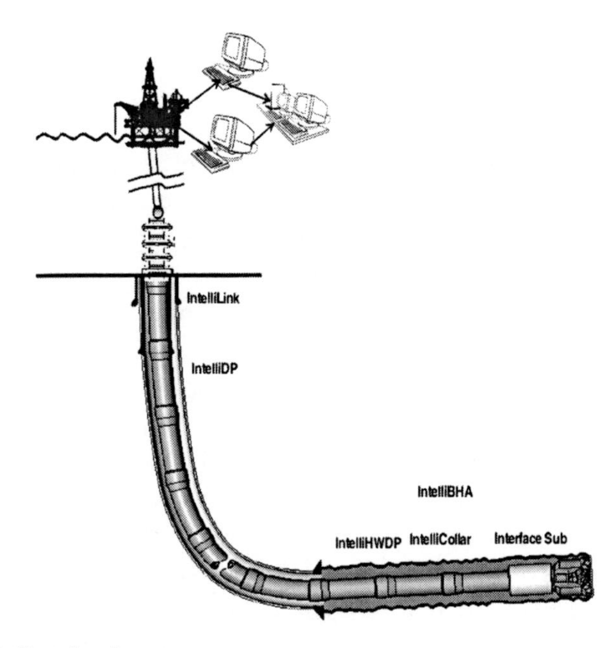

Figure 9.8 Electric telemetry.

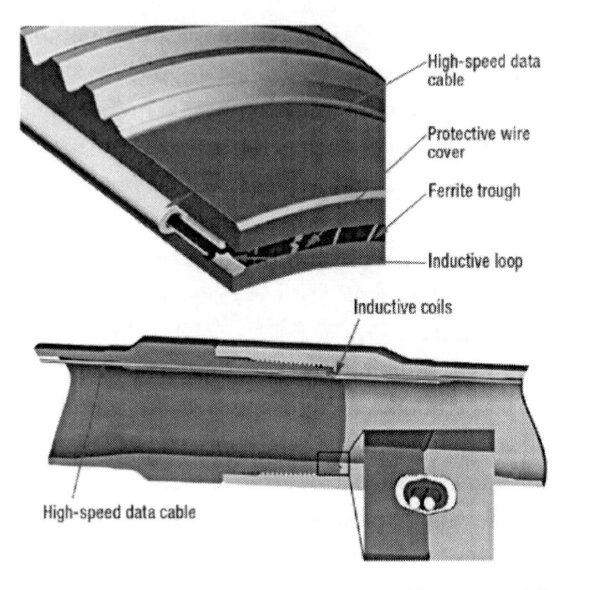

Figure 9.9 Electric cable embedded drill-pipe joints. (Courtesy of *World Oil*.[7])

modification to have continuity in the data transmission. Another disadvantage of this type of wired system is that it needs to withstand the fatigue, impact, bending, and buckling effects.

Borehole Electromagnetic Telemetry

This is wireless telemetry used to transmit data via electromagnetic waves through formation strata, and the data are decoded at the surface through a decoder. Presently, there are two methods: one is through the Earth, and the other uses both the drill string and the Earth as conducting paths. Electromagnetic telemetry can be used for well testing purposes to monitor the reservoir data, reducing the dependency of the wireline cables. It also helps to monitor the production and life of the wells, and especially to remotely operate downhole flow control devices. This type of telemetry can also be used during drilling for optimized drilling conditions.

Traditional mud pulse telemetry has a limited ability to decode pressure pulses when using compressible drilling fluids, whereas electromagnetic telemetry offers a dependable method of data transmission during such drilling operations.[5] The limitations may be due to interferences from some type of high-resistive formation that reduces the signal strength. While using the drill-pipe type method, effectiveness of the transmission is reduced due to electrical rig noise from rig equipment, decreased electrode contacts with the formation and other drill string-related inductive noise.[6] Attenuation is also an associated problem that increases, and data transmission rate decreases with increasing wellbore depth. The data transmission rate of the electromagnetic telemetry is similar to mud-pulse telemetry.

Through-Bit Logging

Through-bit logging, commonly called TBL, is another method of logging in which drilling and open-hole logging operations are performed on demand. The system provides a method through which logging tools are conveyed through the drill string and drill bit for open-hole logging in a single trip for acquiring formation data. The method requires modification to the drilling system, especially the drill bit. The advantages and disadvantages associated with this system should be weighed before considering its use. Some of the advantages are reduced cost of data acquisition, and quality of the data at a time when wellbore conditions are good. The data can be acquired through either logging tools conveyed through wireline or through memory tools.

The modes of operation are:

- Wiper trip mode
- Drilling mode

In the wiper trip mode, the drill string is tripped to the position that required hole condition, and below which the logging is carried out. In the drilling mode, drilling is carried out to the desired depth and a logging run is performed. The drill-string tools should be well suited to allow the deployment of these tools.

The drill bit is usually a polycrystaline diamond compact (PDC) bit and Figure 9.10 shows the through drill bit with a removable central portion, with a latch mechanism and insert tailored to accommodate the logging tools' passage. During drilling, the sleeve along with the insert is latched inside of the bit, forming a seamless bottom at the bit, and resulting in a performance similar to other bits. The running tool, placed at the bottom of the logging tool, can unlatch the sleeve from the bit, and the disengaged sleeve and insert can further be run along with the logging tool. After logging, the sleeve and insert can be latched again to the bit, and the logging tool disengaged from the sleeve.

Figure 9.10 Through bit. (Courtesy of Varel.)

Rotary Steerable Tools

The rotary steerable system, called RSS (shown in Figure 9.11), allows the string to rotate continuously, and at the same time, steer the string to the desired target location. This enables placement of the wellbore with good borehole quality. The latest generation of these can actively steer themselves toward a defined target without intervention from the surface. A big advantage in rotary steerable systems is that by rotating the drill string, axial drag is reduced, increasing the amount of WOB available at the bit. Due to the rotation of the drill string, there is also a considerable improvement of hole cleaning and reduced risk of differential sticking. As mentioned, the process of "guiding" the bit is done while the bit and the drill string rotates. This requires a dynamic evaluation and control of the relative position of the tool with respect to the borehole. The systems must communicate with the surface (using, for example, two-way mud pulse telemetry), sending borehole data to the surface and receiving operational commands from the surface. Alternating between sliding and rotation introduces an undulation in well trajectory, and thereby unwanted doglegs. Continuous monitoring of the trajectory at, or very close to, the bit provides more steering control. Changes to the well trajectory can be done while drilling itself, through surface communication.

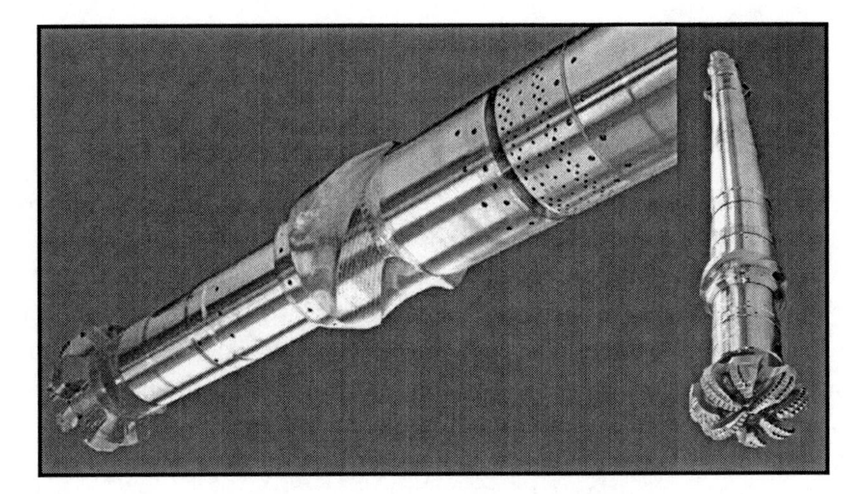

Figure 9.11 Rotary steerable system tool. (Courtesy of Gyrodata.)

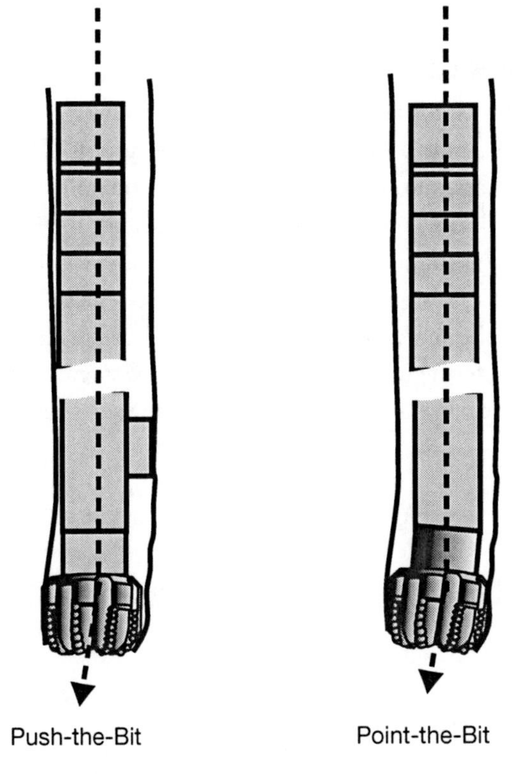

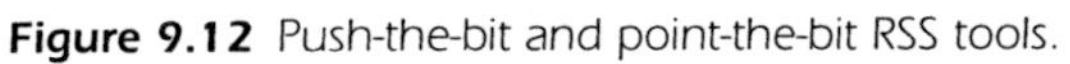

Figure 9.12 Push-the-bit and point-the-bit RSS tools.

There are two types of RSS system (Figure 9.12):

- Point-the-bit
- Push-the-bit

Point-the-bit allows the bit to tilt up, right, down, or left, to the desired direction, using some deflection mechanism, whereas push-the-bit allows the bit to be forced to the desired direction using the bit side force, as shown in Figure 9.12. Due to the nature of the mode, the bit should be compatible to withstand the side forces. The push-the-bit system should have a side-cutting structure, and short gauge vis-à-vis long gauge, for point-the-bit to act as a pivot for deflection. The mechanism involved may have adjustable stabilizers, kickpads, cams, eccentric rings, etc., allowing the various options of tilt angles.

Bit rotational speed is also increased by coupling the rotary steerable tools with the high-power downhole motor, as shown in Figure 9.13.

The flow chart, as shown in Figure 9.14, provides the selection of RSS versus downhole motors.[9]

Hollow Whipstock

A hollow whipstock is a redesigned whipstock, used for a window exit system, which avoids an additional trip to retrieve the system. This couples drilling and production operations. The whipstock can be milled through, or perforated, to establish the communication to the mother bore. The packer at the bottom provides a support for window milling, and can be oriented like any other whipstock. This type of whipstock consists of a hollow core filled with the soft composite material that can be drilled or perforated through. The hard block close to the top of the whipstock prevents milling through the system while sidetracking. Figure 9.15 provides the use of the hollow whipstock.

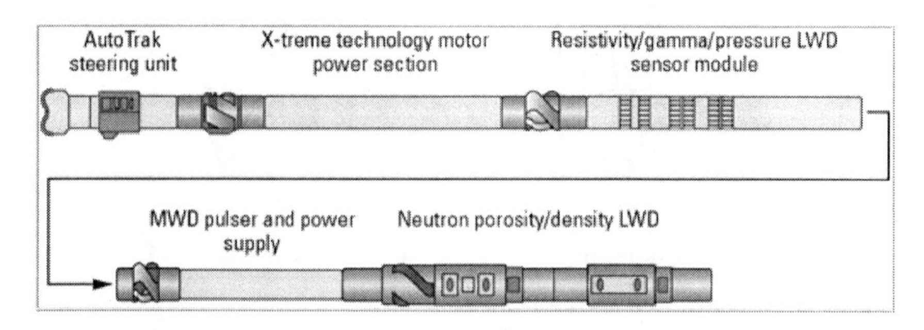

Figure 9.13 RSS tool with downhole motor. (Courtesy of *World Oil*.[8])

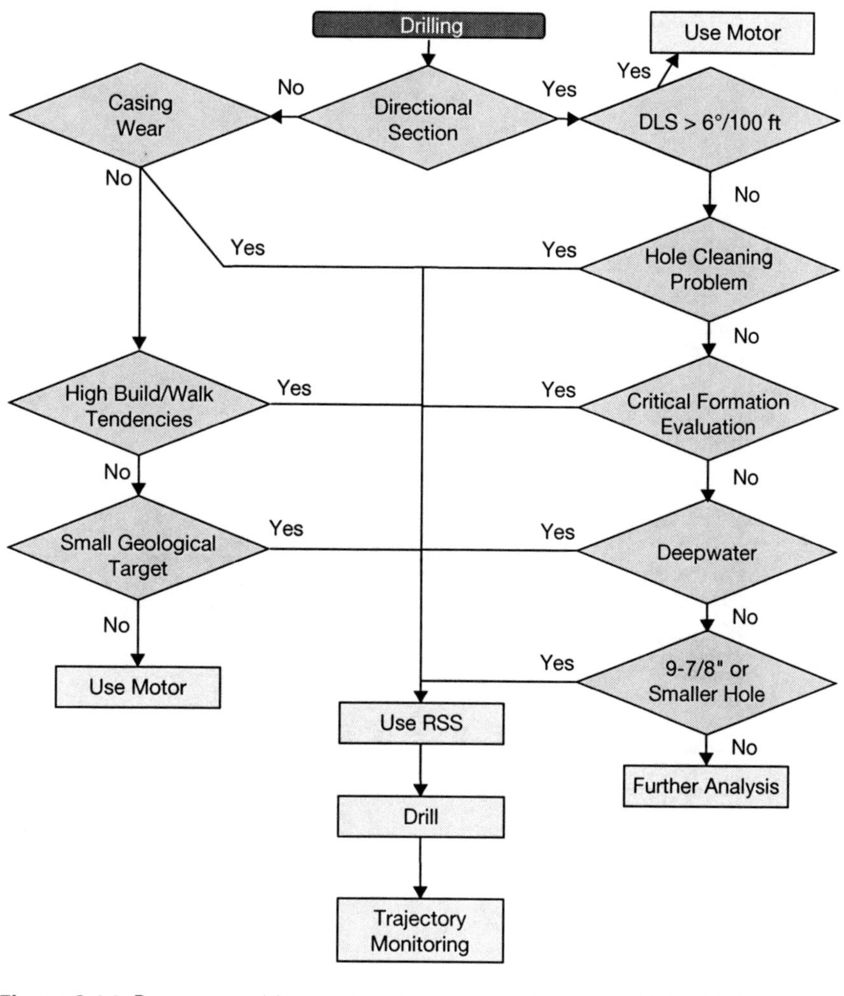

Figure 9.14 Rotary steerable versus motor system. (Courtesy of *World Oil*.[9])

Downhole Deflection Tools

Directional, horizontal drilling, and multilateral or multibranch wells have resulted in the development of several specialized mechanical downhole deviation tools. Tools and instruments associated with the deviation tools, except the tools discussed in this book, are beyond the scope of this book, and readers are advised to refer to *Downhole Drilling Tools—Theory and Practice* for additional information. Choosing the right mechanical downhole deviation tools, and their placement in the drill string, depends on the requirements of the specific operation that are planned or performed.

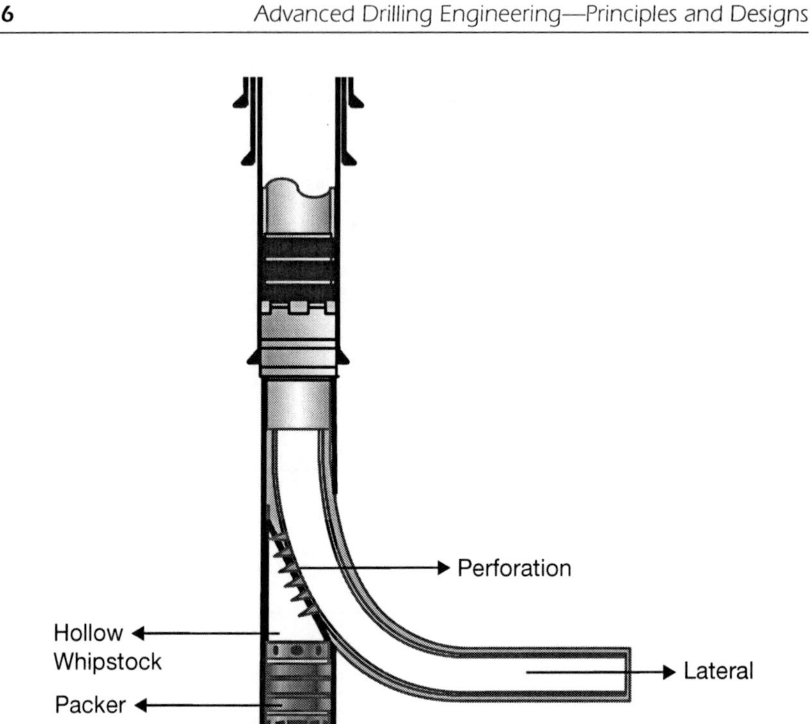

Figure 9.15 Hollow whipstock.

Supplementary Problem

1. An MWD accelerometer's scale factor is 3 mA/g. In a deep well, the accelerometer's readings are $g_x = -2$ mA, $g_y = 2$ mA, $g_z = 2$ mA; and the magnetometer's details are $H_x = -0.1077$ G*, $H_y = 0.2$ G, $H_z = 0.45$ G. Calculate the deviation angle, azimuth angle, and the toolface angle for the defined ranges of scale factors. Draw conclusions.

References

1. Contribution by Jorge Sampaio.
2. Xiushan, L. *Geometry of Wellbore Trajectory*. Beijing, China: Petroleum Industry Press, 2006.
3. Xiushan, L. "Naturalization of Azimuth Angles and Coordinates in Directional Drilling." Renqiu, China. *Oil Drilling and Production Technology* 29, no. 4 (2007): 1–5.
4. Montaron, B. A., J-M. D. Hache, and B. Voisin. "Improvements in MWD Telemetry: 'The Right Data at the Right Time.'" Paper SPE 25356 presented at the SPE Asia Pacific Oil & Gas Conference & Exhibition, Singapore, February 9–10, 1993.
5. Desbrandes, R. "Status Report: MWD Technology, Part 2—Data Transmission." *Petroleum Engineer International* (October 1988): 45–54.
6. Gravley, W. "Review of Down-Hole Measurement-While-Drilling Systems." Paper SPE 10036 presented at the SPE International Petroleum Exhibition and Technical Symposium, Beijing, China, March 18–26, 1982.
7. Fischer, P. A. "Real-Time Drill Pipe Telemetry: A Step-Change in Drilling." *World Oil* (October 2003).
8. Ruszka, J. "Integrating a High-Speed Rotary Steerable System with a High-Power Drilling Motor: Combining Power with Speed Results in Higher Rate of Penetration and Full 3D Deviation Control to Reach Difficult Areas of a Reservoir. Another Advantage Is Reduced Casing and Drill String Wear." *World Oil* 226 (April 2005).
9. Weber, A., I. Gray, R. Neuschaefer, D. Franks, G. Akinniranye, and R. Thomas. "Rotary Steerable Systems in the Gulf of Mexico Gain Acceptance: RSS Is Now Standard Practice for Deepwater GOM Wells." *World Oil* 228 (April 2007).

10

Solid Expansion Technology in Complex Well Construction

This chapter provides a glimpse of solid expandable technology and how it enables drilling, increases production, and lowers well construction costs by better addressing drilling hazards and reducing nonproductive time (NPT), which can typically account for 10–25% of a complex well's construction cost. This chapter provides the fundamental foundation for understanding how expandable products can and are being applied to:

- Drill "undrillable" wells
- Enhance production
- Book additional reserves
- Revitalize mature fields

Knowing how and when to apply expandable products is every bit as important as knowing how the technology works, what products exist, and what value they bring. This chapter provides only a cursory look at drilling and completion engineering, surrounding the appropriate and most efficient application of expandable products. A deeper understanding of the technology can significantly improve the construction of complex wells.

The development of expandable tubular technologies was initiated by the need to reduce drilling costs, increase production of tubing-constrained wells, and enable operators to access reservoirs that could otherwise not be reached economically.

Expansion Technology

The last half of the 20th century saw the age-old process of forming malleable metal into fit-for-purpose shapes transferred to oilfield tubular products, and adapted for downhole applications in the upstream sector of the oil and gas industry. As generally defined by the industry in its simplest form, expandable tubular technology is cold-drawing steel downhole.

Although the first related patent was issued in 1865, it wasn't until the mid-1900s that the industry successfully expanded pipe in situ (e.g., downhole). At this time, operators in the former Soviet Union successfully expanded corrugated pipe with pressure (hydroforming) and roller cones to patch open-hole trouble zones. This transitional system and its relevant application further motivated the evolution of expandable technology.

In the late 1980s and early 1990s the industry began to investigate the possibility of using downhole expandable tubulars to overcome the traditionally tapered wellbore. The idea was to create a technology that would eventually result in a single-diameter wellbore, and drastically reduce overall drilling and completion costs.

The original concept test on what are considered conventional solid expandable tubulars was performed in 1993, by Royal Dutch Shell in The Hague. This early expansion was about 22% of the pipe's original 4-inch diameter. Although the test used joints of pipe welded together, a significant technical advancement occurred in 1998, with the development of expandable threaded connections on expandable Oilfield Country Tubular Goods (OCTG).

The Expansion Process

The expandable tubular technology concept cold-draws steel pipe downhole. This in situ plastic expansion (deformation) of OCTG is done by either pushing or pulling a swage, as shown in Figure 10.1, a tubular with a smaller inside diameter (ID) than the swage. Expansion force depends on the friction coefficient between the swage (cone) and the pipe ID surface. A specially developed lubricated coating applied to the pipe ID provides an environmentally friendly and economical means of reducing friction. To expand pipe plastically, a cone can be driven by the force that is generated by

Figure 10.1 Mandrel used to permanently deform solid expandable tubulars. (Courtesy of Vallouree and Mannesmann.)

applying a pressure deferential across the swage itself, or by a direct pull or push force, if the pipe is anchored. The radial component of the total stress applied to the steel causes plastic deformation of the pipe. The pipe material adheres to the fundamental properties of elasticity and plasticity.

Within the elastic range, material deforms linearly with increasing stress levels, but returns to its original shape when the load is removed. The material reaches its yield point, when the applied stress forces the material past its elastic range into its plastic range. Within the plastic range, deformation no longer increases linearly with increasing stress. Deformation becomes permanent, even if the stress is removed. If stress continues to increase through the plastic range, the material will reach the so-called ultimate strength, where it will eventually fail, as depicted in Figure 10.2.

Hooke's law states that in the elastic range of a material, strain is proportional to stress. Elongation of the pipe is directionally proportional to the tensile force (stress) and the length of the pipe, and inversely proportional to the cross-sectional area and the modulus of elasticity. Hooke's law is given as:

$$\delta = \frac{P\ell}{AE} \tag{10.1}$$

where

P = Force producing an extension of the pipe, lbf
l = Length of pipe, in.

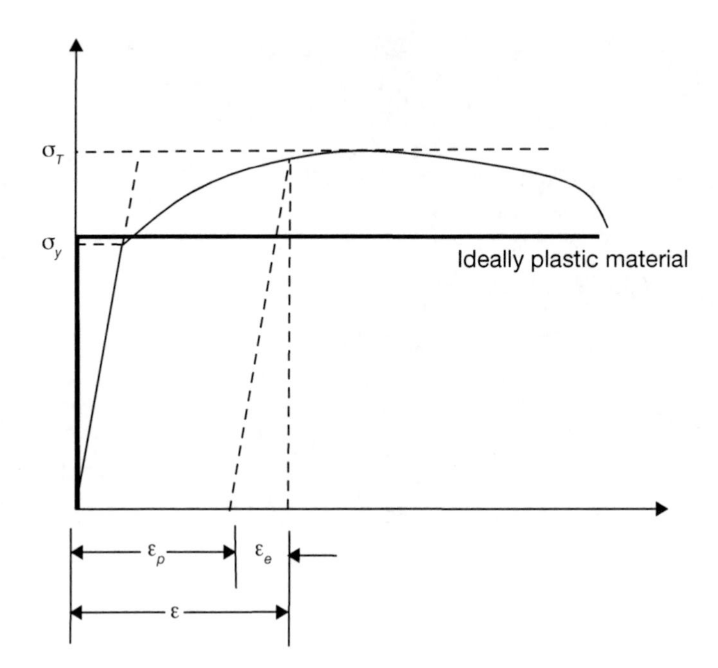

Figure 10.2 Stress-strain curves based on Hooke's law and using Young's modulus.

A = Cross-sectional area of the pipe, in.^{2}
E = Modulus of elasticity, psi

For materials under tension, strain (ε) is proportional to applied stress (σ):

$$\varepsilon = \frac{\sigma}{E}$$

where

σ = Stress, psi
ε = Strain, in./in.

Full strain is a sum of elastic and plastic deformations where ε_e is reversible (disappears when the load is removed) and ε_p is permanent (remains even if the load is removed):

$$\varepsilon = \varepsilon_e + \varepsilon_p \tag{10.2}$$

where

ε = Total strain
ε_e = Elastic strain
ε_p = Plastic strain

σ_y = Yield stress
σ_T = Tensile strength

It is possible to achieve downhole expansions exceeding 20%, but most practical applications use 3½- to 16-inch tubulars and require less than an 18% expansion. Only applications in which no hole size reduction is allowed (open-hole clads and monobore systems) require more than 20% expansion. Plastic deformation of steel does not depend significantly on the rate of deformation. It has been shown in lab/yard tests that it is possible to expand tubulars at a rate exceeding 100 ft/min. In a real rig-floor environment, and taking into account safety considerations, the rate of expansion is usually limited to 50 ft/min.

Expansion Force and Pressures

Expansion is accomplished from the force the mandrel exerts on the pipe when the pressure is applied. The amount of pressure required to initiate and maintain expansion varies according to size, depth, and wellbore conditions. Expandable systems are typically designed to maintain expansion pressure from 40% to 65% of the pipe internal yield. Pressures can range from 1500 psi to 4000 psi, depending on application and tubular size.

During plastic deformation of the pipe, all forces have to be balanced (equilibrium equation), and the volume of the deforming material should remain constant (volumetric incompressibility). These two factors allow derivation of expressions for the magnitude of expansion force and pressure, and the amount of pipe shrinkage and wall reduction.

$$F_{exp} = d_0 \, t_0 \left(1+\mu \cot\alpha\right)\sigma_Y \, \varepsilon \tag{10.3}$$

$$P_{exp} = \frac{4t_0}{d_0}\left(1+\mu\cot\alpha\times\sigma_Y\frac{\varepsilon}{\left(1+\varepsilon\right)^2}\right) \tag{10.4}$$

$$t = t_0\left(1-0.5\varepsilon\right) \tag{10.5}$$

$$L = L_0\left(1-0.5\varepsilon\right) \tag{10.6}$$

where

F_{exp} = Expansion force
P_{exp} = Expansion pressure
d_0 = Original inside diameter
L_0, L = Original and final pipe length

t_0, t = Original and final wall thickness
m = Friction coefficient
α = Half of cone angle

These equations denote that expansion force and pressure are directly proportional to the pipe yield, friction coefficient, and the amount of expansion. Even if the friction coefficient is very small, it is still required to perform a certain amount of mechanical work to deform (expand) the pipe plastically. These equations were derived on the assumption of ideal plastic behavior, which significantly underestimates expansion force and pressure. Ideal plasticity is the simplest form of approximating nonelastic behavior, when it is assumed that after reaching yield stress, material continues to deform plastically under constant stress (see Figure 10.2).

Expansion Effects on Material Properties

Expandable solid tubulars are produced from steels of conventional strength, but ductile enough to withstand a cold-forming operation in which their diameter increases downhole. Expandable pipe initially consisted of seam-welded products, because wall thickness could be controlled within limits required for the process. Seamless casing is now widely used and preferred by some end users. Other applicable materials (case dependent) include tubulars manufactured of nonferrous metals, such as chrome alloys, aluminum, or titanium.

Appropriate pipe materials feature strain-hardening behavior in such a way that the tubulars have similar burst pressures before and after expansion. Collapse of the expanded tubular is negatively affected with the expansion process by a factor of 30–50%, due to the residual stress and the Bauschinger effect. Most of this residual stress dissipates over time, even with low amounts of heat. Reliable engineering with these tubulars is possible because postexpansion values quoted by providers account for this reduced collapse rating.

Wall-thickness imperfections are a potential cause for localization of plastic deformation in areas of minimum wall thickness during expansion. Consequently, necking and ductile failures can occur. Therefore, solid tubular expansion calls for tubular qualities with more restrictive specification of tolerances on the inner and outer diameter. Both surface flaws, and as well as bulk flaws, can be detrimental to the expansion process. Flaws can grow and lead to ductile fracture and catastrophic failure of the pipe. Therefore, solid tubular expansion calls for more stringent tubular inspection practices.

Knowledge of postexpansion mechanical properties is imperative for an accurate service rating of the tubular product under evaluation. Post-expansion strength, ductility, impact toughness, collapse, and burst have been studied. By using lower-yield, carbon-based materials (50 ksi versus 80 ksi) to create expandable casing, low levels of H_2S can be easily tolerated. It was originally thought that the use of the low-yield materials would significantly affect the products' collapse, but it has been proven that, typically, the collapse is only ~100 psi lower for 50-ksi materials than that of casing manufactured from 80-ksi materials. For higher concentrations of H_2S, corrosion-resistant alloy (CRA) materials are required.

As previously mentioned, expandable liners exhibit reduced postexpanded collapse resistance. This reduction is due to the casing being physically larger in diameter after expansion. Also, cold drawing the pipe downhole results in the casing having some residual stress after expansion. Expanding thicker-wall pipe increases collapse resistance, but also increases well slimming.

For example, a liner extension of the 9⅝-inch casing can be run and still offer the possibility of running a flush-joint 7-inch liner. The collapse resistance is then roughly 2800 psi. Expanding a thicker-wall pipe can result in a collapse resistance of 4500 psi but the option of running a 7-inch flush-joint liner through the liner extension is lost. This geometric issue remains the primary reason for expanding thin-walled pipe in certain sizes of casing, but in other sizes, thick-wall casing can be used to add collapse capability to the expanded casing. Increasing the yield strength does not offer any noticeable increase on the collapse resistance, as collapse resistance is mainly a function of wall thickness for thin-walled pipe.

Burst and Collapse

Pipe burst and collapse are proportional to the steel's yield stress. In general, the higher the yield stress, the less plastic deformation pipe can survive before it fails. Balancing postexpansion pipe performance and reliability of the expansion process is critical. The collapse pressure of OCTG is reduced by as much as 50% after expansion.

A lesser effect on the burst pressure is observed. While there is some evidence to suggest that residual stresses are a contributing factor in this reduction in collapse pressure, the preponderance of the evidence suggests that the major cause is the Bauschinger effect.

The Bauschinger effect occurs when plastic flow in one direction (expansion) lowers the applied stress at which plastic flow begins in the reverse direction (collapse). Work is in progress to make an accurate assessment of this effect under simulated downhole conditions (temperatures, stresses, etc.), as well as work on methods to perform in situ remediation (recovery) of the collapse strength.

Tensile

This cold-working method of expanding pipe causes most of the deformation to occur in the hoop direction, with minor wall thinning and some contraction of the pipe. The lowest-strength grade of pipe shows an increase in yield, the higher-strength grades show decreases, and intermediate grades show little change. With expansion, ultimate tensile strength tends to increase and elongation tends to decrease—natural results of cold working on the metal. Experimental difficulties associated with residual stresses in expanded tubulars can cause a high degree of variability in yield-strength test data.

Toughness and Charpy Impact Test

The Charpy test measures the energy absorbed by a standard notched specimen, while breaking under an impact load. The Charpy impact test continues to be used as an economical quality-control method to determine the notch sensitivity and impact toughness of engineering materials.

Of the mechanical properties evaluated, the most significant is the effect of expansion on Charpy impact toughness. Charpy impact toughness is reduced by the cold work of expansion, indicating the need for initially higher than normally required toughness of steels intended for expansion (Figure 10.3).

Expandable Connectors

A family of metal-to-metal seal connectors has been developed beyond the highest API standards, as shown in Figure 10.4. These connections are able to survive extreme stress that may occur if the expandable casing is prevented from shrinking during the system's installation. This condition may occur when the liner becomes differentially stuck while it is being expanded.

No fully qualified, gas-tight expandable connector is currently commercially available; however, water-tight connections have been successfully tested in a gas-tight environment. Ongoing research and development projects include gas-tight connections.

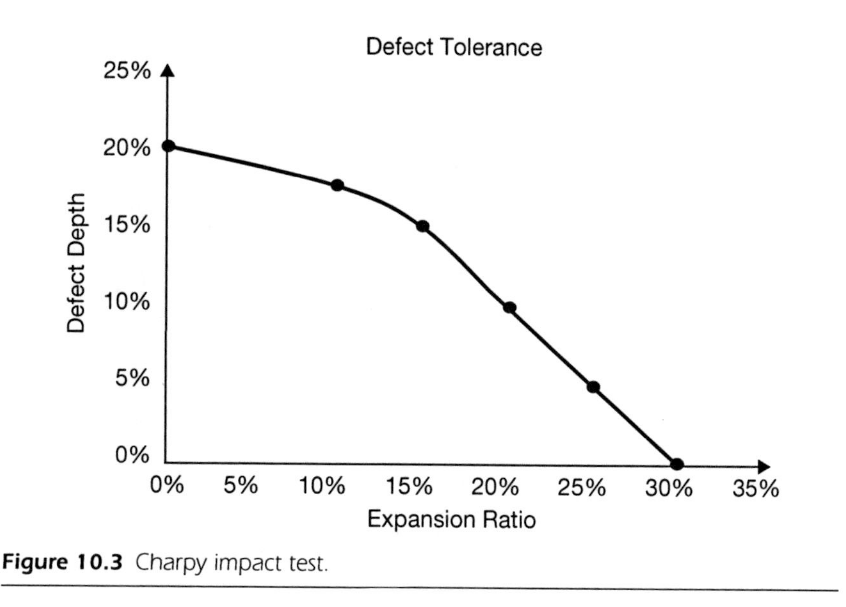

Figure 10.3 Charpy impact test.

Expansion Techniques

With the basic principle defined that expanding pipe requires attaining plastic deformation of the casing to permanently expand the tube, the focus shifts to how best to effect the desired results. Expansion techniques assume different characteristics, depending on dynamic geological conditions and fixed material properties.

This focus describes the methods employed to propagate the technical process that produces an applicable and relevant end result. Current expansion techniques include the following.

Hydroforming Expansion

Hydroforming consists of applying internal hydraulic pressure (movement of liquid under pressure) within the casing to form or reform the pipe (Figure 10.5).

Cone Expansion

Cone expansion forces a swage by pumping, pulling, or pushing through the tubular to deform the casing plastically and allow it to be permanently expanded, as shown in Figure 10.6. Most solid expandable products currently available are expanded through the use of a cone.

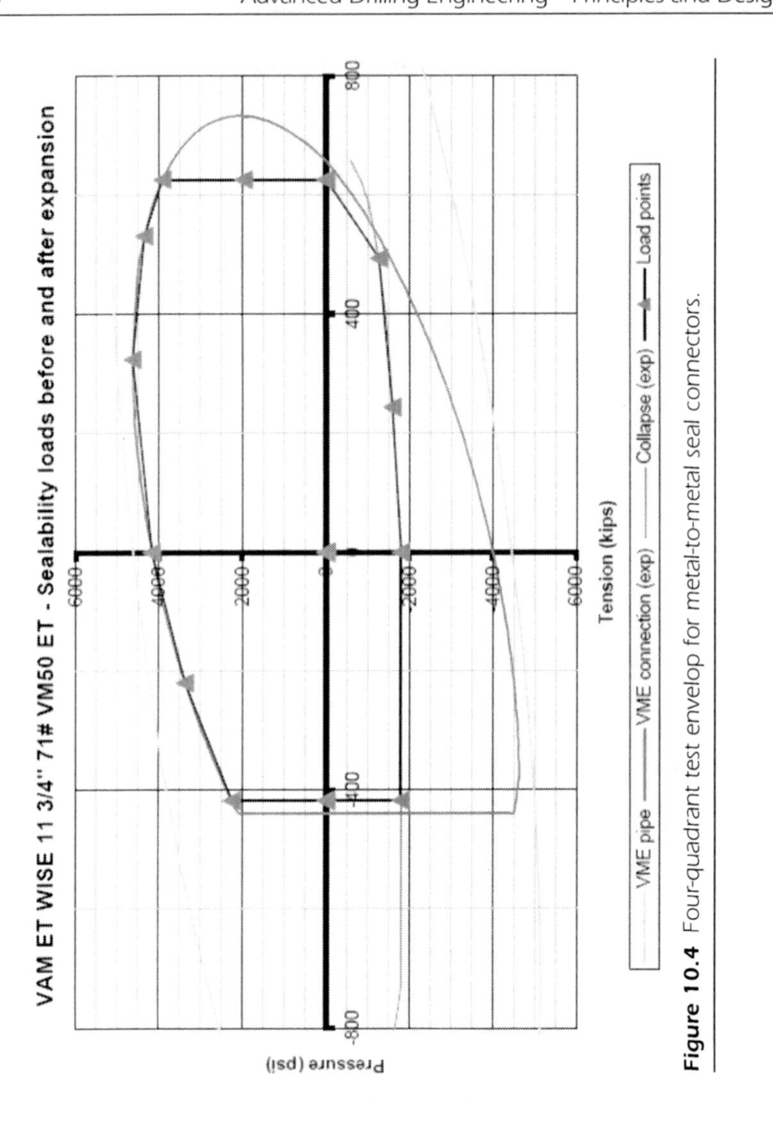

Figure 10.4 Four-quadrant test envelop for metal-to-metal seal connectors.

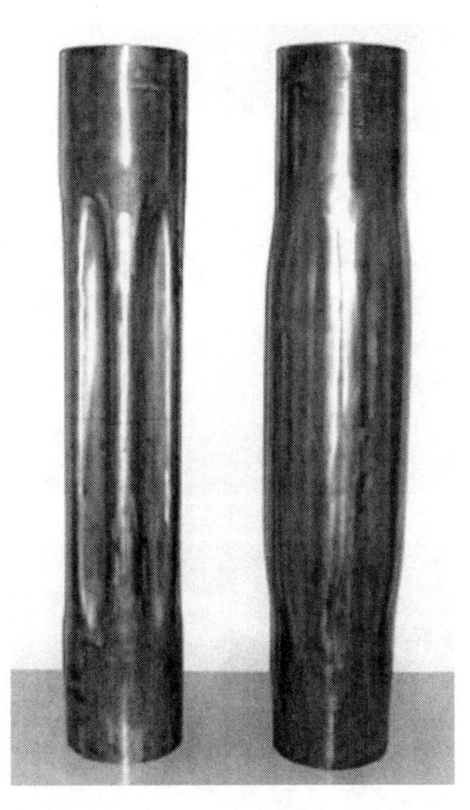

Figure 10.5 Formed casing before (*left*) and after (*right*) hydroforming.

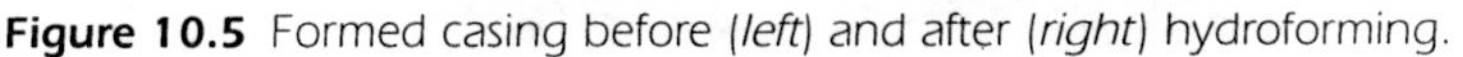

Compliant Rotary Expansion

Compliant rotary expansion combines hydraulic pressure, axial loading, and rotation to deform the casing permanently. The expansion tool is run inside the expandable hanger and rotated from the surface, while simultaneously applying pressure through the work string and expansion tool, forcing a set of rollers out of the expansion tool to perform the expansion. When the rollers of the tool are forced against the inside of the expandable casing from the applied hydraulic pressure, as they are being rotated from the surface of the well, the expandable casing is forced to expand in diameter by a predetermined amount of pressure (Figure 10.6).

Expansion Direction

When casing is expanded in the open hole with a cone, the law of material balance enlarges the diameter of the casing with a slight thinning of the casing

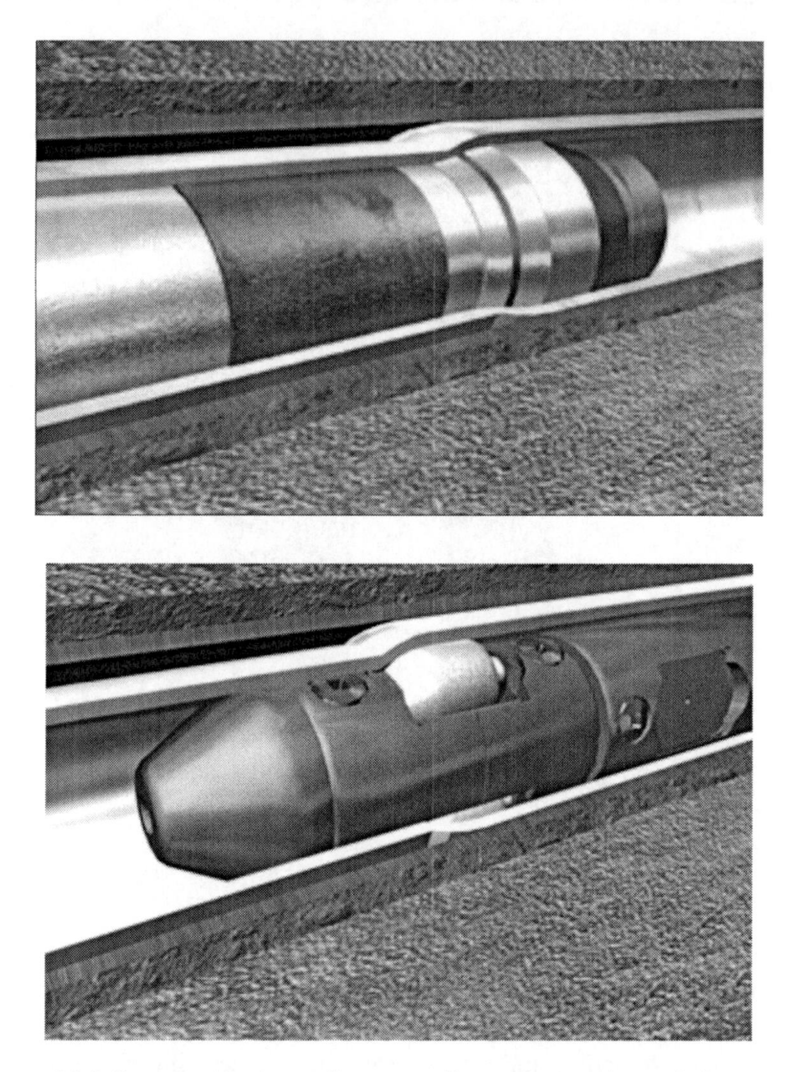

Figure 10.6 Cone (*top*) and compliant rotary (*bottom*) expansion techniques.

wall (~5%), accompanied by a shrinkage in length of 4–7%. Liner length shrinkage depends on the percentage of expansion and the geometry of the expansion device, and must be taken into account.

The physical reaction of material redistribution means that for each 1000 feet of casing expanded, the liner shrinks ~50 ft. If a liner is run to the bottom of the wellbore, anchored at the top, and expanded "top-down," a 1000-ft liner shortens to 950 ft and the bottom 50 ft of wellbore are exposed. Expanding "bottom-up" provides a means to address the exposure issue. As the bottom-up liner expands, the casing string shrinks from the top. Shrinkage can be compensated for by running extra liner footage and allowing it to shorten to the final required length in the overlap of the previous string of casing. This allowance ensures that the bottom of the hole section (typically the most critical section) is cased off.

Expansion Systems

The nature of the wellbore itself dictates what expansion tools and systems are applicable, whether open hole or cased hole. Today, expandable technology is used to construct deeper, slimmer, and more productive wells, and used to repair or seal worn and damaged pipe.

In downhole applications, solid expandable technology reduces or eliminates the telescopic profile of the wellbore, as shown in Figure 10.7. In the open hole, the technology extends casing intervals in preparation for drilling through trouble zones, or when an unplanned event in the wellbore requires sacrificing or compromising a casing point as designed in the drilling plan. In the cased hole, the technology can repair or remedy casing problems with minimal impact of hole size reduction, rather than rendering the completion system size inadequate.

Solid Expandable Systems for the Cased Hole

In a cased-hole environment, solid tubular systems enable remediation of damaged, worn, or corroded casing, and minimize slimming the well profile during repair. Depending on the length of the section, a system may consist of a single joint or multiple joints threaded together. Elastomer seals molded to the outside diameter (OD) of the expandable pipe are spaced to seal above and below each problem area. The elastomers compress between the parent casing and the system as the tubulars expand.

Solid expandable cased-hole liners have been used simply to seal off sets of perforations to remediate entire wellbores. Production from tubing-constrained wells has been increased as much as 50% by removing the conventional production casing, and replacing it with solid expandable casing with an increased ID. Full and partial strings of expandable casing made from CRA material have been installed to "line" wells of which the carbon steel casing suffered from corrosion due to exposure to CO_2, H_2S, or other corrosive formation or injected fluids.

Remediation of tubulars in injection wells are a common application for solid expandable cased-hole liners. Effective repair of these wells can positively

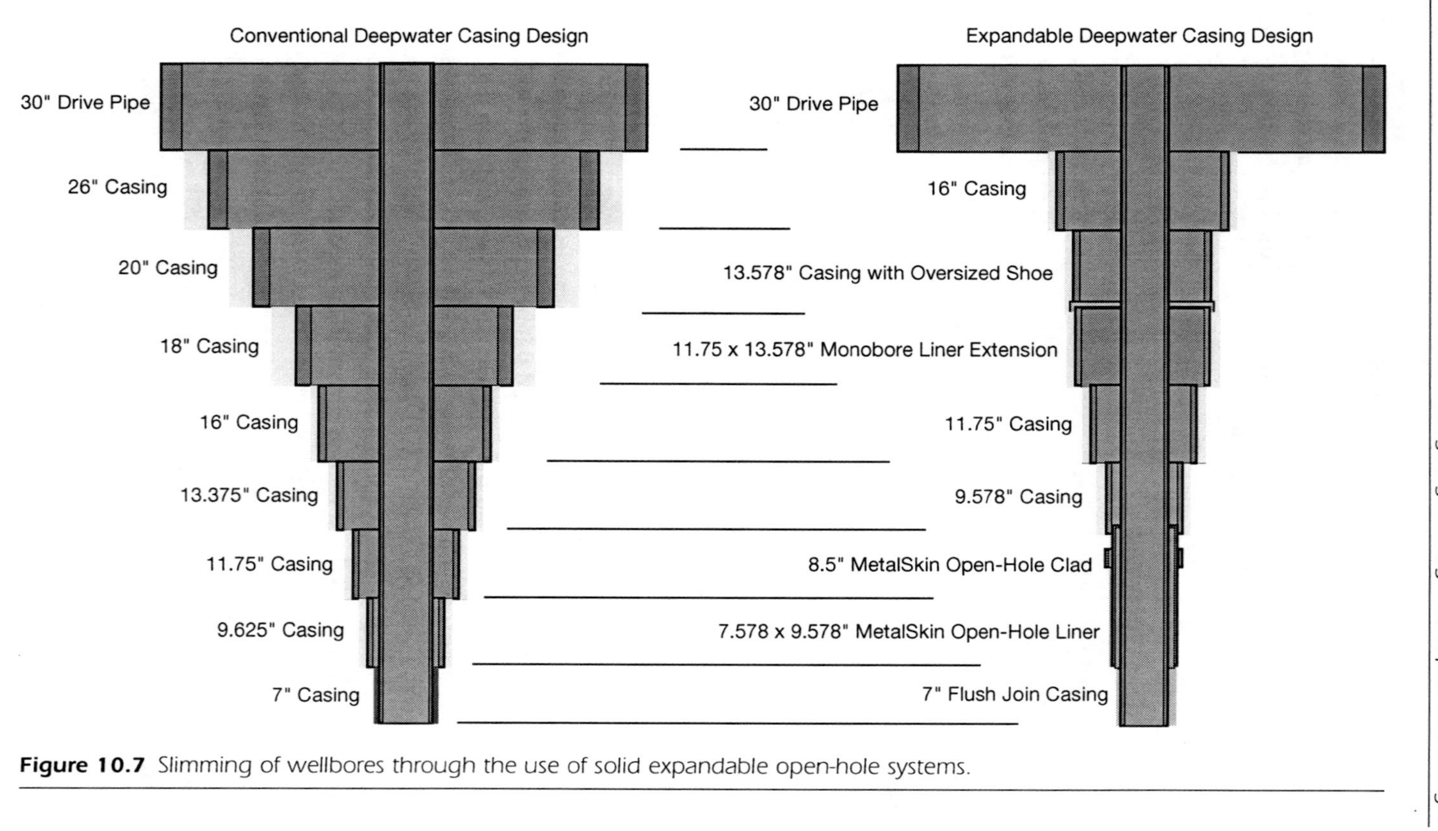

Figure 10.7 Slimming of wellbores through the use of solid expandable open-hole systems.

affect the production of all the wells in their injection pattern, greatly increasing a field's production and increasing the field's recoverable reserves.

Solid Expandable Systems for the Open Hole

In an open-hole environment, the most common application runs a solid expansion system, expands it, and ties it back to the previous casing string. This structural approach minimizes slimming the well profile during well construction. The primary types of open-hole expandable systems consist of the following:

- *A slotted liner system* (Figure 10.8): This liner extension system is used in conjunction with fiber cement and is installed in an underreamed hole to allow zero hole loss. It can be tied back to the previous string of casing, or installed as a stand-alone, non-tied-back liner. The slotted liner is used to reinforce the cement and guide the drill-out assembly, so that it remains in the wellbore rather than inadvertently sidetracking the wellbore. This type of liner/cement system has been applied to mitigate overpressured zones and lost-circulation zones.
- *A solid expandable liner system*, as shown in Figure 10.9, is the most common liner extension system. This system slightly reduces hole ID, but minimizes wellbore slimming.
- *Cladding systems* (Figure 10.10): One cladding system is expanded against a wellbore that has not been underreamed. The ID of the wellbore is reduced by the thickness of the liner wall plus the thickness of

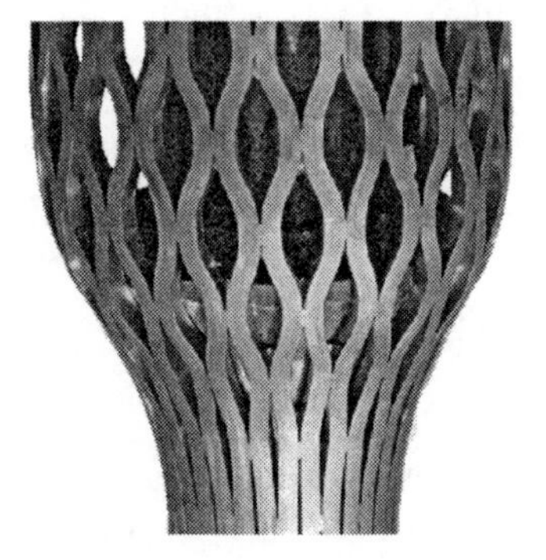

Figure 10.8 Slotted liner expansion system.

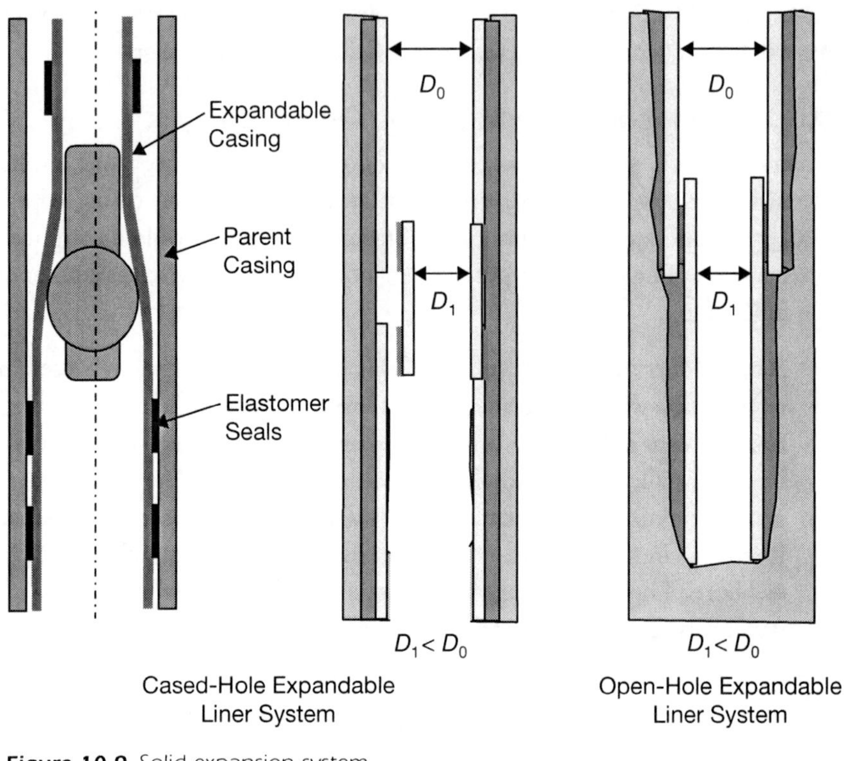

Figure 10.9 Solid expansion system.

the elastomeric sealing elements (if any) formed to the outside of the liner. The second cladding system, called a "clad-thru-clad" liner, is expanded in a slightly underreamed (~1 in.) wellbore. This liner extension system does not reduce ID of the wellbore. The system is not hung off or tied back to the previous casing string and is used to isolate problem zones without having to set a casing or liner.

- *Monobore systems* (Figure 10.11): These single-diameter systems eliminate reduction in the inside diameter.
- *Monobore clad:* This cladding system is used to cover sections of the open hole without sacrificing any hole size.
- *Monobore liner:* This liner extension system typically requires an oversized shoe run on the bottom of the previous string. This system is tied back to the previous casing string.

The monobore liner system can be used as a single-shoe extension that offers the same drift as the previous string of casing. Although not as flexible

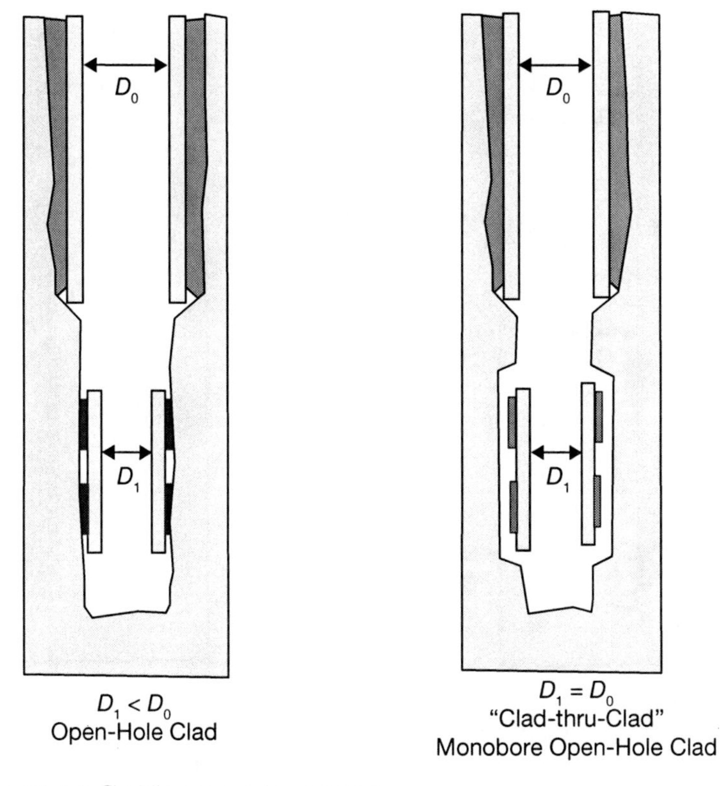

Figure 10.10 Cladding expansion systems.

as conventional expandable systems, shoe-extension systems can extend one or more sized shoes in preparation to deal with one or more trouble zones.

The basic design of the monobore liner comes from a conventional solid expandable open-hole liner system. Using an oversized shoe as part of the bottomhole assembly (BHA) of the previous casing-string configuration facilitates transforming this basic open-hole liner system into a single-diameter system. This oversized shoe has a large enough ID to facilitate a larger-than-normal expansion of the open-hole liner, resulting in a postexpansion ID of the open-hole liner as that of the previous string of casing, thereby extending the previous casing shoe with no tapering of the wellbore. The next hole section can then be drilled with the same size bit. Because the oversized shoe is constructed with casing with flush-joint connections, its running OD is the same as the connections of the conventional casing string, minimizing any equivalent circulating density (ECD) challenges (Figure 10.12).

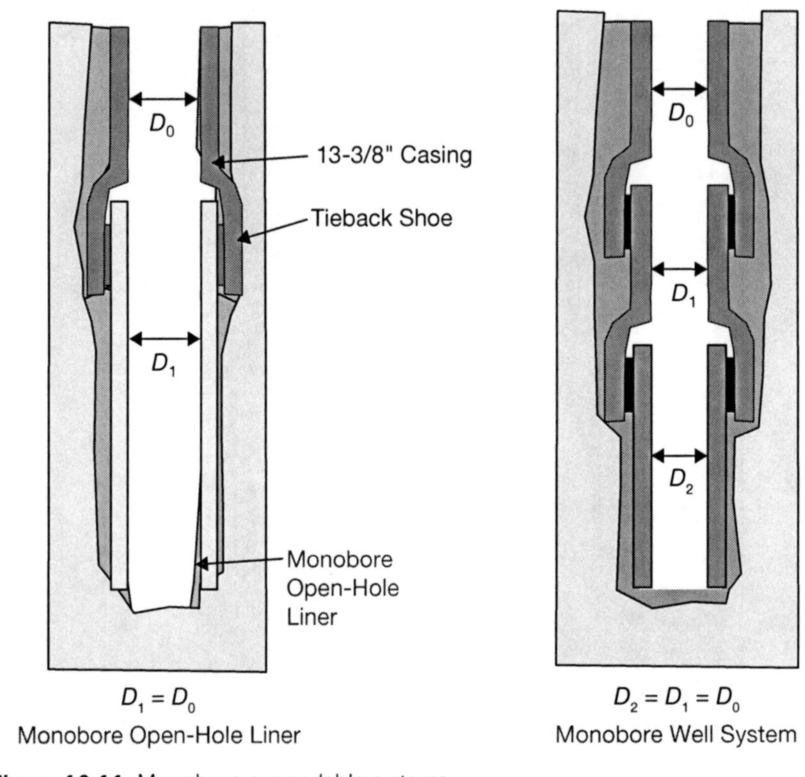

Figure 10.11 Monobore expandable systems.

The monobore liner requires the use of the oversized shoe and the oversized shoe is run on when the previous string of casing is run. These systems must be either used as a planned contingency or as a planned installation.

Monobore liners can be used to downsize the top of a well without downsizing the well's completion. They can also be used as a contingency to ensure that the well's completion is not forced to be downsized, due to drilling challenges that may need to be mitigated (Figure 10.13).

The monobore well system is a single-diameter system that uses sequentially installed, solid expandable products to create a continuous casing string with the same diameter casing, thereby eliminating the telescoping effect. In theory, this system could create a wellbore on a continuous hole size from top to bottom, but is currently used to preserve hole size in sections and is not currently commercially available.

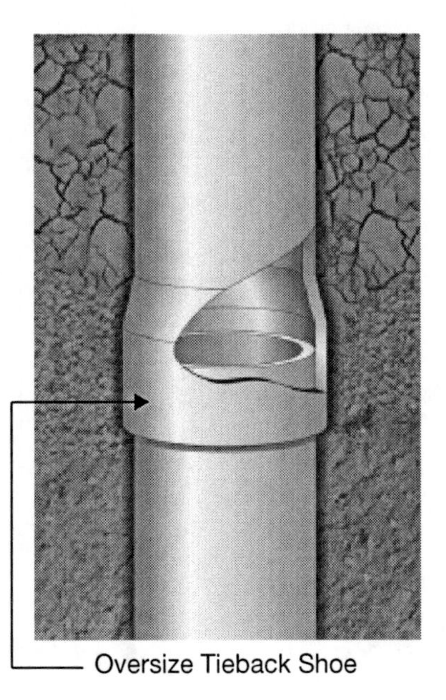

Figure 10.12 Oversized shoe used in the monobore liner.

Using Expansion Technology

Solid expandable technology is usually applied using one of the following approaches:

- One-off contingency, unexpected problems
- Planned-in contingency
- Wellbore construction element
- Planned-in, multiple use

Conditions Mitigated with Expansion Technology

Why the system is used in a project depends on the issues and conditions that demand mitigating. Unexpected problems may require the application of a one-off installation, which is especially common in exploratory wells.

Offset data can identify formation characteristics that may warrant planning in the system as a design contingency. Typical drilling problems that can be mitigated with an expandable liner solution include:

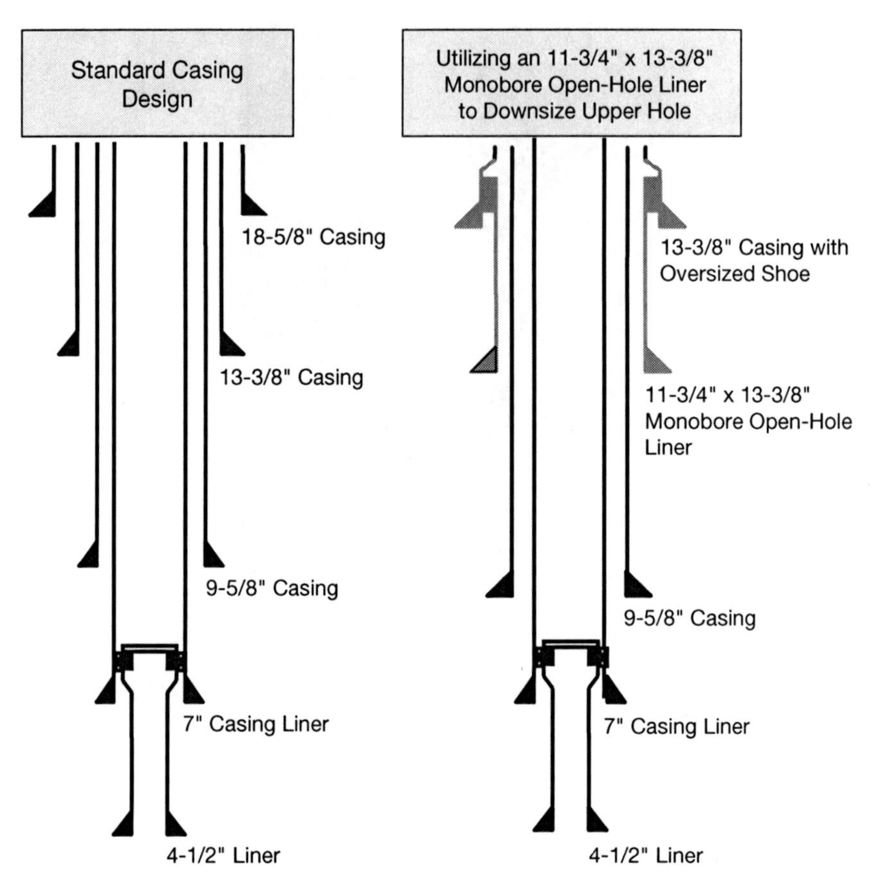

Figure 10.13 Downsizing the top of the wellbore with a monobore liner.

- Inadequate hole stability
- Overexposed hole as a result of drilling issues, equipment failures, prolonged tripping, etc.
- Overpressured formations
- Underpressured formations
- Close fracture gradient/pore pressure tolerances
- Poor isolation across multiple zones
- Remediation for casing that was inadvertently set shallow

In contrast to a last-resort application, expandable systems may be used as a fundamental casing string in the initial wellbore construction. This proactive approach enables the system to be installed before the trouble zone is encountered and before the hole is on the verge of being junked.

Whether an expandable system is used as part of the plan or for contingency purposes, the technology saves the hole size, compensates for unplanned events, and allows for flexibility in the well planning process.

Increasing Production

Expandable technology has enabled operators to reach reservoirs with larger casing sizes than otherwise possible, thus increasing the size of production tubing and the production rate. Comparison of the conventional well plan versus the expandable well plan, as shown in Figure 10.14, illustrates that by using solid expandables as a contingency, or a planned contingency (using only when needed), upsizing a completion is possible.

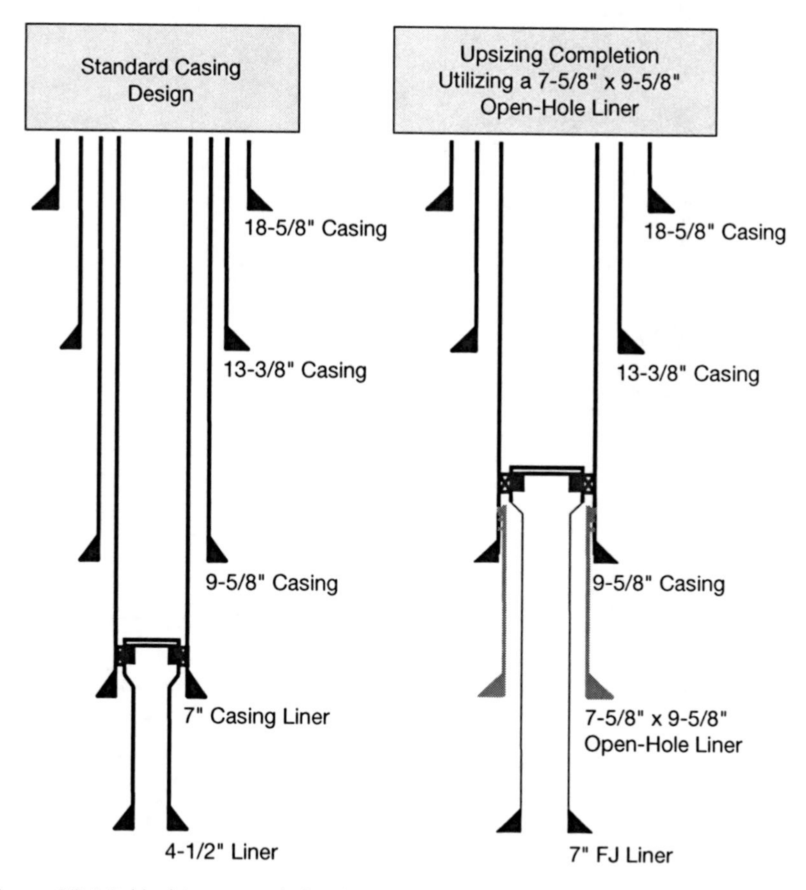

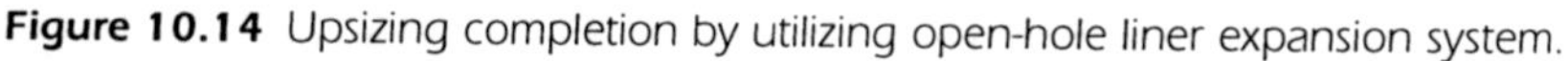
Figure 10.14 Upsizing completion by utilizing open-hole liner expansion system.

Drilling Longer Wells

In some cases it is not possible to reach production zones, because a limit exists on the number of casing points that can be set to isolate different well challenges. With expandable technology, operators have been able to set more casing seats and reach the reservoir with an optimum casing size, thus enabling longer-reach wells. This application is particularly valuable and relevant where the use of monobore technology allows extensive field drainage, or even multiple fields to be developed from one drilling platform through the use of horizontal wells that are 5–15 kilometers (3–10 miles) long.

Traversing Depleted Formations to Reach New Horizons

Solid expandable open-hole liners can economically deepen wells and allow new reservoirs to be exploited, as depicted in Figure 10.15. Without expandables, this objective would require coming up the hole and exiting out

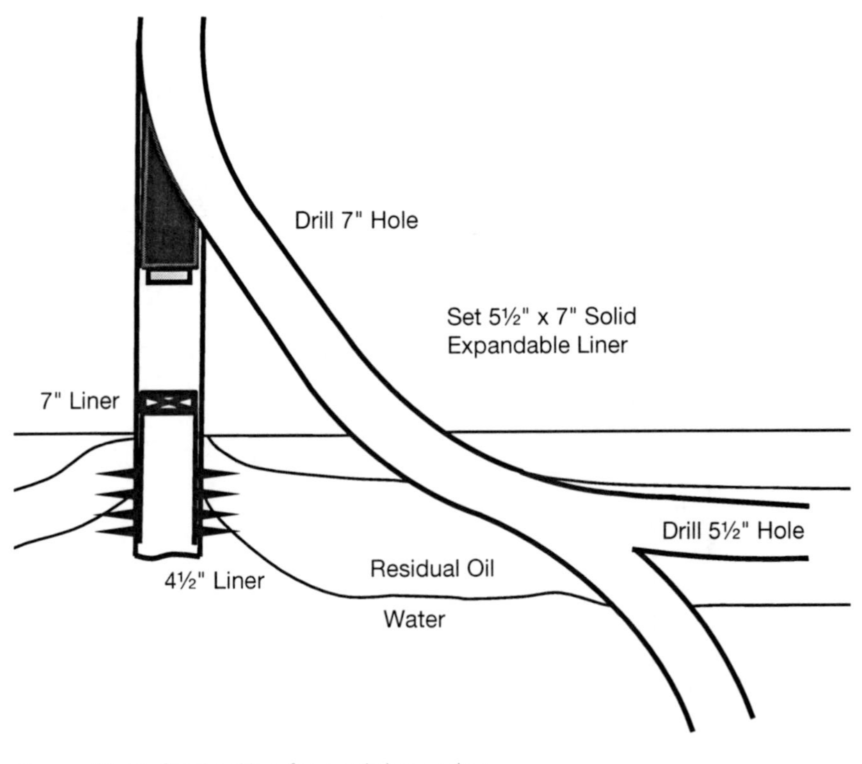

Figure 10.15 Sidetracking from existing casing.

the 9⅝-inch casing, drilling a long hole section to the depleted formation, setting 7-inch or 7⅝-inch casing across the depleted zone, and reaching TD of the well with 5½-inch casing. With expandables, the well can be sidetracked out of the 7⅝-inch casing much deeper in the well, which facilitates drilling a short hole section through the depleted zone. A 6 × 7⅝-inch liner can then be set through the depleted zone, allowing the well to be completed through a 5½-inch completion string.

Downsizing Wells or Preserving Hole Size While Addressing Drilling Hazards

Open-hole clad-thru-clad liners allow for multiple system installation without downsizing the well, and can significantly impact well economics. These liners mitigate drilling hazards, such as depleted zones, overpressure and sloughing formations, or even pore pressure/fracture gradient challenges. One or more open-hole clad liners can be used to mitigate multiple trouble zones (Figure 10.16). However, only open-hole clad liners that allow clad-thru-clad installation (that is, running through and installing beneath an

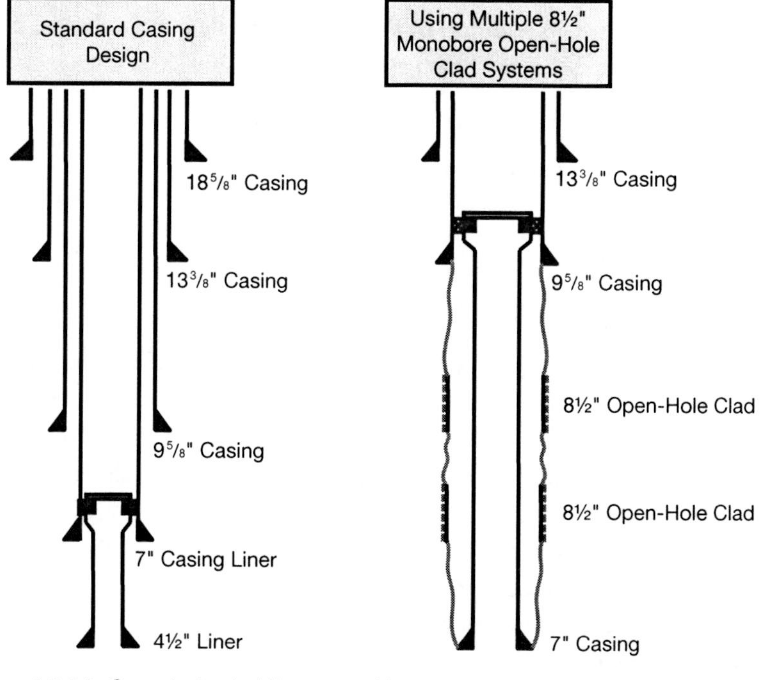

Figure 10.16 Open-hole clad liners to mitigate multiple trouble zones.

already installed open-hole clad liner) enable several liners to be installed as needed.

Casing Exit from an Existing Completion

By planning an expandable into a recompletion project, an operator is able to optimize the equipment and investment. An expandable liner can be set in the window of the liner that enables the operator to drill as large an open hole as possible into the reservoir. With this approach, vertical wells can be converted into horizontal wells and completed with a smart-well system. Sidetracking to recomplete a well with multiple horizontal laterals facilitates better reservoir drainage.

Advantages of Solid Expansion Technology

Combating problematic conditions usually generates extra costs in the rig time needed to battle the problems (NPT), rather than the rig time spent drilling. Mud losses and treatment chemicals result in escalated costs on consumables. Another unplanned expenditure stems from the specialty services and rentals that may be required. Mitigating drilling problems with the installation of a contingency liner at an appropriate juncture or, in catastrophic circumstances, by sidetracking, usually have an even greater impact on the bottom line. In addition to those costs entailed, a number of consequential compromises result, such as:

- Time and cost for invoking the contingency
- Size of eventual completion in reservoir
- Production tubing size
- Production rates
- Life-cycle functionality of the well/completion

Expandable technology used as the contingency can eliminate the compromises normally associated with conventional contingencies. An expandable can be run to mitigate drilling problems without having to downsize the rest of the completion. The time and cost of invoking the contingency will still be an issue; but, if the project entails drilling several wells, and drilling problems are encountered on an irregular basis, the cost of an expandable system can be spread over several wells.

Expandable technology used in lieu of intermediate casing and liners can eliminate the need for larger casing at the surface. Expandable tubulars, particularly single-diameter liner extensions, provide the option of starting with a smaller surface casing, which results in the following:

- Overall casing cost when expandables are run, in conjunction with smaller casings
- Smaller wellheads (Capex) and smaller BOPs (rentals)
- Less overall rig time when drilling smaller holes
- Less cost on mud, cement, and cutting disposal
- Less rig cost by using a smaller rig

The high-level remittance for open-hole expandable liners includes:

- Providing a cost-effective contingency to mitigate knock-on compromises in well functionality and productivity
- Providing the facility for multiple contingencies
- Providing a robust option in casing design to enable true slim-well designs

Solid Expandable System Limitations

If this technology claims all that it does, the question begs asking as to why it is not used in more wells. The answer remains complex, but encompasses a variety of variables and attitudes, the least being metallurgical limitations. Exploration and production companies have limited resources when it comes to research and development budgets, and are reluctant to risk high-profile projects on what some still regard as a radical technology. The high cost of failure is sometimes not worth the risk when projects are scorecard driven. As with any industry, a revolutionary technology requires a paradigm shift in thinking and long-standing conventions. Large service companies that specialize in developing new products and processes sometimes lack incentives when the industry is slow to change. Small innovators, who have proven to be the most ambitious and nimble, lack funding and have limited market access.

As with this technology, a combination of large-company resources and individual effort produced the first commercial solid expandable product. Solid expandable products were widely introduced to the oil industry in 1998, and since then end-users have gone from using them as an emergency "fix" to planning them into drilling projects as contingencies, and as part of the well's basis of design (BOD).

Application of Solid Expandables to Complex Wells

Early applications of solid expandable systems in complex wells were often as a last resort. The operator typically used an expandable open-hole liner when the hole section or well was lost and the alternatives were unacceptable. Although effective in mitigating trouble zones, installation of solid expandables in a well that is already in the process of being lost increases the risk of operation.

"Emergency" installations often result from the need to set casing higher than planned. An unplanned casing point jeopardizes reaching the pay zone with the required hole size. Multiple lost casing points critically reduce the well's diameter. Further "telescoping" (or slimming) often prevents reaching the targeted reservoir with a large enough wellbore to be evaluated effectively or completed optimally. Before solid expandable tubulars, the options to address this situation consisted of sidetracking further up the wellbore, to attempt drilling through the trouble zone, or simply redrilling the well.

Larger-size solid expandable liners can add greater value the sooner they are employed. The option of using the conventional high-collapse casing(s) across the zones they have been designed to cover enables the well's BOD to be maintained. In most complex wells, the critical casing is usually the 13⅜-inch/13⅝-inch string. If this string is set high, or multiple trouble zones are encountered below this string, it becomes very difficult to drill the well to its targeted depth with the planned size hole. As a result, most solid expandable liner providers have focused product development around these sizes to increase the opportunity for a successful resolution.

Many strings of expandable casing have been used to minimize the hole-size reduction, so the targeted section or total depth (TD) can be reached with an appropriate hole size. However, the most effective method to ensure success is to incorporate solid expandable tubulars into the BOD, or as a contingency during the planning stage to address drilling challenges. Figure 10.17 illustrates the dramatic consequences of slimming, using only conventional tubulars in a well requiring ten casing strings. The hole size at TD is compromised, and the resulting reduced production prevents a timely return on investment (ROI), leading to the well's economic failure. Figure 10.18 illustrates the hole-size benefit of planning solid expandable tubulars as a contingency. This installation provides a hole size savings of 2¼ inches. Figure 10.18 illustrates the value of planning solid expandable contingency liners, resulting in adding several strings of casing, while saving a total 3¼-inch hole size at TD.

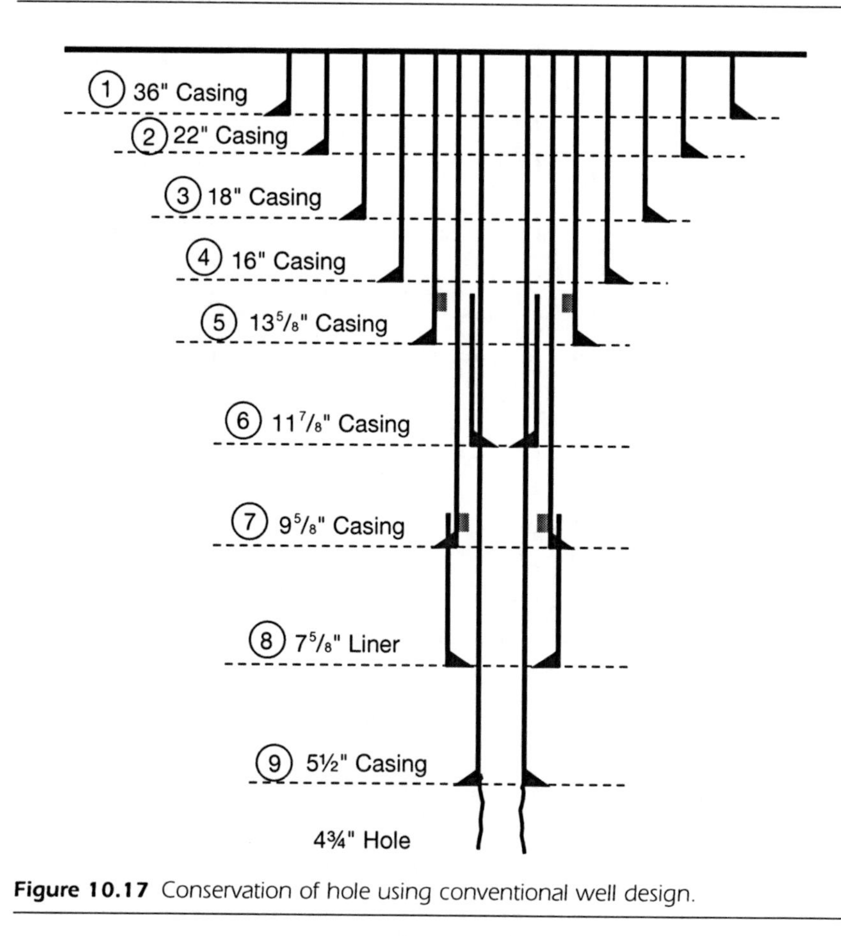

Figure 10.17 Conservation of hole using conventional well design.

Complex Wells in Tight Economic Environments

Economic viability remains a critical element when drilling any prospect. Many projects never come to fruition because the complexity of drilling the prospect drives well construction costs beyond reserve recovery potential, and the ROI is as such to kill the project. These economically "undrillable" prospects are not limited by region, but exist globally in all geologies, environments, and locales. Well designs that "upsize" or apply technologies to address some of these challenges, and enable drilling through the trouble zones, often only add to the well cost. Technically, the operation is successful (the well can be drilled), but excessive well economics prohibit the project.

Since 13⅝-inch casing is rarely used in these wells, trouble zones must be dealt with under smaller strings of casing, while maintaining the size of the well's completion. Managed pressure drilling techniques may be appropriate

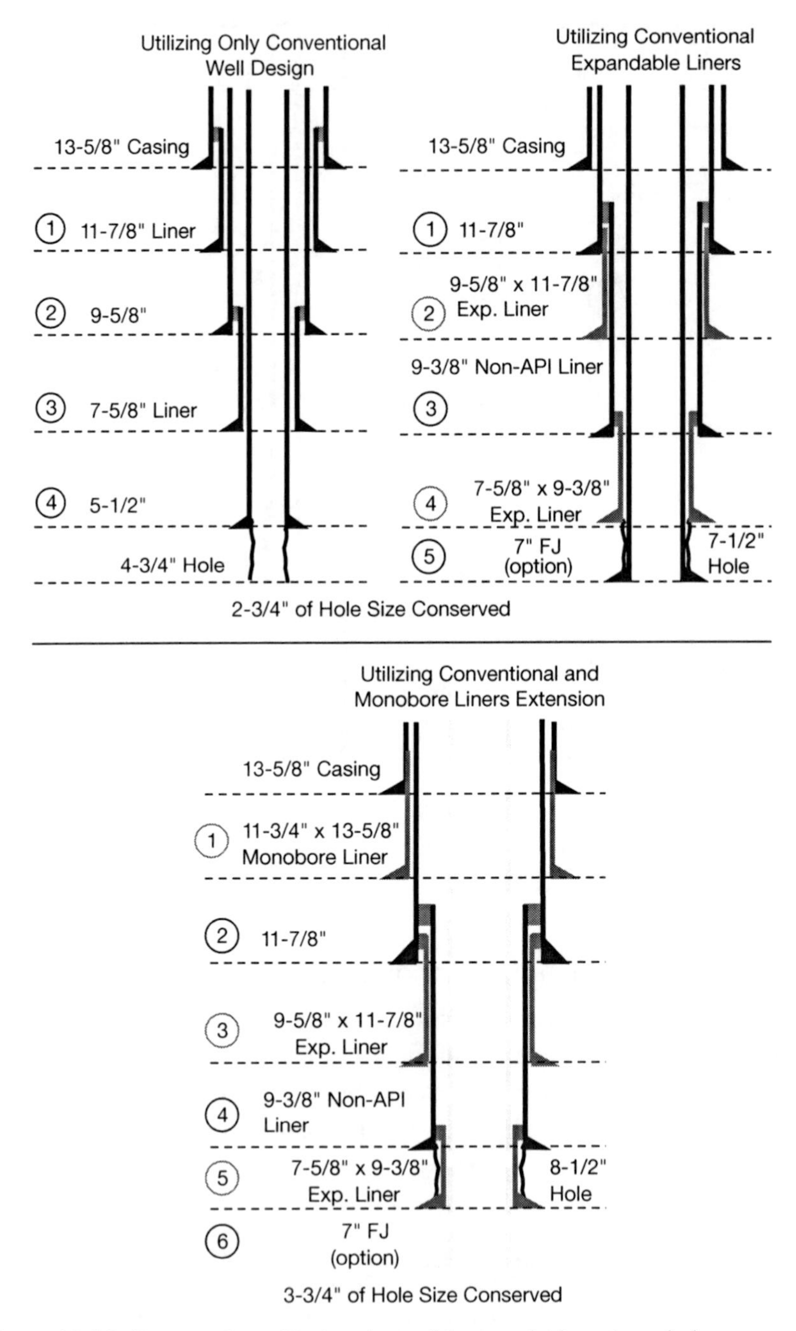

Figure 10.18 Conservation of hole using solid expandable system design.

to address some of these drilling challenges without setting casing. In some instances, shallow flow, sloughing, or overpressured zones may benefit from applying pressure on the backside of the drill string, maintaining a near-balanced environment at the bit and enabling drilling the section to the planned depth. However, if this drilling technique is not planned as part of the well's construction process, it is not easily used as a contingency solution. Alternatively, certain solid expandable products, such as open-hole clad liners, can cover trouble zones and offer a short, cost-effective, non-tied-back casing string with the same size drill-ahead bit used when the trouble zone was encountered. Several open-hole clad-thru-clad liners can address multiple trouble zones in a single hole section without losing any hole size, as shown in Figure 10.19.

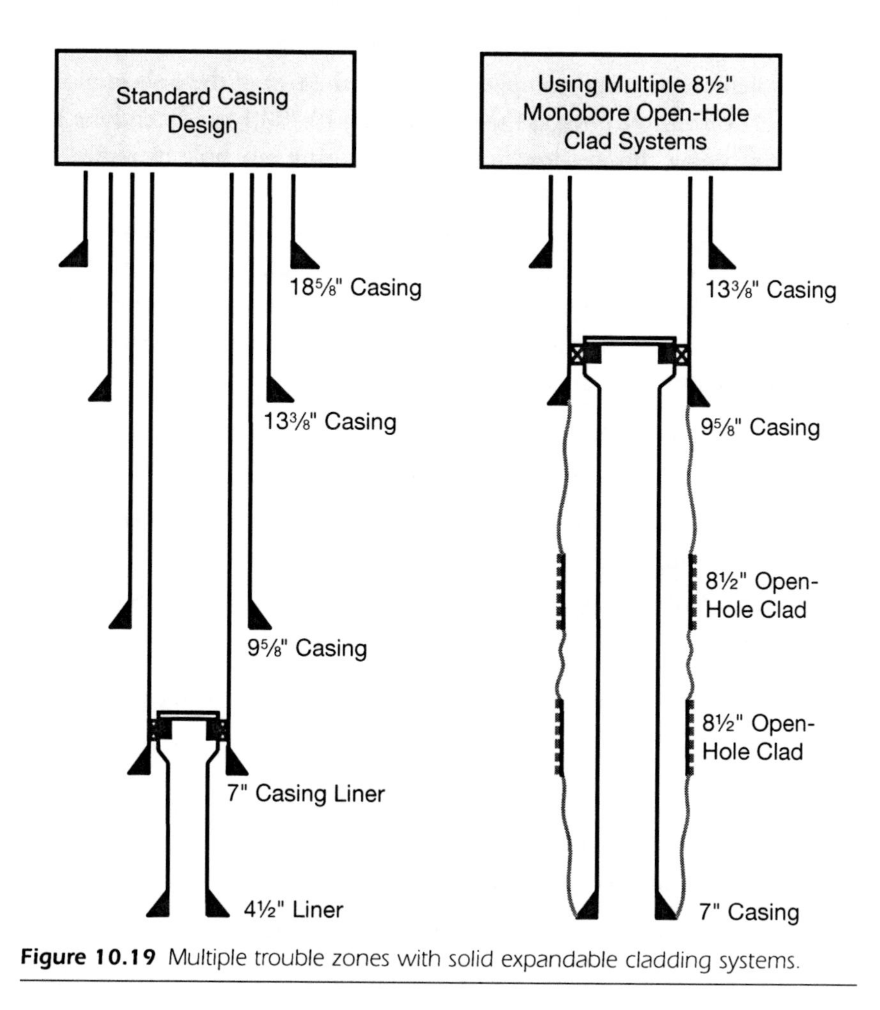

Figure 10.19 Multiple trouble zones with solid expandable cladding systems.

Shallow Flows, Unstable/Sloughing Formations, and Overpressured Gas Zones

To make the economics work, the well's profile must be slimmed down at every opportunity, and still drill through a variety of difficult challenges, such as:

- Shallow flows
- Small gas zones
- Coal stringers
- Sloughing formations

Addressing Trouble Zones and Reducing Well Construction Cost

In some places horizontal wells drilled encounter several formations that challenge the objective of reaching the primary targets, as shown in Figure 10.20(a). The first potential trouble zone, just below the surface casing, can cause a shallow flow and complicate drilling the rest of the hole section. A monobore open-hole liner, as shown in Figure 10.20(b), can extend the 13⅜-inch shoe, cover this shallow flow without losing any hole diameter, and allow drilling ahead with the same size bit.

Although the larger expandable liner addresses the shallow flow drilling challenge, the cost of the liner must also be economically justified. Each potential application to mitigate drilling hazards should be considered in context with the entire well. The cost of a larger monobore liner needs to be evaluated against alternative options, such as:

- Managing the shallow flow-through with effective and conventional drilling techniques and an overbalanced mud system.
- Casing off the shallow flow zone with a conventional liner below the surface casing.
- Implementing managed pressure drilling techniques through the shallow flow.

Considering the needs of the entire well reveals additional trouble zones, as shown in Figure 10.20, deeper in the well that include:

- A potential lost circulation drilling hazard.
- An unstable coal zone.
- A tectonically stressed unstable zone.
- An unstable formation in the build section of the well. This formation also has a high (~28%) H_2S content.

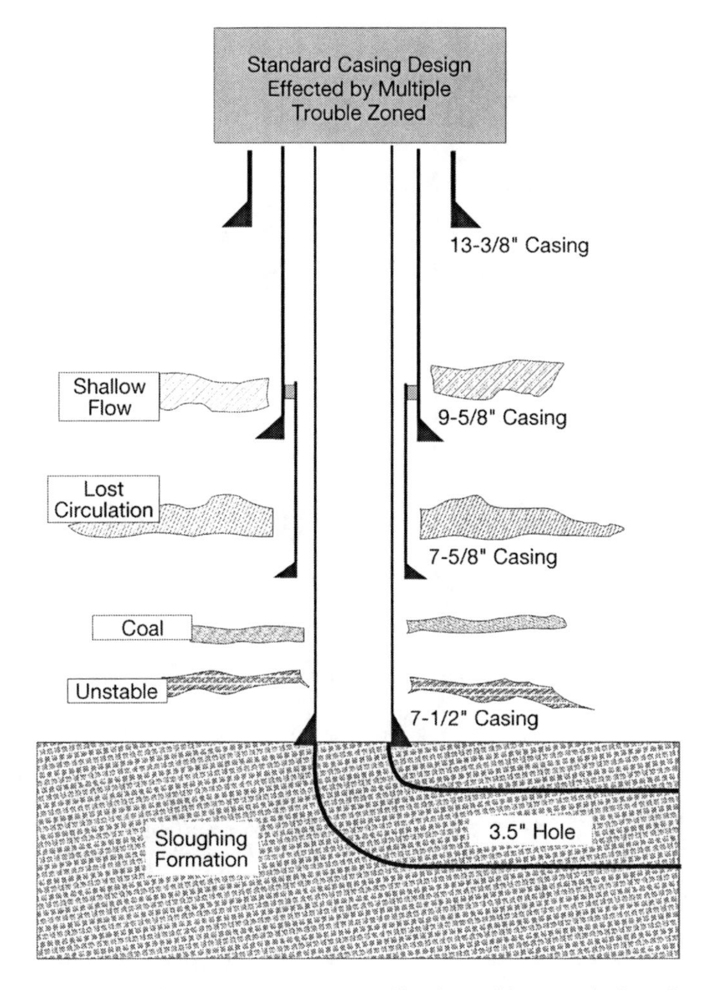

Figure 10.20a Monobore open-hole liner utilized to mitigate a shallow flow and additional problem zones.

Multiple and significant hazards within a tight economic drilling environment are sometimes best mitigated by combining solutions. Evaluating the well's total needs and then applying a combination of managed pressure drilling and solid expandable tubulars provide the most beneficial solution. Figure 10.20 illustrates how, through the use of several different drilling tools, these multiple trouble zones can be mitigated, while maintaining a larger hole size than if only conventional casing strings are used. This resulting large ID allows the horizontal open-hole section to be drilled with a 5⅞-inch bit, enhancing the well's ROI.

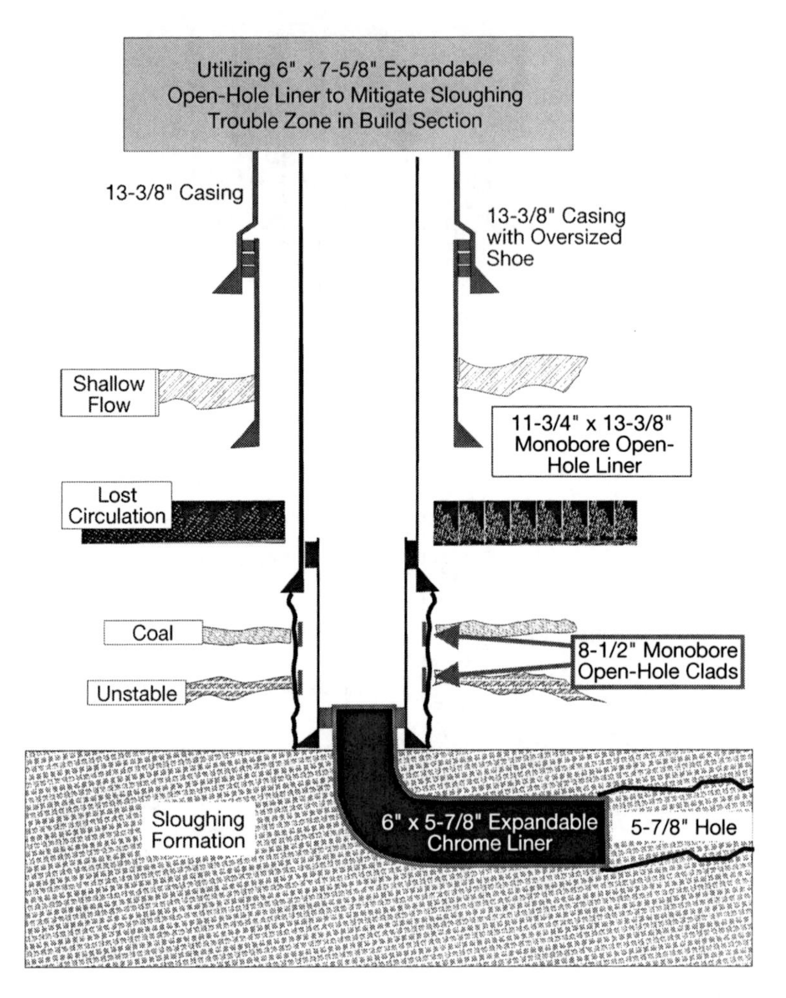

Figure 10.20b Monobore open-hole liner utilized to mitigate a shallow flow and additional problem zones.

The Drilling Solution

Conventional 298-mm (11¾-in.) casing provides a cost-effective method to mitigate the shallow flow, but downsizes the well and prevents drilling the next section with a 311.2-mm (12¼-in.) hole. The larger-size wellbore is critical to setting a 244.5-mm (9⅝-in.) production casing through the first lost-circulation zone. This section requires an underbalanced drilling technique, using nitrogen to lower the mud weight and effectively drill through the lost-circulation zone to the planned setting depth.

The next hole section drilled with a 215.9-mm (8½-in.) bit has to transverse two trouble zones. The first is a lost-circulation zone that can be mitigated with a short (~50-m) section of a 215.9-mm (8½-in.) monobore liner. This non-tied-back open-hole clad has a postexpansion ID of 221 mm (8.7 in.) with a drift of 8.575 inches. Drilling of the hole section can continue with a 215.9-mm (8½-in.) bit and without a special drill-out trip.

The next drilling challenge is an unstable coal. Again, the 215.9-mm (8½-in.) monobore open-hole clad liner provides a barrier to support the unstable coal, so that the remainder of the 215.9-mm hole section can be drilled and then isolated with a 193.7-mm (7⅝-in.) liner. This clad puts one of the primary objective formations behind casing.

At the well's kick-off point (KOP), inclination at this hole section's TD is approximately 50°. After drilling out the 193.7-mm (7.625-in.) casing with a 165.1-mm (6½-in.) bit, the formation entered into typically has a high H_2S content of approximately 25% or more. To exacerbate the drilling challenge, this formation is also unstable after being exposed and begins to slough-in through the well's build section after a few days. This sloughing often requires several sidetracks to successfully drill the 1000-m (2380-ft) horizontal section that is typically open-hole completed.

Key to Drilling Successfully in the Horizontal Section

If the unstable zone is isolated with a conventional 140-mm (5½-in.) liner, the long horizontal section would have to be drilled with a 124-mm (4⅞-in.) bit, adding significant risk to drilling this 1000-m (2380-ft) hole section. Alternatively, a 150-mm (5.905-in.) CRA expandable liner can be run below and tied back into the 193.7-mm (7.625-in.) casing string. The postexpansion ID of the 6 × 7⅝-inch open-hole chrome liner is 151.44 mm (5.962 in.), which facilitates the use of a 149.23-mm (5⅞-in.) bit. Although this well was not drilled in 5000 feet of water, nor required ten casing strings, it was challenging nonetheless, especially when the extreme economic constraints are considered.

Subsalt Rubble Zones

Salt Formations and Deposits

Formations entrapped by salt deposits have been the source of significant hydrocarbons. Salt deposits, as shown in Figure 10.21, are commonly recognized in the Gulf of Mexico (GoM), offshore West Africa, the Middle East, and the North Sea. Some of the salt deposits are quite thick, with wells being drilled through and around these beds in excess of 30,000 feet in true vertical depth (TVD). The first drilling challenge of hydrocarbon objectives that extend deeper has been just to reach the targets. The second challenge

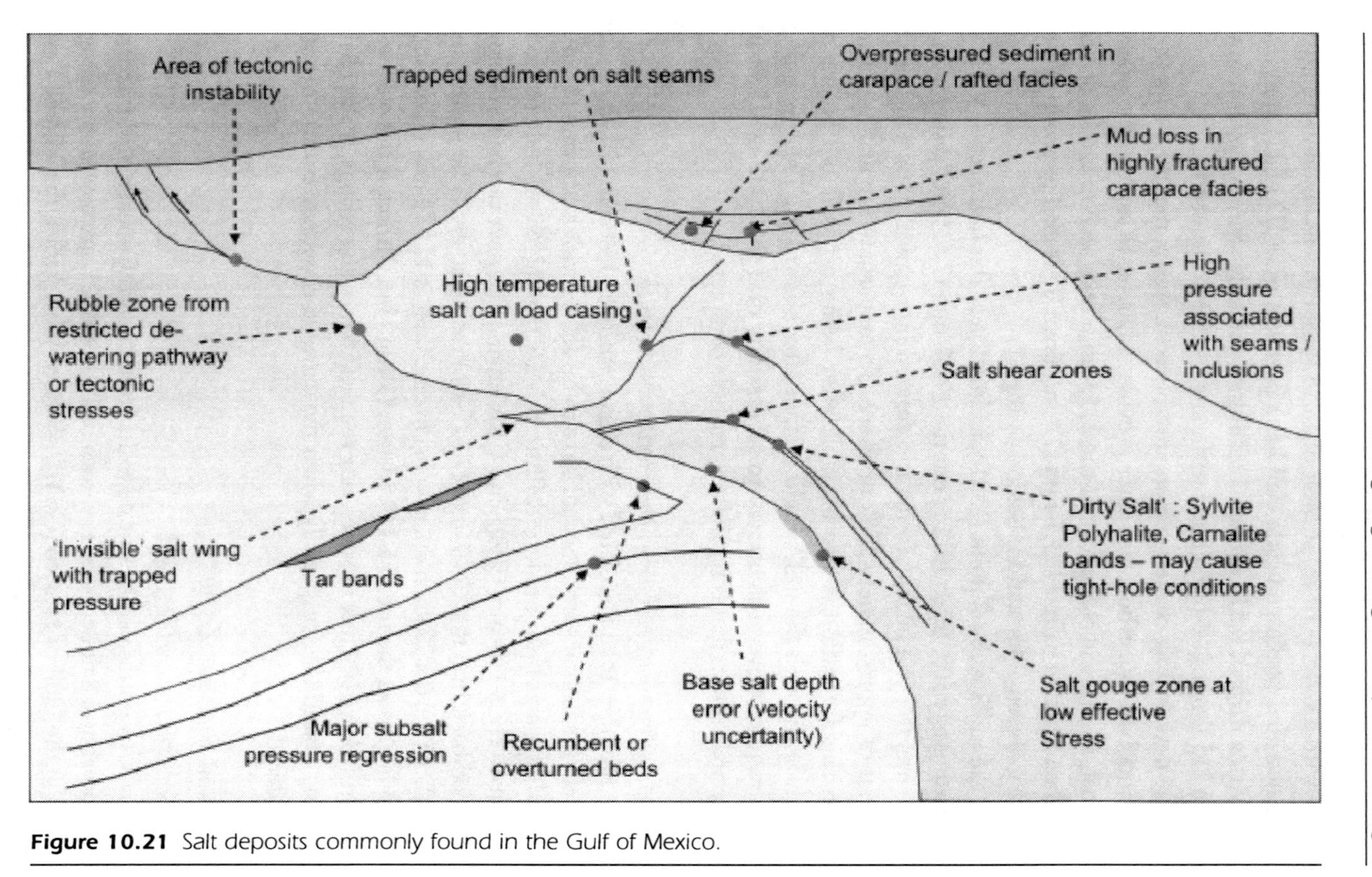

Figure 10.21 Salt deposits commonly found in the Gulf of Mexico.

is to complete these wells with sufficient hole size to make developments economically successful.

Solid expandable liners have helped penetrate these salt formations and facilitated successful drilling operations. Even with years of successful installations, the low collapse of thinner-wall expandable liners has caused significant concern. While expandable liners slow the premature slimming of wellbores when contingency liners are necessary, end users have required an expandable liner that afforded zero hole loss after installation. A solid expandable liner with over twice the collapse of conventional expandable liners, plus zero hole loss postexpansion, actuates successful and effectual drilling into, through, and beneath salt formations.

Addressing Drilling Challenges Surrounding Salt Formations

The majority of nonproductive time (NPT) while drilling for hydrocarbon deposits associated with salt formations is actually drilling the sediments into and out of the salt diaper. Examples of these hazardous drilling environments include:

- Shale sheaths
- Supra-salt "carapace"
- Subsalt "gumbo"
- Subsalt or extrasalt "rubble" or breccia-like zones
- Active or relict translation surfaces along the salt face
- True caprock

In addition to these formation challenges, higher pore pressures and anomalous rock fabric can be present in the sediments surrounding the salt. Given the geological uncertainties around salt bodies, additional contingency casing strings are very advantageous. The hazards associated with drilling in and around salt environments can be divided into many categories, but generally include three main challenges:

- Lost circulation
- Unexpected high-pressure areas
- Loss of hole integrity and wellbore instability

A salt body creates an environment that increases the probability of one or more of these risks manifested while entering, drilling through, or exiting them. This risk potential means that at least two additional strings may be needed to put salt successfully behind casing.

The Drilling Solution

A deepwater well encountered a rubble zone upon drilling out of a salt formation that resulted in the unexpected need to set a casing string. An 11⅞-inch high-collapse conventional liner (Figure 10.22a) had to be set higher than planned to cover the ~1000-ft rubble zone. This lost casing point would put the 9⅝-inch casing being set where the 11⅞-inch was originally planned. Consequently, after encountering another trouble zone, the hole to reach TD would be 4¾ inches and result in the well's potential completion size reduced by as much as 3¾ inches from that originally planned (Figure 10.22b).

To reach TD with a 7½-inch hole, a 3000-ft 9⅝ × 11¾-inch conventional expandable liner was run below the conventional 11⅞-inch liner. Because the expanded ID of the conventional expandable was not large enough to accommodate running conventional 9⅝-inch 53.5 lb/ft casing and set across the subsequent hole section, a non-API 9⅝-inch OD conventional liner was run. To

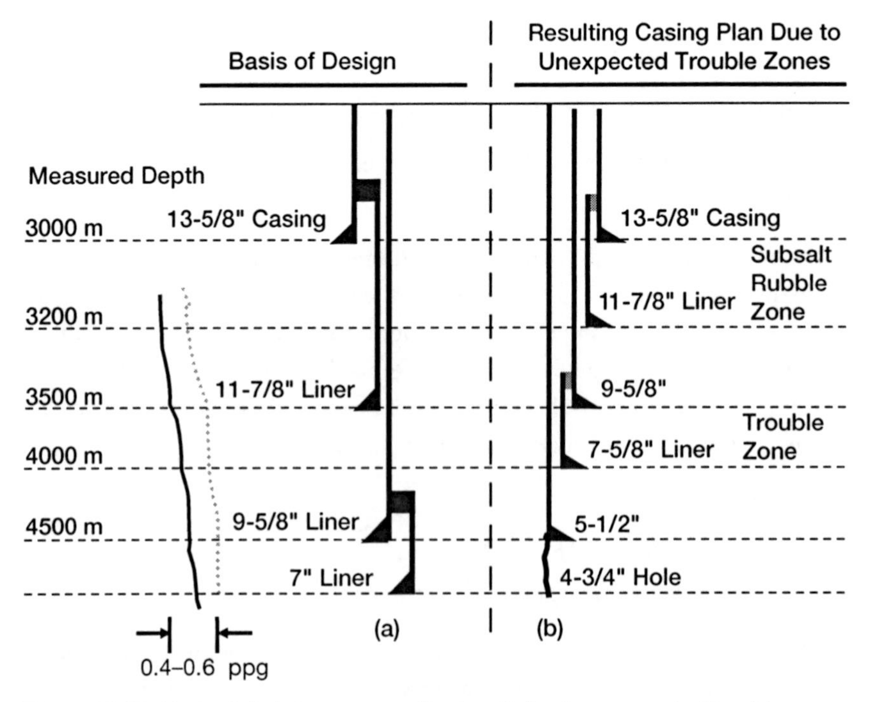

Figure 10.22 Basis of design versus resultant well due to unexpected trouble zones.

get as close as possible back to the original well design, a 7⅝ × 9⅜-inch expandable liner was subsequently run that allowed a 7-inch flush joint liner to be run at TD (Figure 10.23a). Hole-size conservation (2¾-in.) assisted in the management of the ECD and increased the chance to properly evaluate a critical section of the well.

These three liners covered this ~5000-ft interval that had a 0.4–0.6 pound-per-gallon (ppg) pore pressure/fracture gradient window. Using a combination of conventional solid expandable liners and non-API casing preserved ~60% more hole size compared to using only conventional tubulars.

Additional Drilling Solution

Alternately, when the rubble zone was encountered under the salt, a ~1000-ft 11¾ × 13⅝-inch high-collapse, monobore open-hole expandable liner could have been run. The installation of this single monobore expandable

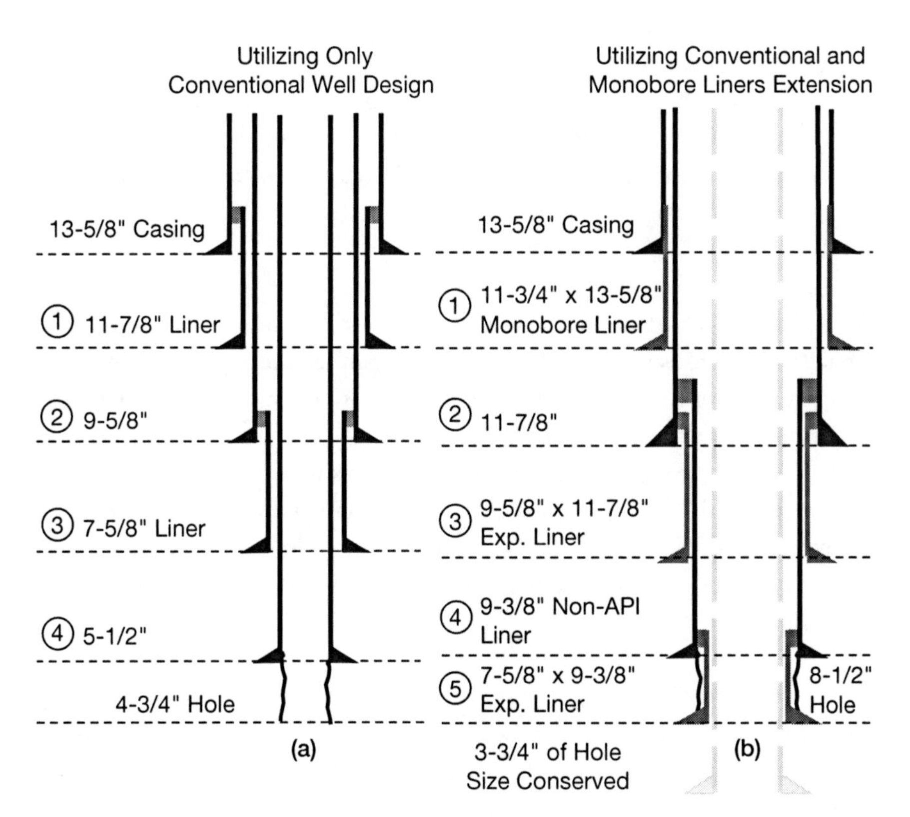

Figure 10.23 Comparison of conventional versus utilizing conventional expandable liners with monobore liner.

liner has the ability to bring the well construction back to the original design, and allow running the 11⅞-inch conventional high-collapse liner at its originally planned depth. The subsequent well construction could have been continued to reach the well's TD with a 7-inch flush-joint liner.

The shorter monobore expandable liner (~1000 ft versus ~3000 ft) doesn't require the subsequent use of the nonconventional-sized well construction equipment. Because the casing used in the 11¾ × 13⅜-inch monobore system has a 0.682-inch wall, its collapse resistance is higher than conventional, thin-walled solid expandable liners.

Extended-Reach Drilling Wells

Shallow-Set Casing and Unstable Sections

Extended-reach drilling (ERD) wells are typically used to "reach" out from a common point to the outer edges of a field development to maximize reservoir drainage. These wells are capable of reaching targets in excess of 10 kilometers and can eliminate the need for permanent drilling facilities/platforms. ERD wells contain long sections of wellbore that transverse several geological horizons, and are exposed for a relatively long time before they are cased off. As illustrated in the following example, unstable or sloughing formations uphole and again near the reservoir can jeopardize the well's economic success, if an optimum-sized completion cannot be facilitated.

Mitigating Shallow Setting of 13⅜-Inch Casing

In an ERD well, setting the 13⅜-inch casing string to cover a set of unstable shale formations is problematic, and can jeopardize the well reaching its primary target with the optimum-size completion. Unstable shale formations must be isolated before the long horizontal well section can be drilled successfully. Setting a conventional casing string is unacceptable, if the well is to use the final 7-inch production liner. If the conventional 13⅜-inch casing is set higher than planned, a monobore open-hole liner can extend the 13⅜-inch casing shoe and isolate the trouble zones without the loss of any hole size (Figure 10.24).

Mitigating Unstable Coal Section above Primary Target

In the casing program for lower in the ERD well, the long 9½ × 12¼-inch hole section has been cased off with a 9⅝-inch liner and an 8½-inch bit has been used to drill to TD. An unstable coal section near TD tends to slough, jeopardizing setting the production string at its planned depth.

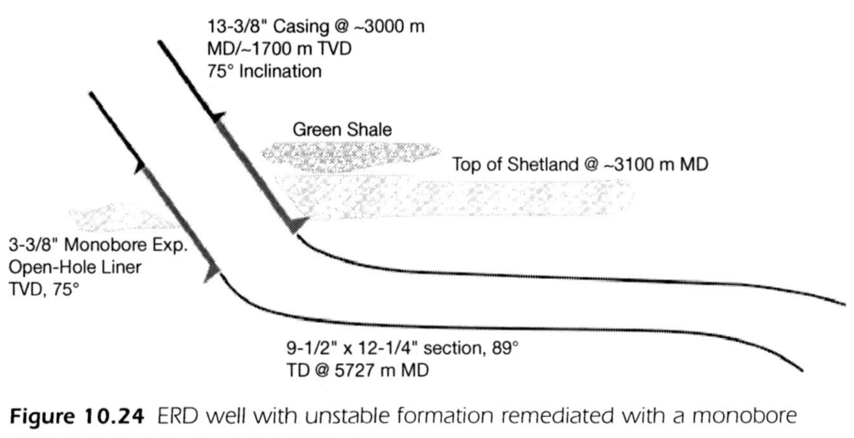

Figure 10.24 ERD well with unstable formation remediated with a monobore open-hole liner in its upper section.

Additional Drilling Solution

The coal hazard can be mitigated using an 8½-inch monobore open-hole clad liner. This liner is not tied back and its length is sized to cover just the drilling hazard. The monobore clad liner is installed in a slightly under-reamed (~1 in. on diameter) hole section and expanded directly against the unstable formation. Effectively isolating the hazard, while maintaining an 8.7-inch ID, allows drilling the remainder of the section and then casing off the entire 8½-inch hole section with the 7-inch production liner (Figure 10.25).

Multiple and Added Contingencies

Extreme Tight Pore Pressure/Fracture Gradient Challenges

Operators are sometimes forced to use nine or more casing strings in their well designs to combat extremely tight pore pressure/fracture gradient challenges as shown in Figure 10.26. Usually, this many strings of conventional casing limit options that can be addressed with conventional casing programs when unexpected trouble zones are encountered. This type of well design has few contingencies, as shown in Figure 10.27, and therefore, can reduce the well's completion and seriously limit the well's productivity or, ultimately, potential field drainage. More forbidding, cost overruns from multiple sidetracks, in some cases, can result in losing the wellbore due to well telescoping and the inability to reach the planned TD.

Solid expandable systems can provide additional contingency strings, as depicted in Figure 10.28, that are best achieved by using these liners in the

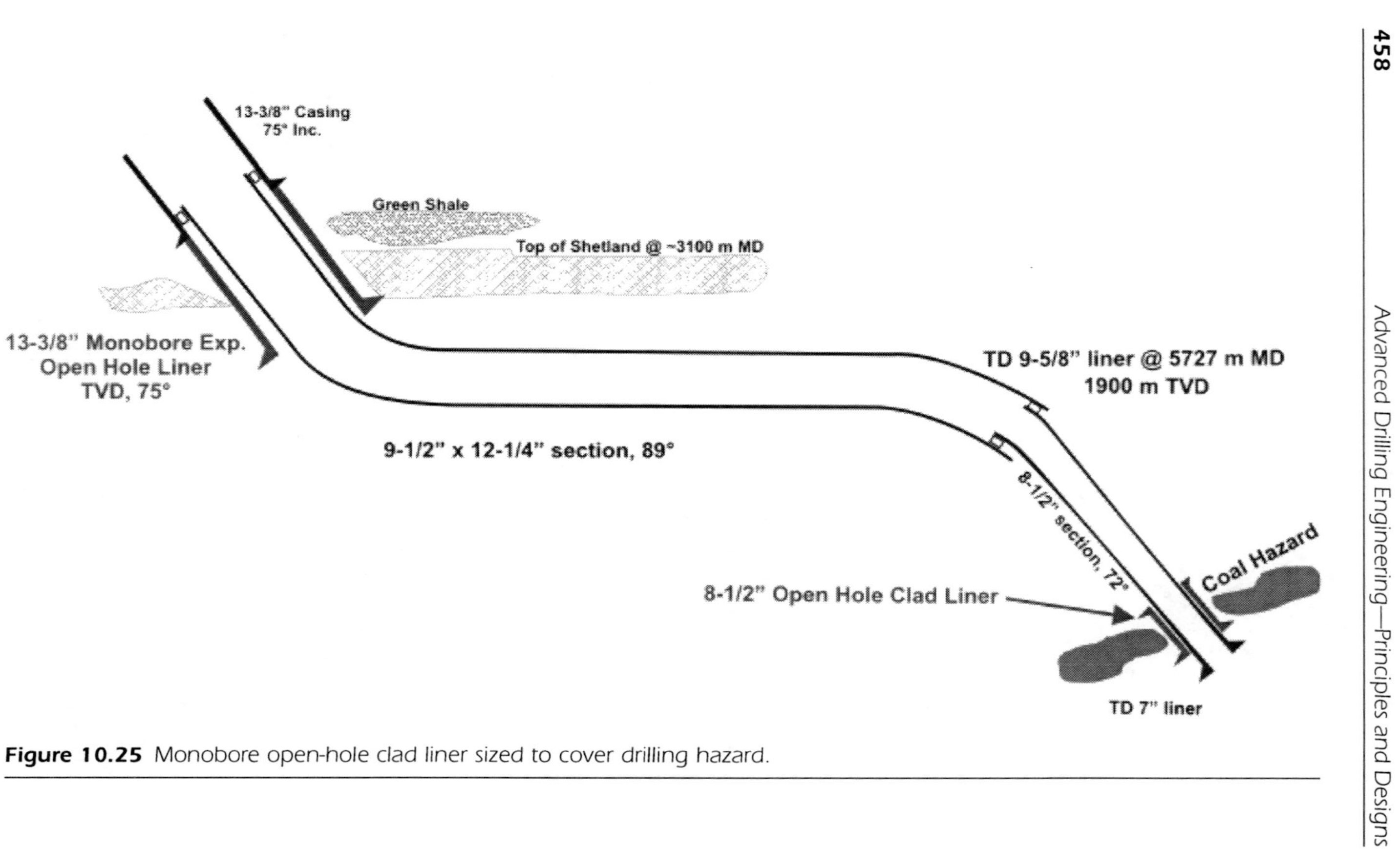

Figure 10.25 Monobore open-hole clad liner sized to cover drilling hazard.

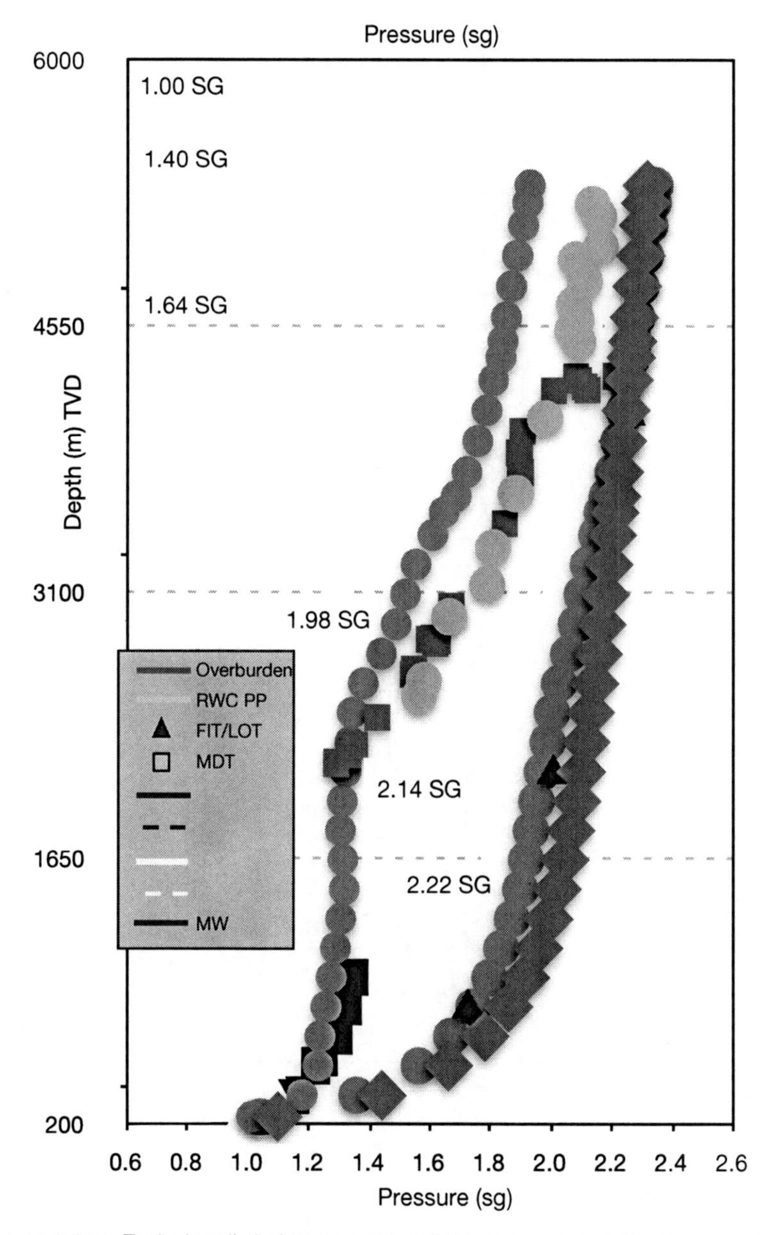

Figure 10.26a Typical well design to combat tight pore pressure/fracture gradient challenges.

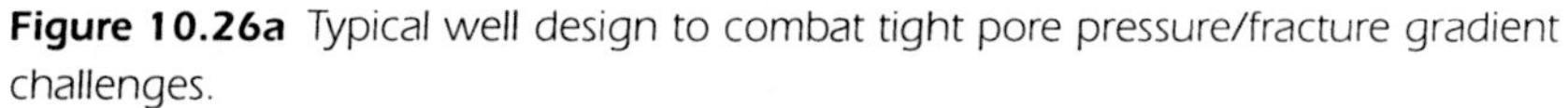

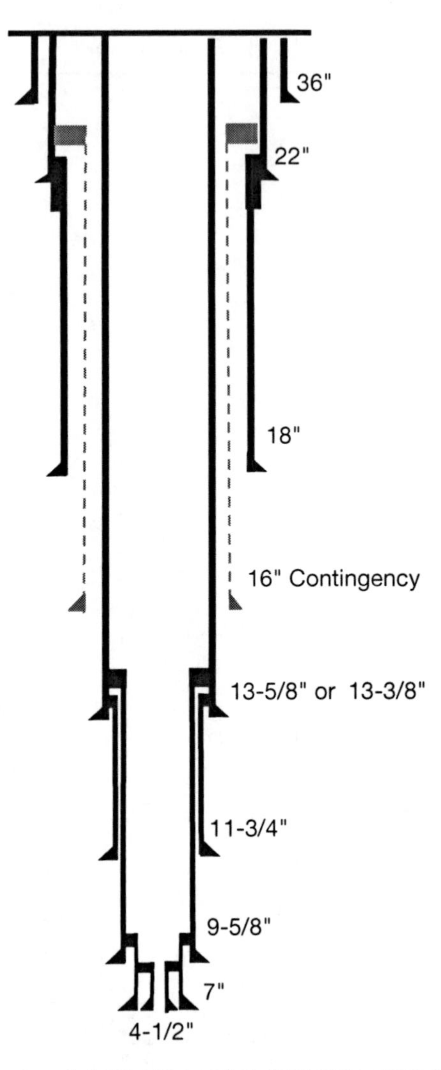

Figure 10.26b Typical well design to combat tight pore pressure/fracture gradient challenges.

first instance in the upper-hole section. A single monobore solid expandable extension provides a means for isolating problematic zones in the upper wellbore. This approach makes it possible to overcome issues created by the weight of long 16-inch strings that can exhaust drilling rig capability and tensile limitations of connections and liner equipment. Testing the draw works' weight capacity can result in prematurely setting this string prior to the pressure event envisaged.

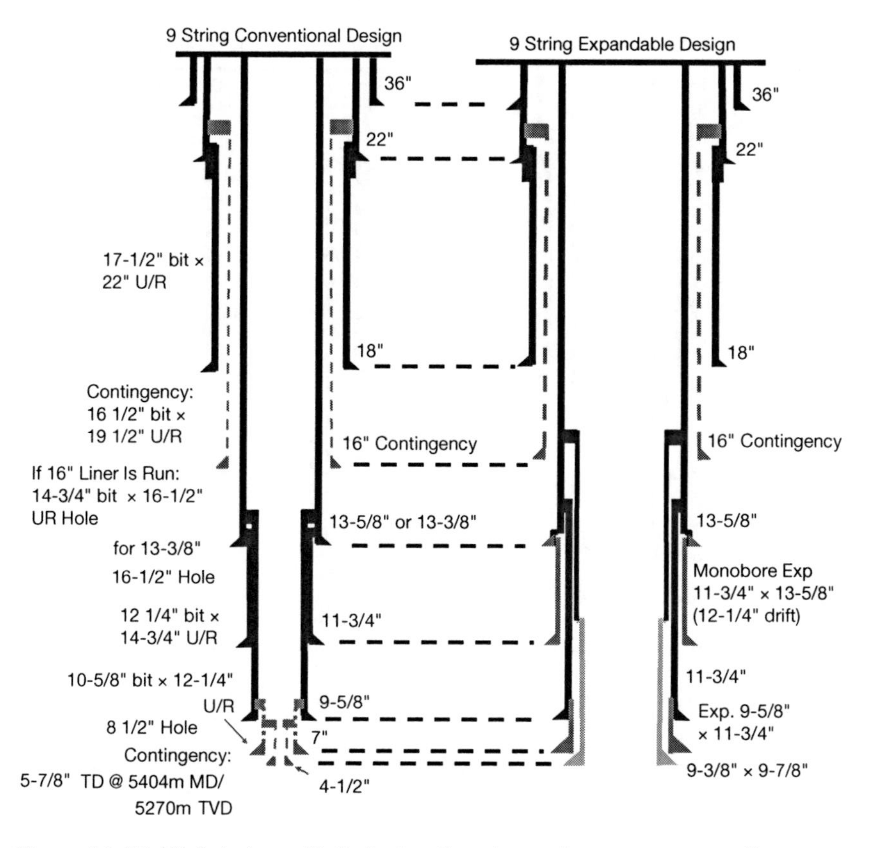

Figure 10.27 Well design with limited options to combat pore pressure/fracture gradient challenges.

The Drilling Solution

After the conventional 13⅜-inch casing string is run affixed with an oversize tie-back shoe, the 11¾ × 13⅝-inch single monobore liner can be run, cemented, and expanded back into the tie-back shoe. The expanded monobore liner has a postexpansion drift of 12¼ inches. This drift enables an additional 11¾-inch liner system to run through the expanded liner and set across a pressure event deeper in the wellbore. Ultimately the monobore/conventional liner configuration keeps the planned casing design on track, while securing the drilled sections behind pipe.

In areas where casing strings are set at short intervals, the necessity to have additional barriers as planned contingencies has proven to be prudent. A monobore tie-back shoe allows for the monobore liner to remain a planned contingency until required. In the event the monobore solution is

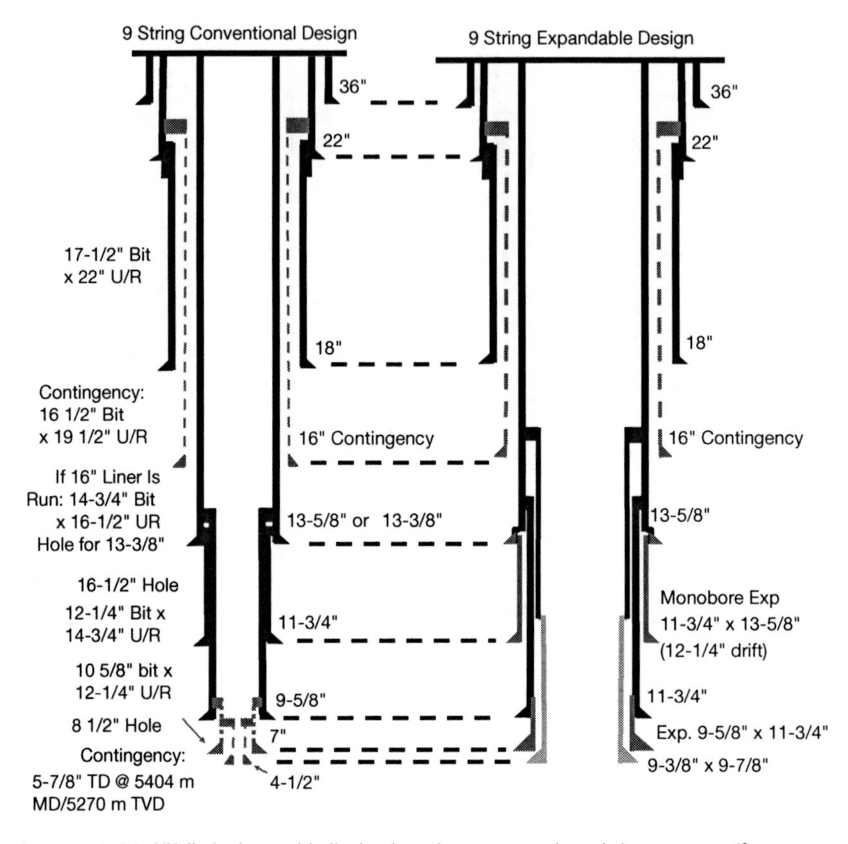

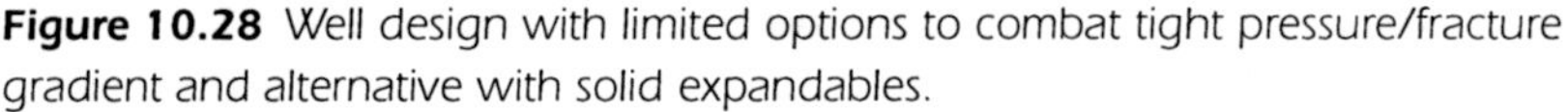
Figure 10.28 Well design with limited options to combat tight pressure/fracture gradient and alternative with solid expandables.

not needed, the next planned casing can be run and tied back to the previous conventional casing string, bypassing the use of the monobore liner. Adding this technology provides the option of another casing string when needed.

Additional Drilling Solution

Additional conventional expandable liners can be used as contingencies as the well increases in depth. A conventional expandable liner can be run below the monobore expandable liner, or below the 11¾- or 11⅞-inch conventional casing string when required. The inclusion of these extra sizes will result in some loss of hole size; however, in many cases, these sizes allow enough pass-through ID to facilitate running a flush-joint casing string.

Tight Drilling Margins

Maximizing Hole Size While Mitigating Drilling Hazards and Fighting Tight Margins

Solid expandables can be planned as contingencies because of pore pressure/fracture gradient challenges, as shown in Figure 10.29. The use of conventional casing designs would require the upper strings in the well to grow in size, driving up the cost significantly. Solid expandable liner systems planned as contingencies could downsize the upper well plans while maintaining completion size.

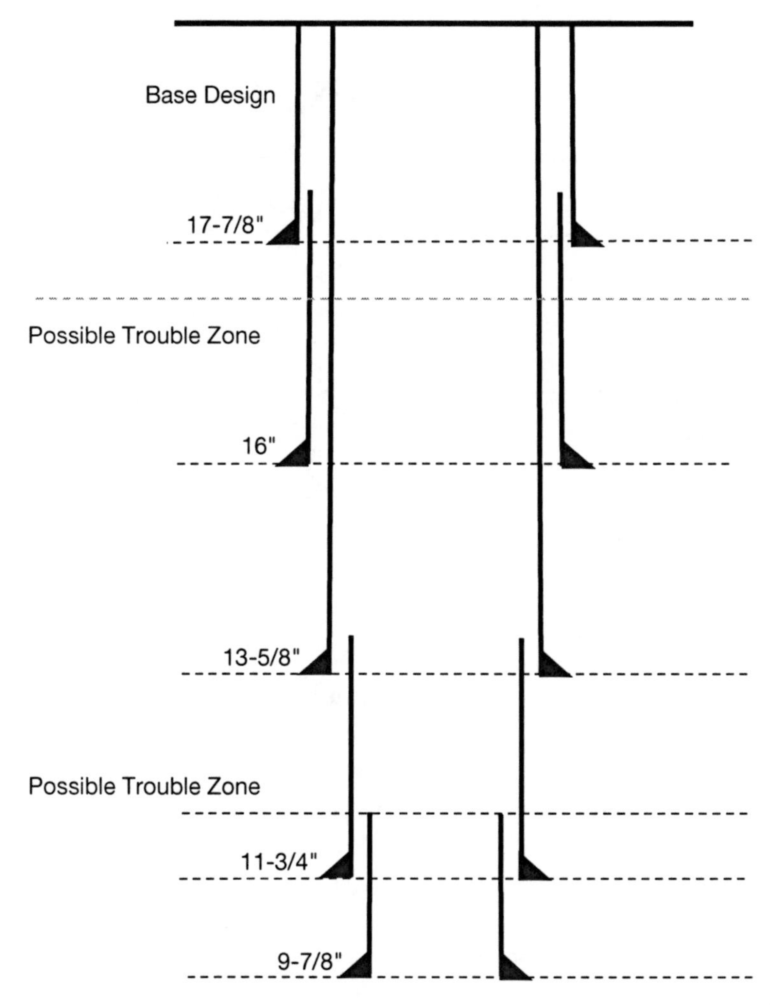

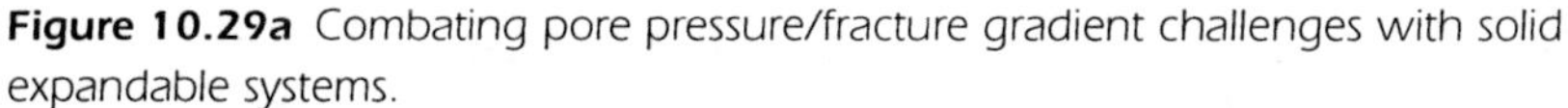

Figure 10.29a Combating pore pressure/fracture gradient challenges with solid expandable systems.

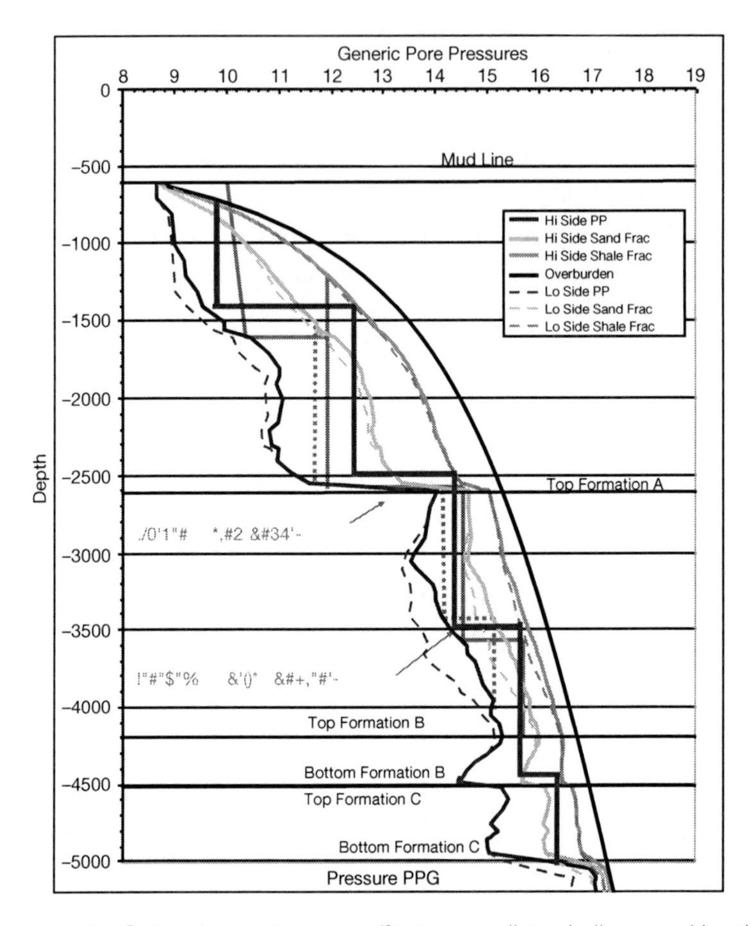

Figure 10.29b Combating pore pressure/fracture gradient challenges with solid expandable systems.

Due to the risks associated with tight-tolerance drilling, several solid expandable liners were identified as contingency options, as shown in Figure 10.30, to maintain hole size and achieve optimum completion sizing within the reservoir. Various drilling hazards and the need to reach the target at ~16,500 ft (~5000 m) measured depth (MD) with long stepouts and eventually TD the well with a 7-inch production casing, necessitated solid expandable liners.

A shallow flow hazard at ~5870 ft (1790 m) is sometimes encountered that requires premature setting of the 16-inch casing. This short landing in turn requires the 13⅝-inch casing string to be set in the hole section originally planned, where the 16-inch liner was to be set. To mitigate setting the 16-inch liner early, due to the shallow trouble zone, an expandable monobore

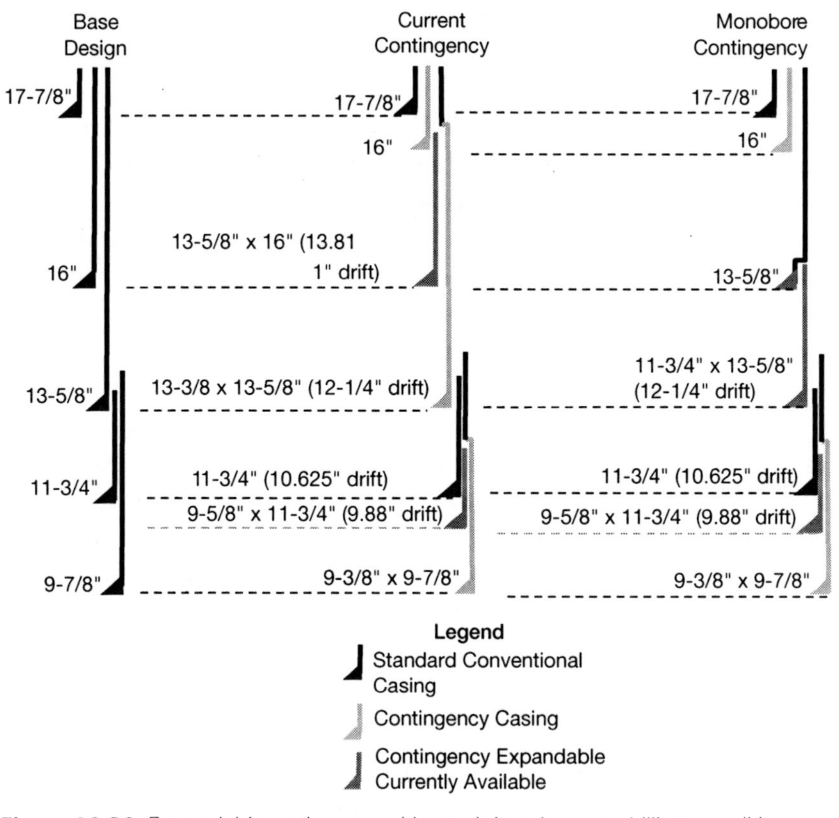

Figure 10.30 Expandable options to mitigate tight-tolerance drilling conditions.

liner could be considered. Approximately 2260 ft (688 m) (2001 ft [610 m] in open hole) of an 11¾ × 13⅝-inch monobore open-hole liner would extend the 13⅝-inch casing shoe to its planned TD. In anticipation of this application, an oversized shoe is run on the bottom of the 13⅝-inch liner, which is set and followed by normal drillout to the next hole section. This configuration allows the monobore system to be run through the tie-back shoe into the 2001 ft (610 m) of the open wellbore, cemented (if required), and expanded back into the tie-back shoe. Effectively, this liner extends the conventional 13⅝-inch casing shoe, maintains hole diameter, and eliminates the need to alter the casing plan of the well.

The design requirement for the solid expandable liner is to provide larger drift IDs, thus rendering more contingency options and improving the chance of reaching TD. Monobore expandable liners used in shallower hole sections, and lower pressure and temperature environments, reduce the operational risks and add contingency options for later in the drilling process, if required.

The monobore system allows conventional drilling products and casing strings to be used above and below this particular solid expandable liner system, while offering the benefit of an "additional" casing string. Following the installation of the 11¾ × 13⅝-inch monobore open-hole liner, drilling ahead can continue with a standard 12¼-inch drill bit. Conventional 11¾- or 9⅝-inch casing or liners can be installed through the expanded liner. Only with upfront planning of these contingency options can the proposed field development be economically feasible.

In the same field development project, for a sand formation at ~12,100-ft (~3690-m) TVD, a 2000-psi pressure change occurs that typically results in the short landing of the 11¾-inch casing string. To minimize telescoping of the well, a 9⅝ × 11¾-inch conventional solid expandable liner can be run, followed by a 9⅞ × 9⅝-inch liner. This combination of liners returns the well profile to a size closer to the original well plan, and facilitates the eventual completion of the well with a 7-inch casing.

Supplementary Problems

Answer true or false to the following questions.

1. The action of running solid expandables has no inherent risk.
2. The first solid expandable–related patent was issued in 1865.
3. The mechanical property of the steel used to make the casing determines the spread between yield and failure.
4. During plastic deformation of the pipe, all forces should remain constant and the volume of the deforming material has to be balanced.

Fill in the blanks to the following questions:

5. With expansion, ultimate tensile strength tends to (a) __________ and elongation tends to (b) __________.
6. Hooke's Law states that in the elastic range of a material, (a) __________ is proportional to (b) __________.
7. Full strain is always a sum of (a) __________ and (b) __________ deformations.
8. In a rig floor environment, the rate of expansion is usually limited to __________ ft/min.

9. In most complex wells, the critical casing size is usually the _________ string.

Answer the following questions with proper reasons:

10. Expansion forces and pressure are directly proportional to what three factors?
11. What methods are used to expand solid tubulars?
12. What is the most common method of expansion used today?
13. What three factors deform the pipe during compliant rotary expansion?
14. Explain how expansion affects material balance.
15. What physical law enlarges the diameter of the casing with a slight thinning of the casing wall accompanied by shrinkage in length?
16. What is the typical amount of pipe shrinkage in percentage?
17. What percentage of complex well costs can typically be attributed to NPT?
18. Why do expandable liners exhibit reduced postexpanded collapse resistance?
19. Why is the "bottom-up" expansion method the most prudent approach?
20. What are the primary types of open-hole expandable systems?
21. Explain the difference between elasticity and plasticity of a material.

References

1. Contribution by Patrick York.
2. Mack, R. D., T. McCoy, and L. Ring. "How In Situ Expansion Affects Casing and Tubing Properties." *World Oil* (July 1999): 69–71.
3. Filippov, A., R. Mack, L. Kendziora, and L. Ring. "In-Situ Expansion of Casing and Tubing—Effect on Mechanical Properties and Resistance to Sulfide Stress Cracking." NACE Standard 00164.
4. Filippov, A., R. Mack, L. Cook, P. York, L. Ring, and Terry McCoy. "Expandable Tubular Solutions." SPE 56500; SPE Annual Technical Conference and Exhibition, Houston, Texas, October 3–6, 1999.
5. Stewart, R. B., F. Marketz, W. C. M. Lohbeck, F. D. Fischer, W. Daves, F. G. Rammerstorfer, and H. J. Böhm. "Expandable Wellbore Tubulars." SPE 60766; SPE Technical Symposium held in Dhahran, Saudi Arabia, October 1999.

6. Lohoefer, C. L., B. Mathis, D. Brisco, K. Waddell, L. Ring, and P. York. "Expandable Liner Hanger Provides Cost-Effective Alternative Solution." IADC/SPE 59151; IADC/SPE Drilling Conference, New Orleans, Louisiana, February 23–25, 2000.
7. Haut, R. C., and Q. Sharif. "Meeting Economic Challenges of Deepwater Drilling With Expandable-Tubular Technology." Deep Offshore Technology Conference, 1999.
8. Mason, D., G. Cales, M. Holland, and J. Jopling. "Using an Engineering Analysis Process to Identify Pragmatic Applications for Solid Expandable Tubular Technology." OTC 17438; Offshore Technology Conference, Houston, Texas, May 2–5, 2005.
9. Dupal, K., C. J. Naquin, C. Daigle, L. Cook, and P. York. "Well Design with Expandable Tubulars Reduces Costs and Increases Success in Deepwater Applications." *Deep Offshore Technology* (2000).
10. York, P., M. Sutherland, D. Stephenson, and L. Ring. "Solid Expandable Monobore Openhole Liner Extends 13⅜ in. Casing Shoe without Hole Size Reduction." OTC 19656-PP, Offshore Technology Conference, Houston, Texas, May 5–8, 2008.
11. Cruz, E. J., R. V. Baker, P. York, and L. Ring. "Mitigating Sub-Salt Rubble Zones Using High Collapse, Cost Effective Solid Expandable Monobore Systems." OTC-19008-PP; Offshore Technology Conference, Houston, Texas, April 30–May 3, 2007.
12. Schmidt, V., J. Dodson, and T. Dodson. "Gulf of Mexico 'Trouble Time' Creates Major Drilling Expenses." *Offshore* 64, no. 1 (January 1, 2004).

11

Directional, Horizontal, and Multilateral Well Economics

Directional, horizontal, and multilateral wells provide more formation exposure in targeted reservoirs by sidetracking at the appropriate kick-off depth, resulting in less total footage drilled and fewer surface wellheads/ locations. Each additional lateral increases reservoir exposure without the necessity of redrilling the vertical distance to reservoir depth. Onshore surface location expenses entail significant costs for permitting, location preparation, land-use rights, road access, and other related activities. Offshore locations further affect costs for platform facilities, bed space, and transportation. To analyze the economics of directional, horizontal, and multilateral wells, one must, therefore, understand the drivers for a given field location and attempt to capture these in the modeling.

Well Planning Considerations

Planning for drilling an oil/gas well requires many detailed studies and evaluations of every aspect that directly, or indirectly, affects the successful and economic outcome of the project. Listed below are some of the more important considerations in well planning and a brief discussion of each[1]:

- Area geology
- Formation pore pressure/fracture gradients

- Logging program
- Casing program
- Mud program
- Cementing program
- Well design and profile
- Bottomhole assemblies (BHAs)
- Hydraulics program
- Drill bit program
- Routine drilling
- Drilling time curve
- Drilling rig

Time Value of Money

The time value of money concept enables one to calculate the future value of an investment made today, and also to calculate the present value of a past investment. It provides the way to compare and adjust the authorization of expenditure to arrive at the realistic budget estimates and calculate the return on investment. It is based on the concept that the present-day dollar is worth more than that of the future. It will have a time value as long as interest can be earned. In a basic sense, this helps to estimate the project cost and financial goals and to evaluate alternate courses of action during the planning and execution of projects.

For example, a company may have a bought a drill bit three years ago at a price of $2000, and may not have used it. The time value of money concept facilitates finding the future or present value, based on the company's average rate of return.

The future value is given as

$$FV = PV\left(1+\frac{r}{n}\right)^{n\times m} \tag{11.1}$$

where

FV = Future value
PV = Present value
r = Periodic interest rate or growth rate in fraction
n = Number of payments per year

m = Number of years

For continuous compounding, the equation can be written as:

$$FV = PVe^{r \times n} \tag{11.2}$$

From the formula, it can be seen that the present value becomes:

- Less as the number of periods becomes larger
- Less as the interest rate period becomes larger

Therefore, the future value is influenced by both the rate of return the company is providing to the investors, as well as the number of payments per year.

PROBLEM 11.1

A company bought a sidetracking tool for a level 2 well five years ago for \$10,000 and did not use it. What would its value be today to break even, assuming an average rate of return for the past five years is 14%.

Solution

Past value = \$10,000
Interest rate = 14% per year
n = 4 periods per year

$$FV = 10{,}000\left(1 + \frac{0.14}{4}\right)^{4\times5} = \$19{,}897$$

Economic Parameters

When evaluating the economics of a project, the criteria include:

- Net present value (NPV)
- Net income (NI)
- Rate of return (ROR)
- Payout time (POT)
- Net cash flow (NCF)

While this book does not intend to provide a comprehensive coverage of economic parameters, this section provides a rudimentary overview of these issues. A fundamental tenet of any economic analysis is the relationship between the sum of future revenues (income), adjusted for the time value of money and capital expenditure (investment). Eq. 11.3, discounting a revenue stream for the time value of money, gives the present value of the income as:

$$V_p = \sum_{n=1}^{N} \frac{R_n}{(1+i)^n} \tag{11.3}$$

where

V_p = Present value of the revenue stream
R_n = Annual revenue, after taxes and depreciation, for each year of n years
n = Number of years the project is considered
i = Discount rate
p = Present value

Eq. 11.4 gives the net present value, symbolized as V_{np}, by setting i (discount rate) and n (project duration) for a given investment I made to generate the revenue flow.

$$V_{np} = \sum_{n=1}^{N} \frac{R_n}{(1+i)^n} - I_0 \tag{11.4}$$

where

V_{np} = Net present value of the revenues stream
R_n = Annual revenue, after taxes and depreciation, for each year of n years
n = Number of years the project is considered
i = Discount rate
p = Present value
I_0 = Initial investment

PROBLEM 11.2

If one expects a net annual revenue of $1 million and applies a discount rate of 10%, the present value of the revenue for five years is (0.91 + 0.83 + 0.75 + 0.68 + 0.62) = $3.79 million. If the initial investment is $2 million, subtracting from the present value of the revenue gives a net present value equal to $1.79 million.

The discount rate reflects a number of considerations:

- Inflation
- Escalation in the value of competing investments (including the interest rate paid by banking institutions)
- Risk (based on weather, geopolitics and infrastructure)

Because of inherent annual compounding, a large discount rate substantially reduces the net present value of future revenue. Using the example above, a discount rate of 30% leads to a present value of the revenue equal to $2.44 million, and a marked reduction of the net present value to $0.44 million.

Eq. 11.4 determines the rate of return and the payout, two frequently used economic criteria, by setting them equal to zero. One can expect to maximize the rate of return among competing investments, or at least have a minimum acceptable value. In this case, one should minimize the payout time.

Also, Eq. 11.4 solves for the rate of return I, by setting the number of years n. In the example above, with $n = 5$, one can calculate the rate of return as 40%, the discount rate at which the future revenue has a present value equal to the investment.

Eq. 11.4 calculates the number of years to pay out the investment, by fixing the discount rate, when the discounted future revenue equals the investment. In the example above, if $i = 20\%$, then the payout time is slightly less than three years.

For the project to be profitable, the initial project investment can be given as[2]:

$$I_0 = \frac{E(r_p)}{1 + r_f\left[E(\tilde{r}_m) - r_f\right]\beta} \tag{11.5}$$

where

$E(r_p)$ = Expected rate of return for a certain period t_p
$E(r_m)$ = Expected rate of return of market portfolio
r_f = Risk-free rate of interest
β = Project risk beta given as

$$\beta = \frac{E(r_p) - r_f}{E(r_m) - r_f} \tag{11.6}$$

PROBLEM 11.3: SINGLE VERSUS DUAL LATERAL WELL

The Austin Chalk, spanning a large part of the state of Texas from northeast to southwest, is one of the main areas of directional well activities. Many consider the formation, a naturally fractured carbonate of very low reservoir permeability of approximately 0.1 md, to be the ideal place for multilateral wells, far superior to hydraulically fractured vertical wells.

Calculations suggest that a 4000-ft single horizontal well in a 50-ft thick structure, with an index of anisotropy equal to 1, provides a dimensionless productivity index.

Eq. 11.7, the steady-state oil inflow calculation, shows that J_D equals 2.35, whereas the J_D is equal to 2 for a well half as long (2000 ft). With k = 0.1 md, h = 50 ft, $B\mu$ = 1, and Δp = 2000 psi, the two well lengths produce initially 162 and 136 STB/d, respectively.

$$q = \frac{kh\Delta p}{141.2B\mu} J_D \tag{11.7}$$

where

J_D = Productivity index
k = Permeability
h = Reservoir formation thickness
B= Formation volume factor
μ = Viscosity
Δp = Pressure drop

Thus, a dual-opposing lateral with two 2000-ft long laterals has an initial production rate of 272 STB/d, compared with 162 b/d initial production from the 4000-ft single lateral. One can decide between these well architectures by balancing expected incremental revenue against incremental costs.

Supposing that for a given drainage, the decline for a single horizontal is 20% per year, and the decline for a dual lateral configuration is 30% per year. One can estimate a monthly production rate schedule, using the following notation in Eq. 11.8:

$$q(t) = q_i e^{-(r_d/12)t} \tag{11.8}$$

where
r_d = Annual decline rate
t = Time in months

Integration of this relationship provides the cumulative recovery for each configuration.

One can also find the answer to a simple question such as "What is the acceptable incremental investment that justifies a dual lateral?" by using Eq. 11.9:

$$\Delta I = \left[\sum_{n=1}^{N} \frac{R_n}{(1+i)^n}\right]_{dual} - \left[\sum_{n=1}^{N} \frac{R_n}{(1+i)^n}\right]_{single} \tag{11.9}$$

Price Elasticity

Price elasticity is used to measure the effect of economic variables, such as demand or supply of rigs or wells drilled, with respect to change in the crude-oil price. It enables discovering how sensitive one variable is with the other one, and also it is independent of units of measurement. It is the ratio of the percentage change of wells and footage drilled, to the percentage change in the crude-oil price. It describes the degree of responsiveness of the rig in demand or rig in supply to the change in the crude-oil price.[1]

Price elasticity of drilling measures how responsive the direction of change of the drilling rigs is when the price of oil changes.

$$\text{Drilling price elasticity} = \frac{\text{Percentage change in drilling wells}}{\text{Percentage change in crude-oil price}} \tag{11.10}$$

The elasticity, E, is given as:

$$E = \frac{\%\Delta R}{\%\Delta P} = \frac{\frac{dR}{R}}{\frac{dP}{P}} = \frac{dR}{dP} \times \frac{P}{R} \tag{11.11}$$

PROBLEM 11.4

If the oil price increases from \$30.00 to \$34.00, and if the drilling rigs increase from 1220 to 1240, calculate the elasticity of drilling rigs.

Solution

The elasticity of drilling rigs would be calculated as:

$$\frac{\frac{(1120-1140)}{1120} \times 100}{\frac{(30-34)}{30} \times 100} = \frac{1.79\%}{13.33\%} = 0.133$$

The midpoint formula is another method, and gives the answer regardless of the direction of change:

$$\text{Price elasticity} = \frac{(R_2 - R_1)/[(R_2 + R_1)/2]}{(P_2 - P_1)/[(P_2 + P_1)/2]}$$

$$E = \frac{\frac{(1140-1120)}{1130}}{\frac{(34-30)}{32}} = 0.14$$

Ranges of Elasticity

In order to compare the calculated elastic values, the elasticity can be classified as follows:

- *Inelastic:* The number of drilling rigs does not respond strongly to the oil price. Elasticity < 1.
- *Elastic:* The number of drilling rigs responds strongly to the oil price. Elasticity > 1.
- *Perfectly inelastic:* The number of drilling rigs does not respond to the oil price change. Elasticity = 0.
- *Perfectly elastic:* The number of drilling rigs responds infinitely to the oil price change. Elasticity > 0.
- *Unit elastic:* The number of drilling rigs responds by the same percentage as the oil price change. Elasticity = 1.

Risk Analysis in Directional, Horizontal, and Multilateral Well Developments

The consideration of risk is very important in assessing directional, horizontal, and multilateral well economics. Drilling and completion investment is concentrated in a limited number of wells, and the inherent risks of failure increase when multiple laterals are drilled.

Comparing the development of a field that has directional, horizontal, and multilateral wells with the development of a field that has multiple single wellbores is the classic example of having "all your eggs in one basket." The economic advantages of directional, horizontal, and multilateral well development can be very high, but failures can be very expensive. Any serious economic analysis of the use of directional, horizontal, and multilateral well technology must therefore include a quantified risk analysis.

A simple method to assess risks involved in a development plan is to use a decision tree analysis. A deterministic decision tree delivers a "go" or "no-go" decision, based on estimated information of project costs, project values (revenues), and risk information. A statistical decision tree considers a range of parameters, and the sensitivity of the parameters used in risk analysis.

A deterministic decision tree, such as the one shown in Figure 11.1, contains three types of nodes:

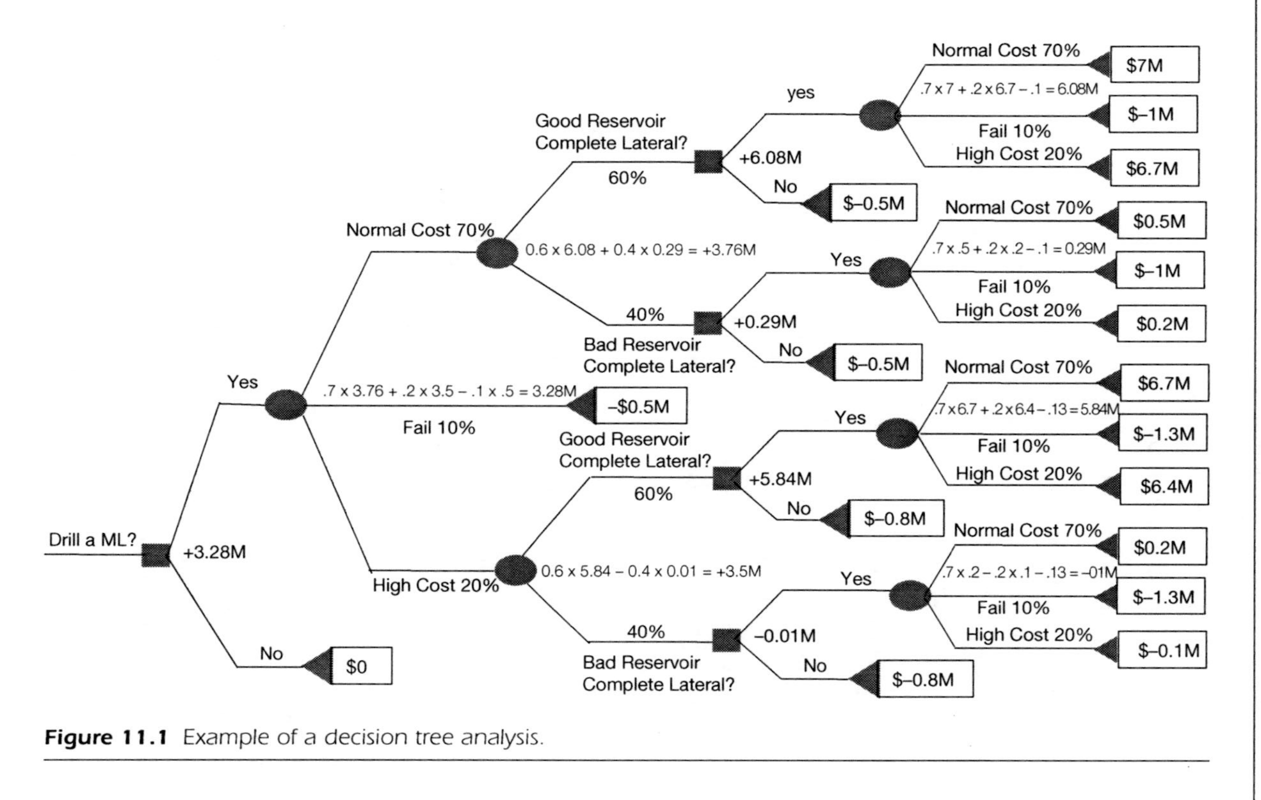

Figure 11.1 Example of a decision tree analysis.

- Terminal—represented by triangles
- Chance—represented by ovals
- Decision—represented by rectangles

Each node in a decision tree has a value. The value of a terminal is its payoff. The value of a chance node changes, since there is more than one outcome associated with the chance. The value of a chance is the expected mean value (V_{em}) at that node, defined as

$$V_{em} = \sum_{i} \left(\text{probability of outcome } X \text{ value of outcome}\right)_{outcome\ i} \tag{11.12}$$

Finally, the value of a decision node is the value of the outcome at the node that has the highest value, which is always selected as the final decision.

The decision-making tree always moves from right to left. Problem 11.5 illustrates how to use the deterministic decision tree to conduct risk analysis.

PROBLEM 11.5

This example analyzes the effect of adding a lateral to an existing well, using a deterministic decision tree. A decision tree helps one make a decision on whether to drill an additional lateral from an existing producing well; in other words, to convert a horizontal well into a dual lateral well.

The cost for drilling the lateral ranges from \$500,000 to \$800,000, and the completion costs range from \$500,000 to \$800,000. There is a 70% chance of a normal drilling cost (\$500,000), a 20% chance of a high drilling cost (\$800,000), and a 10% chance that drilling may fail. Similarly, there is a 70% chance that completion will be a normal cost (\$500,000), 20% chance that completion will be a high cost (\$800,000), and 10% chance that completion may fail. If the reservoir is as expected, the revenue is calculated at \$8,000,000; at the low limit, the revenue is calculated at \$1,500,000.

Assuming that normal drilling and completion costs are used for failure cases, although this may be optimistic, since failures are often at the end of excessive expenditures, what is the sensible decision on the development plan? Figure 11.1 illustrates the decision tree for this problem. From the right, one first decides the value of the chance node: to complete the lateral in a good reservoir. The value of a normal completion cost is the revenue in this case (good reservoir, \$8,000,000, or \$8M), minus the cost of drilling and completion (\$0.5M + \$0.5M), resulting in \$7M. The value for a high completion cost is the revenue in this case (good reservoir, \$8M) minus the cost of drilling and completion (\$0.5 + \$0.8), which is \$6.7M. The value of the failure case is the total cost of drilling and completion, and, using the normal cost case, the value is –\$1M. Thus, the value of the node is:

EMV = (probability of normal completion cost) × (value of normal completion cost)
+ (probability of high completion cost) × (value of high completion cost)
+ (probability of failure) × (value of failure)
= (0.7 × \$7M) + (0.2 × \$6.7M) + (0.1 × –\$1M) = \$6.08M

The value of the not-complete-lateral terminal is \$–0.5M, the cost to drill. This does not include the cost to complete the lateral. Therefore, at the decision node, to complete or not complete the lateral, one can pick the higher value choice to complete the lateral. The value of that node is \$6.08M.

The rest of the decision tree is calculated with a similar approach. The left-most node shows a value of \$3.28M to drill a directional, horizontal, and multilateral well, or \$0 if not, so the decision is to drill the directional, horizontal, or multilateral well.

A convenient way to assess and quantify risk by the method of statistical decision tree analysis is through a Monte Carlo technique. In this technique, all input variables for the economic analysis are described with probabilities of occurrence, not as fixed values. For example, the expected investment is assumed to have an expected value of \$2 million, but the actual investment could follow a Gaussian distribution with this mean value. Similarly, expected reservoir production may range accordingly. The price of oil can be the most wildly fluctuating variable of all.

Analysis of this situation is done by iteration, with a random-number generator picking a value of the variable sampled from a probability density function. The results (for example, NPV, ROR, or POT) are shown in terms of probability of occurrence. The range of these values, especially the fraction above or below an acceptable value, then becomes the criterion of the project attractiveness. Risk-averse companies may take a different decision than risk-taking companies.

Capital Expenditure Reductions

An approach to reservoir exploitation, such as a directional, horizontal, or multilateral well, must be compared to some other next best alternative, in order to estimate its value. As shown previously, a dual lateral well can be compared to a separate single horizontal well of longer length.

In the case of an existing well, abandoning the current production and sidetracking the well to a new target can be compared to another approach that only temporarily shuts in the existing production, adding a new lateral to the current production. In the case of slot-constraints on an offshore platform or a drilling pad, the cost of adding a new slot and drilling and completing can be compared to reconfiguring an existing well as directional, horizontal, or multilateral to drain the newly identified reservoir target.

A method of estimating the value of the technology is to compare the cost of the directional, horizontal, or multilateral wellbore configuration with other configurations, then balance against the revenue streams of each scenario. The initial cost should include the vendor cost of junction-related hardware and installation personnel, the operating rig cost for junction installation, and the normal cost of the number of trips required to install the junction. The junction installation cost depends on the following:

- Depth of the junction
- Tripping time to reach the junction depth
- The "on-bottom" time required to construct the junction
- Rig rate

The deeper the junction, the higher the rig rate, and the lesser the number of trips required for junction installation. The less amount required for on-bottom time, then the larger the benefit of a directional, horizontal, or multilateral approach. One should upscale the total expenses by a factor to account for mechanical risk, or perform a more formal risk analysis. If normal drilling and completion budgets are allowed, for example, a +5% margin, then the directional, horizontal, or multilateral might reflect a +15% increase in lateral costs, depending on the level of maturity of the directional, horizontal, or multilateral system that is considered. Reservoir risk is used to decrease the forecast production to reflect a chance the reservoir may not produce at the rates anticipated.

In conventional directional, horizontal, or multilateral wells, revenue from the production cannot start until all laterals are completed, and the full well is put on production. This means the construction cost benefits of additional laterals must be balanced against delays in production from the already-completed wellbores using the same parent borehole. The total time given to construct the lateral wellbore can be applied to the parent wellbore production as a delay. Each lateral has a cost/revenue worksheet, and totals across worksheets provide the summation of the discounted net present value of the drainage scheme considered.

Reserves Additions

There are many examples of reserves additions enabled by directional, horizontal, or multilateral technology. These are primarily reservoir accumulations that are too small to justify dedicated wells.

Real-Options Valuation

Real-options valuation is a method that addresses uncertainty inherent in any oil and gas development project, due to the inability to know, with certainty, the geology one encounters before drilling, the price of oil or gas due to global events, or the fiscal regimes where governments are less stable. Figure 11.2 shows how this uncertainty was historically managed by ensuring project economics were robust enough to absorb the statistical miscalculations when they occurred. In real-options thinking, uncertainty is exploited: "I do not know the price of oil with certainty two years from now, but I can hold within my portfolio of opportunities an option to spike production upward

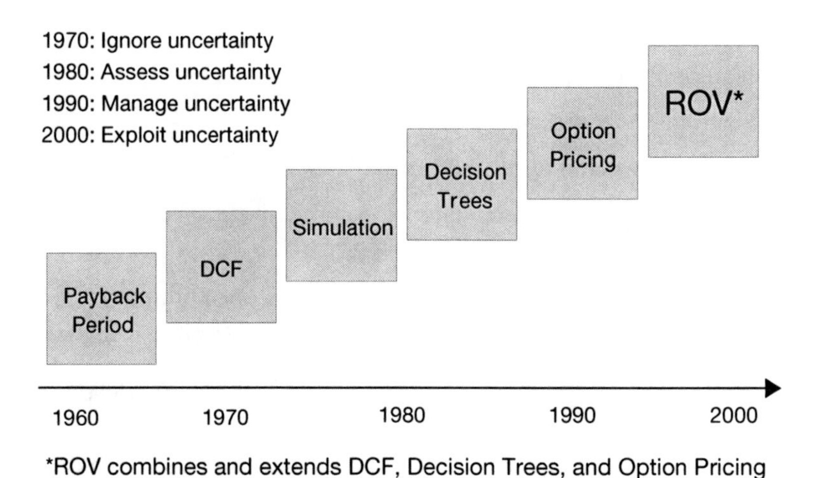

Figure 11.2 Developing methods to exploit uncertainty. (Source: Soussan Faiz, Texaco.)

quickly, in response to increases in oil price." This option to capitalize on the price change has value.[3,4]

Discounted cash-flow analysis versus real-options thinking has been compared to playing the lottery versus playing poker. When playing the lottery, one simply buys a ticket and waits to see the outcome. When playing poker, one makes incremental bets to stay in the game until there is enough data to decide whether to place a larger bet, hold, or fold. Classical discounted cash flow assumes one is playing the lottery, in the sense that there is no control over downstream decisions. Real-options projects are more like poker, where one can actively manage the project to keep options open and reduce uncertainty before irreversible commitments are made.[5]

Real-options valuation has been used in the electric generation industry, in which base load may be satisfied with coal or nuclear plants. During peak consumption hours, however, there is a need for peaking generation. This peaking generation must be easy to activate quickly, and it typically demands a high premium in price. In the realm of oil and gas wells, land assets that can be drilled quickly might be viewed as peaking capacity expansion options. On the other hand, a deepwater development with a billion dollar commitment cannot be easily turned on and off.[3]

The key to a real-options approach is to identify the options inherent in the project under study. Options may include:

- Delaying or deferring decisions, or irreversible capital commitments, until uncertainty is reduced
- Abandoning a project
- Expanding or contracting a project scope
- Shutting down and restarting project later
- Switching input or output materials to take advantage of different market prices
- Growing (for example, hub and spoke designs open options to add production from neighboring properties, once a hub facility is in place)

Setting up a project's analysis in this manner enables active management of the asset as reality evolves.[6]

Real-options value can far exceed the value derived from the more common discounted cash-flow (DCF), net-present-value analysis, and the higher the uncertainty, or volatility, the higher the option value climbs. If one ignores this option value, then corporate officers may actually be destroying corporate value by basing decisions solely on DCF analysis.[3]

Directional, horizontal, and multilateral wells provide flexibility in a project plan:

- By enabling more reservoir drainage points for a given platform slot.
- By enabling an expansion of production by opening more laterals to simultaneous flow.
- By enabling switching options between the flow of oil- or gas-producing laterals.
- By reducing the sensitivity to anisotropy through the addition of laterals in different directions relative to reservoir permeability.[7]

Combining directional, horizontal, or multilateral technology with intelligent well monitoring and control technology effects the options suggested. In the specific case analyzed, an option cost of $13 million for adding intelligent well monitoring and control boosted the conventional development project NPV of $281 million to an improved NPV of $491 million, almost doubling the potential return. This is an example of classic real-option leverage, the ability "to maximize upside gains and minimize downside losses by reacting and responding to uncertainties over time."[8] By integrating learning models together with this option flexibility, investment decisions and project returns can further improve.

Other Methods of Cost Estimation

The cost estimate is an important aspect in the well planning process, especially while drilling horizontal, extended, and ultra-extended-reach, and directional, horizontal, or multilateral wells, because it depends heavily on the technical aspects of the projected well path design. The actual well cost is obtained by integrating expected drilling and completion times with the well design. Two important cost-estimation methods are the deterministic cost-estimation method and the stochastic analysis method (such as Monte Carlo technique or decision tree analysis).

Traditionally, well costs are estimated by using a deterministic calculation procedure with time and cost inputs. In most instances, the well planner makes the "best estimate" of the time required for well completion, based on the historical performance in representative wells called offset wells. The planner calculates well cost by multiplying the drilling time by the anticipated total daily cost of the rig and services combined, then adding an amount for fixed costs. The planner can create a deterministic estimate laying at any point along the continuum of possible well outcomes, producing a value closer to, or rather less than, median outcome.[9-11]

On the other hand, the Monte Carlo method provides a probability distribution based on the offset data. A Monte Carlo trial samples the inputs and combines them together into a possible result. Some of the disadvantages are in defining the minimum and maximum values, which may lead to errors. The errors may further lead to a systematic underestimation of uncertainty in the inputs, and, therefore, in the forecast results. Some of the statistical methods of estimating the well cost are[12,13]:

- Neural network model
- Support-vector machine model
- Econometrics

These models are briefly explained in the following sections.

Neural Network–Based Cost Estimation

Neural networks have a remarkable ability to derive meaning from complicated or imprecise data, and can extract patterns and detect trends that are too complex to be noticed by either humans or other computer techniques. A trained neural network can be thought of as an "expert" in the category of information it has been given to analyze. This expert can then be used to provide projections, given new situations of interest, and answer "What if…?" questions. Artificial neural networks (ANNs) consist of a number of simple processing elements (PEs), or artificial neurons, linked with a set of interconnections, representing neural cells or neurons and their axons or semiconductors in the human brain. The processing elements are arranged and organized in different forms, depending on the type of network and its paradigm. These neurons are organized in the form of layers. In the simplest architecture of an ANN, there is an input layer, one or more hidden layers, and an output layer. Figure 11.3 shows the flow of information in a neural-network model.

Neural networks have the capability of learning from a number of input patterns (representing different problem encounters) and their associated output patterns (representing the conclusions and decisions), seen schematically in Figure 11.4. During the process called "training," the network generalizes the knowledge and becomes capable of providing solutions to new problems.

Once a network is trained using an adequately representative training set, it can be used to classify or to predict the output of the modeled system for a

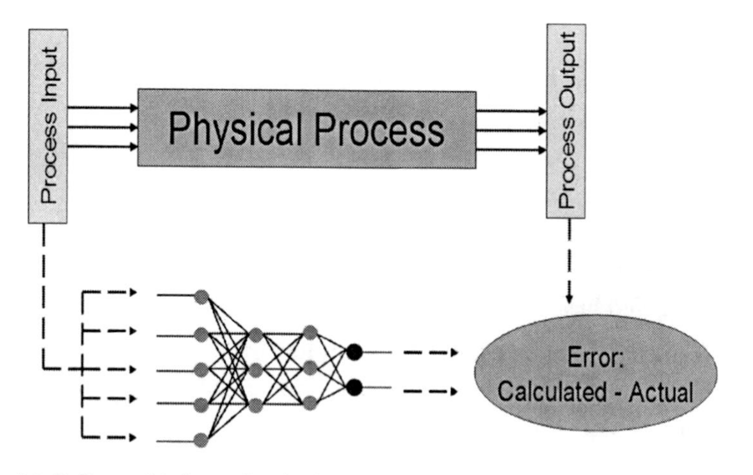

Figure 11.3 Flow of information in the network.

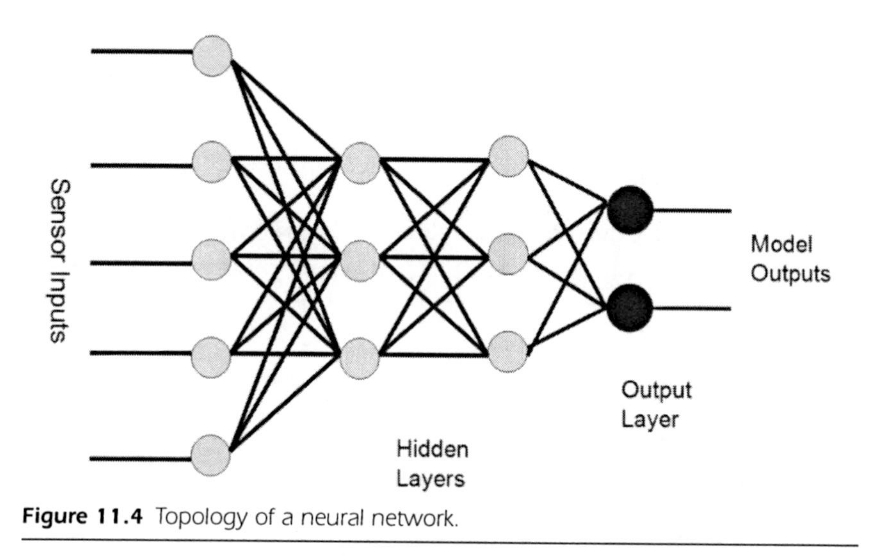

Figure 11.4 Topology of a neural network.

given input pattern. For training, offset well data with their course parameters and well path design types are required. ANNs are very useful in providing the meaningful answers, even when the data to be processed include errors or are incomplete. ANNs can also process information extremely rapidly when applied to real-world problems.

In an ANN, each neuron receives information from other neurons, processes it through an activation function, and produces output to other neurons. The output of a neuron k, Y_k, is given by:

$$Y_k = F\left(\sum_{j=1}^{n} W_{kj}x_j + b_k\right) \tag{11.13}$$

where

F = Activation function
x_j = Inputs to the neuron
w_{kj} = Weights
b_k = Applied bias

Before an ANN can be used to perform its task, it should be trained to do so. This training or learning process determines the weights and biases, using an appropriate learning algorithm.

Weights and biases control the influence of a certain input parameter. The weight represents the correlation between the input and the output. An input parameter that has a large influence on the result is assigned a large-weight magnitude. Transfer functions transform the input of a neuron to an output. Two common transfer functions used for transformation are:

- Nonsigmoid
- Linear function

The selection depends on the complexity of the problem. A neural-network program is designed in such a way that this is performed automatically, and can be given as:

$$E = T - A \tag{11.14}$$

where

E = Error of all output training sets' parameters
T = Target (known output parameters)
A = Output predicted by the neural network

The error between the predicted and actual outputs fine-tunes and evaluates the effectiveness of the model, in order to identify the key differential sticking mechanisms. In this particular case, a generalized feed-forward/back propagation training model can be used to reduce the error by adjusting the weights and biases incorporated in the model. Reduction is achieved by adding this error to the old weights, with new weights calculated as follows:

$$\Delta w = \eta \times O \times \delta + \mu \times \Delta_{prev} \tag{11.15}$$

where

η = Learning rate

μ = Momentum, both between 0 and 1
O = Output from the given layer
δ = Difference between the output and the expected output
Δ_{prev} = Previous change in weights

The number of nodes in the input layers and output layer are determined according to the requirement of the problem. The number of nodes in the hidden layer is maintained at least at three. It is increased according to the weight difference, in order to get the error difference to a minimum value.

Neural-Network Architecture

To estimate the well cost using different well path design or well type, a simple four-layer generalized feed-forward neural-network structure can be used for the ANN model.

The first layer, an input layer, consists of several PEs, including:

- Course parameters
- Target parameters
- Deviation parameters
- Formation properties

The second layer is hidden, and automatically assigns the number of PEs according to the strength of the data. Finally, the output layer consists of two or more PEs, such as well type and well design. The output layer is fully connected to all the units in the hidden layers, as shown in Figure 11.5, where

$f(x_1, ..., x_n)$ = Goal function
$x_n(x_1, x_2, ..., x_i)$= Activation function in the hidden layer of n – units
$x_1, x_2, x_3, ..., x_i$ = Input units
w_i = Weight of the basis function

Support-Vector Machine–Based Cost Estimation

In addition to estimating the cost, it is also important to estimate the percentage of accuracy in the prediction, thereby proving the credibility of the predicted cost estimate. The analysis from such a model is based on the constraints of different well path and formation variables. Traditional neural-network approaches suffer difficulties with generalization, producing models that can overfit the data. This is a consequence of the optimization algorithms used for parameter selection, and the statistical measures used to select the "best" model.[12,13]

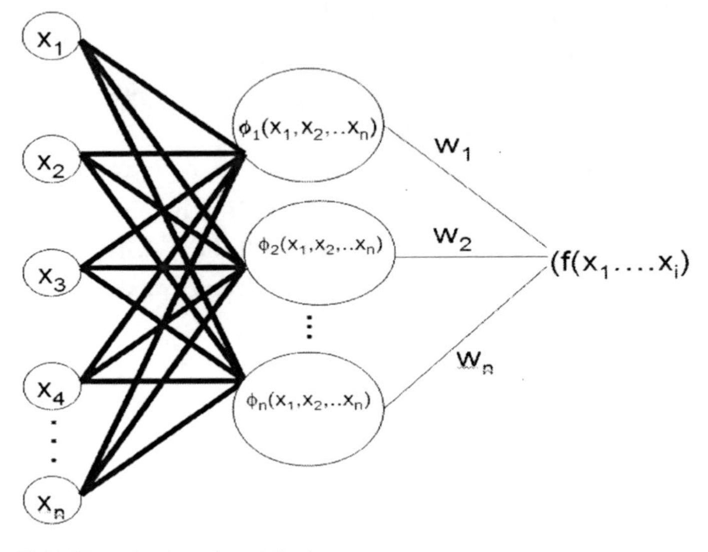

Figure 11.5 Neural-network architecture.

One can track and adjust the well cost by using the weighting functions associated with the operating variables. The problem associated with this type of risk analysis is that the choices are for the "multiple descriptors of risk weights." The best choices are based on reliable, relevant sources describing the uncertainty and variability in the model parameters used in the optimization process. To disassociate the weighting functions and their distributions, this study used a support-vector machine (SVM) model that is based on statistical knowledge of well cost-estimation and prediction using previously available well plan and well path and well trajectory data. One can use an SVM-based algorithm to learn from the analysis of relevant data and to adjust weights, in order to compensate for the current well operations. The algorithm finds the best hyperplane, with the largest margin, that separates instances from different well operations.

What Are Support-Vector Machines?

SVMs are a set of supervised learning methods used for classification and regression. They belong to a family of generalized linear classifiers that can be considered as a special case of Tikhonov regularization. This family of classifiers has a special property: They simultaneously minimize the empirical classification error and maximize the geometric margin. Consequently, they are also known as maximum margin classifiers.

By considering these data points:

$$\{(x_1, c_1), (x_2, c_2), \ldots, (x_i, c_i)\} \tag{11.16}$$

where c_i is either 1 or –1, this constant denotes the class to which the point x_i belongs. Each x_i is a p – (statistics notation), or n – (computer-science notation) dimensional vector of scaled [0, 1] or [–1, 1] values. Scaling is important to guard against variables (attributes) with a larger variance that might otherwise dominate the classification. Samples along the hyperplanes are called the support vectors, as illustrated in Figure 11.6.

This can be viewed as training data, denoting the correct classification, which one wants the SVM to distinguish eventually by means of the dividing hyperplane, using this formula:

$$w \bullet x - b = 0 \tag{11.17}$$

The vector w points perpendicular to the separating hyperplane. Adding the offset parameter b allows one to increase the margin. In its absence, the hyperplane is forced to pass through the origin, restricting the solution. Since one would be interested in the maximum margin, the support vectors and the parallel hyperplanes (to the optimal hyperplane) closest to these support vectors in either class would be of interest. It can be shown that these parallel hyperplanes are described by equations:

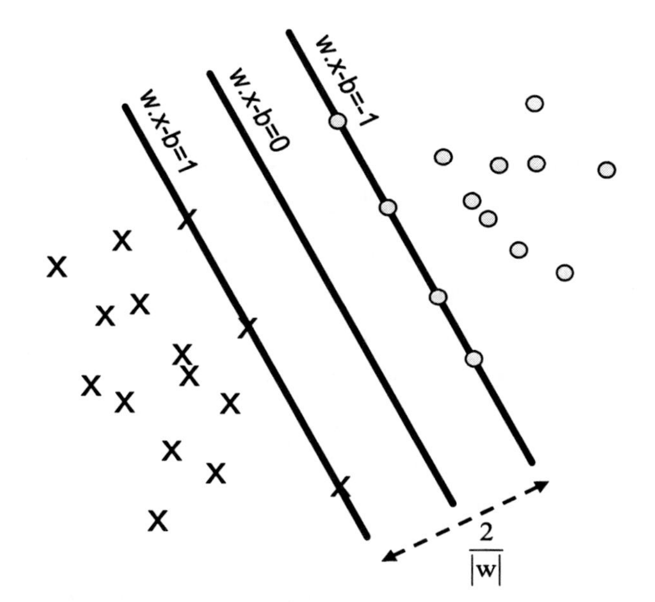

Figure 11.6 Maximum-margin hyperplanes for a SVM trained with samples.

$$w \bullet x - b = 1$$
$$w \bullet x - b = -1 \tag{11.18}$$

If the training data are linearly separable, one can select these hyperlines so that there are no points between them, and then try to maximize their distance. By using geometry, the distance between the hyperplanes is calculated as $2/|w|$. Consequently, $|w|$ needs to be minimized. To exclude data points for all i either:

$$w \bullet x_i - b \geq 1 \text{ or } w \bullet x_i - b \leq -1 \tag{11.19}$$

can be rewritten as:

$$c_i\left(w \bullet x_i - b\right) \geq 1, \quad 1 \leq i \leq n \tag{11.20}$$

The problem now is to minimize $|w|$ subject to the constraint (1). This is a quadratic programming (QP) optimization problem. More clearly, minimize

$$\left(\tfrac{1}{2}\right)\|w\|^2 \text{ subject to } c_i\left(w \bullet x_i - b\right) \geq 1, \ 1 \leq i \leq n \tag{11.21}$$

The factor of ½ is used for mathematical convenience.

Writing the classification rule in its dual form reveals that classification is only a function of the support vectors, the training data on the margin. The dual of the SVM can be shown to be:

$$\max \sum_{i=1}^{n} \alpha_i - \sum_{i,j} \alpha_i \alpha_i c_i c_j x_i^T x_j \tag{11.22}$$

subject to $\alpha_i >= 0$ using linear classifier and nonlinear classifiers.

Methodology

In the actual procedure involved for the economic analysis of various well paths or well types, the following steps may help:

- Define the important attributes that will effect the overall cost of the well directly or indirectly.
- Use these attributes to formulate the input data set for the training samples, based on the existing well data.
- Scale the training sample to normalize it, then use the normalized training sample to train the model.
- Perform predictions using the SVM model generated by the training algorithm.

Special care should be taken during the normalization of the training data set, because this normalization helps reduce a large set of data range into a bounded group, which helps reduce simulation time. The different parameters of the SVM model are optimized, based on the existing data set, by a process called cross validation. Cross validation divides the overall training data set into three sets, and uses each set to predict the result of the other data sets. Therefore, the parameters of the SVM model are tuned to obtain an accurate fit of the data.[14] The main reason for cross validation is to inhibit any kind of biasing in the model, as a result of using only one set of data, therefore, considering only one of the possible scenarios.

Econometrics

Traditionally, well costs are estimated using a deterministic procedure with time and cost inputs. Alternatively, probabilistic methods involving Monte Carlo simulation are used. Uncertainty and variability exist when using these methods, as they involve appropriate weights and distributions for the estimates of various costs.

The econometrics method of well cost estimation is based on the fundamental statistical method used by many econometricians, namely regression analysis.[15] Regression models are used to predict one variable from one or more other variables, with the models built with data from offset wells. The models consider a variety of data, including the poor and good performances of wells, rig performance, problems during drilling, and the performance of personnel during drilling and completing the well. This parametric estimating cost model takes into account all aspects of well operations, as well as anticipated nonlinear behavior of the operating variables involved. It provides a better estimate of the well cost using the historical cost data.

References

1. Azar, J. J., and R. G. Samuel. *Drilling Engineering*. Tulsa, OK: Penwell Publishers, 2007.
2. "Private Finance of Transport Infrastructure Projects." VTT Publication 624, ESPOO, 2007.
3. McCormack, J., R. (Anadarko) LeBlanc, and C. Heiser. "Turning Risk into Shareholder Wealth in the Petroleum Industry." *Journal of Applied Corporate Finance* 15, no. 2 (Fall 2002).
4. Han, J. "There Is Value in Operational Flexibility: An Intelligent Well Application." Paper SPE 82018 presented at the SPE Hydrocarbon Economics and Evaluation Symposium, Dallas, Texas, April 5–8, 2003.
5. Begg S., R. Bratvold, and J. Campbell. "Improving Investment Decisions Using a Stochastic Integrated Asset Model." Paper SPE 71414 presented at the SPE

Annual Technical Conference and Exhibition, New Orleans, Louisiana, September 30–October 3, 2001.
6. Dezen, F., and C. Morooka. "Real Options Applied to Selection of Technological Alternative for Offshore Oilfield Development." Paper SPE 77587 presented at the SPE Annual Technical Conference and Exhibition, San Antonio, Texas, September 29–October 2, 2002.
7. Smith, J., M. Economides, and T. Frick. "Reducing Economic Risk in a Really Anisotropic Formations with Multiple Lateral Horizontal Wells." Paper SPE 30647 presented at the SPE Annual Technical Conference and Exhibition, Dallas, Texas, October 22–25, 1995.
8. Gallant, L., H. Kieffel, R. Chatwin, and J. Smith. "Using Learning Models to Capture Dynamic Complexity in Petroleum Exploration." Paper SPE 52954 presented at the SPE Hydrocarbon Economics and Evaluation Symposium, Dallas, Texas, March 21–23, 1999.
9. Peterson, S. K., J. A. Murtha, and F. F. Schneider. "*Risk Analysis and Monte Carlo Simulation Applied to the Generation of Drilling AFE Estimates.*" Paper SPE 26339, Houston, Texas, June 1995.
10. Williamson, H. S., S. J. Sawaryn, and J. W. Morrison. "Monte Carlo Techniques Applied to Well Forecasting: Some Pitfalls." Paper SPE 89984, 2006.
11. Siruvuri, C., S. Nagarakanti, and R. Samuel. "Stuck Pipe Prediction and Avoidance: A Convolutional Neural Network Approach." Paper IADC/SPE 98378. IADC/SPE Drilling Conference, Miami, Florida, February 21–23, 2006.
12. Bhuddharaju, P., S. Laskar, and R. Samuel. "Robust Well Cost Estimation Using a Support Vector Machine Model." Paper SPE 106577. SPE Digital Energy Conference and Exhibition, Houston, Texas, April 11–12, 2007.
13. Gunn, S. R. "Support Vector Machines for Classification and Regressions." Technical Report, University of Southampton, Southampton, New York, May 1998.
14. Vapnik, V. N. *Statistical Learning Theory*. New York: John Wiley and Sons, 1998.
15. Williamson, H. S., S. J. Sawaryn, and J. W. Morrison. "Monte Carlo Techniques Applied to Well Forecasting: Some Pitfalls." *SPE Drill & Compl.* 21 (no. 3): 216–227.

APPENDIX

Useful Conversion Factors

Length

1 meter = 39.37 in.
1 inch = 2.54 cm
1 feet = 30.48 cm = 0.3048 m
1 mile = 5280 ft = 1720 yard = 1609.344 m
1 nautical mile = 6076 ft

Mass

1 lbm = 453.6 g = 0.4536 kg = 7000 gr (grain)
1 kg = 1000 g = 2.204 6 lbm
1 slug = 1 lbf s^2/ft = 32.174 lbm
1 U.S. ton = 2000 lbm (short ton)
1 long ton = 2240 lbm (British ton)
1 tonne = 1000 kg (metric ton) = 2204.6 lbm
1 kip = 1000 lb

Force

1 lbf = 4.448 N = 4.448×10^5 dynes
1 lbf = 32.174 poundals = 32.174 lbm ft/s^2

Gravitational Acceleration

$g = 32.2\ \text{ft/s}^2 = 9.81\ \text{m/s}^2$

$$g_\lambda = 9.7803267714\left(\frac{1+0.00193185138639\sin^2\lambda}{\sqrt{1-0.00669437999013\sin^2\lambda}}\right)\ \text{m/s}^2$$

Where λ = geographic latitude of Earth ellipsoid measured from the equator in degrees.

Trigonometric Relationships

$$\csc\theta + \cot\theta = \frac{1+\cos\theta}{\sin\theta}$$

$$\csc\theta + \cot\theta = \cot\left(\frac{\theta}{2}\right)$$

$$\sin\theta = 2\sin\left(\frac{\theta}{2}\right)\cos\left(\frac{\theta}{2}\right)$$

$$\cos\theta = \cos^2\left(\frac{\theta}{2}\right) - \sin^2\left(\frac{\theta}{2}\right)$$

$$\sin(a \pm b) = \sin a\cos b \pm \cos a\sin b$$

$$\cos(a \pm b) = \cos a\cos b \mp \sin a\sin b$$

$$\sin\theta = \frac{2e^t}{1+e^t} = \frac{2}{e^t + e^{-t}}$$

$$\cos\theta = \frac{1-e^{2t}}{1+e^{2t}} = \frac{e^{-t} - e^t}{e^t + e^{-t}}$$

$$\sin\theta = \operatorname{sech} t$$

$$\cos\theta = -\tanh t$$

$$\cosh^2 t - \sinh^2 t = 1$$

$$\tanh^2 t + \operatorname{sech}^2 t = 1$$

$$\cosh t = \frac{e^t + e^{-t}}{2}$$

$$\sinh t = \frac{e^t - e^{-t}}{2}$$

$$\sinh^2\theta + \cosh^2\theta = \cosh 2\theta$$

$$\cosh(a \pm b) = \cosh a \cosh b \mp \sinh a \sinh b$$

$$\sin\theta = \frac{2\tan\frac{\theta}{2}}{1+\tan^2\frac{\theta}{2}}$$

$$\cos\theta = \frac{1-\tan^2\frac{\theta}{2}}{1+\tan^2\frac{\theta}{2}}$$

Trigonometric Approximations

$$\sin\theta \cong \theta$$

$$\cos\theta \cong 1$$

For higher-order approximation: $\cos\theta \cong 1 - \frac{\theta^2}{2}$

Useful Relation

$$a^x = e^{x\ln a}$$

Integration Techniques

Integration by parts: $\int_a^b u\,dv = uv\Big|_a^b - \int_a^b v\,du$

Taylor Series

If $f(c)$ is a continuous function in an open interval, its value at neighboring points can be expressed in terms of the Taylor Series as

$$\sum_{n=0}^{\infty}\frac{f^{(n)}(c)}{n!}(x-c) = f(c) + f'(c)(x-c) + f''(c)\frac{(x-c)^2}{2!} + \ldots\ldots$$

$$+ f^{(n)}(c)\frac{(x-c)^n}{n!} + \ldots$$

where f^n denotes the nth derivative. If $c = 0$, then the series is Maclaurin series for f.

SI Metric Conversion Factors

ft × 3.048	E–03 = m
in. × 2.54	E+00 = cm
lbf × 9.869233	E–00 = Nm
lbf/ft^2 × 4.788026	E–02 = kPa
lbm × 4.535924	E–01 = kg
lbm/ft^3 × 1.601846	E+01 = kg/m^3
lbm/gal × 1.198264	E+02 = kg/m^3
md × 6.894757	E–04 = μm^2
psi × 6.894757	E+00 = kPa
psi/ft × 2.262059	E+01 = kPa/m
sq. in. × 6.451	E+00 = cm^2

Index

CPSIA information can be obtained at www.ICGtesting.com
Printed in the USA
LVOW10*1115251013

358585LV00003B/16/P